CONTINUOUS OPTIMIZATION

Current Trends and Modern Applications

Applied Optimization

VOLUME 99

Series Editors:

Panos M. Pardalos
University of Florida, U.S.A.

Donald W. Hearn
University of Florida, U.S.A.

CONTINUOUS OPTIMIZATION

Current Trends and Modern Applications

Edited by

VAITHILINGAM JEYAKUMAR
University of New South Wales, Sydney, Australia

ALEXANDER RUBINOV
University of Ballarat, Ballarat, Australia

 Springer

Library of Congress Cataloging-in-Publication Data

Continuous optimization : current trends and modern applications / edited by
 Vaithilingam Jeyakumar, Alexander Rubinov.
    p. cm. — (Applied optimization ; v. 99)
 Includes bibliographical references.
 ISBN-13: 978-0-387-26769-2 (acid-free paper)
 ISBN-10: 0-387-26769-7 (acid-free paper)
 ISBN-13: 978-0-387-26771-5 (ebook)
 ISBN-10: 0-387-26771-9 (ebook)
   1. Functions, Continuous. 2. Programming (Mathematics). 3. Mathematical models.
 I. Jeyakumar, Vaithilingam. II. Rubinov, Aleksandr Moiseevich. III. Series.

 QA331.C657 2005
 515′.222—dc22

                                                                        2005049900

AMS Subject Classifications: 65Kxx, 90B, 90Cxx, 62H30

© 2005 Springer Science+Business Media, Inc.

All rights reserved. This work may not be translated or copied in whole or in part without the written permission of the publisher (Springer Science+Business Media, Inc., 233 Spring Street, New York, NY 10013, USA), except for brief excerpts in connection with reviews or scholarly analysis. Use in connection with any form of information storage and retrieval, electronic adaptation, computer software, or by similar or dissimilar methodology now know or hereafter developed is forbidden.

The use in this publication of trade names, trademarks, service marks and similar terms, even if the are not identified as such, is not to be taken as an expression of opinion as to whether or not they are subject to proprietary rights.

Printed in the United States of America.

9 8 7 6 5 4 3 2 1   SPIN 11399797

springeronline.com

Contents

Part I Surveys

Linear Semi-infinite Optimization: Recent Advances

Some Theoretical Aspects of Newton's Method for Constrained Best Interpolation

Part II Theory and Numerical Methods

A Numerical Method for Concave Programming Problems

Altannar Chinchuluun, Enkhbat Rentsen, Panos M. Pardalos 251

Convexification and Monotone Optimization

Xiaoling Sun, Jianling Li, Duan Li................................ 275

Part III Applications

Impulsive Control of a Sequence of Rumour Processes

Minimization of the Sum of Minima of Convex Functions and Its Application to Clustering

Analysis of a Practical Control Policy for Water Storage in Two Connected Dams

Phil Howlett, Julia Piantadosi, Charles Pearce.......................435

Preface

Continuous optimization is the study of problems in which we wish to optimize (either maximize or minimize) a continuous function (usually of several variables) often subject to a collection of restrictions on these variables. It has its foundation in the development of calculus by Newton and Leibniz in the 17th century. Nowadys, continuous optimization problems are widespread in the mathematical modelling of real world systems for a very broad range of applications.

Solution methods for large multivariable constrained continuous optimization problems using computers began with the work of Dantzig in the late 1940s on the simplex method for linear programming problems. Recent research in continuous optimization has produced a variety of theoretical developments, solution methods and new areas of applications. It is impossible to give a full account of the current trends and modern applications of continuous optimization. It is our intention to present a number of topics in order to show the spectrum of current research activities and the development of numerical methods and applications.

The collection of 16 refereed papers in this book covers a diverse number of topics and provides a good picture of recent research in continuous optimization. The first part of the book presents substantive survey articles in a number of important topic areas of continuous optimization. Most of the papers in the second part present results on the theoretical aspects as well as numerical methods of continuous optimization. The papers in the third part are mainly concerned with applications of continuous optimization.

We feel that this book will be an additional valuable source of information to faculty, students, and researchers who use continuous optimization to model and solve problems. We would like to take the opportunity to thank the authors of the papers, the anonymous referees and the colleagues who have made direct or indirect contributions in the process of writing this book. Finally, we wish to thank Fusheng Bai for preparing the camera-ready version of this book and John Martindale and Robert Saley for their assistance in producing this book.

Sydney and Ballarat   *Vaithilingam Jeyakumar*
April 2005   *Alexander Rubinov*

List of Contributors

Adil M. Bagirov
CIAO, School of Information
Technology and Mathematical
Sciences
University of Ballarat
Ballarat, VIC 3353
Australia
a.bagirov@ballarat.edu.au

Selma Belen
School of Mathematics
The University of Adelaide
Adelaide, SA, 5005
Australia
sbelen@ankara.baskent.edu.tr

Gleb Beliakov
School of Information Technology
Deakin University
221 Burwood Hwy, Burwood, 3125
Australia
gleb@deakin.edu.au

Altannar Chinchuluun
Department of Industrial and
Systems Engineering
University of Florida
303 Weil Hall, Gainesville, FL, 32611
USA
altannar@ufl.edu

Nguyen Dinh
Department of Mathematics-
Informatics
Ho Chi Minh City University of
Pedagogy
280 An Duong Vuong St., District 5,
HCM city
Vietnam
ndinh@hcmup.edu.vn

Andrew Eberhard
Department of Mathematics
Royal Melbourne University of
Technology
Melbourne, 3001
Australia
andy.eb@rmit.edu.au

Miguel A. Goberna
Dep. de Estadística e Investigación
Operativa
Universidad de Alicante
Spain
mgoberna@ua.es

Semion Gutman
Department of Mathematics
University of Oklahoma
Norman, OK 73019
USA
sgutman@ou.edu

Phil Howlett
Centre for Industrial and Applied
Mathematics
University of South Australia
Mawson Lakes, SA 5095
Australia
phil.howlett@unisa.edu.au

Yalcin Kaya
School of Mathematics and Statistics
University of South Australia
Mawson Lakes, SA, 5095
Australia;
Departamento de Sistemas e
Computação
Universidade Federal do Rio de
Janeiro
Rio de Janeiro
Brazil
Yalcin.Kaya@unisa.edu.au

Duan Li
Department of Systems Engineering
and Engineering
Management
The Chinese University of Hong
Kong
Shatin, N.T., Hong Kong
P.R. China
dli@se.cuhk.edu.hk

Jianling Li
Department of Mathematics
Shanghai University
Shanghai 200436
P.R. China;
College of Mathematics and
Information Science
Guangxi University
Nanning, Guangxi 530004
P.R. China
ljl123@gxu.edu.cn

Gue Myung Lee
Division of Mathematical Sciences

Pukyong National University
599 - 1, Daeyeon-3Dong, Nam-Gu,
Pusan 608 - 737
Korea
gmlee@pknu.ac.kr

Musa Mammadov
CIAO, School of Information
Technology and Mathematical
Sciences
University of Ballarat
Ballarat, VIC 3353
Australia
m.mammadov@ballarat.edu.au

Hossein Mohebi
Department of Mathematics
Shahid Bahonar University of
Kerman
Kerman
Iran
hmohebi@mail.uk.ac.ir

Arkadi Nemirovski
Technion – Israel Institute of
Technology
Haifa 32000
Israel
nemirovs@ie.technion.ac.il

Panos M. Pardalos
Department of Industrial and
Systems Engineering
University of Florida
303 Weil Hall, Gainesville, FL,
32611, USA
pardalos@ufl.edu

Charles Pearce
School of Mathematics
The University of Adelaide
Adelaide, SA 5005
Australia
cpearce@maths.adelaide.edu.au

Julia Piantadosi
Centre for Industrial and Applied
Mathematics
University of South Australia
Mawson Lakes, SA, 5095
Australia
julia.piantadosi@unisa.edu.au

János D. Pintér
Pintér Consulting Services, Inc.
129 Glenforest Drive, Halifax, NS,
B3M 1J2
Canada
jdpinter@hfx.eastlink.ca

Hou-Duo Qi
School of Mathematics
The University of Southampton,
Highfield
Southampton SO17 1BJ
Great Britain
hdqi@soton.ac.uk

Alexander G. Ramm
Department of Mathematics
Kansas State University
Manhattan, Kansas 66506-2602
USA
ramm@math.ksu.edu

Enkhbat Rentsen
Department of Mathematical
Modeling
School of Mathematics and Com-
puter Science
National University of Mongolia
Ulaanbaatar
Mongolia
renkhbat@ses.edu.mn

Alexander Rubinov
CIAO, School of Information
Technology and Mathematical
Sciences
University of Ballarat
Ballarat, VIC 3353
Australia
a.rubinov@ballarat.edu.au

Alexander Shapiro
Georgia Institute of Technology
Atlanta, Georgia 30332-0205
USA
ashapiro@isye.gatech.edu

Nadejda Soukhoroukova
CIAO, School of Information
Technology and Mathematical
Sciences,
University of Ballarat
Ballarat, VIC 3353
Australia
n.soukhoroukova@ballarat.edu.au

Xiaoling Sun
Department of Mathematics
Shanghai University
Shanghai 200444
P. R. China
xlsun@staff.shu.edu.cn

Le Anh Tuan
Ninh Thuan College of Pedagogy
Ninh Thuan
Vietnam
latuan02@yahoo.com

Julien Ugon
CIAO, School of Information
Technology and Mathematical
Sciences
University of Ballarat
Ballarat, VIC 3353
Australia
jugon@students.ballarat.edu.au

Robert Wenczel
Department of Mathematics
Royal Melbourne University of
Technology
Melbourne, VIC 3001
Australia
robert.wenczel@rmit.edu.au

John Yearwood
CIAO, School of Information
Technology and Mathematical
Sciences

University of Ballarat
Ballarat, VIC 3353
Australia
j.yearwood@ballarat.edu.au

Part I

Surveys

Linear Semi-infinite Optimization: Recent Advances

Miguel A. Goberna

Dep. de Estadística e Investigación Operativa
Universidad de Alicante
Spain
mgoberna@ua.es

Summary. Linear semi-infinite optimization (LSIO) deals with linear optimization problems in which either the dimension of the decision space or the number of constraints (but not both) is infinite. This paper overviews the works on LSIO published after 2000 with the purpose of identifying the most active research fields, the main trends in applications, and the more challenging open problems. After a brief introduction to the basic concepts in LSIO, the paper surveys LSIO models arising in mathematical economics, game theory, probability and statistics. It also reviews outstanding real applications of LSIO in semidefinite programming, telecommunications and control problems, in which numerical experiments are reported. In almost all these applications, the LSIO problems have been solved by means of ad hoc numerical methods, and this suggests that either the standard LSIO numerical approaches are not well-known or they do not satisfy the users' requirements. From the theoretical point of view, the research during this period has been mainly focused on the stability analysis of different objects associated with the primal problem (only the feasible set in the case of the dual). Sensitivity analysis in LSIO remains an open problem.

2000 MR Subject Classification. Primary: 90C34, 90C05; Secondary: 15A39, 49K40.

Key words: semi-infinite optimization, linear inequality systems

1 Introduction

Linear semi-infinite optimization (LSIO) deals with linear optimization problems such that either the set of variables or the set of constraints (but not both) is infinite. In particular, LSIO deals with problems of the form

$$(P) \quad \text{Inf } c'x \text{ s.t. } a_t'x \geq b_t, \text{ for all } t \in T,$$

where T is an infinite index set, $c \in \mathbb{R}^n$, $a : T \longmapsto \mathbb{R}^n$, and $b : T \longmapsto \mathbb{R}$, which are called *primal*. The *Haar's dual* problem of (P) is

$$(D) \quad \text{Sup} \sum_{t \in T} \lambda_t b_t, \text{ s.t. } \sum_{t \in T} \lambda_t a_t = c, \quad \lambda \in \mathbb{R}_+^{(T)},$$

where $\mathbb{R}_+^{(T)}$ denotes the positive cone in the space of *generalized finite sequences* $\mathbb{R}^{(T)}$ (the linear space of all the functions $\lambda : T \mapsto \mathbb{R}$ such that $\lambda_t = 0$ for all $t \in T$ except maybe for a finite number of indices). Other dual LSIO problems can be associated with (P) in particular cases, e.g., if T is a compact Hausdorff topological space and a and b are continuous functions, then the *continuous dual* problem of (P) is

$$(D_0) \quad \text{Sup} \int_T b_t \, \mu \, (dt) \text{ s.t. } \int_T a_t \mu \, (dt) = c, \quad \mu \in \mathcal{C}'_+ (T),$$

where $\mathcal{C}'_+ (T)$ represents the cone of nonnegative regular Borel measures on T ($\mathbb{R}_+^{(T)}$ can be seen as the subset of $\mathcal{C}'_+ (T)$ formed by the nonnegative atomic measures). The value of all these dual problems is less or equal to the value of (P) and the equality holds under certain conditions involving either the properties of the constraints system $\sigma = \{a'_t x \geq b_t, t \in T\}$ or some relationship between c and a. Replacing the linear functions in (P) by convex functions we obtain a convex semi-infinite optimization (CSIO) problem. Many results and methods for ordinary linear optimization (LO) have been extended to LSIO, usually assuming that the linear semi-infinite system (LSIS) σ satisfies certain properties. In the same way, LSIO theory and methods have been extended to CSIO and even to nonlinear semi-infinite optimization (NLSIO).

We denote by F, F^* and $v(P)$ the feasible set, the optimal set and the value of (P), respectively (the same notation will be used for NLSIO problems). The boundary and the set of extreme points of F will be denoted by B and E, respectively. We also represent with Λ, Λ^* and $v(D)$ the corresponding objects of (D). We also denote by F the solution set of σ. For the convex analysis concepts we adopt a standard notation (as in [GL98]).

At least three reasons justify the interest of the optimization community in LSIO. First, for its many real life and modeling applications. Second, for providing nontrivial but still tractable optimization problems on which it is possible to check more general theories and methods. Finally, LSIO can be seen as a theoretical model for large scale LO problems.

Section 2 deals with LSISs theory, i.e., with existence theorems (i.e., characterizations of $F \neq \emptyset$) and the properties of the main families of LSISs in the LSIO context. The main purpose of this section is to establish a theoretical frame for the next sections.

Section 3 surveys recent applications of LSIO in a variety of fields. In fact, LSIO models arise naturally in different contexts, providing theoretical tools for a better understanding of scientific and social phenomena. On the other hand, LSIO methods can be a useful tool for the numerical solution of difficult problems. We shall consider, in particular, the connection between LSIO and semidefinite programming (SDP).

Section 4 reviews the last contributions to LSIO numerical methods. We shall also mention some CSIO methods as far as they can be applied, in particular, to linear problems.

Finally, Section 5 deals with the perturbation analysis of LSIO problems. In fact, in many applications, due to either measurement errors or rounding errors occurring during the computation process, the *nominal data* (represented by the triple (a, b, c)) are replaced in practice by approximate data. *Stability* results allow to check whether small perturbations of the data preserve desirable properties of the main objects (as the nonemptiness of F, F^*, Λ and Λ^*, or the boundedness of $v(P)$ and $v(D)$) and, in the affirmative case, allow to know whether small perturbations provoke small variations of these objects. *Sensitivity* results inform about the variation of the value of the perturbed primal and dual problems. Sections 1.2 and 1.5 can be seen as updating the last survey paper on LSIO theory ([GL02]) although for the sake of brevity certain topics are not considered here, e.g., excess of information phenomena in LSIO ([GJR01, GJM03, GJM05]), duality in LSIO ([KZ01, Sha01, Sha04]), inexact LSIO ([GBA05]), etc.

2 Linear semi-infinite systems

Most of the information on σ is captured by its *characteristic cone*,

$$
K = \text{cone} \left\{ \begin{pmatrix} a_t \\ b_t \end{pmatrix}, t \in T; \begin{pmatrix} 0_n \\ -1 \end{pmatrix} \right\}.
$$

The *reference cone* of σ, cl$K$, characterizes the consistency of σ (by the condition $\begin{pmatrix} 0_n \\ 1 \end{pmatrix} \notin \text{cl}K$) as well as the halfspaces containing its solution set, F (if it is nonempty): $a'x \geq b$ is a consequence of σ if and only if $\begin{pmatrix} a \\ b \end{pmatrix} \in \text{cl}K$ (nonhomogeneous Farkas Lemma). Thus, if F_1 and F_2 (K_1 and K_2) are the solution sets (the characteristic cones, respectively) of the consistent systems σ_1 and σ_2, then $F_1 \subset F_2$ if and only if cl$K_2 \subset$ clK_1 (this characterization of set containment is useful in large scale knowledge-based data classification, see [Jey03] and [GJD05]). All these results have been extended from LSISs to linear systems containing strict inequalities ([GJR03, GR05]) and to convex systems possibly containing strict inequalities ([GJD05]). On the other hand, since $F_1 = F_2$ if and only if cl$K_2 =$ clK_1, there exists a one-to-one correspondence between closed convex sets in $\mathbb{R}^n$ and closed convex cones in $\mathbb{R}^{n+1}$ containing $\begin{pmatrix} 0_n \\ -1 \end{pmatrix}$ (the reference cone of their corresponding linear representations). Thus many families of closed convex sets have been characterized by means of the corresponding properties of their corresponding reference cones ([GJR02]). If the index set in σ depends on the variable x, as it happens in generalized semi-infinite optimization (GSIO), F may be nonclosed and even nonconnected ([RS01]).

Let us recall the definition of the main classes of consistent LSIS (which are analyzed in Chapter 5 of [GL98]).

σ is said to be *continuous* (*analytic*, *polynomial*) if T is a compact Hausdorff space (a compact interval, respectively) and the coefficients are continuous (analytic, polynomial, respectively) on T. Obviously,

$$\sigma \text{ polynomial} \to \sigma \text{ analytic} \to \sigma \text{ continuous.}$$

In order to define the remaining three classes of LSISs we associate with $x \in F$ two convex cones. The *cone of feasible directions* at x is

$$D(F;x) = \{d \in \mathbb{R}^n \mid \exists \theta > 0, x + \theta d \in F\}$$

and the *active cone* at x is

$$A(x) := \text{cone}\{a_t \mid a_t'x = b_t, \ t \in T\}$$

(less restrictive definitions of active cone are discussed in [GLT03b] and [GLT03c]).

σ is *Farkas-Minkowsky* (FM) if every consequence of σ is consequence of a finite subsystem (i.e., K is closed). σ is *locally polyhedral* (LOP) if $D(F;x) = A(x)^0$ for all $x \in F$. Finally, σ is *locally Farkas-Minkowsky* (LFM) if every consequence of σ binding at a certain point of F is consequence of a finite subsystem (i.e., $D(F;x)^0 = A(x)$ for all $x \in F$). We have

$$\sigma \text{ continuous \& Slater c.q.} \to \sigma \text{ FM} \to \sigma \text{ LFM} \leftarrow \sigma \text{ LOP.}$$

The statement of two basic theorems and the sketch of the main numerical approaches will show the crucial role played by the above families of LSISs, as constraint qualifications, in LSIO theory and methods (see [GL98] for more details).

Duality theorem: if σ is FM and $F \neq \emptyset \neq \Lambda$, then $v(D) = v(P)$ and (D) is solvable.

Optimality theorem: if $x \in F$ satisfies the KKT condition $c \in A(x)$, then $x \in F^*$, and the converse is true if σ is LFM.

Discretization methods generate sequences of points in $\mathbb{R}^n$ converging to a point of F^* by solving suitable LO problems, e.g., sequences of optimal solutions of the subproblems of (P) which are obtained by replacing T with a sequence of grids. The classical cutting plane approach consists of replacing in (P) the index set T with a finite subset which is formed from the previous one according to certain aggregation and elimination rules. The central cutting plane methods start each step with a polytope containing a sublevel set of (P), calculate a certain "centre" of this polytope by solving a suitable LO problem and then the polytope is updated by aggregating to its defining system either a feasiblity cut (if the center is unfeasible) or an objective cut (otherwise). In order to prove the convergence of any discretization method it is necessary to assume the continuity of σ. The main difficulties with these methods are undesirable jamming (unless (P) has a strongly unique optimal solution) and the increasing size of the auxiliary LO problems (unless efficient elimination rules are implemented).

Reduction methods replace (P) with a nonlinear system of equations (and possibly some inequalities) to be solved by means of a quasi-Newton method. The optimality theorem is the basis of such an approach, so that it requires σ to be LFM. Moreover, some smoothness conditions are required, e.g., σ to be analytic. These methods have a good local behavior provided they start sufficiently close to an optimal solution.

Two-phase methods combine a discretization method (1st phase) and a reduction method (2nd phase). No theoretical result supports the decision to go from phase 1 to phase 2.

Feasible directions (or descent) methods generate a feasible direction at the current iterate by solving a certain LO problem, the next iterate being the result of performing a linear search in this direction. The auxiliary LO problem is well defined assuming that σ is smooth enough, e.g., it is analytic.

Purification methods provide finite sequences of feasible points with decreasing values of the objective functional and the dimension of the corresponding smallest faces containing them, in such a way that the last iterate is an extreme point of either F or Λ (but not necessarily an optimal solution). This approach can only be applied to (P) if the extreme points of F are characterized, i.e., if σ is analytic or LOP.

Hybrid methods (improperly called LSIO simplex method in [AL89]) alternate purification steps (when the current iterate is not an extreme point of F) and descent steps (otherwise).

Simplex methods can be defined for both problems, (P) and (D), and they generate sequences of linked edges of the corresponding feasible set (either F or Λ) in such a way that the objective functional improves on the successive edges under a nondegeneracy assumption. The starting extreme point can be calculated by means of a purification method. Until 2001 the only available simplex method for LSIO problems was conceived for (D) and its convergence status is dubious (recall that the simplex method in [GG83] can be seen as an extension of the classical exchange method for polynomial approximation problems, proposed by Remès in 1934).

Now let us consider the following question: which is the family of solution sets for each class of LSISs?

The answer is almost trivial for continuous, FM and LFM systems. In fact, if

$$T_1 := \left\{ \begin{pmatrix} a \\ b \end{pmatrix} \in \mathbb{R}^n \mid a'x \geq b \,\forall x \in F \right\},$$

$$T_2 := \{ t \in T_1 \mid \|t\| \leq 1 \},$$

and

$$\sigma_i := \left\{ a'x \geq b, \begin{pmatrix} a \\ b \end{pmatrix} \in T_i \right\}, i = 1, 2,$$

it is easy to show that σ_1 and σ_2 are FM (and so LFM) and continuous representations of F, respectively. It is also known that F admits LOP representation if and only if F is quasipolyhedral (i.e., the non-empty intersections of F with polytopes are polytopes).

The problem remains open for analytic and polynomial LSISs. In fact, all we know is that the two families of closed convex sets are different ([GHT05b]) and a list of necessary (sufficient) conditions for F to admit analytic (polynomial) representations. More in detail, it has been shown ([JP04]) that F does not admit analytic representation if either F is a quasi-polyhedral nonpolyhedral set or $F \subset \mathbb{R}^n$, with $n \geq 3$, is smooth (i.e., there exists a unique supporting halfspace at each boundary point of F) and the dimension of the lineality space of F is less than $n - 4$ (e.g., the closed balls in $\mathbb{R}^n$, $n \geq 3$). Between the sets which admit polynomial representation, let us mention the polyhedral convex sets and the plane conic sections, for which it is possible to determine $\deg F$, defined as the minimum of $\deg \sigma := \max \{\deg b; \deg a_i, i = 1, ..., n\}$ (where a_i denotes the ith component of a) for all σ polynomial representation of F ([GHT05a]):

F	$\deg F$
$\{x \in \mathbb{R}^n \mid c_i'x \geq d_i, i = 1, ..., p\}$ (minimal)	$\max\{0, 2p - 3\}$
convex hull of an ellipse	4
convex hull of a parabola	4
convex hull of a branch of hyperbola	2

3 Applications

As the classical applications of LSIO described in Chapters 1 and 2 of [GL98] and in [Gus01b], the new applications could be classified following different criteria as the kind of LSIO problem to be analized or solved $((P), (D), (D_0)$, etc.), the class of constraint system of (P) (continuous, FM, etc.) or the presentation or not of numerical experiments (real or modeling applications, respectively).

Economics

During the 80s different authors formulated and solved risk decision problems as primal LSIO problems without using this name (in fact they solved some examples by means of naive numerical approaches). In the same vein [KM01], instead of using the classical stochastic processes approach to financial mathematics, reformulates and solves dynamic interest rate models as primal LSIO problems where σ is analytical and FM. The chosen numerical approach is two-phase.

Two recent applications in economic theory involve LSIO models where the FM property plays a crucial role. The continuous assignment problem of mathematical economics has been formulated in [GOZ02] as a linear optimization problem over locally convex topological spaces. The discussion involves a certain dual pair of LSIO problems. On the other hand, informational asymmetries generate adverse selection and moral hazard problems. The characterization of economies under asymmetric information (e.g., competitive markets) is a challenging problem. [Jer03] has characterized efficient allocations in this environment by means of LSIO duality theory.

Game Theory

Semi-infinite games arise in those situations (quite frequent in economy) in which one of the players has infinitely many pure strategies whereas the other one only has finitely many alternatives. None of the three reviewed papers reports numerical experiments.

[MM03] deals with transferable utility games, which play a central role in cooperative game theory. The calculus of the linear core is formulated as a primal LSIO problem.

A semi-infinite transportation problem consists of maximizing the profit from the transportation of a certain good from a finite number of suppliers to an infinite number of customers. [SLTT01] uses LSIO duality theory in order to show that the underlying optimization problems have no duality gap and that the core of the game is nonempty. The same authors have considered, in [TTLS01], linear production situations with an infinite number of production techniques. In this context, a LSIO problem arises giving rise to primal and dual games.

Geometry

Different geometrical problems can be formulated and solved with LSIO theory and methods. For instance, the separation and the strong separation of pairs of subsets of a normed space is formulated this way in [GLW01], whereas [JS00] provides a characterization of the minimal shell of a convex body based upon LSIO duality theory (let us recall that the spherical shell of a convex body C with center $x \in C$ is the difference between the smallest closed ball centered at x containing C and the interior of the greatest closed ball centered at x contained in C).

Probability and Statistics

[Dal01] has analyzed the connections between subjective probability theory, maximum likelihood estimation and risk theory with LSIO duality theory and methods. Nevertheless the most promising application field in statistics is Bayesian robustness. Two central problems in this field consist of optimizing posterior functionals over a generalized moment class and calculating minimax decision rules under generalized moment conditions. The first problem has been reformulated as a dual LSIO problem in [BG00]. Concerning the second problem, the corresponding decision rules are obtained by minimizing the maximum of the integrals of the risk function with respect to a given family of distributions on a certain space of parameters. Assuming the compactness of this space, [NS01] proposes a convergence test consisting of solving a certain LSIO problem with continuous constraint system. The authors use duality theory and a discretization algorithm.

Machine Learning

A central problem in machine learning consists of generating a sequence of functions (hypotheses) from a given set of functions which are producible by a base learning algorithm. When this set is infinite, the mentioned problem has been reformulated in [RDB02] as a LSIO one. Certain data classification problems are solved by formulating the problems as linear SDP problems ([JOW05]), so that they can be reformulated and solved as LSIO problems.

Data envelopment analysis

Data Envelopment Analysis (DEA) deals with the comparison of the efficiency of a set of decision making units (e.g., firms, factories, branches or schools) or technologies in order to obtain certain outputs from the available inputs. In the case of a finite set of items to be compared, the efficiency ratios are usually calculated by solving suitable LO problems. In the case of chemical processes which are controlled by means of certain parameters (pressure, temperature, concentrations, etc.) which range on given intervals, the corresponding models can formulated as either LSIO or as bilevel optimization problems. Both approaches are compared in [JJNS01], where a numerical example is provided.

Telecommunication networks

At least three of the techniques for optimizing the capacity of telecommunication systems require the solution of suitable LSIO problems.

In [NNCN01] the capacity of mobile networks is improved by filtering the signal through a beamforming structure. The optimal design of this structure is formulated as an analytic LSIO problem. Numerical results are obtained by means of a hybrid method. The same numerical approach is used in [DCNN01] for the design of narrow-band antennas. Finally, [SAP00] proposes to increase the capacity of cellular systems by means of cell sectorization. A certain technical difficulty arising in this approach can be overcome by solving an associated LSIO problem with continuous σ. Numerical results are provided by means of a discretization procedure.

Control problems

Certain optimal control problems have been formulated as continuous dual LSIO problems. This was done in [Rub00a] for an optimal boundary control problem corresponding to a certain nonlinear diffusion equation with a "rough" initial condition, and in [Rub00b] with two kinds of optimal control problems with unbounded control sets.

On the other hand, in [SIF01] the robust control of certain nonlinear systems with uncertain parameters is obtained by solving a set of continuous primal LSIO problems. Numerical experiments with a discretization procedure are reported.

Optimization under uncertainty

LSIO models arise naturally in inexact LO, when feasibility under any possible perturbation of the nominal problem is required. Thus, the robust counterpart of $\min_x c'x$ subject to $Ax \geq b$, where $(c, A, b) \in \mathcal{U} \subset \mathbb{R}^n \times \mathbb{R}^{mn} \times \mathbb{R}^m$, is formulated in [BN02] as the LSIO problem $\min_{t,x} t$ subject to $t \geq c'x, Ax \geq b \ \forall (c, A, b) \in \mathcal{U}$; the computational tractability of this problem is discussed (in Section 2) for different uncertainty sets $\mathcal{U}$ in a real application (the antenna design problem). On the other hand, [AG01] provides strong duality theorems for inexact LO problems of the form $\min_x \max_{c \in \mathcal{C}} c'x$ subject to $Ax \in \mathcal{B} \ \forall A \in \mathcal{A}$ and $x \in \mathbb{R}_+^n$, where $\mathcal{C}$ and $\mathcal{B}$ are given nonempty convex sets and $\mathcal{A}$ is a given family of matrices. If $Ax \in \mathcal{B}$ can be expressed as $A(t)x = b(t)$, $t \in T$, then this problem admits a continuous dual LSIO formulation.

LSIO also applies to fuzzy systems and optimization. The continuous LSIO (and NLSIO) problems arising in [HF00] are solved with a cutting-plane method. In all the numerical examples reported in [LV01a], the constraint system of the LSIO reformulation is the union of analytic systems (with or without box constraints); all the numerical examples are solved with a hybrid method.

Semidefinite programming

Many authors have analyzed the connections between LSIO and semidefinite programming (see [VB98, Fay02], and references therein, some of them solving SDP problems by means of the standard LSIO methods). In [KZ01] the LSIO duality theory has been used in order to obtain duality theorems for SDP problems. [KK00] and [KGUY02] show that a special class of dual SDP problems can be solved efficiently by means of its reformulation as a continuous LSIO problem which is solved by a cutting-plane discretization method. This idea is also the basis of [KM03], where it is shown that, if the LSIO reformulation of the dual SDP problem has finite value and a FM constraint system, then there exists a low size discretization with the same value. Numerical experiments show that large scale SDP problems which cannot be handled by means of the typical interior point methods (e.g., with more than 3000 dual variables) can be solved applying an ad hoc discretization method which exploits the structure of the problem.

4 Numerical methods

In the previous section we have seen that most of the LSIO problems arising in practical applications in the last years have been solved by means of new methods (usually variations of other already known). Two possible reasons for this phenomenon are the lack of available codes for large classes of LSIO problems (commercial or not) and the computational inefficiency of the known methods (which could fail to exploit the structure of the particular problems). Now we review the specific literature on LSIO methods.

[Bet04] and [WFL01] propose two new central cutting plane methods, taking as center of the current polytope the center of the greatest ball inscribed in the polytope and its analytic center, respectively. [FLW01] proposes a cutting-plane method for solving LSIO and quadratic CSIO problems (an extension of this method to infinite dimensional LSIO can be found in [WFL01]). Several relaxation techniques and their combinations are proposed and discussed. The method in [Bet04], which reports numerical experiments, is an accelerated version of the cutting-plane (Elzinga-Moore) Algorithm 11.4.2 in [GL98] for LSIO problems with continuous σ whereas [WFL01] requires the analiticity of σ. A Kelley cutting-plane algorithm has been proposed in [KGUY02] for a particular class of LSIO problems (the reformulations of dual SDP problems); an extension of this method to SIO problems with nonlinear objective and linear constraints has been proposed in [KKT03].

A reduction approach for LSIO (and CSIO) problems has been proposed in [ILT00], where σ is assumed to be continuous and FM. The idea is to reduce the Wolfe's dual problem to a small number of ordinary non linear optimization problems. The method performs well on a famous test example. This method has been extended to quadratic SIO in [LTW04].

[AGL01] proposes a simplex method (and a reduced gradient method) for LSIO problems such that σ is LOP. These methods are the unique which could be applied to LSIO problems with a countable set of constraints. The proof of the convergence is an open problem.

[LSV00] proposes two hybrid methods to LSIO problems such that σ is a finite union of analytic systems with box constraints. Numerical experiments are reported.

[KM02] considers LSIO problems in which σ is continuous, satisfies the Slater condition and the components of $a_t \in C(T)$ are linearly inpendent. Such kind of problems are reformulated as a linear approximation problem, and then they are solved by means of a classical method of Polya. Convergence proofs are given.

[Kos01] provides a conceptual path-following algorithm for the parametric LSIO problem arising in optimal control consisting of replacing T in (P) with an interval $T(\tau) := [0, \tau]$, where τ ranges on a certain interval. The constraints system of the parametric problem are assumed to be continuous and FM for each τ. An illustrative example is given.

Finally, let us observe that LSIO problems could also be solved by means of numerical methods initially conceived for more general models, as CSIO ([Abb01, TKVB02, ZNF00]), NLSIO ([ZR03, VFG03, GP01, Gus01a]) and GSIO ([Sti01, SS03, Web03] and references therein). The comparison of the particular versions for LSIO problems of these methods with the specific LSIO methods is still to be made.

5 Perturbation analysis

In this section we consider possible any arbitrary perturbation of the nominal data $\pi = (a, b, c)$ which preserve n and T (the constraint system of π is σ). π is *bounded* if $v(P) \neq -\infty$ and it has *bounded data* if a and b are bounded functions. The *parameter space* is $\Pi := (\mathbb{R}^n \times \mathbb{R})^T \times \mathbb{R}^n$, endowed with the *pseudometric of the uniform convergence*:

$$d(\pi_1, \pi) := \max\left\{ \|c^1 - c\|, \sup_{t \in T} \left\| \binom{a_t^1}{b_t^1} - \binom{a_t}{b_t} \right\| \right\},$$

where $\pi_1 = (c^1, a^1, b^1)$ denotes a perturbed data set. The associated problems are (P_1) and (D_1). The sets of consistent (bounded, solvable) perturbed problems are denoted by Π_c (Π_b, Π_s, respectively). Obviously, $\Pi_s \subset \Pi_b \subset \Pi_c \subset \Pi$.

From the primal side, we consider the following set-valued mappings: $\mathcal{F}(\pi_1) := F_1$, $\mathcal{B}(\pi_1) := B_1$, $\mathcal{E}(\pi_1) := E_1$ and $\mathcal{F}^*(\pi_1) := F_1^*$, where F_1, B_1, E_1 and F_1^* denote the feasible set of π_1, its boundary, its set of extreme points and the optimal set of π_1, respectively. The upper and lower semicontinuity (usc and lsc) of these mappings are implicitly understood in the sense of Berge (almost no stability analysis has been made with other semicontinuity concepts). The value function is $\vartheta(\pi_1) := v(P_1)$. Similar mappings can be considered for the dual problem. Some results in this section are new even for LO (i.e., $|T| < \infty$).

Stability of the feasible set

It is easy to prove that $\mathcal{F}$ is closed everywhere whereas the lsc and the usc properties are satisfied or not at a given $\pi \in \Pi_c$ depending on the data a and b.

Chapter 6 of [GL98] provides many conditions which are equivalent to the lsc property of $\mathcal{F}$ at $\pi \in \Pi_c$, e.g., $\pi \in \text{int}\Pi_c$, existence of a strong Slater point $\overline{x}$ (i.e., $a_t'\overline{x} \geq b_t + \varepsilon$ for all $t \in T$, with $\varepsilon > 0$), or

$$0_{n+1} \notin \text{clconv}\left\{ \binom{a_t}{b_t}, t \in T \right\}$$

(a useful condition involving the data).

The characterization of the usc property of $\mathcal{F}$ at $\pi \in \Pi_c$ in [CLP02a] requires some additional notation. Let K^R be the characteristic cone of

$$\sigma^R := \left\{ a'x \geq b, \binom{a}{b} \in \left(\text{conv}\left\{ \binom{a_t}{b_t}, t \in T \right\} \right)_\infty \right\},$$

where $X_\infty := \{\lim_k \mu_k x^k \mid \{x^k\} \subset X, \{\mu_k\} \downarrow 0\}$. If F is bounded, then $\mathcal{F}$ is usc at π. Otherwise two cases are possible:

If F contains at least one line, then $\mathcal{F}$ is usc at π if and only if $K^R = \text{cl}K$.

Otherwise, if w is the sum of a certain basis of $\mathbb{R}^n$ contained in $\{a_t, t \in T\}$, then $\mathcal{F}$ is usc at π if and only if there exists $\beta \in \mathbb{R}$ such that

$$\text{cone}\left(K^R \cup \left\{\begin{pmatrix} w \\ \beta \end{pmatrix}\right\}\right) = \text{cone}\left(\text{cl}K \cup \left\{\begin{pmatrix} w \\ \beta \end{pmatrix}\right\}\right).$$

The stability of the feasible set has been analyzed from the point of view of the dual problem ([GLT01]), for the primal problem with equations and constraint set ([AG05]) and for the primal problem in CSIO ([LV01b] and [GLT02]).

Stability of the boundary of the feasible set

Given $\pi \in \Pi_c$ such that $F \neq \mathbb{R}^n$, then we have ([GLV03], [GLT05]):

$$\mathcal{F} \text{ lsc at } \pi \longleftrightarrow \mathcal{B} \text{ lsc at } \pi$$
$$\searrow {}^{(1)}$$
$$\mathcal{B} \text{ closed at } \pi$$
$$\nearrow {}_{(2)}$$
$$\mathcal{F} \text{ usc at } \pi \longleftarrow \mathcal{B} \text{ usc at } \pi$$

Remarks: (1) the converse holds if $\dim F = n$; (2) the converse statement holds if F is bounded.

Stability of the extreme points set

The following concept is the key of the analysis carried out in [GLV05]: π is *nondegenerate* if $|\{t \in T \mid a_t' x = b_t\}| < n$ for all $x \in B \backslash E$.

Let $\pi_H = (a, 0, c)$. If $|T| \geq n$, $E \neq \emptyset$, and $|F| > 1$ (the most difficult case), then we have:

$$\mathcal{F} \text{ lsc at } \pi \longleftrightarrow \quad \mathcal{E} \text{ lsc at } \pi$$
$$\downarrow {}^{(1)}$$
$$(4)$$
$$\mathcal{E} \text{ closed at } \pi \quad \longrightarrow \quad \pi \text{ nondeg.}$$
$$(2) \downarrow\uparrow \quad {}^{(3)}$$
$$(5)$$
$$\mathcal{E} \text{ usc at } \pi \quad \longrightarrow \quad \pi \ \& \ \pi_H \text{ nondeg.}$$

Remarks: (1) if F is strictly convex; (2) if F is bounded; (3) if $\{a_t, t \in T\}$ is bounded; (4) if $\mathcal{F}$ is lsc at π; the converse holds if $|T| < \infty$; (5) the converse statement holds if $|T| < \infty$.

Stability of the optimal set

In Chapter 10 of [GL98] it is proved that, if $\pi \in \Pi_s$, then the following statements hold:

♦ $\mathcal{F}^*$ is closed at $\pi \longleftrightarrow$ either $\mathcal{F}$ is lsc at π or $F = F^*$.

♦ $\mathcal{F}^*$ is lsc at $\pi \longleftrightarrow \mathcal{F}$ is lsc at π and $|F^*| = 1$ (uniqueness).

♦ If $\mathcal{F}^*$ is usc at π, then $\mathcal{F}^*$ is closed at π (and the converse is true if F^* is bounded).

The following generic result on Π_s has been proved in [GLT03a]: almost every (in a topological sense) solvable LSIO problem with bounded data has a strongly unique solution. Results on the stability of $\mathcal{F}^*$ in CSIO can be found in [GLV03] and [GLT02].

Stability of the value and well-posedness

The following definition of well-posedness is orientated towards the stability of ϑ. $\{x^r\} \subset \mathbb{R}^n$ is an *asymptotically minimazing sequence* for $\pi \in \Pi_c$ associated with $\{\pi_r\} \subset \Pi_b$ if $x^r \in F_r$ for all r, $\lim_r \pi_r = \pi$, and $\lim_r \left[(c^r)' x^r - v(P_r) \right] = 0$. In particular, $\pi \in \Pi_s$ is *Hadamard well-posed* (Hwp) if for every $x^* \in F^*$ and for every $\{\pi_r\} \subset \Pi_b$ such that $\lim_r \pi_r = \pi$ there exists an asymptotically minimazing sequence converging to x^*. The following statements are proved in Chapter 10 of [GL98]:

♦ If $F^* \neq \emptyset$ and bounded, then ϑ is lsc at π. The converse statement holds if $\pi \in \Pi_b$.

♦ ϑ is usc at $\pi \longleftrightarrow \mathcal{F}$ is lsc at π.

♦ If π is Hwp, then $\vartheta|_{\Pi_b}$ is continuous.

♦ If F^* is bounded, π is Hwp $\longleftrightarrow$ either $\mathcal{F}$ is lsc at π or $|F| = 1$.

♦ If F^* is unbounded and π is Hwp, then $\mathcal{F}$ is lsc at π.

A similar analysis has been made in [CLPT01] with other Hwp concepts. Extensions to CSIO can be found in [GLV03]. A generic result on Hwp problems in quadratic SIO can be found [ILR01]. The connection between genericity and Hwp properties is discussed in [Pen01].

Distance to ill-posedness

There exist different concepts of ill-posedness in LSIO: $\mathrm{bd}\Pi_c$ is the set of *ill-posed problems in the feasibility sense*, $\mathrm{bd}\Pi_{si}$ (where Π_{si} denotes the set of problems which have a finite inconsistent subproblem) is the set of *generalized ill-posed problems in the feasibility sense*, and $\mathrm{bd}\Pi_s = \mathrm{bd}\Pi_b$ is the set of *ill-posed problems in the optimality sense*. The following formulae ([CLPT04]) replace the calculus of distances in Π with the calculus of distances in $\mathbb{R}^{n+1}$ involving the so-called *hypographic set*

$$H := \mathrm{conv}\left\{ \begin{pmatrix} a_t \\ b_t \end{pmatrix}, t \in T \right\} + \mathrm{cone}\left\{ \begin{pmatrix} 0_n \\ -1 \end{pmatrix}, t \in T \right\}.$$

♦ If $\pi \in \Pi_c$, then

$$d(\pi, \mathrm{bd}\Pi_{si}) = d(0_{n+1}, \mathrm{bd}H).$$

♦ If $\pi \in (\mathrm{cl}\Pi_s) \cap (\mathrm{int}\Pi_c)$ and $Z^- := \mathrm{conv}\{a_t, t \in T; -c\}$, then

$$d(\pi, \mathrm{bd}\Pi_s) = \min\{d(0_{n+1}, \mathrm{bd}H), d(0_n, \mathrm{bd}Z^-)\}.$$

♦ If $\pi \in (\mathrm{cl}\Pi_s) \cap (\mathrm{bd}\Pi_c)$ and $Z^+ := \mathrm{conv}\{a_t, t \in T; c\}$, then

$$d(\pi, \mathrm{bd}\Pi_s) \geq \min\{d(0_{n+1}, \mathrm{bd}H), d(0_n, \mathrm{bd}Z^+)\}.$$

Error bounds

The *residual function* of π is

$$r(x, \pi) := \sup_{t \in T} (b_t - a_t'x)^+,$$

where $\alpha^+ := \max\{\alpha, 0\}$. Obviously, $x \in F \leftrightarrow r(x, \pi) = 0$. $0 \le \beta < +\infty$ is a *global error bound* for $\pi \in \Pi_c$ if

$$\frac{d(x, F)}{r(x, \pi)} \le \beta \ \forall x \in \mathbb{R}^n \backslash F.$$

If there exists such a β, then the condition number of π is

$$0 \le \tau(\pi) := \sup_{x \in \mathbb{R}^n \backslash F} \frac{d(x, F)}{r(x, \pi)} < +\infty.$$

The following statements hold for any π with bounded data ([Hu00]):

♦ Assume that F is bounded and $\pi \in \operatorname{int} \Pi_c$, and let ρ, x^0 and $\varepsilon > 0$ such that $\|x\| \le \rho \ \forall x \in F$ and $a_t' x^0 \ge b_t + \varepsilon \ \forall t \in T$. Let $0 \le \gamma < 1$. Then, if

$$d(\pi_1, \pi) < \frac{\varepsilon \gamma n^{-\frac{1}{2}}}{2\rho},$$

we have

$$\tau(\pi_1) \le 2\rho\varepsilon^{-1} \left[\frac{1+\gamma}{(1-\gamma)^2} \right].$$

♦ Assume that F is unbounded and $\pi_H \in \operatorname{int} \Pi_c$, and let u and $\eta > 0$ such that $a_t' u \ge \eta \ \forall t \in T$, $\|u\| = 1$. Let $0 < \delta < n^{-\frac{1}{2}}\eta$. Then, if $d(\pi_1, \pi) < \delta$, we have

$$\tau(\pi_1) \le \left(\eta - \delta n^{\frac{1}{2}} \right)^{-1}.$$

Improved error bounds for arbitrary π can be found in [CLPT04]. There exist extensions to CSIO ([Gug00]) and to abstract LSIO ([NY02]).

Sensitivity analysis

The basic problem in sensitivity analysis is to evaluate the impact on the primal and the dual value functions of small perturbations of the data. In the case of perturbations of c, an approximate answer can be obtained from the subdifferentials of these functions (see Chapter 8 in [GL98]). [GGGT05] extends from LO to LSIO the exact formulae in [Gau01] for both value functions under perturbations of c and b (separately). This is done determining neighborhoods of c (b), or at least segments emanating from c (b, respectively), on which the corresponding value function is linear (i.e., finite, convex and concave).

Other perspectives

In the parametric setting the perturbed data depend on a certain parameter $\theta \in \Theta$ (space of parameters), i.e., are expressed as $\pi(\theta) = (a(\theta), b(\theta), c(\theta))$, with T fixed or not, and the nominal problem is $\pi(\overline{\theta})$. The stability of $\mathcal{F}$ in this context has been studied in [MM00, CLP05], where the stability of ϑ and $\mathcal{F}^*$ has been also analyzed. Results on the stability of $\mathcal{F}$ in CSIO in a parametric setting can be found in [CLP02b, CLOP03].

For more information on perturbation analysis in more general contexts the reader is referred to [KH98, BS00] and references therein.

Acknowlegement

This work was supported by DGES of Spain and FEDER, Grant BFM2002-04114-C02-01.

References

[Abb01]   Abbe, L.: Two logarithmic barrier methods for convex semi-infinite problems. In [GL01], 169–195 (2001)

[AG01]   Amaya, J., J.A. Gómez: Strong duality for inexact linear programming problems. Optimization, **49**, 243–369 (2001)

[AG05]   Amaya, J., M.A. Goberna: Stability of the feasible set of linear systems with an exact constraints set. Math. Meth. Oper. Res., to appear (2005)

[AGL01]   Anderson, E.J., Goberna, M.A., López, M.A.: Simplex-like trajectories on quasi-polyhedral convex sets. Mathematics of Oper. Res., **26**, 147–162 (2001)

[AL89]   Anderson, E.J., Lewis, A.S.: An extension of the simplex algorithm for semi-infinite linear programming. Math. Programming (Ser. A), **44**, 247–269 (1989)

[BN02]   Ben-Tal, A., Nemirovski, A.: Robust optimization - methodology and applications. Math. Programming (Ser. B), **92**, 453-480 (2002)

[Bet04]   Betró, B.: An accelerated central cutting plane algorithm for linear semi-infinite linear programming. Math. Programming (Ser. A), **101**, 479–495 (2004)

[BG00]   Betró, B., Guglielmi, A.: Methods for global prior robustness under generalized moment conditions. In: Ríos, D., Ruggeri, F. (ed) Robust Bayesian Analysis, 273–293. Springer, N.Y. (2000)

[BS00]   Bonnans, J.F., Shapiro, A.: Perturbation Analysis of Optimization Problems. Springer Verlag, New York, N.Y. (2000)

[CLOP03]   Cánovas, M.J., López, M.A., Ortega, E.-M., Parra, J.: Upper semicontinuity of closed-convex-valued multifunctions. Math. Meth. Oper. Res., **57**, 409–425 (2003)

[CLP02a]   Cánovas, M.J., López, M.A., Parra, J.: Upper semicontinuity of the feasible set mapping for linear inequality systems. Set-Valued Analysis, **10**, 361–378 (2002)

[CLP02b]   Cánovas, M.J., López, M.A., Parra, J.: Stability in the discretization of a parametric semi-infinite convex inequality system. Mathematics of Oper. Res., **27**, 755–774 (2002)

[CLP05]   Cánovas, M.J., López, M.A. and Parra, J.: Stability of linear inequality systems in a parametric setting, J. Optim. Theory Appl., to appear (2005)

[CLPT01]   Cánovas, M.J., López, M.A., Parra, J., Todorov, M.I.: Solving strategies and well-posedness in linear semi-infinite programming. Annals of Oper. Res., **101**, 171–190 (2001)

[CLPT04]   Cánovas, M.J., López, M.A., Parra, J., F.J. Toledo: Distance to ill-posedness and consistency value of linear semi-infinite inequality systems, Math. Programming (Ser. A), Published online: 29/12/2004, (2004)

[DCNN01] Dahl, M., Claesson, I., Nordebo, S., Nordholm, S.: Chebyshev optimiza-
          tion of circular arrays. In: Yang, X. et al (ed): Optimization Methods and
          Applications, 309–319. Kluwer, Dordrecht, (2001)
[Dal01]   Dall'Aglio: On some applications of LSIP to probability and statistics. In
          [GL01], 237–254 (2001)
[FLW01]   Fang, S.-Ch., Lin, Ch.-J., Wu, S.Y.: Solving quadratic semi-infinite pro-
          gramming problems by using relaxed cutting-plane scheme. J. Comput.
          Appl. Math., **129**, 89–104 (2001)
[Fay02]   Faybusovich, L.: On Nesterov's approach to semi-infinite programming.
          Acta Appl. Math., **74**, 195–215 (2002)
[Gau01]   Gauvin, J.: Formulae for the sensitivity analysis of linear programming
          problems. In Lassonde, M. (ed): Approximation, Optimization and Math-
          ematical Economics, 117–120. Physica-Verlag, Berlin (2001)
[GLV03]   Gayá, V.E., López, M. A., Vera de Serio, V.: Stability in convex semi-
          infinite programming and rates of convergence of optimal solutions of
          discretized finite subproblems. Optimization, **52**, 693–713 (2003)
[GG83]   Glashoff, K., Gustafson, S.-A.: Linear Optimization and Approximation.
          Springer Verlag, Berlin (1983)
[GHT05a] Goberna, M.A., Hernández, L., Todorov, M.I.: On linear inequality sys-
          tems with smooth coefficients. J. Optim. Theory Appl., **124**, 363–386
          (2005)
[GHT05b] Goberna, M.A., Hernández, L., Todorov, M.I.: Separating the solution
          sets of analytical and polynomial systems. Top, to appear (2005)
[GGGT05] Goberna, M.A., Gómez, S., Guerra, F., Todorov, M.I.: Sensitivity analy-
          sis in linear semi-infinite programming: perturbing cost and right-hand-
          side coefficients. Eur. J. Oper. Res., to appear (2005)
[GJD05]   Goberna, M.A., Jeyakumar, V., Dinh, N.: Dual characterizations of set
          containments with strict inequalities. J. Global Optim., to appear (2005)
[GJM03]   Goberna, M.A., Jornet, V., Molina, M.D.: Saturation in linear optimiza-
          tion. J. Optim. Theory Appl., **117**, 327–348 (2003)
[GJM05]   Goberna, M.A., Jornet, V., Molina, M.D.: Uniform saturation. Top, to
          appear (2005)
[GJR01]   Goberna, M.A., Jornet, V., Rodríguez, M.: Directional end of a convex set:
          Theory and applications. J. Optim. Theory Appl., **110**, 389–411 (2001)
[GJR02]   Goberna, M.A., Jornet, V., Rodríguez, M.: On the characterization of
          some families of closed convex sets. Contributions to Algebra and Geom-
          etry, **43**, 153–169 (2002)
[GJR03]   Goberna, M.A., Jornet, V., Rodríguez, M.: On linear systems containing
          strict inequalities. Linear Algebra Appl., **360**, 151–171 (2003)
[GLV03]   Goberna, M.A., Larriqueta, M., Vera de Serio, V.: On the stability of the
          boundary of the feasible set in linear optimization. Set-Valued Analysis,
          11, 203–223 (2003)
[GLV05]   Goberna, M.A., Larriqueta, M., Vera de Serio, V.: On the stability of
          the extreme point set in linear optimization. SIAM J. Optim., to appear
          (2005)
[GL98]   Goberna, M.A., López, M.A.: Linear Semi-Infinite Optimization, Wiley,
          Chichester, England (1998)
[GL01]   Goberna, M.A., López, M.A. (ed): Semi-Infinite Programming: Recent
          Advances. Kluwer, Dordrecht (2001)

[GL02]   Goberna, M.A., López, M.A.: Linear semi-infinite optimization theory: an updated survey. Eur. J. Oper. Res., **143**, 390–415 (2002)

[GLT01]   Goberna, M.A., López, M.A., Todorov, M.I.: On the stability of the feasible set in linear optimization. Set-Valued Analysis, **9**, 75–99 (2001)

[GLT03a]   Goberna, M.A., López, M.A., Todorov, M.I.: A generic result in linear semi-infinite optimization. Applied Mathematics and Optimization, **48**, 181–193 (2003)

[GLT03b]   Goberna, M.A., López, M.A., Todorov, M.I.: A sup-function approach to linear semi-infinite optimization. Journal of Mathematical Sciences, **116**, 3359–3368 (2003)

[GLT03c]   Goberna, M.A., López, M.A., Todorov, M.I.: Extended active constraints in linear optimization with applications. SIAM J. Optim., **14**, 608–619 (2003)

[GLT05]   Goberna, M.A., López, M.A., Todorov, M.I.: On the stability of closed-convex-valued mappings and the associated boundaries, J. Math. Anal. Appl., to appear (2005)

[GLW01]   Goberna, M.A., López, M.A., Wu, S.Y.: Separation by hyperplanes: a linear semi-infinite programming approach. In [GL01], 255–269 (2001)

[GR05]   Goberna, M.A., Rodríguez, M.: Analyzing linear systems containing strict inequalities via evenly convex hulls. Eur. J. Oper. Res., to appear (2005)

[GBA05]   Gómez, J.A., Bosch, P.J., Amaya, J.: Duality for inexact semi-infinite linear programming. Optimization, **54**, 1–25 (2005)

[GLT02]   Gómez, S., Lancho, A., Todorov, M.I.: Stability in convex semi-infinite optimization. C. R. Acad. Bulg. Sci., **55**, 23–26 (2002)

[GOZ02]   Gretsky, N.E., Ostroy, J.M., Zame, W.R.: Subdifferentiability and the duality gap. Positivity, **6**, 261–264 (2002)

[GP01]   Guarino Lo Bianco, C., Piazzi, A.: A hybrid algorithm for infinitely constrained optimization. Int. J. Syst. Sci., **32**, 91–102 (2001)

[Gug00]   Gugat, M.: Error bounds for infinite systems of convex inequalities without Slater's condition. Math. Programing (Ser. B), **88**, 255–275 (2000)

[Gus01a]   Gustafson, S.A.: Semi-infinite programming: Approximation methods. In Floudas, C.A., Pardalos, P.M. (ed) Encyclopedia of Optimization Vol. 5, 96–100. Kluwer, Dordrecht (2001)

[Gus01b]   Gustafson, S.A.: Semi-infinite programming: Methods for linear problems. In Floudas, C.A., Pardalos, P.M. (ed) Encyclopedia of Optimization Vol. 5, 107–112. Kluwer, Dordrecht (2001)

[Hu00]   Hu, H.: Perturbation analysis of global error bounds for systems of linear inequalities. Math. Programming (Ser. B), **88**, 277–284 (2000)

[HF00]   Hu, C. F., Fang, S.-C.: Solving a System of Infinitely Many Fuzzy Inequalities with Piecewise Linear Membership Functions, Comput. Math. Appl., **40**, 721–733 (2000)

[ILR01]   Ioffe, A.D., Lucchetti, R.E., Revalski, J.P.: A variational principle for problems with functional constraints. SIAM J. Optim., **12**, 461–478 (2001)

[ILT00]   Ito, S., Liu, Y., Teo, K.L.: A dual parametrization method for convex semi-infinite programming. Annals of Oper. Res., **98**, 189–213 (2000)

[JP04]   Jaume, D., Puente, R.: Representability of convex sets by analytical linear inequality systems. Linear Algebra Appl., **380**, 135–150 (2004)

[Jer03]   Jerez, B.: A dual characterization of incentive efficiency. J. Econ. Theory, **112**, 1–34 (2003)

20   M.A. Goberna

[JJNS01] Jess, A., Jongen, H.Th., Neralic, L., Stein, O.: A semi-infinite program-
          ming model in data envelopment analysis. Optimization, **49**, 369–385
          (2001)
[Jey03]   Jeyakumar, V.: Characterizing set containments involving infinite convex
          constraints and reverse-convex constraints. SIAM J. Optim., **13**, 947–959
          (2003)
[JOW05]   Jeyakumar, V., Ormerod, J., Womersly, R.S.: Knowledge-based semi-
          definite linear programming classifiers, Optimization Methods and Soft-
          ware, to appear (2005)
[JS00]   Juhnke, F., Sarges, O.: Minimal spherical shells and linear semi-infinite
          optimization. Contributions to Algebra and Geometry, **41**, 93–105 (2000)
[KH98]   Klatte, D., Henrion, R.: Regularity and stability in nonlinear semi-infinite
          optimization. In: Reemtsen, R., Rückmann, J. (ed) Semi-infinite Program-
          ming. Kluwer, Dordrecht, 69–102 (1998)
[KGUY02] Konno, H., Gotho, J., Uno, T., Yuki, A.: A cutting plane algorithm
          for semidefinite programming with applications to failure discriminant
          analysis. J. Comput. and Appl. Math., **146**, 141–154 (2002)
[KKT03]   Konno, H., Kawadai, N. , Tuy, H.: Cutting-plane algorithms for nonlinear
          semi-definite programming problems with applications. J. Global Optim.,
          25, 141–155 (2003)
[KK00]   Konno, H., Kobayashi, H.: Failure discrimination and rating of enterprises
          by semi-definite programming, Asia-Pacific Financial Markets, **7**, 261–273
          (2000)
[KM01]   Kortanek, K.O., Medvedev, V.G.: Building and Using Dynamic Interest
          Rate Models. Wiley, Chichester (2001)
[KZ01]   Kortanek, K.O., Zhang, Q.: Perfect duality in semi-infinite and semidefi-
          nite programming. Math. Programming (Ser. A), **91**, 127–144 (2001)
[KM02]   Kosmol, P., Müller-Wichards, D.: Homotopic methods for semi-infinite
          optimization. J. Contemp. Math. Anal., **36**, 31–48 (2002)
[Kos01]   Kostyukova, O.I.: An algorithm constructing solutions for a family of lin-
          ear semi-infinite problems. J. Optim. Theory Appl., **110**, 585–609 (2001)
[KM03]   Krishnan, K., Mitchel, J.E.: Semi-infinite linear programming approaches
          to semidefinite programming problems. In: Pardalos, P., (ed) Novel Ap-
          proaches to Hard Discrete Optimization, 121–140. American Mathemat-
          ical Society, Providence, RI (2003)
[LSV00]   León, T., Sanmatías, S., Vercher, E.: On the numerical treatment of lin-
          early constrained semi-infinite optimization problems. Eur. J. Oper. Res.,
          121, 78–91 (2000)
[LV01a]   León, T., Vercher, E.: Optimization under uncertainty and linear semi-
          infinite programming: A survey. In [GL01], 327–348 (2001)
[LTW04]   Liu, Y., Teo, K.L., Wu, S.Y.: A new quadratic semi-infinite programming
          algorithm based on dual parametrization. J. Global Optim., **29**, 401–413
          (2004)
[LV01b]   López, M. A., Vera de Serio, V.: Stability of the feasible set mapping in
          convex semi-infinite programming, in [GL01], 101–120 (2001)
[MM03]   Marinacci, M., Montrucchio, L.: Subcalculus for set functions and cores
          of TU games. J. Mathematical Economics, **39**, 1–25 (2003)
[MM00]   Mira, J.A., Mora, G.: Stability of linear inequality systems measured by
          the Hausdorff metric. Set-Valued Analysis, **8**, 253–266 (2000)

[NY02]   Ng, K.F., Yang, W.H.: Error bounds for abstract linear inequality systems. SIAM J. Optim., **13**, 24–43 (2002)

[NNCN01]   Nordholm, S., Nordberg, J., Claesson, I., Nordebo, S.: Beamforming and interference cancellation for capacity gain in mobile networks. Annals of Oper. Res., **98**, 235–253 (2001)

[NS01]   Noubiap, R.F., Seidel, W.: An algorithm for calculating Gamma-minimax decision rules under generalized moment conditions. Ann. Stat., **29**, 1094–1116 (2001)

[Pen01]   Penot, J.-P.: Genericity of well-posedness, perturbations and smooth variational principles. Set-Valued Analysis, **9**, 131–157 (2001)

[RDB02]   Rätsch, G., Demiriz, A., Bennet, K.P.: Sparse regression ensembles in infinite and finite hypothesis spaces. Machine Learning, **48**, 189–218 (2002)

[Rub00a]   Rubio, J.E.: The optimal control of nonlinear diffusion equations with rough initial data. J. Franklin Inst., **337**, 673–690 (2000)

[Rub00b]   Rubio, J.E.: Optimal control problems with unbounded constraint sets. Optimization, **48**, 191–210 (2000)

[RS01]   Rückmann, J.-J., Stein, O.: On linear and linearized generalized semi-infinite optimization problems. Annals Oper. Res., **101**, 191–208 (2001)

[SAP00]   Sabharwal, A., Avidor, D., Potter, L.: Sector beam synthesis for cellular systems using phased antenna arrays. IEEE Trans. on Vehicular Tech., **49**, 1784–1792 (2000)

[SLTT01]   Sanchez-Soriano, J., Llorca, N., Tijs, S., Timmer, J.: Semi-infinite assignment and transportation games. In [GL01], 349–363 (2001)

[Sha01]   Shapiro, A.: On duality theory of conic linear problems. In [GL01], 135–165 (2001)

[Sha04]   Shapiro, A.: On duality theory of convex semi-infinite programming. Tech. Report, School of Industrial and Systems Engineering, Georgia Institute of Technology, Atlanta, GE (2004)

[SIF01]   Slupphaug, O., Imsland, L., Foss, A.: Uncertainty modelling and robust output feedback control of nonlinear discrete systems: A mathematical programming approach. Int. J. Robust Nonlinear Control, **10**, 1129–1152 (2000) [also in: Modeling, Identification and Control, **22**, 29–52 (2001)]

[SS03]   Stein, O., Still, G.: Solving semi-infinite optimization problems with interior point techniques. SIAM J. Control Optim., **42**, 769–788 (2003)

[Sti01]   Still, G.: Discretization in semi-infinite programming: The rate of convergence. Math. Programming (Ser. A), **91**, 53–69 (2001)

[TKVB02]   Tichatschke, R., Kaplan, A., Voetmann, T., Böhm, M.: Numerical treatment of an asset price model with non-stochastic uncertainty. Top, **10**, 1–50 (2002)

[TTLS01]   Tijs, J., Timmer, S., Llorca, N., Sanchez-Soriano, J.: In [GL01], 365–386 (2001)

[VB98]   Vandenberghe, L., Boyd, S.: Connections between semi-infinite and semi-definite programming. In Reemtsen, R., Rückmann, J. (ed) Semi-Infinite Programming, 277–294. Kluwer, Dordrecht (1998)

[VFG03]   Vaz, I., Fernandes, E., Gomes, P.: A sequential quadratic programming with a dual parametrization approach to nonlinear semi-infinite programming. Top **11**, 109–130 (2003)

[Web03]   Weber, G.-W.: Generalized Semi-Infinite Optimization and Related Topics. Heldermann Verlag, Lemgo, Germany (2003)

[WFL01] Wu, S.Y., Fang, S.-Ch., Lin, Ch.-J.: Analytic center based cutting plane method for linear semi-infinite programming. In [GL01], 221–233 (2001)

[ZR03]   Zakovic, S., Rustem, B.: Semi-infinite programming and applications to minimax problems. Annals Oper. Res., **124**, 81–110 (2003)

[ZNF00] Zavriev, S.K., Novikova, N.M., Fedosova, A.V.: Stochastic algorithm for solving convex semi-infinite programming problems with equality and inequality constraints (Russian, English). Mosc. Univ. Comput. Math. Cybern., **2000**, 44–52 (2000)

Some Theoretical Aspects of Newton's Method for Constrained Best Interpolation

Hou-Duo Qi

School of Mathematics, The University of Southampton
Highfield, Southampton SO17 1BJ, Great Britain
hdqi@soton.ac.uk

Summary. The paper contains new results as well as surveys on recent developments on the constrained best interpolation problem, and in particular on the convex best interpolation problem. Issues addressed include theoretical reduction of the problem to a system of nonsmooth equations, nonsmooth analysis of those equations and development of Newton's method, convergence analysis and globalization. We frequently use the convex best interpolation to illustrate the seemingly complex theory. Important techniques such as splitting are introduced and interesting links between approaches from approximation and optimization are also established. Open problems related to polyhedral constraints and strips may be tackled by the tools introduced and developed in this paper.

2000 MR Subject Classification. 49M45, 90C25, 90C33

1 Introduction

The convex best interpolation problem is defined as follows:

$$\text{minimize } \|f''\|_2 \tag{1}$$

$$\text{subject to } f(t_i) = y_i, \quad i = 1, 2, \cdots, n+2,$$
$$f \text{ is convex on } [a, b], \quad f \in W^{2,2}[a, b],$$

where $a = t_1 < t_2 < \cdots < t_{n+2} = b$ and y_i, $i = 1, \ldots, n+2$ are given numbers, $\|\cdot\|_2$ is the Lebesgue $L^2[a, b]$ norm, and $W^{2,2}[a, b]$ denotes the Sobolev space of functions with absolutely continuous first derivatives and second derivatives in $L^2[a, b]$, and equipped with the norm being the sum of the $L^2[a, b]$ norms of the function, its first, and its second derivatives.

Using an integration by parts technique, Favard [Fav40] and, more generally, de Boor [deB78] showed that this problem has an equivalent reformulation as follows:

$$\min \left\{ \|u\| \mid u \in L^2[a,b], \ u \geq 0, \ \langle u, x^i \rangle = d_i, \ i = 1, \ldots, n \right\}, \qquad (2)$$

where the functions $x^i \in L^2[a,b]$ and the numbers d_i can be expressed in terms of the original data $\{t_i, y_i\}$ (in fact, $x^i = B_i(t)$, the B-spline of order 2 defined by the given data and $\{d_i\}$ are the second divided differences of $\{(t_i, y_i)\}_{i=1}^{n+2}$). Under the assumption $d_i > 0$, $i = 1, \ldots, n$ the optimal solution u^* of (2) has the form

$$u^*(t) = \left(\sum_{i=1}^{n} \lambda_i^* B_i(t) \right)_+ \qquad (3)$$

where $\tau_+ := \max\{0, \tau\}$ and $\{\lambda_i^*\}$ satisfy the following interpolation condition:

$$\int_a^b \left(\sum_{i=1}^{n} \lambda_i B_i(t) \right)_+ B_i(t) dt = d_i, \qquad i = 1, \ldots, n. \qquad (4)$$

Once we have the solution u^*, the function required by (1) can be obtained by $f'' = u$. This representation result was obtained first by Hornung [Hor80] and subsequently extended to a much broader circle of problems in [AE87, DK89, IP84, IMS86, MSSW85, MU88]. We briefly discuss below both theoretically and numerically important progresses on those problems.

Theoretically, prior to [MU88] by Micchelli and Utreras, most of research is mainly centered on the problem (1) and its slight relaxations such as f'' is bounded below or above, see [IP84, MSSW85, IMS86, AE87, DK89]. After [MU88] the main focus is on to what degree the solution characterization like (3) and (4) can be extended to a more general problem proposed in Hilbert spaces:

$$\min \left\{ \frac{1}{2} \|x - x^0\|^2 \mid x \in C \text{ and } Ax = b \right\} \qquad (5)$$

where $C \subset X$ is a closed convex set in a Hilbert space X, $A : X \mapsto \mathbb{R}^n$ is a bounded linear operator, $b \in \mathbb{R}^n$. It is easy to see that if we let

$$X = L^2[a,b], \ C = \{x \in X \mid x \geq 0\}, \ Ax = (\langle B_1, x \rangle, \ldots, \langle B_n, x \rangle), \ x^0 = 0, \ b = d \qquad (6)$$

then (5) becomes (2). The abstract interpolation problem (5), initially studied in [MU88], was extensively studied in a series of papers by Chui, Deutsch, and Ward [CDW90, CDW92], Deutsch, Ubhaya, Ward, and Xu [DUWX96], and Deutsch, Li, and Ward [DLW97]. For the complete treatment on this problem in the spirit of those papers, see the recent book by Deutsch [Deu01].

Among the major developments in those papers is an important concept called the *strong CHIP* [DLW97], which is the refinement of the property CHIP [CDW90] (Conical Hull Intersection Property). More studies on the strong CHIP, CHIP and other properties can be found in the two recent papers [BBL99, BBT00]. Roughly speaking, the importance of the strong CHIP is with the following characterization result: The strong CHIP holds for the constraints in (5) if and only the unique solution x^* has the following representation:

$$x^* = P_C(x^0 + A^*\lambda^*), \tag{7}$$

where P_C denotes the projection to the closed convex set C (the closeness and convexity guarantees the existence of P_C), and A^* is the adjoint of A, and $\lambda^* \in \mathbb{R}^n$ satisfies the following nonlinear nonsmooth equation:

$$AP_C(x^0 + A^*\lambda) = b. \tag{8}$$

To see (7) and (8) recover (3) and (4) it is enough to use the fact:

$$P_C = x_+ \quad \text{where } C = \{x \in L^2[a, b] | x \geq 0\}.$$

If the strong CHIP does not hold we still have similar characterization in which P_C is replaced by P_{C_b}, where C_b is an extremal face of C satisfying some properties [Deu01]. However, it is often hard to get enough information to make the calculation of P_{C_b} possible, unless in some particular cases. Hence, we mainly focus on the case where the strong CHIP holds. We will see that the assumption $d_i > 0$, $i = 1, \ldots, n$ for problem (1) is a sufficient condition for the strong CHIP, and much more than that, it ensures the quadratic convergence of Newton's method.

Numerically, problem (1) has been well studied [IP84, IMS86, AE87, MU88, DK89, DQQ01, DQQ03]. As demonstrated in [IMS86] and verified in several other occasions [AE87, DK89], the Newton method is the most efficient compared to many other global methods for solving the equation (4). We delay the description of the Newton method to the end of Section 3, instead we list some difficulties in designing algorithms for (4) and (8). First of all, the equation (4) is generally nonsmooth. The nonsmoothness was a major barrier for Andersson and Elfving [AE87] to establish the convergence of Newton's method (they have to assume that the equation is smooth near the solution (the simple case) in order that the classical convergence result of Newton's method applies). Second, as having been both noticed in [IMS86, AE87], in the simple (i.e., smooth) case, the method presented in [IMS86, AE87] becomes the classical Newton method. More justification is needed to consolidate the name and the use of Newton's method when the equation is nonsmooth. To do this, we appeal to the theory of the generalized Newton method developed by Kummer [Kum88] and Qi and Sun [QS93] for nonsmooth equations. This was done in [DQQ01, DQQ03]. We will review this theory in Section 3. Third, Newton's method is only developed for the conical case, i.e., C is a cone. It is yet to know in what form the Newton method appears even for the polyhedral case (i.e., C is intersection of finitely many halfspaces). We will tackle those difficulties against the problem (5).

The problem (5) can also be studied via a very different approach developed by Borwein and Lewis [BL92] for partially finite convex programming problems:

$$\inf \{f(x) | \ Ax \in b + Q, \ x \in C\}, \tag{9}$$

where $C \in X$ is a closed convex set, X is a topological vector space, $A : X \mapsto \mathbb{R}^n$ is a bounded linear operator, $b \in \mathbb{R}^n$, Q is a polyhedral set in $\mathbb{R}^n$, and $f : X \mapsto (-\infty, \infty]$ is convex. If $f(x) = \|x^0 - x\|^2$ and $Q = \{0\}$, then (9) becomes (5). Under the constraint qualification that there is a feasible point which is in the quasi-relative interior of C, the problem (9) can be solved by its Fenchel-Rockafellar dual problem. We will see in the next section that this approach also leads to the solution characterization (7) and (8). See, e.g., [GT90, Jey92, JW92] for further development of Borwein-Lewis approach.

An interesting aspect of (9) is when $Q = \mathbb{R}^n_+$, the nonnegative orthant of $\mathbb{R}^n$. This yields the following approximation problem:

$$\min \left\{ \frac{1}{2} \|x^0 - x\|^2 | \ Ax \geq b, \ x \in C \right\}. \tag{10}$$

This problem was systematically studied by Deutsch, Li and Ward in [DLW97], proving that the strong CHIP again plays an important role but the sufficient condition ensuring the strong CHIP takes a very different form from that (i.e., $b \in \mathrm{ri} \ AC$) for (5). We will prove in Section 2 that the constraint qualification of Borwein and Lewis also implies the strong CHIP. Nonlinear convex and nonconvex extension of (10) can be found in [LJ02, LN02, LN03].

The paper is organized as follows: The next section contains some necessary background materials. In particular, we review the approach initiated by Micchelli and Utreras [MU88] and all the way to the advent of the strong CHIP and its consequences. We then review the approach of Borwein and Lewis [BL92] and state its implications by establishing the fact that the non-emptiness of the quasi-relative interior of the feasible set implies the strong CHIP. In section 3, we review the theory of Newton's method for nonsmooth equations, laying down the basis for the analysis of the Newton method for (5), which is conducted in Section 4. In the last section, we discuss some extensions to other problems such as interpolation in a strip. Throughout the paper we use the convex best interpolation problem (1) and (4) as an example to illustrate the seemingly complex theory.

2 Constrained Interpolation in Hilbert Space

Since X is a Hilbert space, the bounded linear operator $A : X \mapsto \mathbb{R}^n$ has the following representation: there exist $x_1, \ldots, x_n \in X$ such that

$$Ax = (\langle x_1, x \rangle, \ldots, \langle x_n, x \rangle), \quad \forall x \in X.$$

Defining

$$H_i := \{x \in X | \ \langle x_i, x \rangle = b_i\}, \quad i = 1, \ldots, n$$

the interpolation problem (5) has the following appearance

$$\min\left\{\frac{1}{2}\|x^0 - x\|^2\,\Big|\, x \in K := C \cap \left(\cap_{j=1}^n H_j\right)\right\}. \tag{11}$$

Recall that for any convex set $D \subset X$, the (negative) polar of D, denoted by D°, is defined by

$$D^\circ := \{y \in X|\ \langle y, x\rangle \leq 0,\ \forall x \in D\}.$$

The well-know strong CHIP is now defined as follows.

Definition 1. *[Deu01, Definition 10.2] A collection of closed convex sets $\{C_1, C_2, \ldots, C_m\}$ in X, which has a nonempty intersection, is said to have the strong conical hull intersection property, or the strong CHIP, if*

$$(\cap_1^m C_i - x)^\circ = \sum_1^m (C_i - x)^\circ \qquad \forall x \in \cap_1^m C_i.$$

The concept of the strong CHIP is a refinement of CHIP [DLW97], which requires

$$(\cap_1^m C_i - x)^\circ = \overline{\sum_1^m (C_i - x)^\circ} \qquad \forall x \in \cap_1^m C_i, \tag{12}$$

where $\bar{C}$ denotes the closure of C. It is worth mentioning that one direction of (12) is automatic, that is

$$(\cap_1^m C_i - x)^\circ \supseteq \overline{\sum_1^m (C_i - x)^\circ} \qquad \forall x \in \cap_1^m C_i.$$

Hence, the strong CHIP is actually assuming the other direction. The importance of the strong CHIP is with the following solution characterization of the problem (11).

Theorem 1. *[DLW97, Theorem 3.2] and [Deu01, Theorem 10.13] The set $\{C, \cap_1^n H_j\}$ has the strong CHIP if and only if for every $x^0 \in X$ there exists $\lambda^* \in \mathbb{R}^n$ such that the optimal solution $x^* = P_K(x^0)$ has the representation:*

$$x^* = P_C(x^0 + A^*\lambda^*)$$

and λ^* satisfies the interpolation equation

$$AP_C(x^0 + A^*\lambda) = b.$$

We remark that in general the strong CHIP of the sets $\{C, H_1, \ldots, H_n\}$ implies the strong CHIP of the sets $\{C, \cap_1^n H_j\}$. The following lemma gives a condition that ensures their equivalence.

Lemma 1. *[Deu01, Lemma 10.11] Suppose that X is a Hilbert space and $\{C_0, C_1, \ldots, C_m\}$ is a collection of closed convex subsets such that $\{C_1, \ldots, C_m\}$ has the strong CHIP. Then the following statements are equivalent:*

(i) $\{C_0, C_1, \ldots, C_m\}$ *has the strong CHIP.*
(ii) $\{C_0, \cap_1^m C_j\}$ *has the strong CHIP.*

Since each H_j is a hyperplane, $\{H_1, \ldots, H_n\}$ has the strong CHIP [Deu01, Example 10.9]. It follows from Lemma 1 that the strong CHIP of $\{C, H_1, \ldots, H_n\}$ is equivalent to that of $\{C, A^{-1}(b)\}$. However, it is often difficult to know if $\{C, A^{-1}(b)\}$ has the strong CHIP. Fortunately, there are available easy-to-be-verified sufficient conditions for this property. Given a convex subset $D \subset \mathbb{R}^n$, let ri D denote the relative of D. Note that ri $D \neq \emptyset$ if $D \neq \emptyset$.

Theorem 2. *[Deu01, Theorem 10.32] and [DLW97, Theorem 3.12] If $b \in$ ri AC, then $\{C, A^{-1}(b)\}$ has the strong CHIP.*

Theorem 2 also follows from the approach of Borwein and Lewis [BL92]. The concept of quasi-relative interior of convex sets plays an important role in this approach. We assume temporarily that X be a locally convex topological vector space. Let X^* denote the dual space of X (if X is a Hilbert space then $X^* = X$) and $N_C(\hat{x}) \subset X^*$ denote the normal cone to C at $\hat{x} \in C$, i.e.,

$$N_C(\hat{x}) := \{y \in X^* | \langle y, x - \hat{x} \rangle \leq 0, \quad \forall x \in C\}.$$

The most useful properties of the quasi-relative interiors are contained in the following

Proposition 1. *[BL92] Suppose $C \subset X$ is convex, then*

(i) If X is finite-dimensional then qri $C =$ ri C.
(ii) Let $\hat{x} \in C$ then $\hat{x} \in$ qri C if and only if $N_C(\hat{x})$ is a subspace of X^*.
(iii) Let $A : X \mapsto \mathbb{R}^n$ be a bounded linear map. If qri $C \neq \emptyset$ then $A(\text{qri } C) =$ ri AC.

We note that (ii) serves a definition for the quasi-relative interior of convex sets. One can find several other interesting properties of the quasi-relative interior in [BL92]. Although in finite-dimensional case quasi-relative interior becomes classical relative interior, it is a genuine new concept in infinite-dimensional cases. To see this, let $X = L^p[0,1]$, $(p \geq 1)$, $C := \{x \in X | x \geq 0 \text{ a.e.}\}$. Since C reproduces X (i.e., $X = C - C$), ri $C = \emptyset$, however, qri $C = \{x \in X | x > 0 \text{ a.e.}\}$. One of the basic results in [BL92] is

Theorem 3. *[BL92, Corollary 4.8] Let the assumptions on problem (9) hold. Consider its dual problem*

$$\max \left\{ -(f + \delta(\cdot|C))^*(A^*\lambda) + b^T\lambda | \lambda \in Q^+ \right\}. \qquad (13)$$

If the following constraint qualification is satisfied

$$\text{there exists an } \hat{x} \in \text{qri } C \text{ which is feasible for (9)}, \qquad (14)$$

then the value of (9) and (13) are equal with attainment in (13). Suppose further that $(f + \delta(\cdot|C))$ is closed. If λ^* is optimal for the dual and $(f + \delta(\cdot|C))^*$ is differentiable at $A^*\lambda^*$ with Gateaux derivative $x^* \in X$, then x^* is optimal for (9) and furthermore the unique optimal solution.

In (13), $Q^+ := \{y \in X^* | \langle y, x \rangle \geq 0, \forall x \in Q\}$. We now apply Theorem 3 to problem (11), i.e., we let

$$f(x) = \frac{1}{2}\|x^0 - x\|^2, \ Q = \{0\} \text{ so that } Q^+ = \mathbb{R}^n.$$

Obviously, in this case (9) has a unique solution since $f(x)$ is strongly convex. For $y \in X^*$ we calculate

$$(f + \delta(\cdot|C))^*(y) = \sup_{x \in X} \{\langle y, x \rangle - f(x) - \delta(x|C)\}$$

$$= \sup_{x \in C} \left\{\langle y, x \rangle - \frac{1}{2}\|x - x^0\|^2\right\}$$

$$= \sup_{x \in C} \left\{\langle x, y + x^0 \rangle - \frac{1}{2}\|x\|^2 - \frac{1}{2}\|x^0\|^2\right\}$$

$$= \sup_{x \in C} \left\{\frac{1}{2}\|y + x^0\|^2 - \frac{1}{2}\|y + x^0 - x\|^2 - \frac{1}{2}\|x^0\|^2\right\}$$

$$= \frac{1}{2}\|y + x^0\|^2 - \frac{1}{2}\|y + x^0 - P_C(y + x^0)\|^2 - \frac{1}{2}\|x^0\|^2. \quad (15)$$

It is well known (see, e.g., [MU88, Theorem 3.2]) that the right side of (15) is Gateaux differentiable with

$$\nabla(f + \delta(\cdot|C))^*(y) = P_C(y + x^0).$$

Returning to (13), which is an unconstrained convex optimization problem, we know that the optimal solution λ^* to (13) satisfies

$$AP_C(x^0 + A^*\lambda) = b$$

and the optimal solution to (9) is

$$x^* = P_C(x^0 + A^*\lambda^*).$$

Following Theorem 1 we see that the sets $\{C, A^{-1}(b)\}$ has the strong CHIP. In fact, the qualification (14) is exactly the condition $b \in \text{ri} \, (AC)$ by Proposition 1, except that (14) needs a priori assumption qri $C \neq \emptyset$.

However, for the problem (10), where

$$K = C \cap \{x | Ax \geq b\},$$

the condition $b \in \text{ri} \, AC$ is not suitable as it might happen that $b \notin AC$. It turns out that the strong CHIP again plays an essential role in this case. Let

$$\mathcal{H}_j := \{x | \langle a_j, x \rangle \geq b_j\}.$$

Theorem 4. *[DLW99, Theorem 3.2] The sets $\{C, \cap_1^n \mathcal{H}_j\}$ has the strong CHIP if and only if the optimal solution of (10) $x^* = P_K(x^0)$ has the following representation:*

$$x^* = P_C(x^* + A^*\lambda^*), \tag{16}$$

where λ^* is any solution of the nonlinear complementarity problem:

$$\lambda \geq 0, \; w := AP_C(x^0 + A^*\lambda) - b \geq 0, \; \lambda^T w = 0. \tag{17}$$

The following question was raised in [DLW99] that if the constraint qualification (14) is a sufficient condition for the strong CHIP of $\{C, \cap_1^n \mathcal{H}_j\}$. We give an affirmative answer in the next result.

Theorem 5. *If it holds*

$$qri\ C \cap (\cap_1^n \mathcal{H}_j) \neq \emptyset, \tag{18}$$

then the sets $\{C, \cap_1^n \mathcal{H}_j\}$ has the strong CHIP.

Proof. Suppose (18) is in place, it follows from Theorem 3 with $f(x) = \frac{1}{2}\|x - x^0\|^2$ that there exists an optimal solution λ^* to the problem (13). (15) says that

$$(f + \delta(\cdot|C))^*(y) = \frac{1}{2}\|y + x^0\|^2 - \frac{1}{2}\|y + x^0 - P_C(y + x^0)\|^2 - \frac{1}{2}\|x^0\|^2$$

and it is Gateaux differentiable and convex [MU88, Lemma 3.1]. Then (13) becomes

$$\min\left\{\frac{1}{2}\|A^*\lambda + x^0\|^2 - \frac{1}{2}\|A^*\lambda + x^0 - P_C(A^*\lambda + x^0)\|^2 - b^T\lambda|\ \lambda \geq 0\right\}.$$

It is a finite-dimensional convex optimization problem and the optimal solution is attained. Hence, the optimal solution λ^* is exactly a solution of (17) and the optimal solution of (10) is $x^* = P_C(x^0 + A^*\lambda)$. It then follows from the characterization in Theorem 4 that the sets $\{C, \cap_1^n \mathcal{H}_j\}$ has the strong CHIP. $\qquad\square$

Illustration to problem (2). We recall the problem (2) and the setting in (6). From the fact [BL92, Lemma 7.17]

$$\left\{\left(\int_a^b B_i x dt\right)_1^n \Big|\ x > 0 \text{ a.e. } x \in L^2[a, b]\right\} = \{r \in \mathbb{R}^n|\ r_i > 0, i = 1, \ldots, n\}$$

and the fact $qri\ C = \{x \in L^2[a, b]|\ x > 0 \text{ a.e.}\}$, we have

$$A qri\ C = ri\ AC = int\ AC = \{r \in \mathbb{R}^n|\ r_i > 0, i = 1, \ldots, n\}.$$

It follows from Theorem 2 or Theorem 3 that the solution to (2) is given by (3) and (4), under the assumption that $d_i > 0$ for all i. Moreover, we will see that this assumption implies the uniqueness of the solution λ^*, and eventually guarantees the quadratic convergence of the Newton method.

3 Nonsmooth Functions and Equations

As is well known, if $F : \mathbb{R}^n \mapsto \mathbb{R}^n$ is smooth the classical Newton method for finding a solution x^* of the equation $F(x) = 0$ takes the following form:

$$x^{k+1} = x^k - \left(F'(x^k)\right)^{-1} F(x^k) \tag{19}$$

where F' is the Jacobian of F. If $F'(x^*)$ is nonsingular then (19) is well defined near the solution x^* and is quadratically convergent. However, as we see from the previous sections we are encountered with nonsmooth equations. There is need to develop Newton's method for nonsmooth equation, which is presented below.

Now we suppose that $F : \mathbb{R}^n \mapsto \mathbb{R}^n$ is only locally Lipschitz and we want to find a solution of the equation

$$F(x) = 0. \tag{20}$$

Since F is differentiable almost everywhere according to Redemacher's theorem, the Bouligand differential of F at x, denoted by $\partial_B F(x)$, is defined by

$$\partial_B F(x) := \left\{ V \mid V = \lim_{x^i \to x} F'(x^i), \ F \text{ is differentiable at } x^i \right\}.$$

In other words, $\partial_B F(x)$ is the set of all limits of any sequence $\{F'(x^i)\}$ where F' exists at x^i and $x^i \to x$. The generalized Jacobian of Clark [Cla83] is then the convex hull of $\partial_B F(x)$, i.e.,

$$\partial F(x) = \operatorname{co} \partial_B F(x).$$

The basic properties of ∂F are included in the following result.

Proposition 2. *[Cla83, Proposition 2.6.2]*

(a) ∂F is a nonempty convex compact subset of $\mathbb{R}^{n \times n}$.
(b) ∂F is closed at x; that is, if $x^i \to x$, $M_i \in \partial F(x^i)$, $M_i \to M$, then $M \in \partial F(x)$.
(c) ∂F is upper semicontinuous at x.

Having the object of ∂F, the nonsmooth version of Newton's method for the solution of (20) can be described as follows (see, e.g., [Kum88, QS93]).

$$x^{k+1} = x^k - V_k^{-1} F(x^k), \quad V_k \in \partial F(x^k). \tag{21}$$

We note that different choice of V_k results in different sequence of $\{x^k\}$. Hence, it is more accurate to say that (21) defines a class of Newton-type methods rather than a single method. It is always arguable which element in $\partial F(x^k)$ is the most suitable in defining (21). We will say more about the choice with regard to the convex best interpolation problem. We also note that there are other ways in defining nonsmooth Newton's method, essentially using different definitions $\partial F(x)$, but servicing the same objective as ∂F, see, e.g., [JL98, Xu99, KK02].

Definition 2. *We say that F is regular at x if each element in $\partial F(x)$ is nonsingular.*

If F is regular at x^* it follows from the upper semicontinuity of F at x^* (Prop. 2) that F is regular near x^*, and consequently, (21) is well defined near x^*. Contrasted to the smooth case, the regularity at x^* only is no long a sufficient condition for the convergence of the method (21). It turns out that its convergence also relies on another important property of F, named the *semismoothness*.

Definition 3. *[QS93] We say that F is semismooth at x^* if the following conditions hold:*

(i) F is directionally differentiable at x, and
(ii)it holds

$$F(x + h) - F(x) - Vh = o(\|h\|) \qquad \forall V \in \partial F(x + h) \text{ and } h \in \mathbb{R}^n. \quad (22)$$

Furthermore, if

$$F(x + h) - F(x) - Vh = O(\|h\|^2) \qquad \forall V \in \partial F(x + h) \text{ and } h \in \mathbb{R}^n, \quad (23)$$

F is said strongly semismooth at x. If F is (strongly) semismooth everywhere, we simply say that F is (strongly) semismooth.

The property of semismoothness, as introduced by Mifflin [Mif77] for functionals and scalar-valued functions and further extended by Qi and Sun [QS93] for vector-valued functions, is of particular interest due to the key role it plays in the superlinear convergence of the nonsmooth Newton method (21). It is worth mentioning that in a largely ignored paper [Kum88] by Kummer, the relation (22), being put in a very general form in [Kum88], has been revealed to be essential for the convergence of a class of Newton type methods, which is essentially the same as (21). Nevertheless, Qi and Sun's work [QS93] makes it more accessible to and much easier to use by many researchers (see, e.g., the book [FP03] by Facchinei and Pang). The importance of the semismoothness can be seen from the following convergence result for (21).

Theorem 6. *[QS93, Theorem 3.2] Let x^* be a solution of the equation $F(x) = 0$ and let F be a locally Lipschitz function which is semismooth at x^*. Assume that F is regular at x^*. Then every sequence generated by the method (21) is superlinearly convergent to x^* provided that the starting point x^0 is sufficiently close to x^*. Furthermore, if F is strongly semismooth at x^*, then the convergence rate is quadratic.*

The use of Theorem 6 relies on the availability of the following three elements: (a) availability of an element in $\partial F(x)$ near the solution x^*, (b) regularity of F at x^* and, (c) (strong) semismoothness of F at x^*. We illustrate how the first can be easily calculated below for the convex best interpolation problem and leave the other two tasks to the next section.

Illustration to the convex best interpolation problem. It follow from (3) and (4) that the solution of the convex best interpolation problem can be obtained by solving the following equation:

$$F(\lambda) = d, \tag{24}$$

where $d = (d_1, \ldots, d_n)^T$ and each component of F is given by

$$F_j(\lambda) = \int_a^b \left(\sum_{\ell=1}^n \lambda_\ell B_\ell \right)_+ B_j(t)dt, \quad j = 1, \ldots, n. \tag{25}$$

Irvine, Marin, and Smith [IMS86] developed Newton's method for (24):

$$\lambda^+ = \lambda - (M(\lambda))^{-1} (F(\lambda) - d), \tag{26}$$

where λ and λ^+ denote respectively the old and the new iterate, and $M(\lambda) \in \mathbb{R}^{n \times n}$ is given by

$$(M(\lambda))_{ij} = \int_a^b \left(\sum_{\ell=1}^n \lambda_\ell B_\ell \right)_+^0 B_i(t)B_j(t)dt,$$

and

$$(\tau)_+^0 = \begin{cases} 1 \text{ if } \tau > 0 \\ 0 \text{ if } \tau \le 0. \end{cases}$$

Let e denote the element of all ones in $\mathbb{R}^n$, then it is easy to see that the directional derivative of F at λ along the direction e is

$$F'(\lambda, e) = M(\lambda)e.$$

Moreover, if F is differentiable at λ then $F'(\lambda) = M(\lambda)$. Due to those reasons, the iteration (26) was then called Newton's method, and based on extensive numerical experiments, was observed quadratically convergent in [IMS86]. Independent of [IMS86], partial theoretical results on the convergence of (26) was established by Andersson and Elfving [AE87]. Complete convergence analysis was established by Dontchev, Qi, and Qi [DQQ01, DQQ03] by casting (26) as a particular instance of (21). The convergence analysis procedure verifies exactly the availability of the three elements discussed above, in particular, $M(\lambda) \in \partial F(\lambda)$. We will present in the next section the procedure on the constrained interpolation problem in Hilbert space.

4 Newton's Method and Convergence Analysis

4.1 Newton's Method

We first note that all results in Section 2 assume no other requirements for the set C except being convex and closed. Consequently, we are able to develop (conceptual, at least) Newton's method for the nonsmooth equation (8). However, efficient implementation of Newton's method relies on the assumption that there is an efficient way to calculate the generalized Jacobian of $AP_C(x)$. The most interesting case due to this consideration is when C is a closed convex cone (i.e., the conical case [BL92]), which covers many problems including (1). We recall our setting below

$$X = L^2[a, b], \ C = \{x \in X | x \geq 0\}, Ax = (\langle a_1, x \rangle, \ldots, \langle a_n, x \rangle), \ b \in \mathbb{R}^n$$

where $a_\ell \in X$, $\ell = 1, \ldots, n$ (in fact we may assume that $X = L^p[a, b]$, in this case $a_\ell \in L^q[a, b]$ where $1/p + 1/q = 1$). This setting simplifies our description.

We want to develop Newton's method for the equation:

$$AP_C(x^0 + A^*\lambda) = b.$$

Taking into account of the fact $P_C(x) = x_+$, we let

$$F(\lambda) - b = 0 \tag{27}$$

where each component of $F : \mathbb{R}^n \mapsto \mathbb{R}^n$ is given by

$$F_j(\lambda) := \langle a_j, (x^0 + \sum_{\ell=1}^{n} a_\ell \lambda_\ell)_+ \rangle. \tag{28}$$

We propose a nonsmooth Newton method (in the spirit of Section 3) for nonsmooth equation (27) as follows:

$$V(\lambda)(\lambda^+ - \lambda) = b - F(\lambda), \qquad V(\lambda) \in \partial F(\lambda). \tag{29}$$

One of several difficulties with the Newton method (29) is to select an appropriate matrix $V(\lambda)$ from $\partial F(\lambda)$, which is well defined as F is Lipschitz continuous under Assumption 1 stated later. We will also see the following choice satisfies all the requirements.

$$(V(\lambda))_{ij} := \int_a^b \left(x^0 + \sum_{\ell=1}^{n} \lambda_\ell a_\ell \right)_+^0 a_i a_j dt. \tag{30}$$

We note that for $\beta \in \mathbb{R}^n$

$$\beta^T V(\lambda)\beta = \int_a^b \left(x^0 + \sum_{\ell=1}^{n} \lambda_\ell a_\ell \right)_+^0 \left(\sum_{\ell=1}^{n} \beta_\ell a_\ell \right)^2 dt \geq 0. \tag{31}$$

That is, $V(\lambda)$ is positive semidefinite for arbitrary choice $\lambda \in \mathbb{R}^n$. We need an assumption to make it positive definite. Let the support of a_ℓ be

$$\text{supp}(a_\ell) := \{t \in [a, b] | a_\ell(t) \neq 0\}.$$

Assumption 1. *Each a_ℓ is continuous, and any subset of functions*

$$\{a_\ell, \ell \in \mathcal{I} \subseteq \{1,\ldots,n\} | supp(a_i) \cap supp(a_j) \neq \emptyset \text{ for any pair } i,j \in \mathcal{I}\},$$

are linearly independent on $\cup_{\ell \in \mathcal{I}} supp(a_\ell)$. Moreover,

$$\cup_{\ell=1}^n supp(a_\ell) = [a,b].$$

This assumption is not restrictive. Typical choices of a_ℓ are $\{a_i = t^i\}$ or $\{a_i = B_i\}$. With Assumption 1 we have the following result.

Lemma 2. *Suppose Assumption 1 holds. $V(\lambda)$ is positive definite if and only if $(x^0 + \sum_{\ell=1}^n \lambda_\ell a_\ell)_+$ does not vanish identically on the supporting set of each a_ℓ, $\ell = 1,\ldots,n$.*

Proof. Suppose that $(x^0 + \sum_{\ell=1}^n \lambda_\ell a_\ell)_+$ is nonzero on each $supp(a_\ell)$. Due to the continuity of $(x^0 + \sum_{\ell=1}^n \lambda_\ell a_\ell)$ and a_ℓ, there exists a Borel set $\Omega_\ell \subseteq supp(a_\ell)$ such that $(x^0 + \sum_{\ell=1}^n \lambda_\ell a_\ell)_+^0 = 1$ for all $t \in \Omega_\ell$ and the measure of Ω_ℓ is not zero. Let

$$\mathcal{I}(\Omega_\ell) := \{j | supp(a_j) \cap \Omega_\ell \neq \emptyset\}.$$

Since $\{a_j | j \in \mathcal{I}(\Omega_\ell)\}$ are linearly independent, $\beta^T V(\lambda)\beta = 0$ implies $\beta_j = 0$ for all $j \in \mathcal{I}(\Omega_\ell)$. We also note that

$$\cup_{\ell=1}^n \mathcal{I}(\Omega_\ell) = \{1,\ldots,n\}.$$

We see that $\beta_j = 0$ for all $j = 1,\ldots,n$ if $\beta^T V(\lambda)\beta = 0$. Hence, (31) yields the positive definiteness of $V(\lambda)$. The converse follows from the observation that if $(x^0 + \sum_{\ell=1}^n \lambda_\ell a_\ell)_+ \equiv 0$ on $supp(a_\ell)$ for some ℓ then $\beta^T V(\lambda)\beta = 0$ for $\beta \in \mathbb{R}^n$ with $\beta_\ell = 1$ and $\beta_j = 0$ for $j \neq \ell$. $\square$

Due to the special structure of $V(\lambda)$, Newton's method (29) can be simplified by noticing that

$$F_j(\lambda) = \int_a^b \left(x^0 + \sum_{\ell=1}^n \lambda_\ell a_\ell\right)_+ a_j dt$$

$$= \int_a^b \left(x^0 + \sum_{\ell=1}^n \lambda_\ell a_\ell\right)_+^0 \left(x^0 + \sum_{\ell=1}^n \lambda_\ell a_\ell\right)_+ a_j dt$$

$$= \sum_{\ell=1}^n \lambda_\ell (V(\lambda))_{j\ell} + \int_a^b \left(x^0 + \sum_{\ell=1}^n \lambda_\ell a_\ell\right)_+^0 a_j x^0 d(t).$$

Thus we have

$$F(\lambda) = V(\lambda)\lambda + A\left(\left(x^0 + \sum_{\ell=1}^n \lambda_\ell a_\ell\right)_+^0 x^0\right).$$

Recalling (29) we have

$$V(\lambda)\lambda^+ = b - A\left(\left(x^0 + \sum_{\ell=1}^{n} \lambda_\ell a_\ell\right)_+^0 x^0\right). \tag{32}$$

A very interesting case is when $x^0 = 0$, which implies that no function evaluations are required to implement Newton's method, i.e, (32) takes the form $V(\lambda)\lambda^+ = b$.

Other choices of $V(\lambda)$ are also possible as $\partial F(\lambda)$ usually contains infinitely many elements. For example,

$$\left(\tilde{V}(\lambda)\right)_{ij} := \int_a^b \left(x^0 + \sum_{\ell=1}^{n} \lambda_\ell a_\ell\right)_-^0 a_i a_j dt, \quad \text{and} \quad (\tau)_-^0 := \begin{cases} 1 & \text{if } \tau \geq 0 \\ 0 & \text{if } \tau < 0. \end{cases}$$

It is easy to see that $\beta^T \tilde{V}(\lambda)\beta \geq \beta^T V(\lambda)\beta$ for any $\beta \in \mathbb{R}^n$. This means that $\tilde{V}(\lambda)$ "increases the positivity" of $V(\lambda)$ in the sense that $\tilde{V}(\lambda) - V(\lambda)$ is positive semidefinite. The argument leading to (32) also applies to $\tilde{V}(\lambda)$. We will show below that both $V(\lambda)$ and $\tilde{V}(\lambda)$ are contained in $\partial F(\lambda)$.

4.2 Splitting and Regularity

We now introduce a splitting technique that decomposes the (nonsmooth) function F into two parts, namely F^+ and F^-, satisfying that F^+ is continuously differentiable at the given point and F^- is necessarily nonsmooth nearby. This technique facilitates our arguments that lead to the conclusion that $V(\lambda)$ belongs to $\partial F(\lambda)$ and pave the ways to study the regularity of F at the solution. For the moment, we let $\bar{\lambda}$ be our reference point. Let

$$T(\bar{\lambda}) := \{t \in [a,b] |\, x^0 + \sum_{\ell=1}^{n} \bar{\lambda}_\ell a_\ell = 0\}, \qquad \bar{T}(\bar{\lambda}) := [a,b] \setminus T(\bar{\lambda}).$$

Due to Assumption (1), $T(\bar{\lambda})$ contains closed intervals in $[a,b]$, possibly isolated points. For $j = 1, \ldots, n$, define

$$F_j^+(\lambda) := \int_{\bar{T}(\bar{\lambda})} \left(x^0 + \sum_{\ell=1}^{n} \lambda_\ell a_\ell\right)_+ a_j dt,$$

$$F_j^-(\lambda) := \int_{T(\bar{\lambda})} \left(x^0 + \sum_{\ell=1}^{n} \lambda_\ell a_\ell\right)_+ a_j dt,$$

and

$$F^+(\lambda) := (F_1^+(\lambda), \ldots, F_n^+(\lambda))^T, \quad F^-(\lambda) := (F_1^-(\lambda), \ldots, F_n^-(\lambda))^T.$$

It is easy to see that

$$F(\lambda) = F^+(\lambda) + F^-(\lambda).$$

It is elementary to see that the vector-valued function F^+ is continuous differentiable in a neighborhood $\mathcal{N}(\bar{\lambda})$ of $\bar{\lambda}$. Then from the definition of the generalized Jacobian we obtain that for any $\lambda \in \mathcal{N}(\bar{\lambda})$,

$$\partial F(\lambda) = \nabla F^+(\lambda) + \partial F^-(\lambda), \tag{33}$$

where $\nabla F^+(\lambda)$ denotes the usual Jacobian of F^+ at λ. More precisely,

$$\left(\nabla F^+(\bar{\lambda})\right)_{ij} = \int_{\bar{T}(\bar{\lambda})} \left(x^0 + \sum_{\ell=1}^{n} \bar{\lambda} a_\ell\right)_+^0 a_i a_j dt. \tag{34}$$

Since

$$x^0 + \sum_{\ell=1}^{n} \bar{\lambda} a_\ell = 0 \quad \text{for all } t \in T(\bar{\lambda}),$$

(34) can be written as

$$\left(\nabla F^+(\bar{\lambda})\right)_{ij} = \int_a^b \left(x^0 + \sum_{\ell=1}^{n} \bar{\lambda} a_\ell\right)_+^0 a_i a_j dt = V(\bar{\lambda}). \tag{35}$$

Regarding to F^- we need following assumption:

Assumption 2. *There exists a sequence of $\{\lambda^k\}$ in $\mathbb{R}^n$ converging to zero such that the sum $\sum_{\ell=1}^{n} \lambda_\ell^k a_\ell$ is negative on $[a,b]$ for all λ^k.*

This assumption also holds if each of a_ℓ is nonnegative or nonpositive.

Lemma 3. *For any $\lambda \in \mathbb{R}^n$ every element in $\partial F^-(\lambda)$ is positive semidefinite. Moreover, if Assumption 2 holds then the zero matrix belongs to $\partial F^-(\bar{\lambda})$.*

Proof. We denote

$$y := \left(x^0 + \sum_{\ell=1}^{n} \lambda_\ell a_\ell\right)_+ \chi_{T(\bar{\lambda})},$$

where $\chi_{T(\bar{\lambda})}$ is the characteristic function of the set $T(\bar{\lambda})$. In terms of y, F^- can be written as $F^-(\lambda) = Ay$. Since $T(\bar{\lambda})$ consists of only closed intervals, without loss of generality we assume $T(\bar{\lambda})$ is a closed interval. Let

$$C := \left\{x \in L^2(T(\bar{\lambda})) \mid x \geq 0\right\}.$$

Then we have $L^2[a,b] \subseteq L^2(T(\bar{\lambda}))$ since $(T(\bar{\lambda})) \subseteq [a,b]$. Define

$$\theta(\lambda) := \int_{T(\bar{\lambda})} \left(x^0 + \sum_{\ell=1}^{n} \lambda_\ell a_\ell\right)_+^2 dt = \int_{T(\bar{\lambda})} \left(P_C\left(x^0 + \sum_{\ell=1}^{n} \lambda_\ell a_\ell\right)\right)^2 dt.$$

According to [MU88, Lemma 2.1], $\theta(\lambda)$ is continuously Gateaux differentiable and convex. Moreover,

$$\nabla\theta(\lambda) = Ay = F^-(\lambda).$$

Therefore, any matrix in the generalized Jacobian of the gradient mapping (which is required to be Lipschitz continuous) of a convex function must be positive semidefinite, see, for example, [JQ95, Proposition 2.3]. Now we prove the second part. Suppose Assumption 2 holds for the sequence $\{\lambda^k\}$ which converges to *zero*. Then $F^-(\bar{\lambda} + \lambda^k)$ is differentiable because

$$\sum_{\ell=1}^{n}(\bar{\lambda} + \lambda^k)_\ell a_\ell < 0 \quad \text{for all } t \in T(\bar{\lambda}) \text{ and } \tau > 0.$$

Hence,

$$\lim_{k \to \infty} \nabla F^-(\bar{\lambda} + \lambda^k) = 0 \in \partial F^-(\bar{\lambda}).$$

$\square$

We then have

Corollary 1. *For any* $\lambda \in \mathbb{R}^n$, $V(\lambda) \in \partial F(\lambda)$.

Proof. It follows from Lemma 3 that $0 \in \partial F^-(\bar{\lambda})$ and from (35) that $V(\bar{\lambda}) = \nabla F^+(\bar{\lambda})$. The relation (33) then implies $V(\bar{\lambda}) \in \partial F(\bar{\lambda})$. Since $\bar{\lambda}$ is arbitrary we are done. $\square$

We need another assumption for our regularity result.

Assumption 3. $b_\ell \neq 0$ *for all* $\ell = 1, \ldots, n$.

Lemma 4. *Suppose Assumptions (1), (2) and (3) hold and let* λ^* *be the solution of (27). then every element of* $\partial F(\lambda^*)$ *is positive definite.*

Proof. We have proved that

$$\partial F(\lambda^*) = \partial F^-(\lambda^*) + \nabla F^+(\lambda^*) = \partial F^-(\lambda^*) + V(\lambda^*)$$

and every element in $\partial F^-(\lambda^*)$ is positive semidefinite. It is enough to prove $\nabla F^+(\lambda^*)$ is positive definite. We recall that at the solution

$$b_i = F_i(\lambda^*) = \int_a^b \left(x^0 + \sum_{\ell=1}^{n} \lambda_\ell^* a_\ell \right)_+ a_i dt, \qquad \forall i = 1, \ldots, n.$$

The assumption (3) implies that $\left(x^0 + \sum_{\ell=1}^{n} \lambda_\ell^* a_\ell \right)_+$ does not vanish identically at the support of each a_i. Then Lemma 2 implies that $\nabla F^+(\lambda^*) = V(\lambda^*)$ is positive definite. $\square$

Illustration to problem (2). An essential assumption for problem (2) is that the second divided difference is positive, i.e., $d_i > 0$ for all $i = 1, \ldots, n$. Hence, Assumption (3) is automatically valid. It is easy to see that Assumptions (1) and (2) are also satisfied for B-splines. It follows from the above argument that the Newton method (26) is well defined near the solution. However, to prove its convergence we need the semismoothness property of F, which is addressed below.

4.3 Semismoothness

As we see from Theorem 6 that the property of semismoothness plays an important role in convergence analysis of nonsmooth Newton's method (21). In our application it involves functions of following type:

$$\Phi(\lambda) := \int_0^1 \phi(\lambda, t)dt \tag{36}$$

where $\phi : \mathbb{R}^n \times [a,b] \mapsto \mathbb{R}$ is a locally Lipschitz mapping. The following development is due to D. Ralph [Ral02] and relies on a characterization of semismoothness using the Clarke generalized directional derivative.

Definition 4. *[Cla83] Suppose $\psi : \mathbb{R}^n \mapsto \mathbb{R}$ is locally Lipschitz. The generalized directional derivative of ψ which, when evaluated at λ in the direction h, is given by*

$$\psi^\circ(\lambda; h) := \limsup_{\substack{\beta \to \lambda \\ \delta \downarrow 0}} \frac{\psi(\beta + \delta h) - \psi(\beta)}{\delta}.$$

The different quotient when upper limit is being taken is bounded above in light of Lipschitz condition. So $\psi^\circ(\lambda; h)$ is well defined finite quantity. An important property of ψ° is that for any h,

$$\psi^\circ(\lambda; h) = \max\{\langle \xi, h \rangle | \, \xi \in \partial\psi(\lambda)\}. \tag{37}$$

We now have the following characterization of semismoothness.

Lemma 5. *[Ral02] A locally Lipschitz function $\psi : \mathbb{R}^n \mapsto \mathbb{R}$ is semismooth at $\bar{\lambda}$ if and only if ψ is directionally differentiable and*

$$\begin{aligned}
\psi(\lambda) + \psi^\circ(\lambda; \bar{\lambda} - \lambda) - \psi(\bar{\lambda}) &\leq o(\|\lambda - \bar{\lambda}\|), \text{ and} \\
\psi(\lambda) - \psi^\circ(\lambda; -\bar{\lambda} + \lambda) - \psi(\bar{\lambda}) &\geq o(\|\lambda - \bar{\lambda}\|).
\end{aligned} \tag{38}$$

The equivalence remains valid if the inequalities are replaced by equalities.

Proof. Noticing that (37) implies $-\psi^\circ(\lambda, -h) = \min_{\xi \in \partial\psi(\lambda)} h^T \xi$, the conditions in (38) are equivalent to

$$\psi(\lambda) + [-\psi^\circ(\lambda; -\bar{\lambda} + \lambda), \psi^\circ(\lambda; \bar{\lambda} - \lambda)] - \psi(\bar{\lambda}) = o(\|\lambda - \bar{\lambda}\|).$$

Combining with the directional differentiability of ψ, this set-valued equation clearly implies the semismoothness of ψ at $\bar{\lambda}$ because for any $\xi \in \partial\psi(\lambda)$, we have

$$\xi^T(\bar{\lambda} - \lambda) \in [-\psi^\circ(\lambda; -\bar{\lambda} + \lambda), \psi^\circ(\lambda; \bar{\lambda} - \lambda)].$$

Conversely, if ψ is semismooth at $\bar{\lambda}$ then for any λ we take an element $\xi \in \partial\psi(\lambda)$ (respectively) to obtain

$$\psi^\circ(\lambda, \bar{\lambda} - \lambda) = \xi^T(\bar{\lambda} - \lambda) \quad (\text{respectively } -\psi^\circ(\lambda; -\bar{\lambda} + \lambda) = \xi^T(-\bar{\lambda} + \lambda)).$$

The existence of such ξ follows from compactness of $\partial\psi(\lambda)$. Then the required inequalities follows from the semismoothness of ψ at $\bar{\lambda}$. $\qquad\square$

Now we have our major result concerning the function in (36).

Proposition 3. *[Ral02] Let $\phi : \mathbb{R}^n \times [0,1] \mapsto \mathbb{R}$. Suppose for every $t \in [0,1]$ $\phi(\cdot, t)$ is semismooth at $\lambda \in \mathbb{R}^n$. Then Φ defined in (36) is also semismooth at λ.*

Proof. The directional differentiability of Φ follows from the first part of [DQQ01, Proposition 3.1]. Now we use Lemma 5 to prove the semismoothness of Φ. To this purpose it is enough to establish the following relation:

$$\int_0^1 \left(\phi(\lambda, t) + \phi^\circ((\lambda, t); (\bar{\lambda} - \lambda, 0)) - \phi(\bar{\lambda}, t) \right) dt = o(\|\lambda - \bar{\lambda}\|). \quad (39)$$

This implies

$$\Phi(\lambda) - \Phi^\circ(\lambda; \bar{\lambda} - \lambda) - \Phi(\lambda) \le o(\|\lambda - \bar{\lambda}\|)$$

because the first principles give

$$\Phi^\circ(\lambda; \bar{\lambda} - \lambda) \le \int_0^1 \phi^\circ((\lambda, t); (\bar{\lambda} - \lambda, 0)) dt.$$

If in (39) we replace $\phi^\circ((\lambda, t); (\bar{\lambda} - \lambda, 0))$ by $-\phi^\circ((\lambda, t); (-\bar{\lambda} + \lambda, 0))$ and follow an argument that is almost identical to the subsequent development, we obtain the counter condition

$$\Phi(\lambda) - \Phi^\circ(\lambda; -\bar{\lambda} + \lambda) - \Phi(\lambda) \ge o(\|\lambda - \bar{\lambda}\|)$$

and the proof is sealed in Lemma 5.

Now let U be the closed unit ball in $\mathbb{R}^n$ and

$$e(\cdot, y) = \phi(y) + \phi^\circ(y; \cdot - y) - \phi(\cdot), \qquad y \in \mathbb{R}^n \times [0,1].$$

Let $\epsilon > 0$ we will find $\delta > 0$ such that if $\lambda \in \bar{\lambda} + \delta U$ then

$$\int_0^1 e((\bar{\lambda}, t), (\lambda, t)) dt \le \epsilon \|\lambda - \bar{\lambda}\|.$$

Since ϵ can be made arbitrarily small, verifying existence of δ is equivalent to verifying (39).

For any $\delta > 0$ let

$$\Delta(\delta) := \left\{ t \in [0,1] \mid e((\bar{\lambda}, t), (\lambda, t)) \le \frac{\epsilon}{2} \|\lambda - \bar{\lambda}\|, \ \forall \lambda \in \bar{\lambda} + \delta U \right\}.$$

For each $\lambda \in \mathbb{R}^n$ the mapping $t \mapsto e((\bar{\lambda}, t), (\lambda, t))$ is measurable, hence the set

$$\left\{ t \mid e((\bar{\lambda}, t), (\lambda, t)) \le \frac{\epsilon}{2} \|\lambda - \bar{\lambda}\| \right\}$$

is also measurable. Thus, $\Delta(\delta)$, the interior of measurable sets, is itself measurable. Obviously, $\Delta(\delta) \subseteq \Delta(\delta')$ if $\delta \ge \delta'$. And for fixed $t \in [0,1]$, semismoothness gives, via Lemma 5, that

$$\frac{e((\bar{\lambda}, t), (\lambda, t))}{\|\lambda - \bar{\lambda}\|} \to 0 \text{ as } 0 \neq \lambda - \bar{\lambda} \to 0,$$

i.e., for all small enough $\delta > 0$, $t \in \Delta(\delta)$.

Let $\Omega(\delta) := [0, 1] \setminus \Delta(\delta)$. The properties of $\Delta(\delta)$ yields (a) measurability of $\Omega(\delta)$, (b) $\Omega(\delta) \supseteq \Omega(\delta')$ if $\delta \geq \delta'$, and (c) for each t and all small enough $\delta > 0$, $t \notin \Omega(\delta)$. In particular, $\cap_{\delta > 0} \Omega(\delta) = \emptyset$ and it follows that the measure of $\Omega(\delta)$, meas$(\Omega(\delta))$, converges to 0 as $\delta \to 0_+$.

Let L be the Lipschitz constant of ϕ in a neighborhood of $(\bar{\lambda}, 0)$, so that for each λ near $\bar{\lambda}$,

$$\begin{aligned}
e((\bar{\lambda}, t), (\lambda, t)) &\leq |\phi(\lambda, t) - \phi(\bar{\lambda}, t)| + |\phi^\circ(\lambda, t); (\bar{\lambda} - \lambda, 0))| \\
&\leq 2L\|(\lambda - \bar{\lambda}, 0)\| = 2L\|\lambda - \bar{\lambda}\|
\end{aligned}$$

using the 2-norm. To sum up,

$$\begin{aligned}
\int_0^1 e((\bar{\lambda}, t), (\lambda, t)) dt &= \left(\int_{\Omega(\delta)} + \int_{\Delta(\delta)} \right) e((\bar{\lambda}, t), (\lambda, t)) dt \\
&\leq (2L\|\lambda - \bar{\lambda}\|) \text{meas}(\Omega(\delta)) + (\|\lambda - \bar{\lambda}\|\epsilon/2) \text{meas}(\Delta(\delta)) \\
&\leq \|\lambda - \bar{\lambda}\|(2L \text{meas}(\Omega(\delta)) + \epsilon/2).
\end{aligned}$$

Choose $\delta > 0$ small enough such that meas$(\Omega(\delta)) < \epsilon/(4L)$, and we are done.
□

Corollary 2. *Under Assumption 1, the functions F_j defined in (28) are each semismooth.*

Proof. For each $t \in [a, b]$, the mapping $\phi_j : \mathbb{R}^n \mapsto \mathbb{R}$ by

$$\phi_j(\lambda, t) = a_j(t)(x^0 + \sum_{\ell=1}^n \lambda_\ell a_\ell)_+$$

is piecewise linear with respect to λ, and hence is semismooth. Then Proposition 3 implies that each F_j defined in (28) is semismooth since $F_j(\lambda) = \int_a^b \phi_j(\lambda, t) dt$.
□

Now we are ready to use Theorem 6 of Qi and Sun [QS93] to establish the superlinear convergence of the Newton method (29) for the equation (27).

Theorem 7. *Suppose that Assumptions (1), (2) and (3) hold. Then Newton's method (29) for (27) is superlinearly convergent provided that the initial point λ^0 is close enough to the unique solution λ^*.*

Proof. Three major elements for the use of Theorem 6 have been established: (i) $V(\lambda) \in \partial F(\lambda)$ for any $\lambda \in \mathbb{R}^n$ (see, corollary 1), (ii) F is regular at λ^* (see, Lemma 4), and (iii) F is semismooth since each F_j is semismooth (see, Corollary 2). The result follows the direct application of Theorem 6 to the equation (27).
□

Illustration to (26). The superlinear convergence of the method (26) is a direct consequence of Theorem 7 because all the assumptions for Theorem 7 are satisfied for the convex best interpolation problem (1). This recovers the main result in [DQQ01]. Refinement of some results in [DQQ01] by taking into account of special structures of the B-splines leads to the quadratic convergence analysis conducted in [DQQ03].

4.4 Application to Inequality Constraints

Now we consider the approximation problem given by inequality constraints:

$$K = C \cap \{x | Ax \leq b\}.$$

Under the strong CHIP assumption, we have solution characterization (16) and (17), which we restate below for easy reference.

$$\lambda \geq 0, \ w := AP_C(x^0 + A^*\lambda) - b \geq 0, \ \lambda^T w = 0. \tag{40}$$

Again for computational consideration we assume that C is the cone of positive functions so that $P_C(x) = x_+$. Below we design Newton's method for (40) and study when it is superlinearly convergent. To do this, we use the well-known Fischer-Burmeister NCP function, widely studied in nonlinear complementarity problems [Fis92, SQ99], to reformulate (40) as a system (semismooth) equations.

Recall the Fischer-Burmeister function is given by

$$\phi_{FB}(a, b) := a + b - \sqrt{a^2 + b^2}.$$

Two important properties of ϕ_{FB} are

$$\phi_{FB}(a, b) = 0 \iff a \geq 0, \ b \geq 0, \ ab = 0$$

and the square ϕ_{FB}^2 is continuously differentiable, though ϕ_{FB} is not differentiable. Define

$$\Phi_{FB}(\lambda, w) := \begin{pmatrix} \phi_{FB}(\lambda_1, w_1) \\ \vdots \\ \phi_{FB}(\lambda_n, w_n) \end{pmatrix}$$

and

$$W(\lambda, w) := \begin{pmatrix} AP_C(x^0 + A^*\lambda) - w - b \\ \Phi_{FB}(\lambda, w) \end{pmatrix}.$$

Then it is easy to see that (40) is equivalent to the nonsmooth equation

$$W(\lambda, w) = 0.$$

Since W is locally Lipschitz, direct calculation gives

$$\partial W(\lambda, w) \subseteq \left\{ \begin{pmatrix} V(\lambda) & -I \\ D(\lambda, w) & E(\lambda, w) \end{pmatrix} \middle| \begin{array}{l} V(\lambda) \in \partial F(\lambda) \\ D(\lambda, w), E(\lambda, w) \text{ satisfy (42) and (43)} \end{array} \right\}.$$
(41)

$D(\lambda, w)$ and $E(\lambda, w)$ are diagonal matrices whose ℓth diagonal element is given by

$$D_\ell(\lambda, w) := 1 - \frac{\lambda_\ell}{\|(\lambda_\ell, w_\ell)\|}, \quad E_\ell(\lambda, w) := 1 - \frac{w_\ell}{\|(\lambda_\ell, w_\ell)\|} \tag{42}$$

if $(\lambda_\ell, w_\ell) \neq 0$ and by

$$D_\ell(\lambda, w) = 1 - \xi_\ell, \quad E_\ell(\lambda, w) = 1 - \rho_\ell, \quad \forall (\xi_\ell, \rho_\ell) \in \mathbb{R}^2 \text{ such that } \|(\xi_\ell, \rho_\ell)\| \leq 1 \tag{43}$$

if $(\lambda_\ell, w_\ell) = 0$.

Lemma 6. *Suppose every element $V(\lambda)$ in $\partial F(\lambda)$ is positive definite. Then every element of $\partial W(\lambda, w)$ is nonsingular.*

Proof. Let $M(\lambda, w)$ be an element of the right side set in (41) and let $(y, z) \in \mathbb{R}^{2n}$ be such that $M(y, z) = 0$. Then there exist $V(\lambda) \in \partial F(\lambda)$ and $D(\lambda, w)$ and $E(\lambda, w)$ satisfying (42) and (43) such that

$$V(\lambda)y - z = 0 \quad \text{and} \quad D(\lambda, w)y + E(\lambda, w)z = 0.$$

Since $V(\lambda)$ is nonsingular, it yields that

$$(DV^{-1} + E)z = 0.$$

It is well known from the NCP theory [DFK96, Theorem 21] that the matrix $(DV^{-1} + E)$ is nonsingular because V^{-1} is positive definite according to the assumption. Hence, $z = 0$, implying $y = 0$. This establishes the nonsingularity of all elements in $\partial W(\lambda, w)$. $\qquad\square$

Newton's method for (40) can be developed as follows

$$(\lambda^+, w^+) - (\lambda, w) = -M^{-1}W(\lambda, w), \quad M \in \partial W(\lambda, w). \tag{44}$$

We have proved that each F_j is semismooth (Corollary 2). Using the fact that composite of semismooth functions is semismooth and the Fischer-Burmeister function is strongly semismooth, we know that W is semismooth function. Suppose (λ^*, w^*) is a solution of (40).

Assumption 4. *Each $b_\ell > 0$ for $\ell = 1, \ldots, n$.*

Lemma 7. *Suppose Assumption (1), (2) and (4) hold. Then every element in $\partial W(\lambda^*, w^*)$ is nonsingular.*

Proof. We note that at the solution it holds

$$AP_C(x^0 + A^*\lambda^*) = b + w^*.$$

Since $w_\ell^* \geq 0$, we see that $b_\ell + w_\ell^* > 0$. Following the proof of Lemma 4 we can prove that each element V in $\partial F(\lambda^*)$ is positive definite, and hence each element of $\partial W(\lambda^*, w^*)$ is nonsingular by Lemma 7. $\qquad\square$

All preparation is ready for the use of Theorem 6 to state the superlinear convergence of the method (44). The proof is similar to Theorem 7.

Theorem 8. *Suppose Assumptions (1), (2) and (4) hold. Then the Newton method (44) is superlinearly convergent provided that the initial point (λ^0, w^0) is sufficiently close to (λ^*, w^*).*

We remark that the quadratic convergence is also possible if we could establish the strong semismoothness of W at (λ^*, w^*). A sufficient condition for this property is that each F_j is strongly semismooth since the Fischer-Burmeister function is automatically strongly semismooth.

4.5 Globalization

In the previous subsections, Newton's method is developed for nonsmooth equations arising from constrained interpolation and approximation problems. It is locally superlinearly convergent under reasonable conditions. It is also worth of mentioning it globalization scheme that makes the Newton method globally convergent.

The first issue to be resolved is that we need an objective function for the respective problems. Natural choices for objective functions are briefly described below with outline of an algorithmic scheme, but without global convergence analysis. It is easy to see (following discussion in [MU88, DQQ01]) that the function f given by

$$f(\lambda) := \int_a^b \left(x^0 + \sum_{\ell=1}^n \lambda_\ell a_\ell \right)_+^2 dt - \sum_{\ell=1}^n \lambda_\ell b_\ell$$

severs this purpose because

$$\nabla f(\lambda) = F(\lambda) - b.$$

Since f is convex, $\|\nabla f(\lambda)\| = \|F(\lambda) - b\|$ can be used to monitor the convergence of global methods. We present below a global method, which globalizes the method (29) and has been shown extremely efficient for the convex best interpolation problem (1).

Algorithm 1. (Damped Newton method)

(S.0) (Initialization) Choose $\lambda^0 \in \mathbb{R}^n$, $\rho \in (0,1), \sigma \in (0, 1/2)$, and tolerance tol > 0. $k := 0$.

(S.1) (Termination criterion) If $\epsilon_k = \|F(\lambda^k) - d\| \leq$ tol then stop. Otherwise, go to (S.2).

(S.2) (Direction generation) Let s^k be a solution of the following linear system

$$(V(\lambda^k) + \epsilon_k I)s = -\nabla f(\lambda^k). \tag{45}$$

(S.3)   (Line search) Choose m_k as the smallest nonnegative integer m satisfying

$$f(\lambda^k + \rho^m s^k) - f(\lambda^k) \leq \sigma \rho^m \nabla f(\lambda^k)^T s^k. \qquad (46)$$

(S.4)   (Update) Set $\lambda^{k+1} = \lambda^k + \rho^{m_k} s^k$, $k := k + 1$, return to step (S.1).

Since $V(\lambda)$ is positive semidefinite, the matrix $(V(\lambda) + \epsilon I)$ is positive definite for $\epsilon > 0$. Hence the linear equation (45) is well defined and the direction s^k is a descent direction for the objective function f. The global convergence analysis for Algorithm 1 is standard and can be found in [DQQ03].

Globalized version for the method (44) can be developed as well, but with some notable differences. To this case, the objective function $f(\lambda, w)$ is given by

$$f(\lambda, w) := \int_a^b \left(x^0 + \sum_{\ell=1}^n \lambda_\ell a_\ell \right)_+^2 dt - \sum_{\ell=1}^n \lambda_\ell (b + w) + \|\Phi_{FB}(\lambda, w)\|^2.$$

This function is also continuously differentiable, but not convex because $\|\Phi_{FB}(\lambda, w)\|^2$ is not convex although continuously differentiable. We also note that the gradient of $f(\lambda, w)$ is not $W(\lambda, w)$ any more. A global method based on f can be developed by following the scheme in [DFK96].

5 Open Problems

It is obvious from Section 2 and Section 4 that there is a big gap between theoretical results and Newton-type algorithms for constrained interpolation problems. For example, the solution characterizations appeared in Theorems 1, 3, and 4 are for general convex sets (i.e., C is a closed convex set), however, the Newton method well-developed so far is only on the particular case yet the most important case that C is the cone of positive functions. This is due to the fact that the projection is an essential ingredient when solving the interpolation problem, and that the projection on the cone of positive functions is easy to calculate.

There are many problems that are associated to the projections onto other convex sets including cones. We only discuss two of them which we think are most interesting and likely to be (at least partly) solved by the techniques developed in this paper. The first one is the case that C is a closed polyhedral set in X, i.e.,

$$C := \{x \in X | \langle c_i, x \rangle \leq r_i, \quad i = 1, \ldots, m\}$$

where $c_i \in X$ and $r_i \in \mathbb{R}$. We note that cones are not necessarily polyhedral. It follows from [Deu01, Examples 10.7 and 10.9] that the sets $\{C, \cap H_j\}$ and $\{C, \cap \mathcal{H}_j\}$ both have strong CHIP. Hence the solution characterization theorems are applicable to the polyhedral case. Questions related to P_C include differentiability, directional differentiability, generalized Jacobian and semismoothness of the mapping AP_C, and most importantly how to design Newton's method for this case.

The second is the problem of interpolating a finite set of points with a curve constrained to lie between two piecewise linear splines (with knots at the abscissae of the given points). The objective is to minimize the 2-norm of the second derivative of the interpolant. Let (t_i, y_i) be given data points in $\mathbb{R}^2$ with

$$t_0 < t_1 < \ldots < t_n, \ \phi(t_i) < y_i < \psi(t_i) \text{ for } i = 1, \ldots, n.$$

Hence ϕ and ψ are given piecewise linear functions (or more generally lower and upper semicontinuous functions, respectively) such that

$$\inf_{t \in [t_0, t_n]} (\psi(t) - \phi(t)) > 0.$$

The constraint is

$$\hat{C} := \{x \in W^{2,2}[t_0, t_n] | \ \phi(t) \leq x(t) \leq \psi(t)\}$$

and

$$H := \{x \in W^{2,2}[t_0, t_n] | \ x(t_i) = y_i, i = 1, \ldots, n\}.$$

This problem can be reformulated as a constrained interpolation problem from a convex set in certain Hilbert space [Don93, AE95]. Questions similar to that for the first problem remain unsolved for this interpolation problem from a strip.

Acknowledgement

The author would like to thank Danniel Ralph for his constructive comments on the topic and especially for his kind offer of his material [Ral02] on semi-smoothness of integral functions being included in this survey (i.e., Sec. 4.3). It is also interesting to see how his approach can be extended to cover the strongly semismooth case.

The work was done while the author was with School of Mathematics, The University of New South Wales, Australia, and was supported by Australian Research Council.

References

[AE87]   Andersson, L.-E., Elfving, T.: An algorithm for constrained interpolation. SIAM J. Sci. Statist. Comput., 8, 1012–1025 (1987)

[AE95]   Andersson, L.-E., Elfving, T.: Best constrained approximation in Hilbert
            space and interpolation by cubic splines subject to obstacles. SIAM J. Sci.
            Comput., **16**, 1209–1232 (1995)
[BBT00]   Bauschke, H.H., Borwein, J.M., Tseng, P.: Bounded linear regularity,
            strong CHIP, and CHIP are distinct properties. J. Convex Anal., **7**, 395–
            412 (2000)
[BBL99]   Bauschke, H.H., Borwein, J.M., Li, W.: Strong conical hull intersection
            property, bounded linear regularity, Jameson's property (G), and error
            bounds in convex optimization. Math. Program., **86**, 135–160 (1999)
[BL92]   Borwein, J., Lewis, A.S.: Partially finite convex programming I: Quasi
            relative interiors and duality theory. Math. Program. **57**, 15–48 (1992)
[Cla83]   Clarke, F.H.: Optimization and Nonsmooth Analysis. John Wiley & Sons,
            New York (1983)
[CDW90]   Chui, C.K., Deutsch, F., Ward, J.D.: Constrained best approximation in
            Hilbert space. Constr. Approx., **6**, 35–64 (1990)
[CDW92]   Chui, C.K., Deutsch, F., Ward, J.D.: *Constrained best approximation in
            Hilbert space II*, J. Approx. Theory, 71 (1992), pp. 213–238.
[deB78]   de Boor, C.: A Practical Guide to Splines. Springer-Verlag, New York
            (1978)
[DFK96]   De Luca, T., Facchinei, F., Kanzow, C.: A semismooth equation approach
            to the solution of nonlinear complementarity problems. Math. Program.,
            75, 407–439 (1996)
[Deu01]   Deutsch, F.: Best approximation in inner product spaces. CMS Books in
            Mathematics **7**. Springer-Verlag, New York (2001)
[DLW97]   Deutsch, F., Li, W., Ward, J.D.: A dual approach to constrained inter-
            polation from a convex subset of Hilbert space. J. Approx. Theory, **90**,
            385–414 (1997)
[DLW99]   Deutsch, F., Li, W., Ward, J.D.: Best approximation from the intersection
            of a closed convex set and a polyhedron in Hilbert space, weak Slater
            conditions, and the strong conical hull intersection property. SIAM J.
            Optim., **10**, 252–268 (1999)
[DUWX96]   Deutsch, F., Ubhaya, V.A., Ward, J.D., Xu, Y.: Constrained best ap-
            proximation in Hilbert space. III. Applications to n-convex functions.
            Constr. Approx., **12**, 361–384 (1996)
[Don93]   Dontchev, A.L.: Best interpolation in a strip. J. Approx. Theory, **73** 334–
            342 (1993)
[DK89]   Dontchev, A.L., Kalchev, B.D.: Duality and well-posedness in convex in-
            terpolation. Numer. Funct. Anal. and Optim., **10**, 673–689 (1989)
[DK96]   Dontchev, A.L., Kolmanovsky, I.: Best interpolation in a strip. II. Re-
            duction to unconstrained convex optimization. Comput. Optim. Appl., **5**,
            233–251 (1996)
[DQQ01]   Dontchev, A.L., Qi, H.-D., Qi, L.: Convergence of Newton's method for
            convex best interpolation. Numer. Math., **87** 435–456 (2001)
[DQQ03]   Dontchev, A.L., Qi, H.-D., Qi, L.: Quadratic convergence of Newton's
            method for convex interpolation and smoothing. Constr. Approx., **19**,
            123–143 (2003)
[DQQY02]   Dontchev, A.L., Qi, H.-D., Qi, L., Yin, H.: A Newton method for shape-
            preserving spline interpolation. SIAM J. Optim., **13**, 588–602 (2002)
[FP03]   Facchinei, F., Pang, J.-S.: Finite-dimensional variational inequalities and
            complementarity problems, Vol. I & II. Springer-Verlag, New York (2003)

[Fav40]   Favard, J.: Sur l'interpolation. J. Math. Pures Appl., **19**, 281–306 (1940)

[Fis92]   Fischer, A.: A special Newton-type optimization method. Optimization, **24**, 269-284 (1992)

[GT90]   Gowda, M.S., Teboulle, M.: A comparison of constraint qualifications in infinite-dimensional convex programming. SIAM J. Control Optim., **28**, 925–935 (1990)

[Hor80]   Hornung, U.: Interpolation by smooth functions under restriction on the derivatives. J. Approx. Theory, **28**, 227–237 (1980)

[IP84]   Iliev, G., Pollul, W.: Convex interpolation by functions with minimal L_p norm $(1 < p < \infty)$ of the kth derivative. Mathematics and mathematical education (Sunny Beach, 1984), 31–42, Bulg. Akad. Nauk, Sofia (1984)

[IMS86]   Irvine, L.D., Marin, S.P., Smith, P.W.: Constrained interpolation and smoothing. Constr. Approx., **2**, 129–151 (1986)

[Jey92]   V. Jeyakumar: Infinite-dimensional convex programming with applications to constrained approximation. J. Optim. Theory Appl., **75**, 569–586 (1992)

[JL98]   V. Jeyakumar, D.T. Luc: Approximate Jacobian matrices for nonsmooth continuous maps and C^1-optimization. SIAM J. Control Optim., **36**, 1815–1832 (1998)

[JW92]   V. Jeyakumar, H. Wolkowicz: Generalizations of Slater's constraint qualification for infinite convex programs. Math. Program., **57**, 85–101 (1992)

[JQ95]   Jiang, H., Qi, L.: Local uniqueness and Newton-type methods for nonsmooth variational inequalities. J. Math. Analysis and Appl., **196** 314–331 (1995)

[KK02]   Klatte D., Kummer, B.: Nonsmooth equations in optimization. Regularity, calculus, methods and applications. Nonconvex Optimization and its Applications, **60**. Kluwer Academic Publishers, Dordrecht (2002)

[Kum88]   B. Kummer: Newton's method for nondifferentiable functions. Advances in mathematical optimization, 114–125, Math. Res., **45**, Akademie-Verlag, Berlin (1988)

[LJ02]   Li, C., Jin, X.Q.: Nonlinearly constrained best approximation in Hilbert spaces: the strong chip and the basic constraint qualification. SIAM J. Optim., **13**, 228–239 (2002)

[LN02]   Li, C., Ng, K.F.: On best approximation by nonconvex sets and perturbation of nonconvex inequality systems in Hilbert spaces. SIAM J. Optim., **13**, 726–744 (2002)

[LN03]   Li, C., Ng, K.F.: Constraint qualification, the strong chip, and best approximation with convex constraints in Banach spaces. SIAM J. Optim., **14**, 584–607 (2003)

[MSSW85] Micchelli, C.A., Smith, P.W., Swetits, J., Ward, J.D.: Constrained L_p approximation. Constr. Approx., **1**, 93–102 (1985)

[MU88]   Micchelli, C.A., Utreras, F.I.: Smoothing and interpolation in a convex subset of a Hilbert space. SIAM J. Sci. Statist. Comput., **9**, 728–747 (1988)

[Mif77]   Mifflin, R.: Semismoothness and semiconvex functions in constrained optimization. SIAM J. Control Optim., **15**, 959–972 (1977)

[QS93]   Qi, L., Sun, J.: A nonsmooth version of Newton's method. Math. Program., **58**, 353–367 (1993)

[Ral02]   Ralph, D.: Personal communication. May. (2002)

[SQ99]   Sun, D., Qi, L.: On NCP-functions. Comput. Optim. Appl., **13**, 201–220 (1999)

[Xu99]   Xu, H.: Set-valued approximations and Newton's methods. Math. Program., **84**, 401–420 (1999)

Optimization Methods in Direct and Inverse Scattering

Alexander G. Ramm[1] and Semion Gutman[2]

[1] Department of Mathematics, Kansas State University
Manhattan, Kansas 66506-2602, USA
ramm@math.ksu.edu
[2] Department of Mathematics, University of Oklahoma
Norman OK 73019, USA
sgutman@ou.edu

Summary. In many Direct and Inverse Scattering problems one has to use a parameter-fitting procedure, because analytical inversion procedures are often not available. In this paper a variety of such methods is presented with a discussion of theoretical and computational issues.

The problem of finding small subsurface inclusions from surface scattering data is stated and investigated. This Inverse Scattering problem is reduced to an optimization problem, and solved by the Hybrid Stochastic-Deterministic minimization algorithm. A similar approach is used to determine layers in a particle from the scattering data.

The Inverse potential scattering problem is described and its solution based on a parameter fitting procedure is presented for the case of spherically symmetric potentials and fixed-energy phase shifts as the scattering data. The central feature of the minimization algorithm here is the Stability Index Method. This general approach estimates the size of the minimizing sets, and gives a practically useful stopping criterion for global minimization algorithms.

The 3D inverse scattering problem with fixed-energy data is discussed. Its solution by the Ramm's method is described. The cases of exact and noisy discrete data are considered. Error estimates for the inversion algorithm are given in both cases of exact and noisy data. Comparison of the Ramm's inversion method with the inversion based on the Dirichlet-to-Neumann map is given and it is shown that there are many more numerical difficulties in the latter method than in the Ramm's method.

An Obstacle Direct Scattering problem is treated by a novel Modified Rayleigh Conjecture (MRC) method. MRC's performance is compared favorably to the well known Boundary Integral Equation Method, based on the properties of the single and double-layer potentials. A special minimization procedure allows one to inexpensively compute scattered fields for 2D and 3D obstacles having smooth as well as nonsmooth surfaces.

A new Support Function Method (SFM) is used for Inverse Obstacle Scattering problems. The SFM can work with limited data. It can also be used for Inverse scattering problems with *unknown scattering conditions on its boundary (e.g. soft, or hard scattering)*. Another method for Inverse scattering problems, the Linear Sampling Method (LSM), is analyzed. Theoretical and computational difficulties in using this method are pointed out.

1 Introduction

Suppose that an acoustic or electromagnetic wave encounters an inhomogeneity and, as a consequence, gets scattered. The problem of finding the scattered wave assuming the knowledge of the inhomogeneity (penetrable or not) is the Direct Scattering problem. An impenetrable inhomogeneity is also called an obstacle. On the other hand, if the scattered wave is known at some points outside an inhomogeneity, then we are faced with the Inverse Scattering problem, the goal of which is to identify this inhomogeneity, see [CCM00, CK92, Ram86, Ram92b, Ram94a, Ram05a, Ram05b]

Among a variety of methods available to handle such problems few provide a mathematically justified algorithm. In many cases one has to use a parameter-fitting procedure, especially for inverse scattering problems, because the analytical inversion procedures are often not available. An important part of such a procedure is an efficient global optimization method, see [Flo00, FP01, HPT95, HT93, PRT00, Rub00].

The general scheme for parameter-fitting procedures is simple: one has a relation $B(q) = A$, where B is some operator, q is an unknown function, and A is the data. In inverse scattering problems q is an unknown potential, and A is the known scattering amplitude. If q is sought in a finite-parametric family of functions, then $q = q(x, p)$, where $p = (p_1, ..., p_n)$ is a parameter. The parameter is found by solving a global minimization problem: $\Phi[B(q(x, p)) - A] = \min$, where Φ is some positive functional, and $q \in Q$, where Q is an admissible set of q. In practice the above problem often has many local minimizers, and the global minimizer is not necessarily unique. In [Ram92b, Ram94b] some functionals Φ are constructed which have unique global minimizer, namely, the solution to inverse scattering problem, and the global minimum is zero.

Moreover, as a rule, the data A is known with some error. Thus A_δ is known, such that $\|A - A_\delta\| < \delta$. There are no stability estimates which would show how the global minimizer $q(x, p_{opt})$ is perturbed when the data A are replaced by the perturbed data A_δ. In fact, one can easily construct examples showing that there is no stability of the global minimizer with respect to small errors in the data, in general.

For these reasons there is no guarantee that the parameter-fitting procedures would yield a solution to the inverse problem with a guaranteed accuracy. However, overwhelming majority of practitioners are using parameter-fitting procedures. In dozens of published papers the results obtained by various parameter-fitting procedures look quite good. The explanation, in most of the cases is simple: the authors know the answer beforehand, and it is usually not difficult to parametrize the unknown function so that the exact solution is well approximated by a function from a finite-parametric family, and since the authors know a priori the exact answer, they may choose numerically the values of the parameters which yield a good approximation of the exact solution. *When can one rely on the results obtained by parameter-fitting procedures? Unfortunately, there is no rigorous and complete answer to this question, but some recommendations are given in Section 4.*

In this paper the authors present their recent results which are based on specially designed parameter-fitting procedures. Before describing them, let us mention that usually in a numerical solution of an inverse scattering problem one uses a regularization procedure, e.g. a variational regularization, spectral cut-off, iterative regularization, DSM (the dynamical systems method), quasi-solutions, etc, see e.g. [Ram04a, Ram05a]. This general theoretical framework is well established in the theory of ill-posed problems, of which the inverse scattering problems represent an important class. This framework is needed to achieve a stable method for assigning a solution to an ill-posed problem, usually set in an infinite dimensional space. The goal of this paper is to present optimization algorithms already in a finite dimensional setting of a Direct or Inverse scattering problem.

In Section 2 the problem of finding small subsurface inclusions from surface scattering data is investigated ([Ram97, Ram00a, Ram05a, Ram05b]). This (geophysical) Inverse Scattering problem is reduced to an optimization problem. This problem is solved by the Hybrid Stochastic-Deterministic minimization algorithm ([GR00]). It is based on a genetic minimization algorithm ideas for its random (stochastic) part, and a deterministic minimization without derivatives used for the local minimization part.

In Section 3 a similar approach is used to determine layers in a particle subjected to acoustic or electromagnetic waves. The global minimization algorithm uses Rinnooy Kan and Timmer's Multilevel Single-Linkage Method for its stochastic part.

In Section 4 we discuss an Inverse potential scattering problem appearing in a quantum mechanical description of particle scattering experiments. The central feature of the minimization algorithm here is the Stability Index Method ([GRS02]). This general approach estimates the size of the minimizing sets, and gives a practically useful stopping criterion for global minimization algorithms.

In Section 5 Ramm's method for solving 3D inverse scattering problem with fixed-energy data is presented following [Ram04d], see also [Ram02a, Ram05a]. The cases of exact and noisy discrete data are considered. Error estimates for the inversion algorithm are given in both cases of exact and noisy data. Comparison of the Ramm's inversion method with the inversion based on the Dirichlet-to-Neumann map is given and it is shown that there are many more numerical difficulties in the latter method than in Ramm's method.

In Section 6 an Obstacle Direct Scattering problem is treated by a novel Modified Rayleigh Conjecture (MRC) method. It was introduced in [Ram02b] and applied in [GR02b, GR05, Ram04c, Ram05b]. MRC's performance is compared favorably to the well known Boundary Integral Equation Method, based on the properties of the single and double-layer potentials. A special minimization procedure allows us to inexpensivly compute scattered fields for several 2D and 3D obstacles having smooth as well as nonsmooth surfaces.

In Section 7 a new Support Function Method (SFM) is used to determine the location of an obstacle (cf [GR03, Ram70, Ram86]). Unlike other methods, the SFM can work with limited data. It can also be used for Inverse scattering problems with *unknown scattering conditions on its boundary (e.g. soft or hard obstacles)*.

Finally, in Section 8, we present an analysis of another popular method for Inverse scattering problems, the Linear Sampling Method (LSM), and show that both theoretically and computationally the method fails in many aspects. This section is based on the paper [RG05].

2 Identification of small subsurface inclusions

2.1 Problem description

In many applications it is desirable to find small inhomogeneities from surface scattering data. For example, such a problem arises in ultrasound mammography, where small inhomogeneities are cancer cells. Other examples include the problem of finding small holes and cracks in metals and other materials, or the mine detection. The scattering theory for small scatterers originated in the classical works of Lord Rayleigh (1871). Rayleigh understood that the basic contribution to the scattered field in the far-field zone comes from the dipole radiation, but did not give methods for calculating this radiation. Analytical formulas for calculating the polarizability tensors for homogeneous bodies of arbitrary shapes were derived in [Ram86] (see also references therein). These formulas allow one to calculate the S-matrix for scattering of acoustic and electromagnetic waves by small bodies of arbitrary shapes with arbitrary accuracy. Inverse scattering problems for small bodies are considered in [Ram82] and [Ram94a]. In [Ram97] and [Ram00a] the problem of identification of small subsurface inhomogeneities from surface data was posed and its possible applications were discussed.

In the context of a geophysical problem, let $y \in \mathbb{R}^3$ be a point source of monochromatic acoustic waves on the surface of the earth. Let $u(x, y, k)$ be the acoustic pressure at a point $x \in \mathbb{R}^3$, and $k > 0$ be the wavenumber. The governing equation for the acoustic wave propagation is:

$$\left[\nabla^2 + k^2 + k^2 v(x)\right] u = -\delta(x - y) \text{ in } \mathbb{R}^3, \tag{1}$$

where $x = (x_1, x_2, x_3)$, $v(x)$ is the inhomogeneity in the velocity profile, and $u(x, y, k)$ satisfies the radiation condition at infinity, i.e. it decays sufficiently fast as $|x| \to \infty$.

Let us assume that $v(x)$ is a bounded function vanishing outside of the domain $D = \cup_{m=1}^{M} D_m$ which is the union of M small nonintersecting domains D_m, all of them are located in the lower half-space $\mathbb{R}_-^3 = \{x : x_3 < 0\}$. Smallness is understood in the sense $k\rho \ll 1$, where $\rho := \frac{1}{2}\max_{1 \le m \le M}\{\mathrm{diam}\, D_m\}$, and diam D is the diameter of the domain D. Practically $k\rho \ll 1$ means that $k\rho < 0.1$. In some cases $k\rho < 0.2$ is sufficient for obtaining acceptable numerical results. The background velocity in (1) equals to 1, but we can consider the case of fairly general background velocity [Ram94a].

Denote $\tilde{z}_m$ and $\tilde{v}_m$ the position of the center of gravity of D_m, and the total intensity of the m-th inhomogeneity $\tilde{v}_m := \int_{D_m} v(x)dx$. Assume that $\tilde{v}_m \neq 0$. Let P be the equation of the surface of the earth:

$$P := \{x = (x_1, x_2, x_3) \in \mathbb{R}^3 : x_3 = 0\}. \tag{2}$$

The inverse problem to be solved is:

IP: *Given $u(x, y, k)$ for all source-detector pairs (x, y) on P at a fixed $k > 0$, find the number M of small inhomogeneities, the positions $\tilde{z}_m$ of the inhomogeneities, and their intensities $\tilde{v}_m$.*

Practically, one assumes that a fixed wavenumber $k > 0$, and J source-detector pairs $(x_j, y_j), j = 1, 2, ..., J$, on P are known together with the acoustic pressure measurements $u(x_j, y_j, k)$. Let

$$g(x, y, k) := \frac{\exp(ik|x - y|)}{4\pi|x - y|}, \quad x, y \in P, \tag{3}$$

$$G_j(z) := G(x_j, y_j, z) := g(x_j, z, k)g(y_j, z, k), \quad x_j, y_j \in P, \ z \in \mathbb{R}_-^3, \tag{4}$$

$$f_j := \frac{u(x_j, y_j, k) - g(x_j, y_j, k)}{k^2}, \tag{5}$$

and

$$\Phi(z_1, \ldots, z_M, v_1, \ldots, v_M) := \sum_{j=1}^{J} \left| f_j - \sum_{m=1}^{M} G_j(z_m)v_m \right|^2. \tag{6}$$

The proposed method for solving the (IP) consists of finding the global minimizer of function (6). This minimizer $(\tilde{z}_1, \ldots, \tilde{z}_M, \tilde{v}_1, \ldots, \tilde{v}_M)$ gives the estimates of the positions $\tilde{z}_m$ of the small inhomogeneities and their intensities $\tilde{v}_m$. See [Ram97] and [Ram00a] for a justification of this approach.

The function Φ depends on M unknown points $z_m \in \mathbb{R}_-^3$, and M unknown parameters v_m, $1 \le m \le M$. The number M of the small inhomogeneities is also unknown, and its determination is a part of the minimization problem.

2.2 Hybrid Stochastic-Deterministic Method(HSD)

Let the inhomogeneities be located within the box

$$B = \{(x_1, x_2, x_3) \ : -a < x_1 < a, \ -b < x_2 < b, \ 0 < x_3 < c\}, \qquad (7)$$

and their intensities satisfy

$$0 \leq v_m \leq v_{max}. \qquad (8)$$

The box is located above the earth surface for a computational convenience.

Then, given the location of the points $z_1, z_2, \ldots, z_M$, the minimum of Φ with respect to the intensities $v_1, v_2, \ldots, v_M$ can be found by minimizing the resulting quadratic function in (6) over the region satisfying (8). This can be done using normal equations for (6) and projecting the resulting point back onto the region defined by (8). Denote the result of this minimization by $\tilde{\Phi}$, that is

$$\tilde{\Phi}(z_1, z_2, \ldots, z_M) = \min\{\Phi(z_1, z_2, \ldots, z_M, v_1, v_2, \ldots, v_M) \ : \\ 0 \leq v_m \leq v_{max}, \quad 1 \leq m \leq M\} \qquad (9)$$

Now the original minimization problem for $\Phi(z_1, z_2, \ldots, z_M, v_1, v_2, \ldots, v_M)$ is reduced to the $3M$-dimensional constrained minimization for $\tilde{\Phi}(z_1, z_2, \ldots, z_M)$:

$$\tilde{\Phi}(z_1, z_2, \ldots, z_M) = \min, \quad z_m \in B, \quad 1 \leq m \leq M. \qquad (10)$$

Note, that the dependency of $\tilde{\Phi}$ on its $3M$ variables (the coordinates of the points z_m) is highly nonlinear. In particular, this dependency is complicated by the computation of the minimum in (9) and the consequent projection onto the admissible set B. Thus, an analytical computation of the gradient of $\tilde{\Phi}$ is not computationally efficient. Accordingly, the Powell's quadratic minimization method was used to find local minima. This method uses a special procedure to numerically approximate the gradient, and it can be shown to exhibit the same type of quadratic convergence as conjugate gradient type methods (see [Bre73]).

In addition, *the exact number of the original inhomogeneities M_{orig} is unknown, and its estimate is a part of the inverse problem.* In the HSD algorithm described below this task is accomplished by taking the initial number M sufficiently large, so that

$$M_{orig} \leq M, \qquad (11)$$

which, presumably, can be estimated from physical considerations. After all, our goal is to find only the strongest inclusions, since the weak ones cannot be distinguished from background noise. The Reduction Procedure (see below) allows the algorithm to seek the minimum of $\tilde{\Phi}$ in a lower dimensional subsets

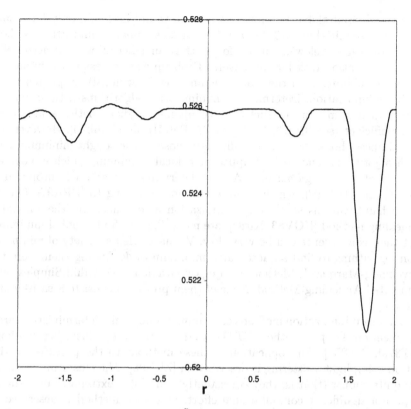

Fig. 1. Objective function $\tilde{\Phi}(z_r, z_2, \tilde{z}_3, \tilde{z}_4, \tilde{z}_5, \tilde{z}_6)$, $\quad -2 \leq r \leq 2$

of the admissible set B, thus finding the estimated number of inclusions M. Still another difficulty in the minimization is a large number of local minima of $\tilde{\Phi}$. This phenomenon is well known for objective functions arising in various inverse problems, and we illustrate this point in Figure 1.

For example, let $M_{orig} = 6$, and the coordinates of the inclusions, and their intensities $(\tilde{z}_1, \ldots, \tilde{z}_6, \tilde{v}_1, \ldots, \tilde{v}_6)$ be as in Table 1. Figure 1 shows the values of the function $\tilde{\Phi}(z_r, z_2, \tilde{z}_3, \tilde{z}_4, \tilde{z}_5, \tilde{z}_6)$, where

$$z_r = (r, 0, 0.520), \quad -2 \leq r \leq 2$$

and

$$z_2 = (-1, 0.3, 0.580).$$

The plot shows multiple local minima and almost flat regions.

A direct application of a gradient type method to such a function would result in finding a local minimum, which may or may not be the sought global one. In the example above, such a method would usually be trapped in a local minimum located at $r = -2$, $r = -1.4$, $r = -0.6$, $r = 0.2$ or $r = 0.9$, and the desired global minimum at $r = 1.6$ would be found only for a sufficiently close initial guess $1.4 < r < 1.9$. Various global minimization methods are known (see below), but we found that an efficient way to accomplish the minimization task for this Inverse Problem was to design a new method (HSD) combining both the stochastic and the deterministic approach to the global minimization. Deterministic minimization algorithms with or without the gradient computation, such as the conjugate gradient methods, are known to be efficient (see [Bre73, DS83, Jac77, Pol71]), and [Rub00]. However, the initial guess should be chosen sufficiently close to the sought minimum. Also such algorithms tend to be trapped at a local minimum, which is not necessarily close to a global one. A new deterministic method is proposed in [BP96] and [BPR97], which is quite efficient according to [BPR97]. On the other hand, various stochastic minimization algorithms, e.g. the simulated annealing method [KGV83, Kir84], are more likely to find a global minimum, but their convergence can be very slow. We have tried a variety of minimization algorithms to find an acceptable minimum of $\tilde{\Phi}$. Among them were the Levenberg-Marquardt Method, Conjugate Gradients, Downhill Simplex, and Simulated Annealing Method. None of them produced consistent satisfactory results.

Among minimization methods combining random and deterministic searches we mention Deep's method [DE94] and a variety of clustering methods [RT87a], [RT87b]. An application of these methods to the particle identification using light scattering is described in [ZUB98]. The clustering methods are quite robust (that is, they consistently find global extrema) but, usually, require a significant computational effort. One such method is described in the next section on the identification of layers in a multilayer particle. The HSD method is a combination of a reduced sample random search method with certain ideas from Genetic Algorithms (see e.g. [HH98]). It is very efficient and seems especially well suited for low dimensional global minimization. Further research is envisioned to study its properties in more detail, and its applicability to other problems.

The steps of the Hybrid Stochastic-Deterministic (HSD) method are outlined below. Let us call a collection of M points (inclusion's centers) $\{z_1, z_2, ..., z_M\}$, $z_i \in B$ a *configuration* Z. Then the minimization problem (10) is the minimization of the objective function $\tilde{\Phi}$ over the set of all configurations.

For clarity, let $P_0 = 1$, $\epsilon_s = 0.5$, $\epsilon_i = 0.25$, $\epsilon_d = 0.1$, be the same values as the ones used in numerical computations in the next section.

Generate a random configuration Z. Compute the best fit intensities v_i corresponding to this configuration. If $v_i > v_{max}$, then let $v_i := v_{max}$. If $v_i < 0$, then let $v_i := 0$. If $\tilde{\Phi}(Z) < P_0\epsilon_s$, then this configuration is a preliminary candidate for the initial guess of a deterministic minimization method (Step 1).

Drop the points $z_i \in Z$ such that $v_i < v_{max}\epsilon_i$. That is, the inclusions with small intensities are eliminated (Step 2).

If two points $z_k, z_j \in Z$ are too close to each other, then replace them with one point of a combined intensity (Step 3).

After completing steps 2 and 3 we would be left with $N \leq M$ points $z_1, z_2, ..., z_N$ (after a re-indexing) of the original configuration Z. Use this reduced configuration Z_{red} as the starting point for the deterministic restraint minimization in the $3N$ dimensional space (Step 4). Let the resulting minimizer be $\tilde{Z}_{red} = (\tilde{z}_1, ..., \tilde{z}_N)$. If the value of the objective function $\tilde{\Phi}(\tilde{Z}_{red}) < \epsilon$, then we are done: $\tilde{Z}_{red}$ is the sought configuration containing N inclusions. If $\tilde{\Phi}(\tilde{Z}_{red}) \geq \epsilon$, then the iterations should continue.

To continue the iteration, randomly generate $M - N$ points in B (Step 5). Add them to the reduced configuration $\tilde{Z}_{red}$. Now we have a new full configuration Z, and the iteration process can continue (Step 1).

This entire iterative process is repeated n_{max} times, and the best configuration is declared to represent the sought inclusions.

2.3 Description of the HSD Method

Let P_0, T_{max}, n_{max}, ϵ_s, ϵ_i, ϵ_d, and ϵ be positive numbers. Let a positive integer M be larger than the expected number of inclusions. Let $N = 0$.

1. Randomly generate $M - N$ additional points $z_{N+1}, \ldots, z_M \in B$ to obtain a full configuration $Z = (z_1, \ldots, z_M)$. Find the best fit intensities v_i, $i = 1, 2, ..., M$. If $v_i > v_{max}$, then let $v_i := v_{max}$. If $v_i < 0$, then let $v_i := 0$. Compute $P_s = \tilde{\Phi}(z_1, z_2 \ldots, z_M)$. If $P_s < P_0\epsilon_s$ then go to step 2, otherwise repeat step 1.

2. Drop all the points with the intensities v_i satisfying $v_i < v_{max}\epsilon_i$. Now only $N \leq M$ points $z_1, z_2 \ldots, z_N$ (re-indexed) remain in the configuration Z.

3. If any two points z_m, z_n in the above configuration satisfy $|z_m - z_n| < \epsilon_d D$, where $D = diam(B)$, then eliminate point z_n, change the intensity of point z_m to $v_m + v_n$, and assign $N := N - 1$. This step is repeated until no further reduction in N is possible. Call the resulting reduced configuration with N points by Z_{red}.

4. Run a constrained deterministic minimization of $\tilde{\Phi}$ in $3N$ variables, with the initial guess Z_{red}. Let the minimizer be $\tilde{Z}_{red} = (\tilde{z}_1, \ldots, \tilde{z}_N)$. If $P = \tilde{\Phi}(\tilde{z}_1, \ldots, \tilde{z}_N) < \epsilon$, then save this configuration, and go to step 6, otherwise let $P_0 = P$, and proceed to the next step 5.

5. Keep intact N points $\tilde{z}_1, \ldots, \tilde{z}_N$. If the number of random configurations has exceeded T_{max} (the maximum number of random tries), then save the configuration and go to step 6, otherwise go to step 1, and use these N points there.

6. Repeat steps 1 through 5 n_{max} times.

7. Find the configuration among the above n_{max} ones, which gives the smallest value to $\tilde{\Phi}$. This is the best fit.

The Powell's minimization method (see [Bre73] for a detailed description) was used for the deterministic part, since this method does not need gradient computations, and it converges quadratically near quadratically shaped minima. Also, in step 1, an idea from the Genetic Algorithm's approach [HH98] is implemented by keeping only the strongest representatives of the population, and allowing a mutation for the rest.

2.4 Numerical results

The algorithm was tested on a variety of configurations. Here we present the results of just two typical numerical experiments illustrating the performance of the method. In both experiments the box B is taken to be

$$B = \{(x_1, x_2, x_3) : -a < x_1 < a, \ -b < x_2 < b, \ 0 < x_3 < c\},$$

with $a = 2$, $b = 1$, $c = 1$. The wavenumber $k = 5$, and the effective intensities v_m are in the range from 0 to 2. The values of the parameters were chosen as follows

$$P_0 = 1, T_{max} = 1000, \ \epsilon_s = 0.5, \ \epsilon_i = 0.25, \ \epsilon_d = 0.1, \epsilon = 10^{-5}, n_{max} = 6$$

In both cases we searched for the same 6 inhomogeneities with the coordinates x_1, x_2, x_3 and the intensities v shown in Table 1.

Table 1. Actual inclusions.

Inclusions	x_1	x_2	x_3	v
1	1.640	-0.510	0.520	1.200
2	-1.430	-0.500	0.580	0.500
3	1.220	0.570	0.370	0.700
4	1.410	0.230	0.740	0.610
5	-0.220	0.470	0.270	0.700
6	-1.410	0.230	0.174	0.600

Parameter M was set to 16, thus the only information on the number of inhomogeneities given to the algorithm was that their number does not exceed 16. This number was chosen to keep the computational time within reasonable limits. Still another consideration for the number M is the aim of the algorithm to find the presence of the most influential inclusions, rather then all inclusions, which is usually impossible in the presence of noise and with the limited amount of data.

Experiment 1. In this case we used 12 sources and 21 detectors, all on the surface $x_3 = 0$. The sources were positioned at $\{(-1.667 + 0.667i, -0.5 + 1.0j, 0), \ i = 0, 1, \ldots, 5, \ j = 0, 1\}$, that is 6 each along two lines $x_2 = -0.5$ and $x_2 = 0.5$. The detectors were positioned at $\{(-2 + 0.667i, -1.0 + 1.0j, 0), \ i = 0, 1, \ldots, 6, \ j = 0, 1, 2\}$, that is seven detectors along each of the three lines

$x_2 = -1$, $x_2 = 0$ and $x_2 = 1$. This corresponds to a mammography search, where the detectors and the sources are placed above the search area. The results for noise level $\delta = 0.00$ are shown in Figure 2 and Table 2. The results for noise level $\delta = 0.05$ are shown in Table 3.

Table 2. Experiment 1. Identified inclusions, no noise, $\delta = 0.00$.

x_1	x_2	x_3	v
1.640	-0.510	0.520	1.20000
-1.430	-0.500	0.580	0.50000
1.220	0.570	0.370	0.70000
1.410	0.230	0.740	0.61000
-0.220	0.470	0.270	0.70000
-1.410	0.230	0.174	0.60000

Table 3. Experiment 1. Identified inclusions, $\delta = 0.05$.

x_1	x_2	x_3	v
1.645	-0.507	0.525	1.24243
1.215	0.609	0.376	0.67626
-0.216	0.465	0.275	0.69180
-1.395	0.248	0.177	0.60747

Experiment 2. In this case we used 8 sources and 22 detectors, all on the surface $x_3 = 0$. The sources were positioned at $\{(-1.75 + 0.5i, 1.5, 0), i = 0, 1, \ldots, 7, j = 0, 1\}$, that is all 8 along the line $x_2 = 1.5$. The detectors were positioned at $\{(-2 + 0.4i, 1.0 + 1.0j, 0), i = 0, 1, \ldots, 10, j = 0, 1\}$, that is eleven detectors along each of the two lines $x_2 = 1$ and $x_2 = 2$. This corresponds to a mine search, where the detectors and the sources must be placed outside of the searched ground. The results of the identification for noise level $\delta = 0.00$ in the data are shown in Figure 3 and Table 4. The results for noise level $\delta = 0.05$ are shown in Table 5.

Table 4. Experiment 2. Identified inclusions, no noise, $\delta = 0.00$.

x_1	x_2	x_3	v
1.656	-0.409	0.857	1.75451
-1.476	-0.475	0.620	0.48823
1.209	0.605	0.382	0.60886
-0.225	0.469	0.266	0.69805
-1.406	0.228	0.159	0.59372

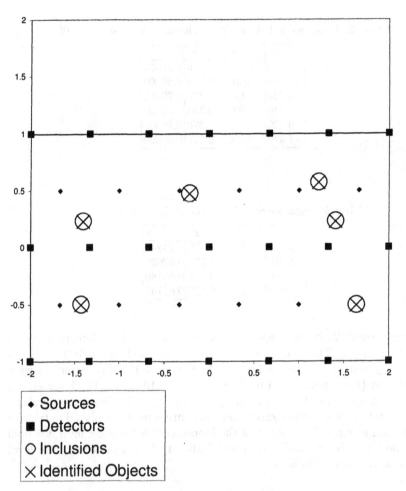

Fig. 2. Inclusions and Identified objects for subsurface particle identification, Experiment 1, $\delta = 0.00$. x_3 coordinate is not shown.

In general, the execution times were less than 2 minutes on a 333MHz PC. As it can be seen from the results, the method achieves a perfect identification in the Experiment #1 when no noise is present. The identification deteriorates in the presence of noise, as well as if the sources and detectors are not located directly above the search area. Still the inclusions with the highest intensity and the closest ones to the surface are identified, while the

Table 5. Experiment 2. Identified inclusions, $\delta = 0.05$.

x_1	x_2	x_3	v
1.575	-0.523	0.735	1.40827
-1.628	-0.447	0.229	1.46256
1.197	0.785	0.578	0.53266
-0.221	0.460	0.231	0.67803

deepest and the weakest are lost. This can be expected, since their influence on the cost functional is becoming comparable with the background noise in the data.

In summary, the proposed method for the identification of small inclusions can be used in geophysics, medicine and technology. It can be useful in the development of new approaches to ultrasound mammography. It can also be used for localization of holes and cracks in metals and other materials, as well as for finding mines from surface measurements of acoustic pressure and possibly in other problems of interest in various applications.

The HSD minimization method is a specially designed low-dimensional minimization method, which is well suited for many inverse type problems. The problems do not necessarily have to be within the range of applicability of the Born approximation. It is highly desirable to apply HSD method to practical problems and to compare its performance with other methods.

3 Identification of layers in multilayer particles.

3.1 Problem Description

Many practical problems require an identification of the internal structure of an object given some measurements on its surface. In this section we study such an identification for a multilayered particle illuminated by acoustic or electromagnetic plane waves. Thus the problem discussed here is an inverse scattering problem. A similar problem for the particle identification from the light scattering data is studied in [ZUB98]. Our approach is to reduce the inverse problem to the best fit to data multidimensional minimization.

Let $D \subset \mathbb{R}^2$ be the circle of a radius $R > 0$,

$$D_m = \{x \in \mathbb{R}^2 : r_{m-1} < |x| < r_m, \quad m = 1, 2, \ldots, N\} \quad (12)$$

and $S_m = \{x \in \mathbb{R}^2 : |x| = r_m\}$ for $0 = r_0 < r_1 < \cdots < r_N < R$. Suppose that a multilayered scatterer in D has a constant refractive index n_m in the region D_m, $m = 1, 2, \ldots, N$. If the scatterer is illuminated by a plane harmonic wave then, after the time dependency is eliminated, the total field $u(x) = u_0(x) + u_s(x)$ satisfies the Helmholtz equation

$$\Delta u + k_0^2 u = 0, \quad |x| > r_N \quad (13)$$

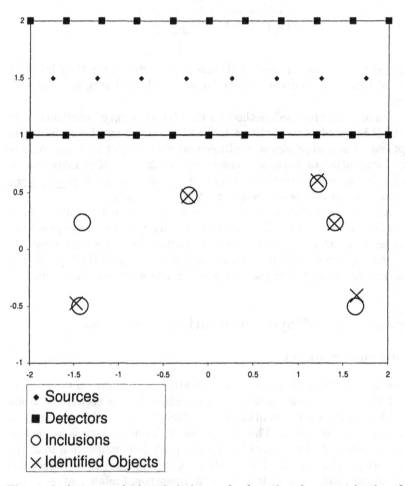

Fig. 3. Inclusions and Identified objects for for subsurface particle identification, Experiment 2, $\delta = 0.00$. x_3 coordinate is not shown.

where $u_0(x) = e^{ik_0 x \cdot \alpha}$ is the incident field and α is the unit vector in the direction of propagation. The scattered field u_s is required to satisfy the radiation condition at infinity, see [Ram86].

Let $k_m^2 = k_0^2 n_m$. We consider the following transmission problem

$$\Delta u_m + k_m^2 u_m = 0 \quad x \in D_m, \tag{14}$$

under the assumption that the fields u_m and their normal derivatives are continuous across the boundaries S_m, $m = 1, 2, \ldots, N$.

In fact, the choice of the boundary conditions on the boundaries S_m depends on the physical model under the consideration. The above model may or may not be adequate for an electromagnetic or acoustic scattering, since the model may require additional parameters (such as the mass density and the compressibility) to be accounted for. However, the basic computational approach remains the same. For more details on transmission problems, including the questions on the existence and the uniqueness of the solutions, see [ARS98, EJP57, RPY00].

The Inverse Problem to be solved is:

IPS: *Given $u(x)$ for all $x \in S = \{x : |x| = R\}$ at a fixed $k_0 > 0$, find the number N of the layers, the location of the layers, and their refractive indices n_m, $m = 1, 2, \ldots, N$ in (14).*

Here the IPS stands for a Single frequency Inverse Problem. Numerical experience shows that there are some practical difficulties in the successful resolution of the IPS even when no noise is present, see [Gut01]. While there are some results on the uniqueness for the IPS (see [ARS98, RPY00]), assuming that the refractive indices are known, and only the layers are to be identified, the stability estimates are few, see [Ram94c, Ram94d, Ram02a]. The identification is successful, however, if the scatterer is subjected to a probe with plane waves of several frequencies. Thus we state the Multifrequency Inverse Problem:

IPM: *Given $u^p(x)$ for all $x \in S = \{x : |x| = R\}$ at a finite number P of wave numbers $k_0^{(p)} > 0$, find the number N of the layers, the location of the layers, and their refractive indices n_m, $m = 1, 2, \ldots, N$ in (14).*

3.2 Best Fit Profiles and Local Minimization Methods

If the refractive indices n_m are sufficiently close to 1, then we say that the scattering is weak. In this case the scattering is described by the Born approximation, and there are methods for the solution of the above Inverse Problems. See [CM90], [Ram86] and [Ram94a] for further details. In particular, the Born inversion is an ill-posed problem even if the Born approximation is very accurate, see [Ram90], or [Ram92b]. When the assumption of the Born approximation is not appropriate, one matches the given observations to a set of solutions for the Direct Problem. Since our interest is in the solution of the IPS and IPM in the non-Born region of scattering, we choose to follow the best fit to data approach. This approach is used widely in a variety of applied problems, see e. g. [Bie97].

Note, that, by the assumption, the scatterer has the rotational symmetry. Thus we only need to know the data for one direction of the incident plane wave. For this reason we fix $\alpha = (1, 0)$ in (13) and define the (complex) functions

$$g^{(p)}(\theta), \quad 0 \le \theta < 2\pi, \quad p = 1, 2, \ldots, P, \tag{15}$$

to be the observations measured on the surface S of the ball D for a finite set of free space wave numbers $k_0^{(p)}$.

Fix a positive integer M. Given a configuration

$$Q = (r_1, r_2, \ldots, r_M, n_1, n_2, \ldots, n_M) \tag{16}$$

we solve the Direct Problem (13)-(14) (for each free space wave number $k_0^{(p)}$) with the layers $D_m = \{x \in \mathbb{R}^2 : r_{m-1} < |x| < r_m, \quad m = 1, 2, \ldots, M\}$, and the corresponding refractive indices n_m, where $r_0 = 0$. Let

$$w^{(p)}(\theta) = u^{(p)}(x)\big|_{x \in S}. \tag{17}$$

Fix a set of angles $\Theta = (\theta_1, \theta_2, \ldots, \theta_L)$ and let

$$\|w\|_2 = (\sum_{l=1}^{L} w^2(\theta_l))^{1/2}. \tag{18}$$

Define

$$\Phi(r_1, r_2, \ldots, r_M, n_1, n_2, \ldots, n_M) = \frac{1}{P} \sum_{p=1}^{P} \frac{\|w^{(p)} - g^{(p)}\|_2^2}{\|g^{(p)}\|_2^2}, \tag{19}$$

where the same set Θ is used for $g^{(p)}$ as for $w^{(p)}$.

We solve the IPM by minimizing the above best fit to data functional Φ over an appropriate set of admissible parameters $A_{adm} \subset \mathbb{R}^{2M}$.

It is reasonable to assume that the underlying physical problem gives some estimate for the bounds n_{low} and n_{high} of the refractive indices n_m as well as for the bound M of the expected number of layers N. Thus,

$$A_{adm} \subset \{(r_1, r_2, \ldots, r_M, n_1, n_2, \ldots, n_M) : 0 \leq r_i \leq R, \; n_{low} \leq n_m \leq n_{high}\}. \tag{20}$$

Note, that the admissible configurations must also satisfy

$$r_1 \leq r_2 \leq r_3 \leq \cdots \leq r_M. \tag{21}$$

It is well known that a multidimensional minimization is a difficult problem, unless the objective function is "well behaved". The most important quality of such a cooperative function is the presence of just a few local minima. Unfortunately, this is, decidedly, not the case in many applied problems, and, in particular, for the problem under the consideration.

To illustrate this point further, let P be the set of three free space wave numbers $k_0^{(p)}$ chosen to be

$$P = \{3.0, \; 6.5, \; 10.0\}. \tag{22}$$

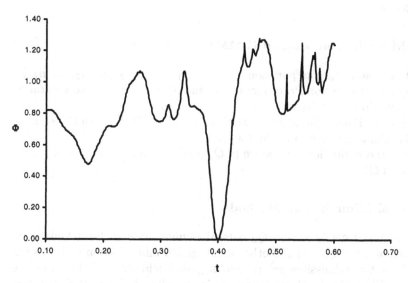

Fig. 4. Best fit profile for the configurations q_t; Multiple frequencies $P = \{3.0, \ 6.5, \ 10.0\}$.

Figure 4 shows the profile of the functional Φ as a function of the variable t, $0.1 \leq t \leq 0.6$ in the configurations q_t with

$$n(x) = \begin{cases} 0.49 & 0 \leq |x| < t \\ 9.0 & t \leq |x| < 0.6 \\ 1.0 & 0.6 \leq |x| \leq 1.0 \end{cases}$$

Thus the objective function Φ has many local minima even along this arbitrarily chosen one dimensional cross-section of the admissible set. There are sharp peaks and large gradients. Consequently, the gradient based methods (see [Bre73, DS83, Fle81, Hes80, Jac77, Pol71]), would not be successful for a significant portion of this region. It is also appropriate to notice that the dependency of Φ on its arguments is highly nonlinear. Thus, the gradient computations have to be done numerically, which makes them computationally expensive. More importantly, the gradient based minimization methods (as expected) perform poorly for these problems.

These complications are avoided by considering conjugate gradient type algorithms which do not require the knowledge of the derivatives at all, for example the Powell's method. Further refinements in the deterministic phase of the minimization algorithm are needed to achieve more consistent performance. They include special line minimization, and Reduction procedures similar to the ones discussed in a previous section on the identification of underground inclusions. We skip the details and refer the reader to [Gut01].

In summary, the entire Local Minimization Method **(LMM)** consists of the following:

Local Minimization Method (LMM)

1. Let your starting configuration be $Q_0 = (r_1, r_2, \ldots, r_M, n_1, n_2, \ldots, n_M)$.
2. Apply the Reduction Procedure to Q_0, and obtain a reduced configuration Q_0^r containing M^r layers.
3. Apply the Basic Minimization Method in $A_{adm} \bigcap \mathbb{R}^{2M^r}$ with the starting point Q_0^r, and obtain a configuration Q_1.
4. Apply the Reduction Procedure to Q_1, and obtain a final reduced configuration Q_1^r.

3.3 Global Minimization Methods

Given an initial configuration Q_0 a local minimization method finds a local minimum near Q_0. On the other hand, global minimization methods explore the entire admissible set to find a global minimum of the objective function. While the local minimization is, usually, deterministic, the majority of the global methods are probabilistic in their nature. There is a great interest and activity in the development of efficient global minimization methods, see e.g. [Bie97],[Bom97]. Among them are the simulated annealing method ([KGV83],[Kir84]), various genetic algorithms [HH98], interval method, TRUST method ([BP96],[BPR97]), etc. As we have already mentioned before, the best fit to data functional Φ has many narrow local minima. In this situation it is exceedingly unlikely to get the minima points by chance alone. Thus our special interest is for the minimization methods, which combine a global search with a local minimization. In [GR00] we developed such a method (the Hybrid Stochastic-Deterministic Method), and applied it for the identification of small subsurface particles, provided a set of surface measurements, see Sections 2.2-2.4. The HSD method could be classified as a variation of a genetic algorithm with a local search with reduction. In this paper we consider the performance of two algorithms: Deep's Method, and Rinnooy Kan and Timmer's Multilevel Single-Linkage Method. Both combine a global and a local search to determine a global minimum. Recently these methods have been applied to a similar problem of the identification of particles from their light scattering characteristics in [ZUB98]. Unlike [ZUB98], our experience shows that Deep's method has failed consistently for the type of problems we are considering. See [DE94] and [ZUB98] for more details on Deep's Method.

Multilevel Single-Linkage Method (MSLM)

Rinnooy Kan and Timmer in [RT87a, RT87b] give a detailed description of this algorithm. Zakovic et. al. in [ZUB98] describe in detail an experience of its application to an inverse light scattering problem. They also discuss different stopping criteria for the MSLM. Thus, we only give here a shortened and an informal description of this method and of its algorithm.

In a pure **Random Search** method a batch H of L trial points is generated in A_{adm} using a uniformly distributed random variable. Then a local search is started from each of these L points. A local minimum with the smallest value of Φ is declared to be the global one.

A refinement of the Random Search is the **Reduced Sample Random Search** method. Here we use only a certain fixed fraction $\gamma < 1$ of the original batch of L points to proceed with the local searches. This reduced sample H_{red} of γL points is chosen to contain the points with the smallest γL values of Φ among the original batch. The local searches are started from the points in this reduced sample.

Since the local searches dominate the computational costs, we would like to initiate them only when it is truly necessary. Given a critical distance d we define a cluster to be a group of points located within the distance d of each other. Intuitively, a local search started from the points within a cluster should result in the same local minimum, and, therefore, should be initiated only once in each cluster.

Having tried all the points in the reduced sample we have an information on the number of local searches performed and the number of local minima found. This information and the critical distance d can be used to determine a statistical level of confidence, that all the local minima have been found. The algorithm is terminated (a stopping criterion is satisfied) if an a priori level of confidence is reached.

If, however, the stopping criterion is not satisfied, we perform another iteration of the MSLM by generating another batch of L trial points. Then it is combined with the previously generated batches to obtain an enlarged batch H^j of jL points (at iteration j), which leads to a reduced sample H^j_{red} of γjL points. According to MSLM the critical distance d is reduced to d_j, (note that $d_j \to 0$ as $j \to \infty$, since we want to find a minimizer), a local minimization is attempted once within each cluster, the information on the number of local minimizations performed and the local minima found is used to determine if the algorithm should be terminated, etc.

The following is an adaptation of the MSLM method to the inverse scattering problem presented in Section 3.1. The LMM local minimization method introduced in the previous Section is used here to perform local searches.

MSLM

(at iteration j).

1. Generate another batch of L trial points (configurations) from a random uniform distribution in A_{adm}. Combine it with the previously generated batches to obtain an enlarged batch H^j of jL points.
2. Reduce H^j to the reduced sample H^j_{red} of γjL points, by selecting the points with the smallest γjL values of Φ in H^j.
3. Calculate the critical distance d_j by

$$d_j^r = \pi^{-1/2} \left(\Gamma \left(1 + \frac{M}{2} \right) R^M \frac{\sigma \ln jL}{jL} \right)^{1/M},$$

$$d_j^m = \pi^{-1/2} \left(\Gamma \left(1 + \frac{M}{2} \right) (n_{high} - n_{low})^M \frac{\sigma \ln jL}{jL} \right)^{1/M}.$$

$$d_j = \sqrt{(d_j^r)^2 + (d_j^n)^2}$$

4. Order the sample points in H^j_{red} so that $\Phi(Q_i) \leq \Phi(Q_{i+1})$, $i = 1, \ldots, \gamma jL$. For each value of i, start the local minimization from Q_i, unless there exists an index $k < i$, such that $\|Q_k - Q_i\| \leq d_j$. Ascertain if the result is a known local minimum.
5. Let K be the number of local minimizations performed, and W be the number of different local minima found. Let

$$W_{tot} = \frac{W(K-1)}{K-W-2}$$

The algorithm is terminated if

$$W_{tot} < W + 0.5. \tag{23}$$

Here Γ is the gamma function, and σ is a fixed constant.

A related algorithm (the Mode Analysis) is based on a subdivision of the admissible set into smaller volumes associated with local minima. This algorithm is also discussed in [RT87a, RT87b]. From the numerical studies presented there, the authors deduce their preference for the MSLM.

The presented MSLM algorithm was successful in the identification of various 2D layered particles, see [Gut01] for details.

4 Potential scattering and the Stability Index method.

4.1 Problem description

Potential scattering problems are important in quantum mechanics, where they appear in the context of scattering of particles bombarding an atom nucleus. One is interested in reconstructing the scattering potential from the results of a scattering experiment. The examples in Section 4 deal with finding a spherically symmetric ($q = q(r)$, $r = |x|$) potential from the fixed-energy scattering data, which in this case consist of the fixed-energy phase shifts. In [Ram96, Ram02a, Ram04d, Ram05a] the three-dimensional inverse scattering problem with fixed-energy data is treated.

Let $q(x)$, $x \in \mathbb{R}^3$, be a real-valued potential with compact support. Let $R > 0$ be a number such that $q(x) = 0$ for $|x| > R$. We also assume that $q \in L^2(B_R)$, $B_R = \{x : |x| \leq R, x \in \mathbb{R}^3\}$. Let S^2 be the unit sphere, and $\alpha \in S^2$. For a given energy $k > 0$ the scattering solution $\psi(x, \alpha)$ is defined as the solution of

$$\Delta\psi + k^2\psi - q(x)\psi = 0, \quad x \in \mathbb{R}^3 \tag{24}$$

satisfying the following asymptotic condition at infinity:

$$\psi = \psi_0 + v, \quad \psi_0 := e^{ik\alpha \cdot x}, \quad \alpha \in S^2, \tag{25}$$

$$\lim_{r \to \infty} \int_{|x|=r} \left| \frac{\partial v}{\partial r} - ikv \right|^2 ds = 0. \tag{26}$$

It can be shown, that

$$\psi(x, \alpha) = \psi_0 + A(\alpha', \alpha, k)\frac{e^{ikr}}{r} + o\left(\frac{1}{r}\right), \quad \text{as } r \to \infty, \quad \frac{x}{r} = \alpha' \quad r := |x|. \tag{27}$$

The function $A(\alpha', \alpha, k)$ is called the scattering amplitude, α and α' are the directions of the incident and scattered waves, and k^2 is the energy, see [New82, Ram94a].

For spherically symmetric scatterers $q(x) = q(r)$ the scattering amplitude satisfies $A(\alpha', \alpha, k) = A(\alpha' \cdot \alpha, k)$. The converse is established in [Ram91]. Following [RS99], the scattering amplitude for $q = q(r)$ can be written as

$$A(\alpha', \alpha, k) = \sum_{l=0}^{\infty} \sum_{m=-l}^{l} A_l(k)Y_{lm}(\alpha')\overline{Y_{lm}(\alpha)}, \tag{28}$$

where Y_{lm} are the spherical harmonics, normalized in $L^2(S^2)$, and the bar denotes the complex conjugate.

The fixed-energy phase shifts $-\pi < \delta_l \leq \pi$ ($\delta_l = \delta(l, k)$, $k > 0$ is fixed) are related to $A_l(k)$ (see e.g., [RS99]) by the formula:

$$A_l(k) = \frac{4\pi}{k}e^{i\delta_l}\sin(\delta_l). \tag{29}$$

Several parameter-fitting procedures were proposed for calculating the potentials from the fixed-energy phase shifts, (by Fiedeldey, Lipperheide, Hooshyar and Razavy, Ioannides and Mackintosh, Newton, Sabatier, May and Scheid, Ramm and others). These works are referenced and their results are described in [CS89, New82]. Recent works [Gut00, Gut01, GR00, GR02a] and [RG01, RS99, RS00], present new numerical methods for solving this problem. In [Ram02d] (also see [Ram04b, Ram05a]) it is proved that the R.Newton-P.Sabatier method for solving inverse scattering problem the fixed-energy phase shifts as the data (see [CS89, New82]) is fundamentally wrong in the sense that its foundation is wrong. In [Ram02c] a counterexample is given to a uniqueness theorem claimed in a modification of the R.Newton's inversion scheme.

Phase shifts for a spherically symmetric potential can be computed by a variety of methods, e.g. by a variable phase method described in [Cal67]. The computation involves solving a nonlinear ODE for each phase shift. However, if the potential is compactly supported and piecewise-constant, then a much simpler method described in [ARS99] and [GRS02] can be used. We refer the reader to these papers for details.

Let $q_0(r)$ be a spherically symmetric piecewise-constant potential, $\{\tilde{\delta}(k,l)\}_{l=1}^{N}$ be the set of its phase shifts for a fixed $k > 0$ and a sufficiently large N. Let $q(r)$ be another potential, and let $\{\delta(k,l)\}_{l=1}^{N}$ be the set of its phase shifts. The best fit to data function $\Phi(q, k)$ is defined by

$$\Phi(q, k) = \frac{\sum_{l=1}^{N} |\delta(k,l) - \tilde{\delta}(k,l)|^2}{\sum_{l=1}^{N} |\tilde{\delta}(k,l)|^2}. \tag{30}$$

The phase shifts are known to decay rapidly with l, see [RAI98]. Thus, for sufficiently large N, the function Φ is practically the same as the one which would use all the shifts in (30). The inverse problem of the reconstruction of the potential from its fixed-energy phase shifts is reduced to the minimization of the objective function Φ over an appropriate admissible set.

4.2 Stability Index Minimization Method

Let the minimization problem be

$$\min\{\Phi(q) \ : \ q \in A_{adm}\} \tag{31}$$

Let $\tilde{q}_0$ be its global minimizer. Typically, the structure of the objective function Φ is quite complicated: this function may have many local minima. Moreover, the objective function in a neighborhood of minima can be nearly flat resulting in large minimizing sets defined by

$$S_\epsilon = \{q \in A_{adm} \ : \ \Phi(q) < \Phi(\tilde{q}_0) + \epsilon\} \tag{32}$$

for an $\epsilon > 0$.

Given an $\epsilon > 0$, let D_ϵ be the diameter of the minimizing set S_ϵ, which we call the **Stability Index** D_ϵ of the minimization problem (31).

Its usage is explained below.

One would expect to obtain stable identification for minimization problems with small (relative to the admissible set) stability indices. Minimization problems with large stability indices have distinct minimizers with practically the same values of the objective function. If no additional information is known, one has an uncertainty of the minimizer's choice. The stability index provides a quantitative measure of this uncertainty or instability of the minimization.

If $D_\epsilon < \eta$, where η is an a priori chosen treshold, then one can solve the global minimization problem stably. In the above general scheme it is not discussed in detail what are possible algorithms for computing the Stability Index.

One idea to construct such an algorithm is to iteratively estimate stability indices of the minimization problem, and, based on this information, to conclude if the method has achieved a stable minimum.

One such algorithm is an Iterative Reduced Random Search (IRRS) method, which uses the Stability Index for its stopping criterion. Let a batch H of L trial points be randomly generated in the admissible set A_{adm}. Let γ be a certain fixed fraction, e.g., $\gamma = 0.01$. Let S_{min} be the subset of H containing points $\{p_i\}$ with the smallest γL values of the objective function Φ in H. We call S_{min} the minimizing set. If all the minimizers in S_{min} are close to each other, then the objective function Φ is not flat near the global minimum. That is, the method identifies the minimum consistently. Let $\| \cdot \|$ be a norm in the admissible set.

Let

$$\epsilon = \max_{p_j \in S_{min}} \Phi(p_j) - \min_{p_j \in S_{min}} \Phi(p_j)$$

and

$$\tilde{D}_\epsilon = diam(S_{min}) = \max\{\|p_i - p_j\| \ : \ p_i, p_j \in S_{min}\}. \tag{33}$$

Then $\tilde{D}_\epsilon$ can be considered an estimate for the **Stability Index** D_ϵ of the minimization problem. The Stability Index reflects the size of the minimizing sets. Accordingly, it is used as a self-contained stopping criterion for an iterative minimization procedure. The identification is considered to be stable if the Stability Index $D_\epsilon < \eta$, for an a priori chosen $\eta > 0$. Otherwise, another batch of L trial points is generated, and the process is repeated. We used $\beta = 1.1$ as described below in the stopping criterion to determine if subsequent iterations do not produce a meaningful reduction of the objective function.

More precisely

Iterative Reduced Random Search (IRRS)

(at the j−th iteration).
   Fix $0 < \gamma < 1$, $\beta > 1$, $\eta > 0$ and N_{max}.

1. Generate another batch H^j of L trial points in A_{adm} using a random distribution.

2. Reduce H^j to the reduced sample H^j_{min} of γL points by selecting the points in H^j with the smallest γL values of Φ.

3. Combine H^j_{min} with H^{j-1}_{min} obtained at the previous iteration. Let S^j_{min} be the set of γL points from $H^j_{min} \cup H^{j-1}_{min}$ with the smallest values of Φ. (Use H^1_{min} for $j = 1$).

4. Compute the Stability Index (diameter) D^j of S^j_{min} by $D^j = \max\{\|p_i - p_k\| : p_i, p_k \in S_{min}\}$.

5. Stopping criterion.

Let $p \in S^j_{min}$ be the point with the smallest value of Φ in S^j_{min} (the global minimizer).

If $D^j \leq \eta$, then stop. The global minimizer is p. The minimization is stable.

If $D^j > \eta$ and $\Phi(q) \leq \beta\Phi(p) : q \in S^j_{min}$, then stop. The minimization is unstable. The Stability Index D^j is the measure of the instability of the minimization.

Otherwise, return to step 1 and do another iteration, unless the maximum number of iterations N_{max} is exceeded.

One can make the stopping criterion more meaningful by computing a normalized stability index. This can be achieved by dividing D^j by a fixed normalization constant, such as the diameter of the entire admissible set A_{adm}. To improve the performance of the algorithm in specific problems we found it useful to modify (IRRS) by combining the stochastic (global) search with a deterministic local minimization. Such Hybrid Stochastic-Deterministic (HSD) approach has proved to be successful for a variety of problems in inverse quantum scattering (see [Gut01, GRS02, RG01]) as well as in other applications (see [Gut00, GR00]). A somewhat different implementation of the Stability Index Method is described in [GR02a].

We seek the potentials $q(r)$ in the class of piecewise-constant, spherically symmetric real-valued functions. Let the admissible set be

$$A_{adm} \subset \{(r_1, r_2, \ldots, r_M, q_1, q_2, \ldots, q_M) : 0 \leq r_i \leq R, \; q_{low} \leq q_m \leq q_{high}\}, \tag{34}$$

where the bounds q_{low} and q_{high} for the potentials, as well as the bound M on the expected number of layers are assumed to be known.

A configuration $(r_1, r_2, \ldots, r_M, q_1, q_2, \ldots, q_M)$ corresponds to the potential

$$q(r) = q_m, \quad \text{for} \quad r_{m-1} \leq r < r_m, \quad 1 \leq m \leq M, \tag{35}$$

where $r_0 = 0$ and $q(r) = 0$ for $r \geq r_M = R$.

Note, that the admissible configurations must also satisfy

$$r_1 \leq r_2 \leq r_3 \leq \cdots \leq r_M. \tag{36}$$

We used $\beta = 1.1$, $\epsilon = 0.02$ and $j_{max} = 30$. The choice of these and other parameters ($L = 5000$, $\gamma = 0.01$, $\nu = 0.16$) is dictated by their meaning in the algorithm and the comparative performance of the program at their different values. As usual, some adjustment of the parameters, stopping criteria, etc., is needed to achieve the optimal performance of the algorithm. The deterministic part of the IRRs algorithm was based on the Powell's minimization method, one-dimensional minimization, and a Reduction procedure similar to ones described in the previous section 3, see [GRS02] for details.

4.3 Numerical Results

We studied the performance of the algorithm for 3 different potentials $q_i(r)$, $i = 1, 2, 3$ chosen from the physical considerations.

The potential $q_3(r) = -10$ for $0 \leq r < 8.0$ and $q_3 = 0$ for $r \geq 8.0$ and a wave number $k = 1$ constitute a typical example for elastic scattering of neutral particles in nuclear and atomic physics. In nuclear physics one measures the length in units of fm $= 10^{-15}$m, the quantity q_3 in units of $1/\text{fm}^2$, and the wave number in units of $1/\text{fm}$. The physical potential and incident energy are given by $V(r) = \frac{\hbar^2}{2\mu} q_3(r)$ and $E = \frac{\hbar^2 k^2}{2\mu}$, respectively. here $\hbar := \frac{h}{2\pi}$, $h = 6.62510^{-27}$ erg·s is the Planck constant, $\hbar c = 197.32$ MeV·fm, $c = 3 \cdot 10^6$ m/sec is the velocity of light, and μ is the mass of a neutron. By choosing the mass μ to be equal to the mass of a neutron $\mu = 939.6$ MeV/c^2, the potential and energy have the values of $V(r) = -207.2$ MeV for $0 \leq r < 8.0$ fm and $E(k = 1/\text{fm}) = 20.72$ MeV. In atomic physics one uses atomic units with the Bohr radius $a_0 = 0.529 \cdot 10^{-10}$m as the unit of length. Here, r, k and q_3 are measured in units of $a_0, 1/a_0$ and $1/a_0^2$, respectively. By assuming a scattering of an electron with mass $m_0 = 0.511$ MeV/c^2, we obtain the potential and energy as follows: $V(r) = -136$ eV for $0 \leq r < 8a_0 = 4.23 \cdot 10^{-10}$m and $E(k = 1/a_0) = 13.6$ eV. These numbers give motivation for the choice of examples applicable in nuclear and atomic physics.

The method used here deals with finite-range (compactly supported) potentials. One can use this method for potentials with the Coulomb tail or other potentials of interest in physics, which are not of finite range. This is done by using the phase shifts transformation method which allows one to transform the phase shifts corresponding to a potential, not of finite range, whose behavior is known for $r > a$, where a is some radius, into the phase shifts corresponding to a potential of finite range a (see [Apa97], p.156).

In practice differential cross section is measured at various angles, and from it the fixed-energy phase shifts are calculated by a parameter-fitting procedure. Therefore, we plan in the future work to generalize the stability index method to the case when the original data are the values of the differential cross section, rather than the phase shifts.

For the physical reasons discussed above, we choose the following three potentials:

$$q_1(r) = \begin{cases} -2/3 & 0 \leq r < 8.0 \\ 0.0 & r \geq 8.0 \end{cases}$$

$$q_2(r) = \begin{cases} -4.0 & 0 \le r < 8.0 \\ 0.0 & r \ge 8.0 \end{cases}$$

$$q_3(r) = \begin{cases} -10.0 & 0 \le r < 8.0 \\ 0.0 & r \ge 8.0 \end{cases}$$

In each case the following values of the parameters have been used. The radius R of the support of each q_i was chosen to be $R = 10.0$. The admissible set A_{adm} (34) was defined with $M = 2$. The Reduced Random Search parameters: $L = 5000$, $\gamma = 0.01$, $\nu = 0.16$, $\epsilon = 0.02$, $\beta = 1.10$, $j_{max} = 30$. The value $\epsilon_r = 0.1$ was used in the Reduction Procedure during the local minimization phase. The initial configurations were generated using a random number generator with seeds determined by the system time. A typical run time was about 10 minutes on a 333 MHz PC, depending on the number of iterations in IRRS. The number N of the shifts used in (30) for the formation of the objective function $\Phi(q)$ was 31 for all the wave numbers. It can be seen that the shifts for the potential q_3 decay rapidly for $k = 1$, but they remain large for $k = 4$. The upper and lower bounds for the potentials $q_{low} = -20.0$ and $q_{high} = 0.0$ used in the definition of the admissible set A_{adm} were chosen to reflect a priori information about the potentials.

The identification was attempted with 3 different noise levels h. The levels are $h = 0.00$ (no noise), $h = 0.01$ and $h = 0.1$. More precisely, the noisy phase shifts $\delta_h(k, l)$ were obtained from the exact phase shifts $\delta(k, l)$ by the formula

$$\delta_h(k, l) = \delta(k, l)(1 + (0.5 - z) \cdot h),$$

where z is the uniformly distributed on $[0, 1]$ random variable.

The distance $d(p_1(r), p_2(r))$ for potentials in step 5 of the IRRS algorithm was computed as

$$d(p_1(r), p_2(r)) = \|p_1(r) - p_2(r)\|$$

where the norm is the L_2-norm in $\mathbb{R}^3$.

The results of the identification algorithm (the Stability Indices) for different iterations of the IRRS algorithm are shown in Tables 6-8.

For example, Table 8 shows that for $k = 2.5$, $h = 0.00$ the Stability Index has reached the value 0.013621 after 2 iteration. According to the Stopping criterion for IRRS, the program has been terminated with the conclusion that the identification was stable. In this case the potential identified by the program was

$$p(r) = \begin{cases} -10.000024 & 0 \le r < 7.999994 \\ 0.0 & r \ge 7.999994 \end{cases}$$

which is very close to the original potential

Table 6. Stability Indices for $q_1(r)$ identification at different noise levels h.

k Iteration	$h = 0.00$	$h = 0.01$	$h = 0.10$
1.00	1 1.256985	0.592597	1.953778
	2 0.538440	0.133685	0.799142
	3 0.538253	0.007360	0.596742
	4 0.014616		0.123247
	5		0.015899
2.00	1 0.000000	0.020204	0.009607
2.50	1 0.000000	0.014553	0.046275
3.00	1 0.000000	0.000501	0.096444
4.00	1 0.000000	0.022935	0.027214

$$q_3(r) = \begin{cases} -10.0 & 0 \leq r < 8.0 \\ 0.0 & r \geq 8.0 \end{cases}$$

On the other hand, when the phase shifts of $q_3(r)$ were corrupted by a 10% noise ($k = 2.5$, $h = 0.10$), the program was terminated (according to the Stopping criterion) after 4 iterations with the Stability Index at 0.079241. Since the Stability Index is greater than the a priori chosen threshold of $\epsilon = 0.02$ the conclusion is that the identification is unstable. A closer look into this situation reveals that the values of the objective function $\Phi(p_i)$, $p_i \in S_{min}$ (there are 8 elements in S_{min}) are between 0.0992806 and 0.100320. Since we chose $\beta = 1.1$ the values are within the required 10% of each other. The actual potentials for which the normalized distance is equal to the Stability Index 0.079241 are

$$p_1(r) = \begin{cases} -9.997164 & 0 \leq r < 7.932678 \\ -7.487082 & 7.932678 \leq r < 8.025500 \\ 0.0 & r \geq 8.025500 \end{cases}$$

and

$$p_2(r) = \begin{cases} -9.999565 & 0 \leq r < 7.987208 \\ -1.236253 & 7.987208 \leq r < 8.102628 \\ 0.0 & r \geq 8.102628 \end{cases}$$

with $\Phi(p_1) = 0.0992806$ and $\Phi(p_2) = 0.0997561$. One may conclude from this example that the threshold $\epsilon = 0.02$ is too tight and can be relaxed, if the above uncertainty is acceptable.

Finally, we studied the dependency of the Stability Index from the dimension of the admissible set A_{adm}, see (34). This dimension is equal to $2M$, where M is the assumed number of layers in the potential. More precisely, $M = 3$, for example. means that the search is conducted in the class of potentials having 3 or less layers. The experiments were conducted for the identification of the original potential $q_2(r)$ with $k = 2.0$ and no noise present in the data.

Table 7. Stability Indices for $q_2(r)$ identification at different noise levels h.

k Iteration		$h = 0.00$	$h = 0.01$	$h = 0.10$
1.00	1	0.774376	0.598471	0.108902
	2	0.773718	1.027345	0.023206
	3	0.026492	0.025593	0.023206
	4	0.020522	0.029533	0.024081
	5	0.020524	0.029533	0.024081
	6	0.000745	0.029533	
	7		0.029533	
	8		0.029533	
	9		0.029533	
	10		0.029533	
	11		0.029619	
	12		0.025816	
	13		0.025816	
	14		0.008901	
2.00	1	0.863796	0.799356	0.981239
	2	0.861842	0.799356	0.029445
	3	0.008653	0.000993	0.029445
	4			0.029445
	5			0.026513
	6			0.026513
	7			0.024881
2.50	1	1.848910	1.632298	0.894087
	2	1.197131	1.632298	0.507953
	3	0.580361	1.183455	0.025454
	4	0.030516	0.528979	
	5	0.016195	0.032661	
3.00	1	1.844702	1.849016	1.708201
	2	1.649700	1.782775	1.512821
	3	1.456026	1.782775	1.412345
	4	1.410253	1.457020	1.156964
	5	0.624358	0.961263	1.156964
	6	0.692080	0.961263	0.902681
	7	0.692080	0.961263	0.902681
	8	0.345804	0.291611	0.902474
	9	0.345804	0.286390	0.159221
	10	0.345804	0.260693	0.154829
	11	0.043845	0.260693	0.154829
	12	0.043845	0.260693	0.135537
	13	0.043845	0.260693	0.135537
	14	0.043845	0.260693	0.135537
	15	0.042080	0.157024	0.107548
	16	0.042080	0.157024	
	17	0.042080	0.157024	
	18	0.000429	0.157024	
	19		0.022988	
4.00	1	0.000000	0.000674	0.050705

Table 8. Stability Indices for $q_3(r)$ identification at different noise levels h.

k	Iteration	$h = 0.00$	$h = 0.01$	$h = 0.10$
1.00	1	0.564168	0.594314	0.764340
	2	0.024441	0.028558	0.081888
	3	0.024441	0.014468	0.050755
	4	0.024684		
	5	0.024684		
	6	0.005800		
2.00	1	0.684053	1.450148	0.485783
	2	0.423283	0.792431	0.078716
	3	0.006291	0.457650	0.078716
	4		0.023157	0.078716
	5			0.078716
	6			0.078716
	7			0.078716
	8			0.078716
	9			0.078716
	10			0.078716
	11			0.078716
2.50	1	0.126528	0.993192	0.996519
	2	0.013621	0.105537	0.855049
	3		0.033694	0.849123
	4		0.026811	0.079241
3.00	1	0.962483	1.541714	0.731315
	2	0.222880	0.164744	0.731315
	3	0.158809	0.021775	0.072009
	4	0.021366		
	5	0.021366		
	6	0.001416		
4.00	1	1.714951	1.413549	0.788434
	2	0.033024	0.075503	0.024482
	3	0.018250	0.029385	
	4		0.029421	
	5		0.029421	
	6		0.015946	

The results are shown in Table 9. Since the potential q_2 consists of only one layer, the smallest Stability Indices are obtained for $M = 1$. They gradually increase with M. Note, that the algorithm conducts the global search using random variables, so the actual values of the indices are different in every run. Still the results show the successful identification (in this case) for the entire range of the a priori chosen parameter M. This agrees with the theoretical consideration according to which the Stability Index corresponding to an ill-posed problem in an infinite-dimensional space should be large. Reducing the original ill-posed problem to a one in a space of much lower dimension regularizes the original problem.

Table 9. Stability Indices for $q_2(r)$ identification for different values of M.

Iteration	$M = 1$	$M = 2$	$M = 3$	$M = 4$
1	0.472661	1.068993	1.139720	1.453076
2	0.000000	0.400304	0.733490	1.453076
3		0.000426	0.125855	0.899401
4			0.125855	0.846117
5			0.033173	0.941282
6			0.033173	0.655669
7			0.033123	0.655669
8			0.000324	0.948816
9				0.025433
10				0.025433
11				0.012586

5 Inverse scattering problem with fixed-energy data.

5.1 Problem description

In this Section we continue a discussion of the Inverse potential scattering with a presentation of Ramm's method for solving inverse scattering problem with fixed-energy data, see [Ram04d]. The method is applicable to both exact and noisy data. Error estimates for this method are also given. An inversion method using the Dirichlet-to-Neumann (DN) map is discussed, the difficulties of its numerical implementation are pointed out and compared with the difficulties of the implementation of the Ramm's inversion method. See the previous Section on the potential scattering for the problem set up.

5.2 Ramm's inversion method for exact data

The results we describe in this Section are taken from [Ram94a] and [Ram02a]. Assume $q \in Q := Q_a \cap L^\infty(\mathbb{R}^3)$, where $Q_a := \{q : q(x) = \overline{q(x)}, \quad q(x) \in L^2(B_a), \quad q(x) = 0 \text{ if } |x| \geq a\}$, $B_a := \{x : |x| \leq a\}$. Let $A(\alpha', \alpha)$ be the corresponding scattering amplitude at a fixed energy k^2, $k = 1$ is taken without loss of generality. One has:

$$A(\alpha', \alpha) = \sum_{\ell=0}^{\infty} A_\ell(\alpha) Y_\ell(\alpha'), \quad A_\ell(\alpha) := \int_{S^2} A(\alpha', \alpha) \overline{Y_\ell(\alpha')} d\alpha', \qquad (37)$$

where S^2 is the unit sphere in $\mathbb{R}^3$, $Y_\ell(\alpha') = Y_{\ell,m}(\alpha')$, $-\ell \leq m \leq \ell$, are the normalized spherical harmonics, summation over m is understood in (37) and in (44) below. Define the following algebraic variety:

$$M := \{\theta : \theta \in \mathbb{C}^3, \theta \cdot \theta = 1\}, \quad \theta \cdot w := \sum_{j=1}^{3} \theta_j w_j. \qquad (38)$$

This variety is non-compact, intersects $\mathbb{R}^3$ over S^2, and, given any $\xi \in \mathbb{R}^3$, there exist (many) $\theta, \theta' \in M$ such that

$$\theta' - \theta = \xi, \quad |\theta| \to \infty, \quad \theta, \theta' \in M. \tag{39}$$

In particular, if one chooses the coordinate system in which $\xi = te_3$, $t > 0$, e_3 is the unit vector along the x_3-axis, then the vectors

$$\theta' = \frac{t}{2}e_3 + \zeta_2 e_2 + \zeta_1 e_1, \quad \theta = -\frac{t}{2}e_3 + \zeta_2 e_2 + \zeta_1 e_1, \quad \zeta_1^2 + \zeta_2^2 = 1 - \frac{t^2}{4}, \tag{40}$$

satisfy (39) for any complex numbers ζ_1 and ζ_2 satisfying the last equation (40) and such that $|\zeta_1|^2 + |\zeta_2|^2 \to \infty$. There are infinitely many such $\zeta_1, \zeta_2 \in \mathbb{C}$. Consider a subset $M' \subset M$ consisting of the vectors $\theta = (\sin\vartheta\cos\varphi, \sin\vartheta\sin\varphi, \cos\vartheta)$, where ϑ and φ run through the whole complex plane. Clearly $\theta \in M$, but M' is a proper subset of M. Indeed, any $\theta \in M$ with $\theta_3 \neq \pm 1$ is an element of M'. If $\theta_3 = \pm 1$, then $\cos\vartheta = \pm 1$, so $\sin\vartheta = 0$ and one gets $\theta = (0, 0, \pm 1) \in M'$. However, there are vectors $\theta = (\theta_1, \theta_2, 1) \in M$ which do not belong to M'. Such vectors one obtains choosing $\theta_1, \theta_2 \in \mathbb{C}$ such that $\theta_1^2 + \theta_2^2 = 0$. There are infinitely many such vectors. The same is true for vectors $(\theta_1, \theta_2, -1)$. Note that in (39) one can replace M by M' for any $\xi \in \mathbb{R}^3$, $\xi \neq 2e_3$.

Let us state two estimates proved in [Ram94a]:

$$\max_{\alpha \in S^2} |A_\ell(\alpha)| \leq c \left(\frac{a}{\ell}\right)^{\frac{1}{2}} \left(\frac{ae}{2\ell}\right)^{\ell+1}, \tag{41}$$

where $c > 0$ is a constant depending on the norm $\|q\|_{L^2(B_a)}$, and

$$|Y_\ell(\theta)| \leq \frac{1}{\sqrt{4\pi}} \frac{e^{r|Im\theta|}}{|j_\ell(r)|}, \quad \forall r > 0, \quad \theta \in M', \tag{42}$$

where

$$j_\ell(r) := \left(\frac{\pi}{2r}\right)^{\frac{1}{2}} J_{\ell+\frac{1}{2}}(r) = \frac{1}{2\sqrt{2}} \frac{1}{\ell} \left(\frac{er}{2\ell}\right)^\ell [1 + o(1)] \text{ as } \ell \to \infty, \tag{43}$$

and $J_\ell(r)$ is the Bessel function regular at $r = 0$. Note that $Y_\ell(\alpha')$, defined above, admits a natural analytic continuation from S^2 to M by taking ϑ and φ to be arbitrary complex numbers. The resulting $\theta' \in M' \subset M$.

The series (37) converges absolutely and uniformly on the sets $S^2 \times M_c$, where M_c is any compact subset of M.

Fix any numbers a_1 and b, such that $a < a_1 < b$. Let $\| \cdot \|$ denote the $L^2(a_1 \leq |x| \leq b)$-norm. If $|x| \geq a$, then the scattering solution is given analytically:

$$u(x, \alpha) = e^{i\alpha \cdot x} + \sum_{\ell=0}^{\infty} A_\ell(\alpha) Y_\ell(\alpha') h_\ell(r), \quad r := |x| > a, \quad \alpha' := \frac{x}{r}, \tag{44}$$

where $A_\ell(\alpha)$ and $Y_\ell(\alpha')$ are defined above,

$$h_\ell(r) := e^{i\frac{\pi}{2}(\ell+1)}\sqrt{\frac{\pi}{2r}}H^{(1)}_{\ell+\frac{1}{2}}(r),$$

$H^{(1)}_\ell(r)$ is the Hankel function, and the normalizing factor is chosen so that $h_\ell(r) = \frac{e^{ir}}{r}[1 + o(1)]$ as $r \to \infty$. Define

$$\rho(x) := \rho(x; \nu) := e^{-i\theta \cdot x} \int_{S^2} u(x, \alpha)\nu(\alpha, \theta)d\alpha - 1, \quad \nu \in L^2(S^2). \qquad (45)$$

Consider the minimization problem

$$\|\rho\| = \inf := d(\theta), \qquad (46)$$

where the infimum is taken over all $\nu \in L^2(S^2)$, and (39) holds.

It is proved in [Ram94a] that

$$d(\theta) \le c|\theta|^{-1} \text{ if } \theta \in M, \quad |\theta| \gg 1. \qquad (47)$$

The symbol $|\theta| \gg 1$ means that $|\theta|$ is sufficiently large. The constant $c > 0$ in (47) depends on the norm $\|q\|_{L^2(B_a)}$ but not on the potential $q(x)$ itself.

An algorithm for computing a function $\nu(\alpha, \theta)$, which can be used for inversion of the exact, fixed-energy, three-dimensional scattering data, is as follows:

a) Find an approximate solution to (46) in the sense

$$\|\rho(x, \nu)\| < 2d(\theta), \qquad (48)$$

where in place of the factor 2 in (48) one could put any fixed constant greater than 1.

b) Any such $\nu(\alpha, \theta)$ generates an estimate of $\tilde{q}(\xi)$ with the error $O\left(\frac{1}{|\theta|}\right)$, $|\theta| \to \infty$. This estimate is calculated by the formula

$$\hat{q} := -4\pi \int_{S^2} A(\theta', \alpha)\nu(\alpha, \theta)d\alpha, \qquad (49)$$

where $\nu(\alpha, \theta) \in L^2(S^2)$ is any function satisfying (48).

Our basic result is:

Theorem 1. *Let (39) and (48) hold. Then*

$$\sup_{\xi \in \mathbb{R}^3} |\hat{q} - \tilde{q}(\xi)| \le \frac{c}{|\theta|}, \quad |\theta| \to \infty, \qquad (50)$$

The constant $c > 0$ in (50) depends on a norm of q, but not on a particular q.

The norm of q in the above Theorem can be any norm such that the set $\{q : \|q\| \leq const\}$ is a compact set in $L^\infty(B_a)$.

In [Ram94a, Ram02a] an inversion algorithm is formulated also for noisy data, and the error estimate for this algorithm is obtained. Let us describe these results.

Assume that the scattering data are given with some error: a function $A_\delta(\alpha', \alpha)$ is given such that

$$\sup_{\alpha', \alpha \in S^2} |A(\alpha', \alpha) - A_\delta(\alpha', \alpha)| \leq \delta. \tag{51}$$

We emphasize that $A_\delta(\alpha', \alpha)$ is not necessarily a scattering amplitude corresponding to some potential, it is an arbitrary function in $L^\infty(S^2 \times S^2)$ satisfying (51). It is assumed that the unknown function $A(\alpha', \alpha)$ is the scattering amplitude corresponding to a $q \in Q$.

The problem is: *Find an algorithm for calculating $\widehat{q}_\delta$ such that*

$$\sup_{\xi \in \mathbb{R}^3} |\widehat{q}_\delta - \widetilde{q}(\xi)| \leq \eta(\delta), \quad \eta(\delta) \to 0 \text{ as } \delta \to 0, \tag{52}$$

and estimate the rate at which $\eta(\delta)$ tends to zero.

An algorithm for inversion of noisy data will now be described.

Let

$$N(\delta) := \left[\frac{|\ln \delta|}{\ln |\ln \delta|} \right], \tag{53}$$

where $[x]$ is the integer nearest to $x > 0$,

$$\widehat{A}_\delta(\theta', \alpha) := \sum_{\ell=0}^{N(\delta)} A_{\delta \ell}(\alpha) Y_\ell(\theta'), \quad A_{\delta \ell}(\alpha) := \int_{S^2} A_\delta(\alpha', \alpha) \overline{Y_\ell(\alpha')} d\alpha', \tag{54}$$

$$u_\delta(x, \alpha) := e^{i\alpha \cdot x} + \sum_{\ell=0}^{N(\delta)} A_{\delta \ell}(\alpha) Y_\ell(\alpha') h_\ell(r), \tag{55}$$

$$\rho_\delta(x; \nu) := e^{-i\theta \cdot x} \int_{S^2} u_\delta(x, \alpha) \nu(\alpha) d\alpha - 1, \quad \theta \in M, \tag{56}$$

$$\mu(\delta) := e^{-\gamma N(\delta)}, \quad \gamma = \ln \frac{a_1}{a} > 0, \tag{57}$$

$$a(\nu) := \|\nu\|_{L^2(S^2)}, \quad \kappa := |Im\theta|. \tag{58}$$

Consider the variational problem with constraints:

$$|\theta| = \sup := \vartheta(\delta), \tag{59}$$

$$|\theta| \left[\|\rho_\delta(\nu)\| + a(\nu) e^{\kappa b} \mu(\delta) \right] \leq c, \quad \theta \in M, \quad |\theta| = \sup := \vartheta(\delta), \tag{60}$$

the norm is defined above (44), and it is assumed that (39) holds, where $\xi \in \mathbb{R}^3$ is an arbitrary fixed vector, $c > 0$ is a sufficiently large constant, and the supremum is taken over $\theta \in M$ and $\nu \in L^2(S^2)$ under the constraint (60). By c we denote various positive constants.

Given $\xi \in \mathbb{R}^3$ one can always find θ and θ' such that (39) holds. We prove that $\vartheta(\delta) \to \infty$, more precisely:

$$\vartheta(\delta) \geq c \frac{|\ln \delta|}{(\ln |\ln \delta|)^2}, \quad \delta \to 0. \tag{61}$$

Let the pair $\theta(\delta)$ and $\nu_\delta(\alpha, \theta)$ be any approximate solution to problem (59)-(60) in the sense that

$$|\theta(\delta)| \geq \frac{\vartheta(\delta)}{2}. \tag{62}$$

Calculate

$$\widehat{q}_\delta := -4\pi \int_{S^2} \widehat{A}_\delta(\theta', \alpha)\nu_\delta(\alpha, \theta)d\alpha. \tag{63}$$

Theorem 2. *If* (39) *and* (62) *hold, then*

$$\sup_{\xi \in \mathbb{R}^3} |\widehat{q}_\delta - \widetilde{q}(\xi)| \leq c \frac{(\ln |\ln \delta|)^2}{|\ln \delta|} \quad as \; \delta \to 0, \tag{64}$$

where $c > 0$ is a constant depending on a norm of q.

In [Ram94a] estimates (50) and (64) were formulated with the supremum taken over an arbitrary large but fixed ball of radius ξ_0. Here these estimates are improved: $\xi_0 = \infty$. The key point is: the constant $c > 0$ in the estimate (47) does not depend on θ.

Remark. In [Ram96] (see also [Ram92a, Ram02a]) an analysis of the approach to ISP, based on the recovery of the DN (Dirichle-to-Neumann) map from the fixed-energy scattering data, is given. This approach is discussed below.

The basic numerical difficulty of the approach described in Theorems 1 and 2 comes from solving problems (46) for exact data, and problem (59)-(60) for noisy data. Solving (46) amounts to finding a global minimizer of a quadratic form of the variables c_ℓ, if one takes ν in (45) as a linear combination of the spherical harmonics: $\nu = \sum_{\ell=0}^{L} c_\ell Y_\ell(\alpha)$. If one uses the necessary condition for a minimizer of a quadratic form, that is, a linear system, then the matrix of this system is ill-conditioned for large L. This causes the main difficulty in the numerical solution of (46). On the other hand, there are methods for global minimization of the quadratic functionals, based on the gradient descent, which may be more efficient than using the above necessary condition.

5.3 Discussion of the inversion method which uses the DN map

In [Ram96] the following inversion method is discussed:

$$\tilde{q}(\xi) = \lim_{|\theta| \to \infty} \int_S \exp(-i\theta' \cdot s)(\Lambda - \Lambda_0)\psi ds, \qquad (65)$$

where (39) is assumed, Λ is the Dirichlet-to-Neumann (DN) map, ψ is found from the equation:

$$\psi(s) = \psi_0(s) - \int_S G(s-t)B\psi dt, \quad B := \Lambda - \Lambda_0, \quad \psi_0(s) := e^{i\theta \cdot s}, \qquad (66)$$

and G is defined by the formula:

$$G(x) = \exp(i\theta \cdot x)\frac{1}{(2\pi)^3}\int_{\mathbb{R}^3} \frac{\exp(i\xi \cdot x)d\xi}{\xi^2 + 2\xi \cdot \theta}. \qquad (67)$$

The DN map is constructed from the fixed-energy scattering data $A(\alpha', \alpha)$ by the method of [Ram96] (see also [Ram94a]).

Namely, given $A(\alpha', \alpha)$ for all $\alpha', \alpha \in S^2$, one finds Λ using the following steps.

Let $f \in H^{3/2}(S)$ be given, S is a sphere of radius a centered at the origin, f_ℓ are its Fourier coefficients in the basis of the spherical harmonics,

$$w = \sum_{l=0}^{\infty} f_l Y_l(x^0)\frac{h_l(r)}{h_l(a)}, \quad r \geq a, \quad x^0 := \frac{x}{r}, \quad r := |x|. \qquad (68)$$

Let

$$w = \int_S g(x, s)\sigma(s)ds, \qquad (69)$$

where σ is some function, which we find below, and g is the Green function (resolvent kernel) of the Schroedinger operator, satisfying the radiation condition at infinity. Then

$$w_N^+ = w_N^- + \sigma, \qquad (70)$$

where N is the outer normal to S, so N is directed along the radius-vector. We require $w = f$ on S. Then w is given by (68) in the exterior of S, and

$$w_N^- = \sum_{l=0}^{\infty} f_l Y_l(x^0)\frac{h_l'(a)}{h_l(a)}. \qquad (71)$$

By formulas (70) and (71), finding Λ is equivalent to finding σ. By (69), asymptotics of w as $r := |x| \to \infty$, $x/|x| := x^0$, is (cf [Ram94a], p.67):

$$w = \frac{e^{ir}}{r}\frac{u(y, -x^0)}{4\pi} + o(\frac{1}{r}), \qquad (72)$$

where u is the scattering solution,

$$u(y, -x^0) = e^{-ix^0 \cdot y} + \sum_{\ell=0}^{\infty} A_\ell(-x^0)Y_\ell(y^0)h_\ell(|y|). \qquad (73)$$

From (68), (72) and (73) one gets an equation for finding σ ([Ram96], eq. (23), see also [Ram94a], p. 199):

$$\frac{f_l}{h_l(a)} = \frac{1}{4\pi} \int_S ds\sigma(s)\, (u(s, -\beta), Y_l(\beta))_{L^2(S^2)}, \tag{74}$$

which can be written as a linear system:

$$\frac{4\pi f_l}{h_l(a)} = a^2(-1)^l \sum_{l'=0}^{\infty} \sigma_{l'} [4\pi i^l j_l(a)\delta_{ll'} + A_{l'l}h_{l'}(a)], \tag{75}$$

for the Fourier coefficients σ_ℓ of σ. The coefficients

$$A_{l'l} := ((A(\alpha', \alpha), Y_\ell(\alpha'))_{L^2(S^2)}, Y_\ell(\alpha))_{L^2(S^2)}$$

are the Fourier coefficients of the scattering amplitude. Problems (74) and (75) are very ill-posed (see [Ram96] for details).

   This approach faces many difficulties:

   1) The construction of the DN map from the scattering data is a very ill-posed problem,

   2) The construction of the potential from the DN map is a very difficult problem numerically, because one has to solve a Fredholm-type integral equation (equation (66)) whose kernel contains G, defined in (67). This G is a tempered distribution, and it is very difficult to compute it,

   3) One has to calculate a limit of an integral whose integrand grows exponentially to infinity if a factor in the integrand is not known exactly. The solution of equation (66) is one of the factors in the integrand. It cannot be known exactly in practice because it cannot be calculated with arbitrary accuracy even if the scattering data are known exactly. Therefore the limit in formula (65) cannot be calculated accurately.

   No error estimates are obtained for this approach.

   In contrast, in Ramm's method, there is no need to compute G, to solve equation (66), to calculate the DN map from the scattering data, and to compute the limit (65). The basic difficulty in Ramm's inversion method for exact data is to minimize the quadratic form (46), and for noisy data to solve optimization problem (59)-(60). The error estimates are obtained for the Ramm's method.

6 Obstacle scattering by the Modified Rayleigh Conjecture (MRC) method.

6.1 Problem description

In this section we present a novel numerical method for Direct Obstacle Scattering Problems based on the Modified Rayleigh Conjecture (MRC). The basic theoretical foundation of the method was developed in [Ram02b]. The MRC has the appeal of an easy implementation for obstacles of complicated geometry, e.g. having edges and corners. A special version of the MRC method was used in [GR05] to compute the scattered field for 3D obstacles. In our numerical experiments the method has shown itself to be a competitive alternative to the BIEM (boundary integral equations method), see [GR02b]. Also, unlike the BIEM, one can apply the algorithm to different obstacles with very little additional effort.

We formulate the obstacle scattering problem in a 3D setting with the Dirichlet boundary condition, but the discussed method can also be used for the Neumann and Robin boundary conditions.

Consider a bounded domain $D \subset \mathbb{R}^3$, with a boundary S which is assumed to be Lipschitz continuous. Denote the exterior domain by $D' = \mathbb{R}^3 \backslash D$. Let $\alpha, \alpha' \in S^2$ be unit vectors, and S^2 be the unit sphere in $\mathbb{R}^3$.

The acoustic wave scattering problem by a soft obstacle D consists in finding the (unique) solution to the problem (76)-(77):

$$\left(\nabla^2 + k^2 \right) u = 0 \text{ in } D', \quad u = 0 \text{ on } S, \tag{76}$$

$$u = u_0 + A(\alpha', \alpha)\frac{e^{ikr}}{r} + o\left(\frac{1}{r}\right), \quad r := |x| \to \infty, \quad \alpha' := \frac{x}{r}. \tag{77}$$

Here $u_0 := e^{ik\alpha \cdot x}$ is the incident field, $v := u - u_0$ is the scattered field, $A(\alpha', \alpha)$ is called the scattering amplitude, its k-dependence is not shown, $k > 0$ is the wavenumber. Denote

$$A_\ell(\alpha) := \int_{S^2} A(\alpha', \alpha)\overline{Y_\ell(\alpha')}d\alpha', \tag{78}$$

where $Y_\ell(\alpha)$ are the orthonormal spherical harmonics, $Y_\ell = Y_{\ell m}, -\ell \le m \le \ell$. Let $h_\ell(r)$ be the spherical Hankel functions, normalized so that $h_\ell(r) \sim \frac{e^{ikr}}{r}$ as $r \to +\infty$.

Informally, the Random Multi-point MRC algorithm can be described as follows.

Fix a $J > 0$. Let $x_j, j = 1, 2, ..., J$ be a batch of points randomly chosen inside the obstacle D. For $x \in D'$, let

$$\alpha' = \frac{x - x_j}{|x - x_j|}, \quad \psi_\ell(x, x_j) = Y_\ell(\alpha')h_\ell(k|x - x_j|). \tag{79}$$

Let $g(x) = u_0(x)$, $x \in S$, and minimize the discrepancy

$$\Phi(\mathbf{c}) = \left\| g(x) + \sum_{j=1}^{J} \sum_{\ell=0}^{L} c_{\ell,j}\psi_\ell(x, x_j) \right\|_{L^2(S)}, \tag{80}$$

over $\mathbf{c} \in \mathbb{C}^N$, where $\mathbf{c} = \{c_{\ell,j}\}$. That is, the total field $u = g(x) + v$ is desired to be as close to zero as possible at the boundary S, to satisfy the required condition for the soft scattering. If the resulting residual $r^{min} = \min \Phi$ is smaller than the prescribed tolerance ϵ, than the procedure is finished, and the sought scattered field is

$$v_\epsilon(x) = \sum_{j=1}^{J} \sum_{\ell=0}^{L} c_{\ell,j} \psi_\ell(x, x_j), \quad x \in D',$$

(see Lemma 1 below).

If, on the other hand, the residual $r^{min} > \epsilon$, then we continue by trying to improve on the already obtained fit in (80). Adjust the field on the boundary by letting $g(x) := g(x) + v_\epsilon(x)$, $x \in S$. Create another batch of J points randomly chosen in the interior of D, and minimize (80) with this new $g(x)$. Continue with the iterations until the required tolerance ϵ on the boundary S is attained, at the same time keeping the track of the changing field v_ϵ.

Note, that the minimization in (80) is always done over the same number of points J. However, the points x_j are sought to be different in each iteration to assure that the minimal values of Φ are decreasing in consequent iterations. Thus, computationally, the size of the minimization problem remains the same. This is the new feature of the Random multi-point MRC method, which allows it to solve scattering problems untreatable by previously developed MRC methods, see [GR02b].

Here is the precise description of the algorithm.

Random Multi-point MRC.

For $x_j \in D$, and $\ell \geq 0$ functions $\psi_\ell(x, x_j)$ are defined as in (79).

1. **Initialization.** Fix $\epsilon > 0$, $L \geq 0$, $J > 0$, $N_{max} > 0$. Let $n = 0$, $v_\epsilon = 0$ and $g(x) = u_0(x)$, $x \in S$.
2. **Iteration.**
    a) Let $n := n + 1$. Randomly choose J points $x_j \in D$, $j = 1, 2, \ldots, J$.
    b) Minimize

$$\Phi(\mathbf{c}) = \|g(x) + \sum_{j=1}^{J} \sum_{\ell=0}^{L} c_{\ell,j} \psi_\ell(x, x_j)\|_{L^2(S)}$$

    over $\mathbf{c} \in \mathbb{C}^N$, where $\mathbf{c} = \{c_{\ell,j}\}$.
    Let the minimal value of Φ be r^{min}.
    c) Let

$$v_\epsilon(x) := v_\epsilon(x) + \sum_{j=1}^{J} \sum_{\ell=0}^{L} c_{\ell,j} \psi_\ell(x, x_j), \quad x \in D'.$$

3. **Stopping criterion.**
    a) If $r^{min} \leq \epsilon$, then stop.
    b) If $r^{min} > \epsilon$, and $n \neq N_{max}$, let

$$g(x) := g(x) + \sum_{j=1}^{J} \sum_{\ell=0}^{L} c_{\ell,j} \psi_\ell(x, x_j), \quad x \in S$$

and repeat the iterative step (2).

c) If $r^{min} > \epsilon$, and $n = N_{max}$, then the procedure failed.

6.2 Direct scattering problems and the Rayleigh conjecture.

Let a ball $B_R := \{x : |x| \leq R\}$ contain the obstacle D. In the region $r > R$ the solution to (76)-(77) is:

$$u(x, \alpha) = e^{ik\alpha \cdot x} + \sum_{\ell=0}^{\infty} A_\ell(\alpha)\psi_\ell, \quad \psi_\ell := Y_\ell(\alpha')h_\ell(kr), \quad r > R, \quad \alpha' = \frac{x}{r},$$

(81)

where the sum includes the summation with respect to m, $-\ell \leq m \leq \ell$, and $A_\ell(\alpha)$ are defined in (78).

The Rayleigh conjecture (RC) is: the series (81) converges up to the boundary S (originally RC dealt with periodic structures, gratings). This conjecture is false for many obstacles, but is true for some ([Bar71, Mil73, Ram86]). For example, if $n = 2$ and D is an ellipse, then the series analogous to (81) converges in the region $r > a$, where $2a$ is the distance between the foci of the ellipse [Bar71]. In the engineering literature there are numerical algorithms, based on the Rayleigh conjecture. Our aim is to give a formulation of a *Modified Rayleigh Conjecture* (MRC) which holds for any Lipschitz obstacle and can be used in numerical solution of the direct and inverse scattering problems (see [Ram02b]). We discuss the Dirichlet condition but similar argument is applicable to the Neumann boundary condition, corresponding to acoustically hard obstacles.

Fix $\epsilon > 0$, an arbitrary small number.

Lemma 1. *There exist $L = L(\epsilon)$ and $c_\ell = c_\ell(\epsilon)$ such that*

$$\left\| u_0 + \sum_{\ell=0}^{L(\epsilon)} c_\ell(\epsilon)\psi_\ell \right\|_{L^2(S)} \leq \epsilon.$$

(82)

If (82) and the boundary condition (76) hold, then

$$\|v_\epsilon - v\|_{L^2(S)} \leq \epsilon, \quad v_\epsilon := \sum_{\ell=0}^{L(\epsilon)} c_\ell(\epsilon)\psi_\ell.$$

(83)

Lemma 2. *If (83) holds then*

$$\||v_\epsilon - v\|| = O(\epsilon), \quad \epsilon \to 0,$$

(84)

where $\||\cdot\|| := \|\cdot\|_{H^m_{loc}(D')} + \|\cdot\|_{L^2(D';(1+|x|)^{-\gamma})}$, $\gamma > 1$, $m > 0$ is an arbitrary integer, H^m is the Sobolev space, and v_ϵ, v in (84) are functions defined in D'.

In particular, (84) implies

$$\|v_\epsilon - v\|_{L^2(S_R)} = O(\epsilon), \quad \epsilon \to 0, \tag{85}$$

where S_R is the sphere centered at the origin with radius R.

Lemma 3. *One has:*

$$c_\ell(\epsilon) \to A_\ell(\alpha), \quad \forall \ell, \quad \epsilon \to 0. \tag{86}$$

The Modified Rayleigh Conjecture (MRC) is formulated as a theorem, which follows from the above three lemmas:

Theorem 3. *For an arbitrary small $\epsilon > 0$ there exist $L(\epsilon)$ and $c_\ell(\epsilon)$, $0 \le \ell \le L(\epsilon)$, such that (82), (84) and (86) hold.*

See [Ram02b] for a proof of the above statements.

The difference between RC and MRC is: (83) does not hold if one replaces v_ϵ by $\sum_{\ell=0}^{L} A_\ell(\alpha)\psi_\ell$, and lets $L \to \infty$ (instead of letting $\epsilon \to 0$). Indeed, the series $\sum_{\ell=0}^{\infty} A_\ell(\alpha)\psi_\ell$ diverges at some points of the boundary for many obstacles. Note also that the coefficients in (83) depend on ϵ, so (83) is *not* a partial sum of a series.

For the Neumann boundary condition one minimizes

$$\left\| \frac{\partial[u_0 + \sum_{\ell=0}^{L} c_\ell \psi_\ell]}{\partial N} \right\|_{L^2(S)}$$

with respect to c_ℓ. Analogs of Lemmas 1-3 are valid and their proofs are essentially the same.

See [Ram04c] for an extension of these results to scattering by periodic structures.

6.3 Numerical Experiments.

In this section we desribe numerical results obtained by the Random Multi-point MRC method for 2D and 3D obstacles. We also compare the 2D results to the ones obtained by our earlier method introduced in [GR02b]. The method that we used previously can be described as a Multi-point MRC. Its difference from the Random Multi-point MRC method is twofold: It is just the first iteration of the Random method, and the interior points x_j, $j = 1, 2, ..., J$ were chosen deterministically, by an *ad hoc* method according to the geometry of the obstacle D. The number of points J was limited by the size of the resulting numerical minimization problem, so the accuracy of the scattering solution (i.e. the residual r^{min}) could not be made small for many obstacles. The method was not capable of treating 3D obstacles. These limitations were removed by using the Random Multi-point MRC method. As we mentioned previously, [GR02b] contains a favorable comparison of the Multi-point MRC method with the BIEM, inspite in spite of the fact that the numerical implementation of the MRC method in [GR02b] is considerably less efficient than the one presented in this paper.

A numerical implementation of the Random Multi-point MRC method follows the same outline as for the Multi-point MRC, which was described in [GR02b]. Of course, in a 2D case, instead of (79) one has

$$\psi_l(x, x_j) = H_l^{(1)}(k|x - x_j|)e^{il\theta_j},$$

where $(x - x_j)/|x - x_j| = e^{i\theta_j}$.

For a numerical implementation choose M nodes $\{t_m\}$ on the surface S of the obstacle D. After the interior points x_j, $j = 1, 2, ..., J$ are chosen, form N vectors

$$\mathbf{a}^{(n)} = \{\psi_l(t_m, x_j)\}_{m=1}^M,$$

$n = 1, 2, ..., N$ of length M. Note that $N = (2L + 1)J$ for a 2D case, and $N = (L + 1)^2 J$ for a 3D case. It is convenient to normalize the norm in $\mathbb{R}^M$ by

$$\|\mathbf{b}\|^2 = \frac{1}{M} \sum_{m=1}^M |b_m|^2, \quad \mathbf{b} = (b_1, b_2, ..., b_M).$$

Then $\|u_0\| = 1$.

Now let $\mathbf{b} = \{g(t_m)\}_{m=1}^M$, in the Random Multi-point MRC (see section 1), and minimize

$$\Phi(\mathbf{c}) = \|\mathbf{b} + A\mathbf{c}\|, \tag{87}$$

for $\mathbf{c} \in \mathbb{C}^N$, where A is the matrix containing vectors $\mathbf{a}^{(n)}$, $n = 1, 2, ..., N$ as its columns.

We used the Singular Value Decomposition (SVD) method (see e.g. [PTVF92]) to minimize (87). Small singular values $s_n < w_{min}$ of the matrix A are used to identify and delete linearly dependent or almost linearly dependent combinations of vectors $\mathbf{a}^{(n)}$. This spectral cut-off makes the minimization process stable, see the details in [GR02b].

Let r^{min} be the residual, i.e. the minimal value of $\Phi(\mathbf{c})$ attained after N_{max} iterations of the Random Multi-point MRC method (or when it is stopped). For a comparison, let r_{old}^{min} be the residual obtained in [GR02b] by an earlier method.

We conducted 2D numerical experiments for four obstacles: two ellipses of different eccentricity, a kite, and a triangle. The M=720 nodes t_m were uniformly distributed on the interval $[0, 2\pi]$, used to parametrize the boundary S. Each case was tested for wave numbers $k = 1.0$ and $k = 5.0$. Each obstacle was subjected to incident waves corresponding to $\alpha = (1.0, 0.0)$ and $\alpha = (0.0, 1.0)$.

The results for the Random Multi-point MRC with $J = 1$ are shown in Table 10, in the last column r^{min}. In every experiment the target residual $\epsilon = 0.0001$ was obtained in under 6000 iterations, in about 2 minutes run time on a 2.8 MHz PC.

In [GR02b], we conducted numerical experiments for the same four 2D obstacles by a Multi-point MRC, as described in the beginning of this section. The interior points x_j were chosen differently in each experiment. Their choice is indicated in the description of each 2D experiment. The column J shows the number of these interior points. Values $L = 5$ and $M = 720$ were used in all the experiments. These results are shown in Table 10, column r_{old}^{min}.

Thus, the Random Multi-point MRC method achieved a significant improvement over the earlier Multi-point MRC.

Table 10. Normalized residuals attained in the numerical experiments for 2D obstacles, $\|u_0\| = 1$.

Experiment	J	k	α	r_{old}^{min}	r^{min}
I	4	1.0	$(1.0, 0.0)$	0.000201	0.0001
	4	1.0	$(0.0, 1.0)$	0.000357	0.0001
	4	5.0	$(1.0, 0.0)$	0.001309	0.0001
	4	5.0	$(0.0, 1.0)$	0.007228	0.0001
II	16	1.0	$(1.0, 0.0)$	0.003555	0.0001
	16	1.0	$(0.0, 1.0)$	0.002169	0.0001
	16	5.0	$(1.0, 0.0)$	0.009673	0.0001
	16	5.0	$(0.0, 1.0)$	0.007291	0.0001
III	16	1.0	$(1.0, 0.0)$	0.008281	0.0001
	16	1.0	$(0.0, 1.0)$	0.007523	0.0001
	16	5.0	$(1.0, 0.0)$	0.021571	0.0001
	16	5.0	$(0.0, 1.0)$	0.024360	0.0001
IV	32	1.0	$(1.0, 0.0)$	0.006610	0.0001
	32	1.0	$(0.0, 1.0)$	0.006785	0.0001
	32	5.0	$(1.0, 0.0)$	0.034027	0.0001
	32	5.0	$(0.0, 1.0)$	0.040129	0.0001

Experiment 2D-I. The boundary S is an ellipse described by

$$\mathbf{r}(t) = (2.0\cos t, \ \sin t), \quad 0 \le t < 2\pi. \tag{88}$$

The Multi-point MRC used $J = 4$ interior points $x_j = 0.7\mathbf{r}(\frac{\pi(j-1)}{2})$, $j = 1, \ldots, 4$. Run time was 2 seconds.

Experiment 2D-II. The kite-shaped boundary S (see [CK92], Section 3.5) is described by

$$\mathbf{r}(t) = (-0.65 + \cos t + 0.65 \cos 2t, \ 1.5 \sin t), \quad 0 \le t < 2\pi. \tag{89}$$

The Multi-point MRC used $J = 16$ interior points $x_j = 0.9\mathbf{r}(\frac{\pi(j-1)}{8})$, $j = 1, \ldots, 16$. Run time was 33 seconds.

Experiment 2D-III. The boundary S is the triangle with vertices $(-1.0, 0.0)$ and $(1.0, \pm1.0)$. The Multi-point MRC used the interior points $x_j = 0.9\mathbf{r}(\frac{\pi(j-1)}{8})$, $j = 1, \ldots, 16$. Run time was about 30 seconds.

Experiment 2D-IV. The boundary S is an ellipse described by

$$\mathbf{r}(t) = (0.1\cos t, \ \sin t), \quad 0 \le t < 2\pi. \tag{90}$$

The Multi-point MRC used $J = 32$ interior points $x_j = 0.95\mathbf{r}(\frac{\pi(j-1)}{16})$, $j = 1, \ldots, 32$. Run time was about 140 seconds.

The 3D numerical experiments were conducted for 3 obstacles: a sphere, a cube, and an ellipsoid. We used the Random Multi-point MRC with $L = 0$, $w_{min} = 10^{-12}$, and $J = 80$. The number M of the points on the boundary S is indicated in the description of the obstacles. The scattered field for each obstacle was computed for two incoming directions $\alpha_i = (\theta, \phi)$, $i = 1, 2$, where ϕ was the polar angle. The first unit vector α_1 is denoted by (1) in Table 11, $\alpha_1 = (0.0, \pi/2)$. The second one is denoted by (2), $\alpha_2 = (\pi/2, \pi/4)$. A typical number of iterations N_{iter} and the run time on a 2.8 MHz PC are also shown in Table 11. For example, in experiment I with $k = 5.0$ it took about 700 iterations of the Random Multi-point MRC method to achieve the target residual $r^{min} = 0.001$ in 7 minutes.

Experiment 3D-I. The boundary S is the sphere of radius 1, with $M = 450$.

Experiment 3D-II. The boundary S is the surface of the cube $[-1, 1]^3$ with $M = 1350$.

Experiment 3D-III. The boundary S is the surface of the ellipsoid $x^2/16 + y^2 + z^2 = 1$ with $M = 450$.

Table 11. Normalized residuals attained in the numerical experiments for 3D obstacles, $\|\mathbf{u}_0\| = 1$.

Experiment	k	α_i	r^{min}	N_{iter}	run time
I	1.0		0.0002	1	1 sec
	5.0		0.001	700	7 min
II	1.0	(1)	0.001	800	16 min
	1.0	(2)	0.001	200	4 min
	5.0	(1)	0.0035	2000	40 min
	5.0	(2)	0.002	2000	40 min
III	1.0	(1)	0.001	3600	37 min
	1.0	(2)	0.001	3000	31 min
	5.0	(1)	0.0026	5000	53 min
	5.0	(2)	0.001	5000	53 min

In the last experiment the run time could be reduced by taking a smaller value for J. For example, the choice of $J = 8$ reduced the running time to about 6-10 minutes.

Numerical experiments show that the minimization results depend on the choice of such parameters as J, w_{min}, and L. They also depend on the choice of the interior points x_j. It is possible that further versions of the MRC could be made more efficient by finding a more efficient rule for their placement. Numerical experiments in [GR02b] showed that the efficiency of the minimization greatly depended on the deterministic placement of the interior points, with better results obtained for these points placed sufficiently close to the boundary S of the obstacle D, but not very close to it. The current choice of a random placement of the interior points x_j reduced the variance in the obtained results, and eliminated the need to provide a justified algorithm for their placement. The random choice of these points distributes them in the entire interior of the obstacle, rather than in a subset of it.

6.4 Conclusions.

For 3D obstacle Rayleigh's hypothesis (conjecture) says that the acoustic field u in the exterior of the obstacle D is given by the series convergent up to the boundary of D:

$$u(x, \alpha) = e^{ik\alpha \cdot x} + \sum_{\ell=0}^{\infty} A_\ell(\alpha)\psi_\ell, \quad \psi_\ell := Y_\ell(\alpha')h_\ell(kr), \quad \alpha' = \frac{x}{r}. \tag{91}$$

While this conjecture (RC) is false for many obstacles, it has been modified in [Ram02b] to obtain a valid representation for the solution of (76)-(77). This representation (Theorem 3) is called the Modified Rayleigh Conjecture (MRC), and is, in fact, not a conjecture, but a Theorem.

Can one use this approach to obtain solutions to various scattering problems? A straightforward numerical implementation of the MRC may fail, but, as we show here, it can be efficiently implemented and allows one to obtain accurate numerical solutions to obstacle scattering problems.

The Random Multi-point MRC algorithm was successfully applied to various 2D and 3D obstacle scattering problems. This algorithm is a significant improvement over previous MRC implementation described in [GR02b]. The improvement is achieved by allowing the required minimizations to be done iteratively, while the previous methods were limited by the problem size constraints. In [GR02b], such MRC method was presented, and it favorably compared to the Boundary Integral Equation Method.

The Random Multi-point MRC has an additional attractive feature, that it can easily treat obstacles with complicated geometry (e.g. edges and corners). Unlike the BIEM, it is easily modified to treat different obstacle shapes.

Further research on MRC algorithms is conducted. It is hoped that the MRC in its various implementation can emerge as a valuable and efficient alternative to more established methods.

7 Support Function Method for inverse obstacle scattering problems.

7.1 Support Function Method (SFM)

The Inverse Scattering Problem consists of finding the obstacle D from the Scattering Amplitude, or similarly observed data. The Support Function Method (SFM) was originally developed in a 3-D setting in [Ram70], see also [Ram86, pp. 94–99]. It is used to approximately locate the obstacle D. The method is derived using a high-frequency approximation to the scattered field for smooth, strictly convex obstacles. It turns out that this inexpensive method also provides a good localization of obstacles in the resonance region of frequencies. If the obstacle is not convex, then the SFM yields its convex hull.

One can restate the SFM in a 2-D setting as follows (see [GR03]). Let $D \subset \mathbb{R}^2$ be a smooth and strictly convex obstacle with the boundary Γ. Let $\nu(\mathbf{y})$ be the unique outward unit normal vector to Γ at $\mathbf{y} \in \Gamma$. Fix an incident direction $\alpha \in S^1$. Then the boundary Γ can be decomposed into the following two parts:

$$\Gamma_+ = \{\mathbf{y} \in \Gamma \ : \ \nu(\mathbf{y}) \cdot \alpha < 0\}, \text{ and } \quad \Gamma_- = \{\mathbf{y} \in \Gamma \ : \ \nu(\mathbf{y}) \cdot \alpha \geq 0\}, \quad (92)$$

which are, correspondingly, the illuminated and the shadowed parts of the boundary for the chosen incident direction α.

Given $\alpha \in S^1$, its **specular point** $\mathbf{s}_0(\alpha) \in \Gamma_+$ is defined from the condition:

$$\mathbf{s}_0(\alpha) \cdot \alpha = \min_{\mathbf{s} \in \Gamma_+} \mathbf{s} \cdot \alpha \qquad (93)$$

Note that the equation of the tangent line to Γ_+ at $\mathbf{s}_0$ is

$$< x_1, x_2 > \cdot \, \alpha = \mathbf{s}_0(\alpha) \cdot \alpha, \qquad (94)$$

and

$$\nu(\mathbf{s}_0(\alpha)) = -\alpha. \qquad (95)$$

The **Support function** $d(\alpha)$ is defined by

$$d(\alpha) = \mathbf{s}_0(\alpha) \cdot \alpha. \qquad (96)$$

Thus $|d(\alpha)|$ is the distance from the origin to the unique tangent line to Γ_+ perpendicular to the incident vector α. Since the obstacle D is assumed to be convex

$$D = \cap_{\alpha \in S^1} \{\mathbf{x} \in \mathbb{R}^2 \ : \ \mathbf{x} \cdot \alpha \geq d(\alpha)\}. \qquad (97)$$

The boundary Γ of D is smooth, hence so is the Support Function. The knowledge of this function allows one to reconstruct the boundary Γ using the following procedure.

Parametrize unit vectors $\mathbf{l} \in S^1$ by $\mathbf{l}(t) = (\cos t, \sin t)$, $0 \leq t < 2\pi$ and define

$$p(t) = d(\mathbf{l}(t)), \quad 0 \leq t < 2\pi. \tag{98}$$

Equation (94) and the definition of the Support Function give

$$x_1 \cos t + x_2 \sin t = p(t). \tag{99}$$

Since Γ is the envelope of its tangent lines, its equation can be found from (99) and

$$-x_1 \sin t + x_2 \cos t = p'(t). \tag{100}$$

Therefore the parametric equations of the boundary Γ are

$$x_1(t) = p(t) \cos t - p'(t) \sin t, \quad x_2(t) = p(t) \sin t + p'(t) \cos t. \tag{101}$$

So, the question is how to construct the Support function $d(\mathbf{l})$, $\mathbf{l} \in S^1$ from the knowledge of the Scattering Amplitude. In 2-D the Scattering Amplitude is related to the total field $u = u_0 + v$ by

$$A(\alpha', \alpha) = -\frac{e^{i\frac{\pi}{4}}}{\sqrt{8\pi k}} \int_\Gamma \frac{\partial u}{\partial \nu(\mathbf{y})} e^{-ik\alpha' \cdot \mathbf{y}} \, ds(\mathbf{y}). \tag{102}$$

In the case of the "soft" boundary condition (i.e. the pressure field satisfies the Dirichlet boundary condition $u = 0$) the Kirchhoff (high frequency) approximation gives

$$\frac{\partial u}{\partial \nu} = 2 \frac{\partial u_0}{\partial \nu} \tag{103}$$

on the illuminated part Γ_+ of the boundary Γ, and

$$\frac{\partial u}{\partial \nu} = 0 \tag{104}$$

on the shadowed part Γ_-. Therefore, in this approximation,

$$A(\alpha', \alpha) = -\frac{ike^{i\frac{\pi}{4}}}{\sqrt{2\pi k}} \int_{\Gamma_+} \alpha \cdot \nu(\mathbf{y}) \, e^{ik(\alpha - \alpha') \cdot \mathbf{y}} \, ds(\mathbf{y}). \tag{105}$$

Let L be the length of Γ_+, and $\mathbf{y} = \mathbf{y}(\zeta)$, $0 \leq \zeta \leq L$ be its arc length parametrization. Then

$$A(\alpha', \alpha) = -\frac{i\sqrt{k} \, e^{i\frac{\pi}{4}}}{\sqrt{2\pi}} \int_0^L \alpha \cdot \nu(\mathbf{y}(\zeta)) \, e^{ik(\alpha - \alpha') \cdot \mathbf{y}(\zeta)} \, d\zeta. \tag{106}$$

Let $\zeta_0 \in [0, L]$ be such that $\mathbf{s}_0 = \mathbf{y}(\zeta_0)$ is the specular point of the unit vector $\mathbf{l}$, where

$$\mathbf{l} = \frac{\alpha - \alpha'}{|\alpha - \alpha'|}. \tag{107}$$

Then $\nu(\mathbf{s}_0) = -\mathbf{l}$, and $d(\mathbf{l}) = \mathbf{y}(\zeta_0) \cdot \mathbf{l}$. Let

$$\varphi(\zeta) = (\alpha - \alpha') \cdot \mathbf{y}(\zeta).$$

Then $\varphi(\zeta) = \mathbf{1} \cdot \mathbf{y}(\zeta)|\alpha - \alpha'|$. Since $\nu(\mathbf{s}_0)$ and $\mathbf{y}'(\zeta_0)$ are orthogonal, one has

$$\varphi'(\zeta_0) = \mathbf{1} \cdot \mathbf{y}'(\zeta_0)|\alpha - \alpha'| = 0.$$

Therefore, due to the strict convexity of D, ζ_0 is also the unique non-degenerate stationary point of $\varphi(\zeta)$ on the interval $[0, L]$, that is $\varphi'(\zeta_0) = 0$, and $\varphi''(\zeta_0) \neq 0$.

According to the Stationary Phase method

$$\int_0^L f(\zeta)e^{ik\varphi(\zeta)}d\zeta = f(\zeta_0)\exp\left[ik\varphi(\zeta_0) + \frac{i\pi}{4}\frac{\varphi''(\zeta_0)}{|\varphi''(\zeta_0)|}\right]\sqrt{\frac{2\pi}{k|\varphi''(\zeta_0)|}}\left[1 + O\left(\frac{1}{k}\right)\right],$$
(108)

as $k \to \infty$.

By the definition of the curvature $\kappa(\zeta_0) = |\mathbf{y}''(\zeta_0)|$. Therefore, from the collinearity of $\mathbf{y}''(\zeta_0)$ and $\mathbf{1}$, $|\varphi''(\zeta_0)| = |\alpha - \alpha'|\kappa(\zeta_0)$. Finally, the strict convexity of D, and the definition of $\varphi(\zeta)$, imply that ζ_0 is the unique point of minimum of φ on $[0, L]$, and

$$\frac{\varphi''(\zeta_0)}{|\varphi''(\zeta_0)|} = 1.$$
(109)

Using (108)-(109), expression (106) becomes:

$$A(\alpha', \alpha) = -\frac{\mathbf{1} \cdot \alpha}{\sqrt{|\alpha - \alpha'|\kappa(\zeta_0)}}e^{ik(\alpha - \alpha')\cdot \mathbf{y}(\zeta_0)}\left[1 + O\left(\frac{1}{k}\right)\right], \quad k \to \infty. \quad (110)$$

At the specular point one has $\mathbf{1} \cdot \alpha' = -\mathbf{1} \cdot \alpha$. By the definition $\alpha - \alpha' = \mathbf{1}|\alpha - \alpha'|$. Hence $\mathbf{1} \cdot (\alpha - \alpha') = |\alpha - \alpha'|$ and $2\mathbf{1} \cdot \alpha = |\alpha - \alpha'|$. These equalities and $d(\mathbf{1}) = \mathbf{y}(\zeta_0) \cdot \mathbf{1}$ give

$$A(\alpha', \alpha) = -\frac{1}{2}\sqrt{\frac{|\alpha - \alpha'|}{\kappa(\zeta_0)}}e^{ik|\alpha - \alpha'|d(\mathbf{1})}\left[1 + O\left(\frac{1}{k}\right)\right], \quad k \to \infty. \quad (111)$$

Thus, the approximation

$$A(\alpha', \alpha) \approx -\frac{1}{2}\sqrt{\frac{|\alpha - \alpha'|}{\kappa(\zeta_0)}}e^{ik|\alpha - \alpha'|d(\mathbf{1})} \quad (112)$$

can be used for an approximate recovery of the curvature and the support function (modulo $2\pi/k|\alpha - \alpha'|$) of the obstacle, provided one knows that the total field satisfies the Dirichlet boundary condition. The uncertainty in the support function determination can be remedied by using different combinations of vectors α and α' as described in the numerical results section.

Since it is also of interest to localize the obstacle in the case when the boundary condition is not a priori known, one can modify the SFM as shown in [RG04], and obtain

$$A(\alpha', \alpha) \sim \frac{1}{2} \sqrt{\frac{|\alpha - \alpha'|}{\kappa(\zeta_0)}} \, e^{i(k|\alpha - \alpha'|d(1) - 2\gamma_0 + \pi)}, \tag{113}$$

where

$$\gamma_0 = \arctan \frac{k}{h},$$

and

$$\frac{\partial u}{\partial n} + hu = 0$$

along the boundary Γ of the sought obstacle.

Now one can recover the Support Function $d(1)$ from (113), and the location of the obstacle.

7.2 Numerical results for the Support Function Method.

In the first numerical experiment the obstacle is the circle

$$D = \{(x_1, x_2) \in \mathbb{R}^2 \; : \; (x_1 - 6)^2 + (x_2 - 2)^2 = 1\}. \tag{114}$$

It is reconstructed using the Support Function Method for two frequencies in the resonance region: $k = 1.0$, and $k = 5.0$. Table 12 shows how well the approximation (112) is satisfied for various pairs of vectors α and α' all representing the same vector $1 = (1.0, 0.0)$ according to (107). The Table shows the ratios of the approximate Scattering Amplitude $A_a(\alpha', \alpha)$ defined as the right hand side of the equation (112) to the exact Scattering Amplitude $A(\alpha', \alpha)$. Note, that for a sphere of radius a, centered at $\mathbf{x}_0 \in \mathbb{R}^2$, one has

$$A(\alpha', \alpha) = -\sqrt{\frac{2}{\pi k}} \, e^{-i\frac{\pi}{4}} e^{ik(\alpha - \alpha') \cdot \mathbf{x}_0} \sum_{l=-\infty}^{\infty} \frac{J_l(ka)}{H_l^{(1)}(ka)} \, e^{il(\theta - \beta)}, \tag{115}$$

where $\alpha' = \mathbf{x}/|\mathbf{x}| = e^{i\theta}$, and $\alpha = e^{i\beta}$. Vectors α and α' are defined by their polar angles shown in Table 12.

Table 12 shows that only vectors α close to the vector 1 are suitable for the Scattering Amplitude approximation. This shows the practical importance of the backscattering data. Any single combination of vectors α and α' representing 1 is not sufficient to uniquely determine the Support Function $d(1)$ from (112) because of the phase uncertainty. However, one can remedy this by using more than one pair of vectors α and α' as follows.

Let $1 \in S^1$ be fixed. Let

$$R(1) = \{\alpha \in S^1 \; : \; |\alpha \cdot 1| > 1/\sqrt{2}\}.$$

Table 12. Ratios of the approximate and the exact Scattering Amplitudes $A_a(\alpha',\alpha)/A(\alpha',\alpha)$ for $l = (1.0, 0.0)$.

α'	α	$k = 1.0$	$k = 5.0$
π	0	0.88473 - 0.17487i	0.98859 - 0.05846i
$23\pi/24$	$\pi/24$	0.88272 - 0.17696i	0.98739 - 0.06006i
$22\pi/24$	$2\pi/24$	0.87602 - 0.18422i	0.98446 - 0.06459i
$21\pi/24$	$3\pi/24$	0.86182 - 0.19927i	0.97977 - 0.07432i
$20\pi/24$	$4\pi/24$	0.83290 - 0.22411i	0.96701 - 0.08873i
$19\pi/24$	$5\pi/24$	0.77723 - 0.25410i	0.95311 - 0.10321i
$18\pi/24$	$6\pi/24$	0.68675 - 0.27130i	0.92330 - 0.14195i
$17\pi/24$	$7\pi/24$	0.57311 - 0.25360i	0.86457 - 0.14959i
$16\pi/24$	$8\pi/24$	0.46201 - 0.19894i	0.81794 - 0.22900i
$15\pi/24$	$9\pi/24$	0.36677 - 0.12600i	0.61444 - 0.19014i
$14\pi/24$	$10\pi/24$	0.28169 - 0.05449i	0.57681 - 0.31075i
$13\pi/24$	$11\pi/24$	0.19019 + 0.00075i	0.14989 - 0.09479i
$12\pi/24$	$12\pi/24$	0.00000 + 0.00000i	0.00000 + 0.00000i

Define $\Psi : \mathbb{R} \to \mathbb{R}^+$ by

$$\Psi(t) = \left\| \frac{A(\alpha',\alpha)}{|A(\alpha',\alpha)|} + e^{ik|\alpha - \alpha'|t} \right\|^2_{L^2(R(1))},$$

where $\alpha' = \alpha'(\alpha)$ is defined by l and α according to (107), and the integration is done over $\alpha \in R(1)$.

If the approximation (112) were exact for any $\alpha \in R(1)$, then the value of $\Psi(d(l))$ would be zero. This justifies the use of the minimizer $t_0 \in \mathbb{R}$ of the function $\Psi(t)$ as an approximate value of the Support Function $d(l)$. If the Support Function is known for sufficiently many directions $l \in S^1$, the obstacle can be localized using (97) or (101). The results of such a localization for $k = 1.0$ together with the original obstacle D is shown on Figure 5. For $k = 5.0$ the identified obstacle is not shown, since it is practically the same as D. The only a priori assumption on D was that it was located inside the circle of radius 20 with the center in the origin. The Support Function was computed for 16 uniformly distributed in S^1 vectors l. The program run takes about 80 seconds on a 333 MHz PC.

In another numerical experiment we used $k = 1.0$ and a kite-shaped obstacle. Its boundary is described by

$$\mathbf{r}(t) = (5.35 + \cos t + 0.65 \cos 2t, \; 2.0 + 1.5 \sin t), \quad 0 \leq t < 2\pi. \tag{116}$$

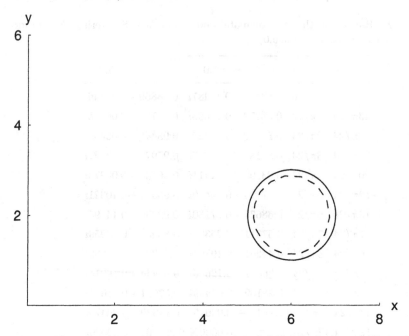

Fig. 5. Identified (dotted line), and the original (solid line) obstacle D for $k = 1.0$.

Numerical experiments using the boundary integral equation method (BIEM) for the direct scattering problem for this obstacle centered in the origin are described in [CK92, Section 3.5]. Again, the Dirichlet boundary conditions were assumed. We computed the scattering amplitude for 120 directions α using the MRC method with about 25% performance improvement over the BIEM, see [GR02b].

The Support Function Method (SFM) was used to identify the obstacle D from the synthetic scattering amplitude with no noise added. The only a priori assumption on D was that it was located inside the circle of radius 20 with the center in the origin. The Support Function was computed for 40 uniformly distributed in S^1 vectors l in about 10 seconds on a 333 MHz PC. The results of the identification are shown in Figure 6. The original obstacle is the solid line. The points were identified according to (101). As expected, the method recovers the convex part of the boundary Γ, and fails for the concave part. The same experiment but with $k = 5.0$ achieves a perfect identification of the convex part of the boundary. In each case the convex part of the obstacle was successfully localized. Further improvements in the obstacle localization using the MRC method are suggested in [Ram02b], and in the next section.

For the identification of obstacles with unknown boundary conditions let

$$A(t) = A(\alpha', \alpha) = |A(t)|e^{i\psi(t)}$$

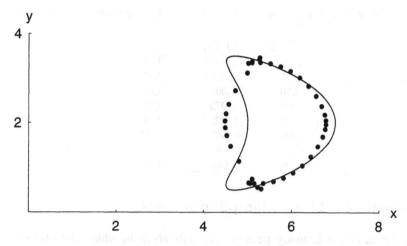

Fig. 6. Identified points and the original obstacle D (solid line); $k = 1.0$.

where, given t, the vectors α and α' are chosen as above, and the phase function $\psi(t)$, $\sqrt{2} < t \leq 2$ is continuous. Similarly, let $A_a(t)$, $\psi_a(t)$ be the approximate scattering amplitude and its phase defined by formula (113).

If the approximation (113) were exact for any $\alpha \in R(1)$, then the value of

$$|\psi_a(t) - ktd(1) + 2\gamma_0 - \pi|$$

would be a multiple of 2π.

This justifies the following algorithm for the determination of the Support Function $d(1)$:

Use a linear regression to find the approximation

$$\psi(t) \approx C_1 t + C_2$$

on the interval $\sqrt{2} < t \leq 2$. Then

$$d(1) = \frac{C_1}{k}. \tag{117}$$

Also

$$h = -k \tan \frac{C_2}{2}.$$

However, the formula for h did not work well numerically. It could only determine if the boundary conditions were or were not of the Dirichlet type. Table 13 shows that the algorithm based on (117) was successful in the identification of the circle of radius 1.0 centered in the origin for various values of h with no a priori assumptions on the boundary conditions. For this circle the Support Function $d(1) = -1.0$ for any direction 1.

Table 13. Identified values of the Support Function for the circle of radius 1.0 at $k = 3.0$.

h	Identified $d(1)$	Actual $d(1)$
0.01	-0.9006	-1.00
0.10	-0.9191	-1.00
0.50	-1.0072	-1.00
1.00	-1.0730	-1.00
2.00	-0.9305	-1.00
5.00	-1.3479	-1.00
10.00	-1.1693	-1.00
100.00	-1.0801	-1.00

8 Analysis of a Linear Sampling method.

During the last decade many papers were published, in which the obstacle identification methods were based on a numerical verification of the inclusion of some function $f := f(\alpha, z)$, $z \in \mathbb{R}^3$, $\alpha \in S^2$, in the range $R(B)$ of a certain operator B. Examples of such methods include [CCM00, CK96, Kir98]. However, one can show that the methods proposed in the above papers have essential difficulties, see [RG05]. Although it is true that $f \notin R(B)$ when $z \notin D$, it turns out that in any neighborhood of f there are elements from $R(B)$. Also, although $f \in R(B)$ when $z \in D$, there are elements in every neighborhood of f which do not belong to $R(B)$ even if $z \in D$. Therefore it is quite difficult to construct a stable numerical method for the identification of D based on the verification of the inclusions $f \notin R(B)$, and $f \in R(B)$. Some published numerical results were intended to show that the method based on the above idea works practically, but it is not clear how these conclusions were obtained.

Let us introduce some *notations* : $N(B)$ and $R(B)$ are, respectively, the null-space and the range of a linear operator B, $D \in \mathbb{R}^3$ is a bounded domain (obstacle) with a smooth boundary S, $D' = \mathbb{R}^3 \setminus D$, $u_0 = e^{ik\alpha \cdot x}$, $k = const > 0$, $\alpha \in S^2$ is a unit vector, N is the unit normal to S pointing into D', $g = g(x, y, k) := g(|x - y|) := \frac{e^{ik|x-y|}}{4\pi|x-y|}$, $f := e^{-ik\alpha' \cdot z}$, where $z \in \mathbb{R}^3$ and $\alpha' \in S^2$, $\alpha' := xr^{-1}$, $r = |x|$, $u = u(x, \alpha, k)$ is the scattering solution:

$$(\Delta + k^2)u = 0 \quad in \quad D', u|_S = 0, \tag{118}$$

$$u = u_0 + v, \quad v = A(\alpha', \alpha, k)e^{ikr}r^{-1} + o(r^{-1}), \quad as \quad r \to \infty, \tag{119}$$

where $A := A(\alpha', \alpha, k)$ is called the scattering amplitude, corresponding to the obstacle D and the Dirichlet boundary condition. Let $G = G(x, y, k)$ be the resolvent kernel of the Dirichlet Laplacian in D':

$$(\Delta + k^2)G = -\delta(x - y) \quad in \quad D', G|_S = 0, \tag{120}$$

and G satisfies the outgoing radiation condition.

If

$$(\Delta + k^2)w = 0 \quad in \quad D', w|_S = h, \tag{121}$$

and w satisfies the radiation condition, then ([Ram86]) one has

$$w(x) = \int_S G_N(x, s)h(s)ds, \quad w = A(\alpha', k)e^{ikr}r^{-1} + o(r^{-1}), \tag{122}$$

as $r \to \infty$, and $xr^{-1} = \alpha'$. We write $A(\alpha')$ for $A(\alpha', k)$, and

$$A(\alpha') := Bh := \frac{1}{4\pi} \int_S u_N(s, -\alpha')h(s)ds, \tag{123}$$

as follows from Ramm's lemma:

Lemma 1. ([Ram86, p. 46]) *One has:*

$$G(x, y, k) = g(r)u(y, -\alpha', k) + o(r^{-1}), \quad as \quad r = |x| \to \infty, \quad xr^{-1} = \alpha', \tag{124}$$

where u is the scattering solution of (118)-(119).

One can write the scattering amplitude as:

$$A(\alpha', \alpha, k) = -\frac{1}{4\pi} \int_S u_N(s, -\alpha')e^{ik\alpha \cdot s}ds. \tag{125}$$

The following claim follows easily from the results in [Ram86], [Ram92b] (cf [Kir98]):

Claim: $f := e^{-ik\alpha' \cdot z} \in R(B)$ if and only if $z \in D$.

Proof. If $e^{-ik\alpha' \cdot z} = Bh$, then Lemma 1 and (12.6) imply

$$g(y, z) = \int_S G_N(s, y)h ds \quad for \quad |y| > |z|.$$

Thus $z \in D$, because otherwise one gets a contradiction: $\lim_{y \to z} g(y, z) = \infty$ if $z \in \overline{D'}$, while $\lim_{y \to z} \int_S G_N(s, y)h ds < \infty$ if $z \in \overline{D'}$. Conversely, if $z \in D$, then Green's formula yields $g(y, z) = \int_S G_N(s, y)g(s, z)ds$. Taking $|y| \to \infty$, $\frac{y}{|y|} = \alpha'$, and using Lemma 1, one gets $e^{-ik\alpha' \cdot z} = Bh$, where $h = g(s, z)$. The claim is proved. □

Consider $B : L^2(S) \to L^2(S^2)$, and $A : L^2(S^2) \to L^2(S^2)$, where B is defined in (123) and $Aq := \int_{S^2} A(\alpha', \alpha)q(\alpha)d\alpha$. Then one proves (see [RG05]):

Theorem 1. *The ranges $R(B)$ and $R(A)$ are dense in $L^2(S^2)$.*

Remark 1. In [CK96] the 2D inverse obstacle scattering problem is considered. It is proposed to solve the equation (1.9) in [CK96]:

$$\int_{S^1} A(\alpha, \beta)\gamma d\beta = e^{-ik\alpha \cdot z}, \tag{126}$$

where A is the scattering amplitude at a fixed $k > 0$, S^1 is the unit circle, $\alpha \in S^1$, and z is a point on $\mathbb{R}^2$. If $\gamma = \gamma(\beta, z)$ is found, the boundary S of the obstacle is to be found by finding those z for which $\|\gamma\| := \|\gamma(\beta, z)\|_{L^2(S^1)}$ is maximal. Assuming that k^2 is not a Dirichlet or Neumann eigenvalue of the Laplacian in D, that D is a smooth, bounded, simply connected domain, the authors state Theorem 2.1 [CK96, p. 386], which says that for every $\epsilon > 0$ there exists a function $\gamma \in L^2(S^1)$, such that

$$\lim_{z \to S} \|\gamma(\beta, z)\| = \infty, \tag{127}$$

and (see [CK96, p. 386]),

$$\left\| \int_{S^1} A(\alpha, \beta)\gamma d\beta - e^{-ik\alpha \cdot z} \right\| < \epsilon. \tag{128}$$

There are several questions concerning the proposed method.

First, equation (126), in general, is not solvable. The authors propose to solve it approximately, by a regularization method. The regularization method applies for stable solution of solvable ill-posed equations (with exact or noisy data). If equation (126) is not solvable, it is not clear what numerical "solution" one seeks by a regularization method.

Secondly, since the kernel of the integral operator in (126) is smooth, one can always find, for any $z \in \mathbb{R}^2$, infinitely many γ with arbitrary large $\|\gamma\|$, such that (128) holds. Therefore it is not clear how and why, using (127), one can find S numerically by the proposed method.

A numerical implementation of the Linear Sampling Method (LSM) suggested in [CK96] consists of solving a discretized version of (126)

$$F\mathbf{g} = \mathbf{f}, \tag{129}$$

where $F = \{A\alpha_i, \beta_j\}$, $i = 1, ..., N$, $j = 1, ..., N$ be a square matrix formed by the measurements of the scattering amplitude for N incoming, and N outgoing directions. In 2-D the vector $\mathbf{f}$ is formed by

$$\mathbf{f}_n = \frac{e^{i\frac{\pi}{4}}}{\sqrt{8\pi k}} e^{-ik\alpha_n \cdot z}, \quad n = 1, ..., N,$$

see [BLW01] for details.

Denote the Singular Value Decomposition of the far field operator by $F = USV^H$. Let s_n be the singular values of F, $\rho = U^H\mathbf{f}$, and $\mu = V^H\mathbf{f}$. Then the norm of the sought function g is given by

$$\|\gamma\|^2 = \sum_{n=1}^{N} \frac{|\rho_n|^2}{s_n^2}. \tag{130}$$

A different LSM is suggested by A. Kirsch in [Kir98]. In it one solves

$$(F^*F)^{1/4}\mathbf{g} = \mathbf{f} \tag{131}$$

instead of (129). The corresponding expression for the norm of γ is

$$\|\gamma\|^2 = \sum_{n=1}^{N} \frac{|\mu_n|^2}{s_n}. \tag{132}$$

A detailed numerical comparison of the two LSMs and the linearized tomographic inverse scattering is given in [BLW01].

The conclusions of [BLW01], as well as of our own numerical experiments are that the method of Kirsch (131) gives a better, but a comparable identification, than (129). The identification is significantly deteriorating if the scattering amplitude is available only for a limited aperture, or the data are corrupted by noise. Also, the points with the *smallest* values of the $\|\gamma\|$ are the best in locating the inclusion, and not the *largest* one, as required by the theory in [CK96, Kir98]. In Figures 7 and 8 the implementation of the Colton-Kirsch LSM (130) is denoted by *gnck*, and of the Kirsch method (132) by *gnk*. The Figures show a contour plot of the logarithm of the $\|\gamma\|$. In all the cases the original obstacle was the circle of radius 1.0 centered at the point (10.0, 15.0). A similar circular obstacle that was identified by the Support Function Method (SFM) is discussed in Section 10. Note that the actual radius of the circle is 1.0, but it cannot be seen from the LSM identification. The LSM does not require any knowledge of the boundary conditions on the obstacle. The use of the SFM for unknown boundary conditions is discussed in the previous section. The LSM identification was performed for the scattering amplitude of the circle computed analytically with no noise added. In all the experiments the value for the parameter N was chosen to be 128.

References

[ARS99] Airapetyan, R., Ramm, A.G., Smirnova, A.: Example of two different potentials which have practically the same fixed-energy phase shifts. Phys. Lett. A, **254**, N3-4, 141–148(1999).

[Apa97] Apagyi, B. et al (eds): Inverse and algebraic quantum scattering theory. Springer, Berlin (1997)

[ARS98] Athanasiadis, C., Ramm A.G., Stratis I.G.: Inverse Acoustic Scattering by a Layered Obstacle. In: Ramm A. (ed) Inverse Problems, Tomography, and Image Processing. Plenum Press, New York, 1–8 (1998)

[Bar71] Barantsev, R.: Concerning the Rayleigh hypothesis in the problem of scattering from finite bodies of arbitrary shapes. Vestnik Lenungrad Univ., Math., Mech., Astron., **7**, 56–62 (1971)

[BP96] Barhen, J., Protopopescu, V.: Generalized TRUST algorithm for global optimization. In: Floudas C. (ed) State of The Art in Global Optimization. Kluwer, Dordrecht (1996)

[BPR97] Barhen, J., Protopopescu, V., Reister, D.: TRUST: A deterministic algorithm for global optimization. Science, **276**, 1094–1097 (1997)

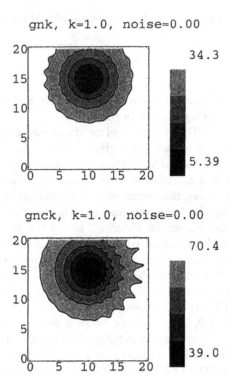

Fig. 7. Identification of a circle at $k = 1.0$.

[Bie97]   Biegler, L.T. (ed): Large-scale Optimization With Applications. In: IMA volumes in Mathematics and Its Applications, **92–94**. Springer-Verlag, New York (1997)

[BR87]   Boender, C.G.E., Rinnooy Kan, A.H.G.: Bayesian stopping rules for multistart global optimization methods. Math. Program., **37**, 59–80 (1987)

[Bom97]   Bomze, I.M. (ed): Developments in Global Optimization. Kluwer Academia Publ., Dordrecht (1997)

[BLW01]   Brandfass, M., Lanterman A.D., Warnick K.F.: A comparison of the Colton-Kirsch inverse scattering methods with linearized tomographic inverse scattering. Inverse Problems, **17**, 1797–1816 (2001)

[Bre73]   Brent, P.: Algorithms for minimization without derivatives. Prentice-Hall, Englewood Cliffs, NJ (1973)

[Cal67]   Calogero, F.: Variable Phase Approach to Potential Scattering. Academic Press, New York and London (1967)

[CS89]   Chadan, K., Sabatier, P.: Inverse Problems in Quantum Scattering Theory. Springer, New York (1989)

[CCM00]   Colton, D., Coyle, J., Monk, P.: Recent developments in inverse acoustic scattering theory. SIAM Rev., **42**, 369–414 (2000)

[CK96]   Colton, D., Kirsch, A.: A simple method for solving inverse scattering problems in the resonance region. Inverse Problems **12**, 383–393 (1996)

gnk, k=5.0, noise=0.00

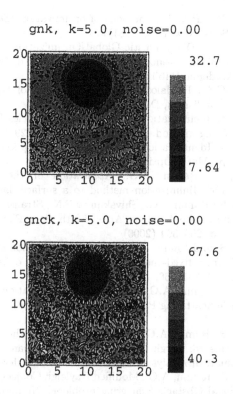

Fig. 8. Identification of a circle at $k = 5.0$.

[CK92]   Colton, D., Kress, R.: Inverse Acoustic and Electromagnetic Scattering
         Theory. Springer-Verlag, New York (1992)
[CM90]   Colton, D., Monk, P.: The Inverse Scattering Problem for acoustic waves
         in an Inhomogeneous Medium. In: Colton D., Ewing R., Rundell W. (eds)
         Inverse Problems in Partial Differential Equations. SIAM Publ. Philadel-
         phia, 73–84 (1990)
[CT70]   Cox, J., Thompson, K.: Note on the uniqueness of the solution of an
         equation of interest in the inverse scattering problem. J. Math. Phys., **11**,
         815–817 (1970)
[DE94]   Deep, K., Evans, D.J.: A parallel random search global optimization
         method. Technical Report **882**, Computer Studies, Loughborough Uni-
         versity of Technology (1994)
[DS83]   Dennis, J.E., Schnabel, R.B.: Numerical methods for unconstrained op-
         timization and nonlinear equations. Prentice-Hall, Englewood Cliffs, NJ
         (1983)
[DJ93]   Dixon, L.C.W., Jha, M.: Parallel algorithms for global optimization. J.
         Opt. Theor. Appl., **79**, 385–395 (1993)
[EJP57] Ewing, W.M, Jardetzky, W.S., Press, F.: Elastic Waves in Layered Media.
         McGraw-Hill, New York (1957)

[Fle81]   Fletcher, R. : Practical methods of optimization (Second edition). John Wiley & Sons, New York (1981)

[Flo00]   Floudas, C.A.: Deterministic Global Optimization-Theory, Methods and Applications. In: Nonconvex Optimization and Its Applications, **37**, Kluwer Academic Publishers, Dordrecht (2000)

[FP01]   Floudas, C.A., Pardalos, P.M.: Encyclopedia of Optimization. Kluwer Academic Publishers, Dordrecht (2001)

[Gut00]   Gutman, S. Identification of multilayered particles from scattering data by a clustering method. J. Comp. Phys., **163**, 529–546 (2000)

[Gut01]   Gutman, S.: Identification of piecewise-constant potentials by fixed-energy shifts. Appl. Math. Optim., **44**, 49–65 (2001)

[GR00]   Gutman, S., Ramm, A.G.: Application of the Hybrid Stochastic-deterministic Minimization method to a surface data inverse scattering problem. In: Ramm A.G., Shivakumar P.N., Strauss A.V. (eds) Operator Theory and Its Applications, Amer. Math. Soc., Fields Institute Communications, **25**, 293–304 (2000)

[GR02a]   Gutman, S. and Ramm, A.G.: Stable identification of piecewise-constant potentials from fixed-energy phase shifts. Jour. of Inverse and Ill-Posed Problems, **10**, 345–360.

[GR02b]   Gutman, S., Ramm, A.G.: Numerical Implementation of the MRC Method for obstacle Scattering Problems. J. Phys. A: Math. Gen., **35**, 8065–8074 (2002)

[GR03]   Gutman, S., Ramm, A.G.: Support Function Method for Inverse Scattering Problems. In: Wirgin A. (ed) Acoustics, Mechanics and Related Topics of Mathematical Analysis. World Scientific, New Jersey, 178–184 (2003)

[GR05]   Gutman, S., Ramm, A.G.: Modified Rayleigh Conjecture method for Multidimensional Obstacle Scattering problems. Numerical Funct. Anal and Optim., **26** (2005)

[GRS02]   Gutman, S., Ramm, A.G., Scheid, W.: Inverse scattering by the stability index method. Jour. of Inverse and Ill-Posed Problems, **10**, 487–502 (2002)

[HH98]   Haupt, R.L., Haupt, S.E.: Practical genetic algorithms. John Wiley and Sons Inc., New York (1998)

[Hes80]   Hestenes, M.: Conjugate direction methods in optimization. In: Applications of mathematics **12**. Springer-Verlag, New York (1980)

[HPT95]   Horst, R., Pardalos, P.M., Thoai, N.V.: Introduction to Global Optimization. Kluwer Academic Publishers, Dordrecht (1995)

[HT93]   Horst, R., Tuy, H.: Global Optimization: Deterministic Approaches, second edition. Springer, Heidelberg (1993)

[Hu95]   Hu, F.Q.: A spectral boundary integral equation method for the 2D Helmholtz equation. J. Comp. Phys., **120**, 340–347 (1995)

[Jac77]   Jacobs, D.A.H. (ed): The State of the Art in Numerical Analysis. Academic Press, London (1977)

[Kir84]   Kirkpatrick, S.: Optimization by simulated annealing: quantitative studies. Journal of Statistical Physics, **34**, 975–986 (1984)

[KGV83]   Kirkpatrick, S., Gelatt, C.D., Vecchi, M.P.: Optimization by Simulated Annealing. Science, **220**, 671–680 (1983)

[Kir96]   Kirsch, A.: An Introduction to the Mathematical Theory of Inverse Problems. Springer-Verlag, New York (1996)

[Kir98]   Kirsch, A.: Characterization of the shape of a scattering obstacle using the spectral data for far field operator. Inverse Probl., **14**, 1489–1512.

[Mil73]   Millar, R.: The Rayleigh hypothesis and a related least-squares solution to the scattering problems for periodic surfaces and other scatterers. Radio Sci., **8**, 785–796 (1973)

[New82]   Newton R.: Scattering Theory of Waves and Particles. Springer, New York (1982)

[PRT00]   Pardalos, P.M., Romeijn, H.E., Tuy, H.: Recent developments and trends in global optimization. J. Comput. Appl. Math., **124**, 209–228 (2000)

[Pol71]   Polak, E.: Computational methods in optimization. Academic Press, New York (1971)

[PTVF92]   Press, W.H., Teukolsky, S.A., Vetterling, W.T., Flannery, B.P.: Numerical Recepies in FORTRAN, Second Edition, Cambridge University Press (1992)

[Ram70]   Ramm, A.G.: Reconstruction of the shape of a reflecting body from the scattering amplitude. Radiofisika, **13**, 727–732 (1970)

[Ram82]   Ramm A.G.: Iterative methods for calculating the static fields and wave scattering by small bodies. Springer-Verlag, New York, NY (1982)

[Ram86]   Ramm A.G.: Scattering by Obstacles. D. Reidel Publishing, Dordrecht, Holland (1986)

[Ram88]   Ramm, A.G. Recovery of the potential from fixed energy scattering data. Inverse Problems, **4**, 877–886.

[Ram90]   Ramm, A.G.: Is the Born approximation good for solving the inverse problem when the potential is small? J. Math. Anal. Appl., **147**, 480–485 (1990)

[Ram91]   Ramm, A.G.: Symmetry properties for scattering amplitudes and applications to inverse problems. J. Math. Anal. Appl., **156**, 333–340 (1991)

[Ram92a]   Ramm, A.G.: Stability of the inversion of 3D fixed-frequency data. J. Math. Anal. Appl., **169**, 329–349 (1992)

[Ram92b]   Ramm, A.G.: Multidimensional Inverse Scattering Problems. Longman/Wiley, New York (1992)

[Ram94a]   Ramm, A.G.: Multidimensional Inverse Scattering Problems. Mir, Moscow (1994) (expanded Russian edition of [Ram92b])

[Ram94b]   Ramm, A.G.: Numerical method for solving inverse scattering problems. Doklady of Russian Acad. of Sci., **337**, 20–22 (1994)

[Ram94c]   Ramm, A.G.: Stability of the solution to inverse obstacle scattering problem. J. Inverse and Ill-Posed Problems, **2**, 269–275 (1994)

[Ram94d]   Ramm, A.G.: Stability estimates for obstacle scattering. J. Math. Anal. Appl., **188**, 743-751 (1994)

[Ram96]   Ramm, A.G.: Finding potential from the fixed-energy scattering data via D-N map. J. of Inverse and Ill-Posed Problems, **4**, 145–152 (1996)

[Ram97]   Ramm, A.G.: A method for finding small inhomogeneities from surface data. Math. Sci. Research Hot-Line, **1**, 10 , 40–42 (1997)

[Ram00a]   Ramm A.G.: Finding small inhomogeneities from scattering data. Jour. of inverse and ill-posed problems, **8**, 1–6 (2000)

[Ram00b]   Ramm, A.G.: Property C for ODE and applications to inverse problems. In: Operator Theory and Its Applications, Amer. Math. Soc., Fields Institute Communications, Providence, RI, **25**, 15–75 (2000)

[Ram02a]   Ramm, A.G.: Stability of the solutions to 3D inverse scattering problems. Milan Journ of Math **70**, 97–161 (2002)

[Ram02b]   Ramm, A.G.: Modified Rayleigh Conjecture and applications. J. Phys. A: Math. Gen., **35**, 357–361.

[Ram02c] Ramm, A.G.: A counterexample to the uniqueness result of Cox and Thompson. Appl. Anal., **81**, 833–836 (2002)

[Ram02d] Ramm, A.G.: Analysis of the Newton-Sabatier scheme for inverting fixed-energy phase shifts. Appl. Anal., **81**, 965–975 (2002)

[Ram04a] Ramm, A.G.: Dynamical systems method for solving operator equations. Communic. in Nonlinear Science and Numer. Simulation, **9**, 383–402 (2004)

[Ram04b] Ramm, A.G.: One-dimensional inverse scattering and spectral problems. Cubo a Mathem. Journ., **6**, 313–426 (2004)

[Ram04c] Ramm, A.G., Gutman, S.: Modified Rayleigh Conjecture for scattering by periodic structures. Intern. Jour. of Appl. Math. Sci., **1**, 55–66 (2004)

[Ram04d] Ramm, A.G.: Inverse scattering with fixed-energy data. Jour. of Indones. Math. Soc., **10**, 53–62 (2004)

[Ram05a] Ramm, A.G.: Inverse Problems. Springer, New York (2005)

[Ram05b] Ramm, A.G.: Wave Scattering by Small Bodies of Arbitrary Shapes. World Sci. Publishers, Singapore (2005)

[RAI98] Ramm, A.G., Arredondo, J.H., Izquierdo, B.G.: Formula for the radius of the support of the potential in terms of the scattering data. Jour. Phys. A, **31**, 39–44 (1998)

[RG01] Ramm, A.G., Gutman, S.: Piecewise-constant positive potentials with practically the same fixed-energy phase shifts. Appl. Anal., **78**, 207–217 (2001)

[RG04] Ramm, A.G., Gutman, S.: Numerical solution of obstacle scattering problems. Internat. Jour. of Appl. Math. and Mech., **1**, 71–102 (2004)

[RG05] Ramm, A.G., Gutman, S.: Analysis of a linear sampling method for identification of obstacles. Acta Math. Appl. Sinica, **21**, 1–6 (2005)

[RPY00] Ramm, A.G, Pang, P., Yan, G.: A uniqueness result for the inverse transmission problem. Internat. Jour. of Appl. Math., **2**, 625–634 (2000)

[RS99] Ramm, A.G., Scheid, W.: An approximate method for solving inverse scattering problems with fixed-energy data. Jour. of Inverse and Ill-posed Problems, **7**, 561-571 (1999)

[RS00] Ramm, A.G., Smirnova, A.: A numerical method for solving the inverse scattering problem with fixed-energy phase shifts. Jour. of Inverse and Ill-Posed Problems, **3**, 307–322.

[RT87a] Rinnooy Kan, A.H.G., Timmer, G.T.: Stochastic global optimization methods, part I: clustering methods. Math. Program., **39**, 27–56 (1987)

[RT87b] Rinnooy Kan, A.H.G., Timmer, G.T.: Stochastic global optimization methods, part II: multi level methods. Math. Prog., **39**, 57–78 (1987)

[Rub00] Rubinov, A.M.: Abstract Convexity and Global Optimization. Kluwer Acad. Publ., Dordrecht (2000)

[Sch90] Schuster, G.T.: A fast exact numerical solution for the acoustic response of concentric cylinders with penetrable interfaces. J. Acoust. Soc. Am., **87**, 495–502 (1990)

[ZUB98] Zakovic, S., Ulanowski, Z., Bartholomew-Biggs, M.C.: Application of global optimization to particle identification using light scattering. Inverse Problems, **14**, 1053–1067 (1998)

On Complexity of Stochastic Programming Problems

Alexander Shapiro[1] and Arkadi Nemirovski[2]

[1] Georgia Institute of Technology
   Atlanta, Georgia 30332-0205, USA
   ashapiro@isye.gatech.edu
[2] Technion – Israel Institute of Technology
   Haifa 32000, Israel
   nemirovs@ie.technion.ac.il

Summary. The main focus of this paper is in a discussion of complexity of stochastic programming problems. We argue that two-stage (linear) stochastic programming problems with recourse can be solved with a reasonable accuracy by using Monte Carlo sampling techniques, while multi-stage stochastic programs, in general, are intractable. We also discuss complexity of chance constrained problems and multi-stage stochastic programs with linear decision rules.

Key words: stochastic programming, complete recourse, chance constraints, Monte Carlo sampling, SAA method, large deviations bounds, convex programming, multi-stage stochastic programming.

1 Introduction

In real life we constantly have to make decisions under uncertainty and, moreover, we would like to make such decisions in a reasonably optimal way. Then for a specified objective function $F(x, \xi)$, depending on decision vector $x \in \mathbb{R}^n$ and vector $\xi \in \mathbb{R}^d$ of uncertain parameters, we are faced with the problem of optimizing (say minimizing) $F(x, \xi)$ over x varying in a permissible (feasible) set $X \subset \mathbb{R}^n$. Of course, such an optimization problem is not well defined since our objective depends on an unknown value of ξ. A way of dealing with this is to optimize the objective on *average*. That is, it is assumed that ξ is a random vector[3], with known probability distribution P having support $\Xi \subset \mathbb{R}^d$, and the following optimization problem is formulated

---

[3] Sometimes, in the sequel, ξ denotes a random vector and sometimes its particular realization (numerical value). Which one of these two meanings is used will be clear from the context.

$$\underset{x \in X}{\text{Min}} \left\{ f(x) := \mathbb{E}_P[F(x, \xi)] \right\}. \tag{1}$$

We assume throughout the paper that considered expectations are well defined, e.g., $F(x, \cdot)$ is measurable and P-integrable.

In particular, the above formulation can be applied to two-stage stochastic programming problem with recourse, pioneered by Beale [Bea55] and Dantzig [Dan55]. That is, an optimization problem is divided into two stages. At the first stage one has to make a decision on the basis of some available information. At the second stage, after a realization of the uncertain data becomes known, an optimal second stage decision is made. Such stochastic programming problem can be written in the form (1) with $F(x, \xi)$ being the optimal value of the second stage problem.

It should be noted that in the formulation (1) all uncertainties are concentrated in the objective function while the feasible set X is supposed to be known (deterministic). Quite often the feasible set itself is defined by constraints which depend on uncertain parameters. In some cases one can reasonably formulate such problems in the form (1) by introducing penalties for possible infeasibilities. Alternatively one can try to optimize the objective subject to satisfying constraints for *all* values of unknown parameters in a chosen (uncertain) region. This is the approach of robust optimization (cf., Ben-Tal and Nemirovski [BN01]). Satisfying the constraints for all possible realizations of random data may be too conservative and, more reasonably, one may try to satisfy the constraints with a high (close to one) probability. This leads to the chance, or probabilistic, constraints formulation which is going back to Charnes and Cooper [CC59].

There are several natural questions which arise with respect to formulation (1).

(i) How do we know the probability distribution P? In some cases one has historical data which can be used to obtain a reasonably accurate estimate of the corresponding probability distribution. However, this happens in rather specific situations and often the probability distribution either cannot be accurately estimated or changes with time. Even worse, in many cases one deals with *scenarios* (i.e., possible realizations of the random data) with the associated probabilities assigned by a subjective judgment.

(ii) Why, at the first stage, do we optimize the *expected* value of the second stage optimization problem? If the optimization procedure is repeated many times, with the same probability distribution of the data, then it could be argued by employing the Law of Large Numbers that this gives an optimal decision on average. However, if in the process, because of the variability of the data one looses all its capital, it does not help that the decisions were optimal on average.

(iii)How difficult is it to solve the stochastic programming problem (1)? Evaluation of the expected value function $f(x)$ involves calculation of the corresponding multivariate integrals. Only in rather specific cases it can be done analytically. Therefore, typically, one employs a finite discretization of the random data which allows to write the expectation in a form of summation. Note, however, that if random vector ξ has d elements each with just 3 possible realizations independent of each other, then the total number of scenarios is 3^d, i.e., the number of scenarios grows exponentially fast with dimension d of the data vector.

(iv) Finally, what can be said about multi-stage stochastic programming, when decisions are made in several stages based on available information at the time of making the sequential decisions?

It turns out that there is a close relation between questions (i) and (ii). As far as question (i) is concerned, one can approach it from the following point of view. Suppose that a plausible family $\mathfrak{P}$ of probability distributions, of the random data vector ξ, can be identified. Consequently, the "worst-case-distribution" minimax problem

$$\operatorname*{Min}_{x \in X} \left\{ f(x) := \sup_{P \in \mathfrak{P}} \mathbb{E}_P[F(x, \xi)] \right\} \tag{2}$$

is formulated. The worst-case approach to decision analysis, of course, is not new. It was also discussed extensively in the stochastic programming literature (e.g., [Dup79, Dup87, EGN85, Gai91, SK00, Zac66]).

Again we are facing the question of how to choose the set $\mathfrak{P}$ of possible distributions. Traditionally this problem is approached by assuming knowledge of certain moments of the involved random parameters. This leads to the so-called Problem of Moments, where the set $\mathfrak{P}$ is formed by probability measures P satisfying moment constraints $\mathbb{E}_P[\psi_i(\xi)] = b_i$, $i = 1, ..., m$ (see, e.g., [Lan87]). In that case the extreme (worst case) distributions are measures with a finite support of at most $m + 1$ points.

On the other hand, it often happens in applications that one is given a deterministic value μ of the uncertain data vector ξ and does not have an idea what a corresponding distribution may be. For example, ξ could represent an uncertain demand and μ is viewed as its mean vector given by a forecast. It is well recognized now that solving a corresponding optimization problem for the deterministic value $\xi = \mu$ may give a poor solution from a robustness point of view. It is natural then to introduce random perturbations to the deterministic vector μ and to solve the obtained stochastic program. For instance, one can assume that components ξ_i of the uncertain data vector are independent and have a certain type (say, log-normal if ξ_i should be nonnegative) distribution with means μ_i and standard deviations σ_i which are defined within a certain percentage of μ_i, $i = 1, ..., d$. Often this quickly stabilizes optimal solutions of the corresponding stochastic programs irrespective of the underlying distribution (cf., [SAGS05]). Furthermore, we can approach this setup from the minimax point of view by considering a worst distribution supported on, say, a box region around vector μ. If, moreover, we consider unimodal type families of distributions, then the worst case distribution is uniform (cf., [Sha04]). For a given x, even unimodal distributions and $F(x, \cdot) := -\mathbb{1}_S(\cdot)$, where $\mathbb{1}_S(\cdot)$ is the indicator function of a symmetric convex set S, this result was first established by Barmish and Lagoa [BL97], where it was called the "Uniformity Principle".

Question (ii) has also a long history. One can optimize a weighted sum of the expected value and a term representing variability of the second stage objective function. For example, we can try to minimize

$$f(x) := \mathbb{E}[F(x, \xi)] + c\mathrm{Var}[F(x, \xi)], \tag{3}$$

where $c \geq 0$ is a chosen constant. This approach goes back to Markowitz [Mar52]. The additional (variance) term in (3) can be viewed as a risk measure of the second stage (optimal) outcome. It could be noted, however, that adding the variance term may destroy convexity of the function $f(\cdot)$ even if $F(\cdot, \xi)$ is convex for all realizations of ξ (cf., [TA04]). An axiomatic approach to a mathematical theory of risk measures was suggested recently by Artzner et al. [ADEH99]. That is, value of a random variable Z is measured by a function $\rho(Z)$ satisfying certain axioms. An example of such function $\rho(Z)$, called coherent risk measure, is the mean-semideviation

$$\rho(Z) := \mathbb{E}[Z] + c\left\{\mathbb{E}\left([Z - \mathbb{E}[Z]]_+^2\right)\right\}^{1/2},$$

where $c \in [0, 1]$.

It turns out that $\rho(Z)$ is a coherent risk measure if and only if it can be represented in the form $\rho(Z) = \sup_{P \in \mathfrak{P}} \mathbb{E}_P[Z]$, where $\mathfrak{P}$ is a set of probability measures. In different frameworks this dual representation was derived in [ADEH99, FS02, RUZ02, RS04a]. Therefore, the min-max problem (2) and the problem of minimization of a coherent risk measure, of $F(x, \xi)$, in fact are equivalent. We may refer to [ADEHK03, ER05, Rie03, RS04b] for extensions of this approach to a multi-stage setting.

2 Complexity of two-stage stochastic programs

In this section we discuss question (iii) mentioned in the introduction, that is, how difficult is to solve a stochastic program. Problem (1) is a problem of minimizing a deterministic *implicitly given* objective $f(x)$. We should expect that this problem is at least as difficult as minimizing $f(x)$, $x \in X$, in the case where $f(x)$ is given explicitly, say by a "closed form analytic expression", or, more general, by an "oracle" capable to compute the values and the derivatives of $f(x)$ at every given point. As far as problems of minimization of $f(x)$, $x \in X$, with explicitly given objective, are concerned, the "solvable case" is known, this is the Convex Programming case. That is, X is a closed convex set and $f : X \to \mathbb{R}$ is a convex function. It is known that generic Convex Programming problems satisfying mild computability and boundedness assumptions can be solved in polynomial time. In contrast to this, typical nonconvex problems turn out to be NP-hard[4]. It follows that when speaking about conditions under which the stochastic program (1) is efficiently solvable, it makes sense to assume that X is a closed convex set, and $f(\cdot)$ is convex on X. We gain from a technical point (and do not lose much from practical viewpoint) by assuming X to be bounded. These assumptions, plus mild technical conditions, would be sufficient to make (1) easy, if $f(x)$ were given explicitly. However, in Stochastic Programming it makes no sense to assume that we can compute efficiently the expectation in (1), thus arriving at an explicit representation of $f(x)$. Would it be the case, there would be no necessity to treat (1) as a stochastic program.

We argue now that stochastic programming problems of the form (1) can be solved reasonably efficiently by using Monte Carlo sampling techniques provided that the probability distribution of the random data is not "too bad" and certain general conditions are met. In this respect we should explain what do we mean by "solving" stochastic programming problems. Let us consider, for example, two-stage linear stochastic programming problems with recourse. Such problems can be written in the form (1) with[5]

$$X := \{x : Ax = b, \ x \geq 0\} \text{ and } F(x, \xi) := \langle c, x \rangle + Q(x, \xi),$$

where $Q(x, \xi)$ is the optimal value of the second stage problem:

$$\underset{y \geq 0}{\text{Min}} \ \langle q, y \rangle \text{ subject to } Tx + Wy \geq h. \tag{4}$$

Here T and W are matrices of an appropriate order and $\xi \in \mathbb{R}^d$ is a vector whose elements are composed from elements of vectors q and h and matrices T and W which, in a considered problem, are assumed to be random. If we assume that the random data vector has a finite number of realizations (scenarios) $\xi_k = (q_k, W_k, T_k, h_k)$ with respective probabilities p_k, $k = 1, ..., K$, then the obtained two-stage problem can be written as one large linear programming problem:

$$
\begin{aligned}
\underset{x, y_1, ..., y_K}{\text{Min}} \quad & \langle c, x \rangle + \sum_{k=1}^{K} p_k \langle q_k, y_k \rangle \\
\text{s.t} \quad & Ax = b, \ T_k x + W_k y_k \geq h_k, \ k = 1, ..., K, \\
& x \geq 0, \ y_k \geq 0, \ k = 1, ..., K.
\end{aligned} \tag{5}
$$

---

[4] It is beyond the scope of this paper to give a detailed explanation of what "polynomial time solvability" and "NP-hardness" mean. Informally speaking, the former property of a problem P means that P is "easy to solve" – it admits a computationally efficient solution algorithm. NP-hardness of P means that no efficient solution algorithms for P are known, and there are strong theoretical reasons to believe that they do not exist. For formal treatment of these issues in Continuous Optimization, see, e.g. [BN01, Chapter 5].

We should also stress that a claim "such and such problem is difficult" relates to a *generic* problem in question and does *not* imply that the problem has no solvable particular cases.

[5] By $\langle x, y \rangle$ we denote the standard scalar product of two vectors $x, y \in \mathbb{R}^n$.

If the number of scenarios K is not "too large", then the above linear programming problem (5) can be solved accurately in a reasonable time. However, even a crude discretization of the probability distribution of ξ typically results in an exponential growth of the number of scenarios with increase of the number d of random parameters. Suppose, for example, that components of the random vector ξ are mutually independently distributed each having a small number r of possible realizations. Then the size of the corresponding input data grows linearly in d (and r) while the number of scenarios $K = r^d$ grows exponentially. Yet in some cases problem (5) can be solved numerically in a reasonable time. For example, suppose that matrices T and W are constant (deterministic) and only h is random and, moreover, $Q(x, \xi)$ decomposes into the sum $Q(x, \xi) = Q_1(x_1, h_1) + \ldots + Q_n(x_n, h_n)$. This happens in the case of the so-called simple recourse with

$$Q_i(x_i, h_i) = q_i^+ [x_i - h_i]_+ + q_i^- [h_i - x_i]_+, \ i = 1, \ldots, n,$$

where q_i^+ and q_i^- are some positive numbers. Then $\mathbb{E}[Q(x, \xi)] = \mathbb{E}[Q_1(x_1, h_1)] + \ldots + \mathbb{E}[Q_n(x_n, h_n)]$, i.e., calculation of the multidimensional expectation is reduced to calculations of one dimensional expectations. Of course, the above is a rather specific case and in a general situation there is no hope to solve problem (5) accurately (say with machine precision) even for moderate values of r and d (cf., [DS03]).

It should be said at this point that from a practical point of view, typically, it does not make sense to try to solve a stochastic programming problem with a high precision. A numerical error resulting from an inaccurate estimation of the involved probability distributions, modeling errors, etc., can be far bigger than such an optimization error. We argue now that two-stage stochastic problems can be solved efficiently with a reasonable accuracy provided that the following conditions are met:

(a) The feasible set X is fixed (deterministic).
(b) For all $x \in X$ and $\xi \in \Xi$ the objective function $F(x, \xi)$ is real valued.
(c) The considered stochastic programming problem can be solved efficiently (by a deterministic algorithm) if the number of scenarios is not "too large".

When applied to two-stage stochastic programming, the above conditions (a) and (b) mean that the recourse is relatively complete[6] and the second stage problem is bounded from below. The above condition (c) certainly holds in the case of two-stage *linear* stochastic programming with recourse.

In order to proceed let us consider the following Monte Carlo sampling approach. Suppose that we can generate an iid (independent identically distributed) random sample $\xi^1, \ldots, \xi^N$ of N realizations of the considered random vector. Then we can estimate the expected value function $f(x)$ by the sample

---

[6] It is said that the recourse is *relatively complete* if for every $x \in X$ and every possible realization of random data, the second stage problem is feasible.

average[7]

$$\hat{f}_N(x) := \frac{1}{N} \sum_{j=1}^{N} F(x, \xi^j). \tag{6}$$

Consequently, we approximate the true problem (1) by the problem:

$$\min_{x \in X} \hat{f}_N(x). \tag{7}$$

We refer to (7) as the Sample Average Approximation (SAA) problem. The optimal value $\hat{v}_N$ and the set $\hat{S}_N$ of optimal solutions of the SAA problem (7) provide estimates of their true counterparts of problem (1). It should be noted that once the sample is generated, $\hat{f}_N(x)$ becomes a deterministic function and problem (7) becomes a stochastic programming problem with N scenarios $\xi^1, ..., \xi^N$ taken with equal probabilities $1/N$. It also should be mentioned that the SAA method is *not* an algorithm. One still has to solve the obtained problem (7) by employing an appropriate (deterministic) algorithm.

By the Law of Large Numbers we have that $\hat{f}_N(x)$ converges (pointwise in x) w.p.1 to $f(x)$ as N tends to infinity. Therefore it is reasonable to expect for $\hat{v}_N$ and $\hat{S}_N$ to converge to their counterparts of the true problem (1) with probability one (w.p.1) as N tends to infinity. And, indeed, such convergence can be proved under mild regularity conditions. However, for a fixed $x \in X$, convergence of $\hat{f}_N(x)$ to $f(x)$ is notoriously slow. By the Central Limit Theorem it is of order $O_p(N^{-1/2})$. The rate of convergence can be improved, sometimes significantly, by variance reduction methods. However, by using Monte Carlo (Quasi-Monte Carlo) techniques one cannot evaluate the expected value $f(x)$ very accurately.

The following analysis is based on exponential bounds of the Large Deviations (LD) theory (see, e.g., [DZ98] for a general discussion of LD theory). Denote by S^ε and $\hat{S}_N^\varepsilon$ the sets of ε-optimal solutions of the true and SAA problems, respectively, i.e., $\bar{x} \in S^\varepsilon$ iff $\bar{x} \in X$ and $f(\bar{x}) \leq \inf_{x \in X} f(x) + \varepsilon$. Choose accuracy constants $\varepsilon > 0$ and $0 \leq \delta < \varepsilon$, and significance level $\alpha \in (0, 1)$. Suppose for the moment that the set X is finite although its cardinality $|X|$ can be very large. Then by using Cramér's LD theorem it is not difficult to show that the sample size

$$N \geq \frac{1}{\eta(\varepsilon, \delta)} \log \left(\frac{|X|}{\alpha} \right) \tag{8}$$

guarantees that probability of the event $\{\hat{S}_N^\delta \subset S^\varepsilon\}$ is at least $1 - \alpha$ (see [KSH01],[Sha03b, section 3.1]). That is, for any N bigger than the right hand side of (8) we are guaranteed that any δ-optimal solution of the corresponding SAA problem provides an ε-optimal solution of the true problem with probability at least $1 - \alpha$, in other words, solving the SAA problem with accuracy δ

---

[7] In order to simplify notation we only write in the subscript the sample size N while actually $\hat{f}_N(\cdot)$ depends on the generated sample, and in that sense is random.

guarantees solving the true problem with accuracy ε with probability at least $1 - \alpha$.

The number $\eta(\varepsilon, \delta)$ in the estimate (8) is defined as follows. Consider a mapping $u : X \setminus S^\varepsilon \to X$ such that $f(u(x)) \leq f(x) - \varepsilon$ for all $x \in X \setminus S^\varepsilon$. Such mappings do exist, although not unique. For example, any mapping $u : X \setminus S^\varepsilon \to S$ satisfies this condition. Choice of such a mapping gives a certain flexibility to the corresponding estimate of the sample size. For $x \in X$, consider random variable

$$Y_x := F(u(x), \xi) - F(x, \xi),$$

its moment generating function $M_x(t) := \mathbb{E}\left[e^{tY_x}\right]$ and the LD rate function[8]

$$I_x(z) := \sup_{t \in \mathbb{R}} \left\{ tz - \log M_x(t) \right\}.$$

Note that, by construction of mapping $u(x)$, the inequality

$$\mu_x := \mathbb{E}[Y_x] = f(u(x)) - f(x) \leq -\varepsilon \qquad (9)$$

holds for all $x \in X \setminus S^\varepsilon$. Finally, we define

$$\eta(\varepsilon, \delta) := \min_{x \in X \setminus S^\varepsilon} I_x(-\delta). \qquad (10)$$

Because of (9) and since $\delta < \varepsilon$, the number $I_x(-\delta)$ is positive provided that the probability distribution of Y_x is not "too bad". Specifically, if we assume that the moment generating function $M_x(t)$, of Y_x, is finite valued for all t in a neighborhood of 0, then the random variable Y_x has finite moments and $I_x(\mu_x) = I'(\mu_x) = 0$, and $I''(\mu_x) = 1/\sigma_x^2$ where $\sigma_x^2 := \mathrm{Var}\,[Y_x]$. Consequently, $I_x(-\delta)$ can be approximated, by using second order Taylor expansion, as follows

$$I_x(-\delta) \approx \frac{(-\delta - \mu_x)^2}{2\sigma_x^2} \geq \frac{(\varepsilon - \delta)^2}{2\sigma_x^2}.$$

This suggests that one can expect the constant $\eta(\varepsilon, \delta)$ to be of order of $(\varepsilon - \delta)^2$. And, indeed, this can be ensured by various conditions. Consider the following condition.

(A1) There exists constant $\sigma > 0$ such that for any $x', x \in X$, the moment generating function $M^*(t)$ of $F(x', \xi) - F(x, \xi) - \mathbb{E}\left[F(x', \xi) - F(x, \xi)\right]$ satisfies:

$$M^*(t) \leq \exp\left(\tfrac{1}{2}\sigma^2 t^2\right), \quad \forall t \in \mathbb{R}. \qquad (11)$$

Note that random variable $F(x', \xi) - F(x, \xi) - \mathbb{E}\left[F(x', \xi) - F(x, \xi)\right]$ has zero mean. Moreover, if it has a normal distribution, with variance σ^2, then

---

[8] That is, $I_x(\cdot)$ is the conjugate of the function $\log M_x(\cdot)$ in the sense of convex analysis.

its moment generating function is equal to the right hand side of (11). Condition (11) means that tail probabilities $\mathrm{Prob}\big(|F(x',\xi) - F(x,\xi)| > t\big)$ are bounded from above[9] by $O(1)\exp\left(-\frac{t^2}{2\sigma^2}\right)$. This condition certainly holds if the distribution of the considered random variable has a bounded support.

For $x' = u(x)$, random variable $F(x',\xi) - F(x,\xi)$ coincides with Y_x, and hence (11) implies that $M_x(t) \leq \exp(\mu_x t + \sigma^2 t^2/2)$. It follows that

$$I_x(z) \geq \sup_{t \in \mathbb{R}} \left\{zt - \mu_x t - \sigma^2 t^2/2\right\} = \frac{(z - \mu_x)^2}{2\sigma^2}, \tag{12}$$

and hence for any $\varepsilon > 0$ and $\delta \in [0, \varepsilon)$:

$$\eta(\varepsilon, \delta) \geq \frac{(-\delta - \mu_x)^2}{2\sigma^2} \geq \frac{(\varepsilon - \delta)^2}{2\sigma^2}. \tag{13}$$

It follows that, under assumption (A1), the estimate (8) can be written as

$$N \geq \frac{2\sigma^2}{(\varepsilon - \delta)^2} \log\left(\frac{|X|}{\alpha}\right). \tag{14}$$

Remark 1. Condition (11) can be replaced by a more general condition

$$M^*(t) \leq \exp(\psi(t)), \quad \forall t \in \mathbb{R}, \tag{15}$$

where $\psi(t)$ is a convex even function with $\psi(0) = 0$. Then $\log M_x(t) \leq \mu_x t + \psi(t)$ and hence $I_x(z) \geq \psi^*(z - \mu_x)$, where ψ^* is the conjugate of the function ψ. It follows then that

$$\eta(\varepsilon, \delta) \geq \psi^*(-\delta - \mu_x) \geq \psi^*(\varepsilon - \delta). \tag{16}$$

For example, instead of assuming that the bound (11) holds for all $t \in \mathbb{R}$, we can assume that it holds for all t in a finite interval $[-a, a]$, where $a > 0$ is a given constant. That is, we can take $\psi(t) := \frac{1}{2}\sigma^2 t$ if $|t| \leq a$, and $\psi(t) := +\infty$ otherwise. In that case $\psi^*(z) = z^2/(2\sigma^2)$ for $|z| \leq a\sigma^2$, and $\psi^*(z) = a|z| - \frac{1}{2}a^2\sigma^2$ for $|z| > a\sigma^2$.

Now let X be a bounded, not necessary finite, subset of $\mathbb{R}^n$ of diameter

$$D := \sup_{x', x \in X} \|x' - x\|.$$

Then for $\tau > 0$ we can construct a set $X_\tau \subset X$ such that for any $x \in X$ there is $x' \in X_\tau$ satisfying $\|x - x'\| \leq \tau$, and $|X_\tau| = O(1)(D/\tau)^n$. Suppose that condition (A1) holds. Then by (14), for $\varepsilon' > \delta$, we can estimate the corresponding sample size required to solve the reduced optimization problem, obtained by replacing X with X_τ, as

---

[9] By $O(1)$ we denote generic absolute constants.

$$N \geq \frac{2\sigma^2}{(\varepsilon' - \delta)^2} \left[n \left(\log D - \log \tau \right) + \log \left(O(1)/\alpha \right) \right]. \qquad (17)$$

Suppose, further, that there exists a (measurable) function $\kappa : \Xi \to \mathbb{R}_+$ and $\gamma > 0$ such that

$$|F(x', \xi) - F(x, \xi)| \leq \kappa(\xi)\|x' - x\|^\gamma \qquad (18)$$

holds for all $x', x \in X$ and all $\xi \in \Xi$. It follows by (18) that

$$|\hat{f}_N(x') - \hat{f}_N(x)| \leq N^{-1} \sum_{j=1}^{N} |F(x', \xi^j) - F(x, \xi^j)| \leq \hat{\kappa}_N \|x' - x\|^\gamma, \qquad (19)$$

where $\hat{\kappa}_N := N^{-1} \sum_{j=1}^N \kappa(\xi^j)$.

Let us assume, further, the following:

(A2) The moment generating function $M_\kappa(t) := \mathbb{E}\left[e^{t\kappa(\xi)}\right]$ of $\kappa(\xi)$ is finite valued for all t in a neighborhood of 0.

It follows then that the expectation $L := \mathbb{E}[\kappa(\xi)]$ is finite, and moreover, by Cramér's LD Theorem that for any $L' > L$ there exists a positive constant $\beta = \beta(L')$ such that

$$P\left(\hat{\kappa}_N > L'\right) \leq e^{-N\beta}. \qquad (20)$$

Let $\hat{x}_N$ be a δ-optimal solution of the SAA problem and $\tilde{x}_N \in X_\tau$ be a point such that $\|\hat{x}_N - \tilde{x}_N\| \leq \tau$. Let us take $N \geq \beta^{-1}\log(2/\alpha)$, so that by (20) we have that

$$\text{Prob}\left(\hat{\kappa}_N > L'\right) \leq \alpha/2. \qquad (21)$$

Then with probability at least $1 - \alpha/2$, the point $\tilde{x}_N$ is a $(\delta + L'\tau^\gamma)$-optimal solution of the reduced SAA problem. Setting

$$\tau := [(\varepsilon - \delta)/(2L')]^{1/\gamma},$$

we obtain that with probability at least $1 - \alpha/2$, the point $\tilde{x}_N$ is an ε'-optimal solution of the reduced SAA problem with $\varepsilon' := (\varepsilon + \delta)/2$. Moreover, by taking a sample size satisfying (17), we obtain that $\tilde{x}_N$ is an ε'-optimal solution of the reduced expected value problem with probability at least $1 - \alpha/2$. It follows that $\hat{x}_N$ is an ε''-optimal solution of the SAA problem (1) with probability at least $1 - \alpha$ and $\varepsilon'' = \varepsilon' + L\tau^\gamma \leq \varepsilon$. We obtain the following estimate

$$N \geq \frac{4\sigma^2}{(\varepsilon - \delta)^2} \left[n \left(\log D + \gamma^{-1} \log \frac{2L'}{\varepsilon - \delta} \right) + \log \left(\frac{O(1)}{\alpha} \right) \right] \vee \left[\beta^{-1} \log \left(2/\alpha \right) \right] \qquad (22)$$

for the sample size (cf., [Sha03b, section 3.2]).

The above result is quite general and does not involve the assumption of convexity. Estimate (22) of the sample size contains various constants and is too conservative for practical applications. However, in a sense, it gives an estimate of complexity of two-stage stochastic programming problems. We will discuss this in the next section. In typical applications (e.g., in the convex case) the constant $\gamma = 1$, in which case condition (18) means that $F(\cdot, \xi)$ is Lipschitz continuous on X with constant $\kappa(\xi)$. However, there are also some applications where γ could be less than 1 (cf., [Sha05a]).

We obtain the following basic positive result.

Theorem 1. *Suppose that assumptions* (A1) *and* (A2) *hold and* X *has a finite diameter* D. *Then for* $\varepsilon > 0$, $0 \leq \delta < \varepsilon$ *and sample size* N *satisfying* (22), *we are guaranteed that any* δ-*optimal solution of the SAA problem is an* ε-*optimal solution of the true problem with probability at least* $1 - \alpha$.

Let us also consider the following simplified variant of Theorem 1. Suppose that:

(A3) There is a positive constant C such that $|F(x', \xi) - F(x, \xi)| \leq C$ for all $x', x \in X$ and $\xi \in \Xi$.

Under assumption (A3) we have that for any $\varepsilon > 0$ and $\delta \in [0, \varepsilon]$:

$$I_x(-\delta) \geq O(1)\frac{(\varepsilon - \delta)^2}{C^2}, \quad \text{for all } x \in X \setminus S^\varepsilon, \ \xi \in \Xi, \tag{23}$$

and hence $\eta(\varepsilon, \delta) \geq O(1)(\varepsilon - \delta)^2/C^2$. Consequently, the bound (8) for the sample size which is required to solve the true problem with accuracy $\varepsilon > 0$ and probability at least $1 - \alpha$, by solving the SAA problem with accuracy $\delta := \varepsilon/2$, takes the form

$$N \geq O(1) \left(\frac{C}{\varepsilon}\right)^2 \log \left(\frac{|X|}{\alpha}\right). \tag{24}$$

The estimate (24) can be also derived by using Hoeffding's inequality[10] instead of Cramér's LD bound.

In particular, if we assume that $\gamma = 1$ and $\kappa(\xi) = L$ for all $\xi \in \Xi$, i.e., $F(\cdot, \xi)$ is Lipschitz continuous on X with constant L independent of $\xi \in \Xi$, then we can take $C = DL$ and remove the term $\beta^{-1} \log(2/\alpha)$ in the right hand side of (22). By taking, further, $\delta := \varepsilon/2$ we obtain in that case the following estimate of the sample size

$$N \geq O(1) \left(\frac{DL}{\varepsilon}\right)^2 \left[n \log \left(\frac{DL}{\varepsilon}\right) + \log \left(\frac{O(1)}{\alpha}\right)\right]. \tag{25}$$

We can write the following simplified version of Theorem 1.

---

[10] Recall that Hoeffding's inequality states that if $Z_1, ..., Z_N$ is an iid random sample from a distribution supported on a bounded interval $[a, b]$, then for any $t > 0$,

$$\text{Prob}\left(\bar{Z} - \mu \geq t\right) \leq e^{-2t^2 N/(b-a)^2},$$

where $\bar{Z}$ is the sample average and $\mu = \mathbb{E}[Z_i]$.

Theorem 2. *Suppose that X has a finite diameter D and condition (18) holds with $\gamma = 1$ and $\kappa(\xi) = L$ for all $\xi \in \Xi$. Then with sample size N satisfying (25) we are guaranteed that every $(\varepsilon/2)$-optimal solution of the SAA problem is an ε-optimal solution of the true problem with probability at least $1 - \alpha$.*

In the next section we compare complexity estimates implied by the bound (25) with complexity of "deterministic" convex programming.

3 What is easy and what is difficult in stochastic programming?

Since, generically, nonconvex problems are difficult already in the deterministic case, when discussing the question of what is easy and what is not in Stochastic Programming, it makes sense to restrict ourselves with convex problems (1). Thus, in the sequel it is assumed by default that X is a closed and bounded convex set, and $f : X \to \mathbb{R}$ is convex. These assumptions, plus mild technical conditions, would be sufficient to make (1) easy, provided that $f(x)$ were given explicitly, but the latter is *not* what we assume in SP. What we usually (and everywhere below) do assume in SP is that:

(i) The function $F(x, \xi)$ is given explicitly, so that we can compute efficiently its value (and perhaps the derivatives in x) at every given pair $(x, \xi) \in X \times \Xi$.

(ii) We have access to a mechanism which is capable of sampling from the distribution P, that is, we can generate a sample ξ^1, ξ^2,... of independent realizations of ξ.

For the sake of discussion to follow we assume in this section that we are under the premise of Theorem 2 and that problem (1) is *convex*. To proceed, let us compare the complexity bound given by Theorem 2 with a typical result on the "black box" complexity of the usual (deterministic) Convex Programming.

Theorem 3. *Consider a convex problem*

$$\operatorname*{Min}_{x \in X} f(x), \tag{CP}$$

where $X \subset \mathbb{R}^n$ is a closed convex set which is contained in a centered at the origin ball of diameter D and contains a ball of given diameter $d > 0$, and that $f : X \to \mathbb{R}$ is convex Lipschitz continuous, with constant L. Assume that X is given by a Separation Oracle which, given on input a point $x \in \mathbb{R}^n$, reports whether $x \in X$, and if it is not the case, returns $e \in \mathbb{R}^n$ which separates x and X: such that $\langle e, x \rangle > \max_{y \in X} \langle e, y \rangle$. Assume, further, that f is given by a First Order oracle which, given on input $x \in X$, returns on output the value $f(x)$ and a subgradient $\nabla f(x)$, $\|\nabla f(x)\|_2 \leq L$, of f at x.

In this framework, for every $\varepsilon > 0$ one can find an ε-solution to (CP) by an algorithm which requires at most

$$M = O(1)n^2 \left[\log\left(\frac{DL}{\varepsilon}\right) + \log\left(\frac{D}{d}\right)\right] \tag{26}$$

calls to the Separation and First Order oracles, with a call accompanied by $O(n^2)$ arithmetic operations to process oracle's answer.

In our context, the role of Theorem 3 is twofold. First, it can be viewed as a necessary follow-up to Theorem 2 which reduces solving (1) to solving the corresponding SAA problem and says nothing on how difficult is the latter task. This question is answered by Theorem 3 in the convex case[11]. However, the main role of Theorem 3 in our context is the one of a benchmark for the SP complexity results. Let us use this benchmark to evaluate the result stated in Theorem 2.

Observation 1. In contrast to Theorem 3, Theorem 2 provides us with no more than probabilistic quality guarantees. That is, the random approximate solution to (1) implied by the outlined SAA approach, being ε-solution to (1) with probability $1 - \alpha$, can be very bad with the remaining probability α. In our "black box" informational environment (the distribution of ξ is not given in advance, all we have is an access to a black box generating independent realizations of ξ), this "shortcoming" is unavoidable. Note, however, that the sample size N as given by (22) is "nearly independent" of α, i.e., to reduce unreliability from 10^{-2} to 10^{-12} requires at most 6-fold increase in the sample size. Note that unreliability as small as 10^{-12} is, for all practical purposes, the same as 100% reliability.

Observation 2. To proceed with our comparison, it makes sense to measure the complexity of the SAA method merely by the number of scenarios N required to get an ε-solution with probability at least $1 - \alpha$, and to measure the complexity of deterministic convex optimization as presented in Theorem 3 by the number M of oracle calls required to get an ε-solution. The rationale behind is that "very large" N definitely makes the SAA method impractical, while with a "moderate" N, the method becomes practical, provided that $F(\cdot, \cdot)$ and X are not too complicated, and similarly for M in the context of Theorem 3.

When comparing bounds (25) and (26), our first observation is that both of them depend polynomially on the design dimension n of the problem, which is nice. What does make difference between these bounds, is their dependence

---

[11] In our context, Theorem 3 allows to handle the most general "black box" situation – no assumptions on $F(\cdot, \xi)$ and X except for convexity and computability. When $F(\cdot, \xi)$ possesses appropriate analytic structure, the complexity of solving the SAA problem can be reduced by using a solver adjusted to this structure.

on the required accuracy ε, or, better to say, on the relative accuracy[12] $\nu :=$ $\varepsilon/(DL)$. In contrast to bound (26) which is polynomial in $\log(1/\nu)$, bound (25) is polynomial (specifically, quadratic) in $1/\nu$. In reality this means that the SAA method could solve in a reasonable time to a moderate relative accuracy, like $\nu = 10\%$ or even $\nu = 1\%$, stochastic problems involving an astronomically large, or even infinite, number of scenarios. This was verified in a number of numerical experiments (e.g., [LSW05, MMW99, SAGS05, VAKNS03]). On the other hand, in general, the SAA method does *not* allow to solve, even simply-looking, problems to high relative accuracy[13]: according to (25), the *estimated* sample size N required to achieve $\nu = 10^{-3}$ ($\nu = 10^{-5}$) is at least of order of millions (respectively, tens of billions). In sharp contrast to this, bound (26) says that in the deterministic case, relative accuracy $\nu = 10^{-5}$ is just by factor 5 "more costly" than $\nu = 0.1$.

It should be stressed that in our general setting the outlined phenomenon is not a shortcoming of the SAA method – it is unavoidable. Indeed, given positive constants L, D and ε such that $\nu = \varepsilon/(LD) \leq 0.1$, consider the pair of stochastic problems:

$$\underset{x \in [0,D]}{\text{Min}} \left\{ f_\chi(x) := \mathbb{E}_{P_\chi}[x\xi] \right\} \tag{SP_χ}$$

indexed by $\chi = \pm 1$, and with distribution P_χ of ξ supported on the two-point set $\{-L; L\}$ on the axis. Specifically, P_1 assigns the mass $1/2 - 4\nu$ to the point $-L$ and the mass $1/2 + 4\nu$ to the point L, while P_{-1} assigns to the same points $-L, L$ the masses $1/2 + 4\nu$, $1/2 - 4\nu$, respectively. Of course, $f_1(x) = 4\varepsilon D^{-1}x$, $f_{-1}(x) = -4\varepsilon D^{-1}x$, the solution to ($SP_1$) is $x_1 = 0$, while the solution to (SP_{-1}) is $x_{-1} = D$. Note, however, that the situation is that trivial *only when we know in advance what is the distribution P_χ we deal with*. If it is not the case and all we can see is a sample of N independent realizations of ξ, the situation changes dramatically: an algorithm capable of solving with accuracy ε and reliability $1 - \alpha = 0.9$ every one of the problems ($SP_{\pm 1}$) using sample of size N, would, as a byproduct, imply a procedure which, given the sample, decides, with the same reliability, which one of the two possible distributions $P_{\pm 1}$ underlies the sample. The laws of Statistics say that such a reliable identification of the underlying distribution is possible only when $N \geq O(1)\frac{D^2 L^2}{\varepsilon^2}$ (compare with bound (25)). Note that both stochastic problems in question satisfy all the assumptions in Theorem 2, so that in the

---

[12] Recall that, under assumptions of Theorem 2, DL gives an upper bound on the variation of the objective on the feasible domain. While using bound (22) we can take $\nu := \varepsilon/\sigma$. Passing from ε to ν, means quantifying inaccuracies as fractions of the variation, which is quite natural.

[13] It is possible to solve true problem (1) by the SAA method with high (machine) accuracy in some specific situations, for example, in some cases of linear two-stage stochastic programming with a finite (although very large) number of scenarios, see [SH00, SHK02].

situation considered in this statement the bound (25) is the best possible (up to logarithmic term) as far as the dependence on D, L and ε is concerned.

To make our presentation self-contained, we explain here what are the "laws of Statistics" which underlie the above conclusions. First, an algorithm $\mathcal{A}$ capable of solving within accuracy ε and reliability 0.9 every one of the problems $(\text{SP}_{\pm 1})$, given an N-element sample drawn from the corresponding distribution, indeed implies a "0.9-reliable" procedure which decides, based on the same sample, what is the distribution; this procedure accepts hypothesis I stating that the sample is drawn from distribution P_1 if and only if the approximate solution generated by $\mathcal{A}$ is in $[0, D/2]$; if it is not the case, the procedure accepts hypothesis II "the sample is drawn from P_{-1}". Note that if the first of the hypotheses is true and the outlined procedure accepts the second one, the approximate solution produced by $\mathcal{A}$ is *not* and ε-solution to (SP_1), so that the probability p^{I} to accept the second hypothesis when the first is true is $\leq 1 - 0.9 = 0.1$. Similarly, probability p^{II} for the procedure to accept the first hypothesis when the second is true is ≤ 0.1. The announced lower bound on N is given by the following observation: *Consider a decision rule which, given on input a sequence ξ^N of N independent realizations of ξ known in advance to be drawn either from the distribution P_1, or from the distribution P_{-1}, decides which one of these two options takes place, and let p^{I}, p^{II} be the associated probabilities of wrong decisions. Then*

$$\max\{p^{\text{I}}, p^{\text{II}}\} \leq 0.1 \text{ implies that } N \geq O(1)\nu^{-2}, \qquad (27)$$

where $O(1)$ is a positive absolute constant.

Indeed, a candidate decision rule can be identified with a subset S of $\mathcal{L}^N$; this set is comprised of all realizations ξ^N resulting, via the decision rule in question, in acceptance of hypothesis I. Let P_1^N, P_{-1}^N be the distributions of ξ^N corresponding to hypotheses I, II. We clearly have

$$p^{\text{I}} = \sum_{\xi^N \notin S} P_1^N(\xi^N), \; p^{\text{II}} = \sum_{\xi^N \in S} P_{-1}^N(\xi^N).$$

Now consider the Kullback distance from P_1^N to P_{-1}^N:

$$\mathcal{K} = \sum_{\xi^N \in \mathcal{L}^N} \log \left(\frac{P_1^N(\xi^N)}{P_{-1}^N(\xi^N)} \right) P_1^N(\xi^N).$$

the function $p \log \frac{p}{q}$ of two positive variables p, q is jointly convex; denoting by $\bar{S}$ the complement of S in $\mathcal{L}^N$ and by $|A|$ the cardinality of a finite set A, it follows that

$$|\bar{S}|^{-1} \sum_{\xi^N \in \bar{S}} \log\left(\frac{P_1^N(\xi^N)}{P_{-1}^N(\xi^N)}\right) P_1^N(\xi^N)$$

$$\geq \left(|\bar{S}|^{-1} \sum_{\xi^N \in \bar{S}} P_1^N(\xi^N)\right) \log\left(\frac{|\bar{S}|^{-1} \sum_{\xi^N \in \bar{S}} P_1^N(\xi^N)}{|\bar{S}|^{-1} \sum_{\xi^N \in \bar{S}} P_{-1}^N(\xi^N)}\right),$$

whence

$$\sum_{\xi^N \in \bar{S}} \log\left(\frac{P_1^N(\xi^N)}{P_{-1}^N(\xi^N)}\right) P_1^N(\xi^N) \geq p^{\mathrm{I}} \log\left(\frac{p^{\mathrm{I}}}{1 - p^{\mathrm{II}}}\right),$$

and similarly

$$\sum_{\xi^N \in S} \log\left(\frac{P_1^N(\xi^N)}{P_{-1}^N(\xi^N)}\right) P_1^N(\xi^N) \geq (1 - p^{\mathrm{I}}) \log\left(\frac{1 - p^{\mathrm{I}}}{p^{\mathrm{II}}}\right),$$

whence

$$\mathcal{K} \geq p^{\mathrm{I}} \log\left(\frac{p^{\mathrm{I}}}{1 - p^{\mathrm{II}}}\right) + (1 - p^{\mathrm{I}}) \log\left(\frac{1 - p^{\mathrm{I}}}{p^{\mathrm{II}}}\right).$$

For every $p \in (0, 1/2)$, the minimum of the left hand side in the latter inequality in $p^{\mathrm{I}}, p^{\mathrm{II}} \in (0, p]$ is achieved when $p^{\mathrm{I}} = p^{\mathrm{II}} = p$ and is equal to $p \log \frac{p}{1-p} + (1 - p) \log \frac{1-p}{p} \geq 4(p - 1/2)^2$. Thus,

$$p := \max[p^{\mathrm{I}}, p^{\mathrm{II}}] \leq 1/2 \text{ implies that } \mathcal{K} \geq (2p - 1)^2. \qquad (28)$$

On the other hand, taking into account the product structure of $P_{\pm 1}^N$, we have

$$\mathcal{K} = N \left[P_1(-L) \log \frac{P_1(-L)}{P_{-1}(-L)} + P_1(L) \log \frac{P_1(L)}{P_{-1}(L)} \right]$$

$$= N \left[(1/2 - 4\nu) \log\left(\frac{1 - 8\nu}{1 + 8\nu}\right) + (1/2 + 4\nu) \log\left(\frac{1 + 8\nu}{1 - 8\nu}\right) \right] = 8N\nu \log\left(\frac{1 + 8\nu}{1 - 8\nu}\right).$$

The concluding quantity is $\leq O(1)N\nu^2$, provided that $\nu \leq 0.1$. Combining this observation and (28), we arrive at (27).

Observation 3. One can argue that the phenomenon discussed in Observation 2 is not too dangerous from the practical viewpoint. In reality, especially in an "uncertain one", treated in stochastic models, relative accuracy like 1% or 5% is more than satisfactory. This indeed is true in numerous applications, which, in our opinion, is the intrinsic reason for Stochastic Programming to be of significant practical value. At the same time, there are some unpleasant exceptions; the most disturbing, from applied viewpoint, is the one related to problems without *relatively complete recourse*. This is the issue we are consider next.

The above analysis, summarized in Theorem 2, implicitly depends on the assumptions (i) and (ii) formulated in the beginning of this section (which are parallel to the assumptions (a)-(c) specified in the previous section). When applied to two-stage stochastic programming with recourse these assumptions imply that the recourse is relatively complete, i.e., for every $x \in X$ and every possible realization of ξ, the second stage problem is feasible. If, on the other hand, for some $x \in X$ and $\xi \in \Xi$ the second stage problem is infeasible, we can formally set the value $F(x, \xi)$ of the second stage problem to be $+\infty$. In order to avoid such infinite penalizations and to restore the applicability of Theorem 2 one can introduce a finite penalty for infeasibility. In some cases this can reasonably solve the problem. However, in some situations the infeasibility may result in a catastrophic event. In that case the penalty could be huge. Translated into the sample size bounds considered in the previous section, this means huge variances in the estimate (22) or huge Lipschitz constant in (25), which makes these estimates useless. In a sense, in such situation "nothing works".

It is NP-hard even to check whether a given first-stage decision $x \in X$ leads to feasible, with probability 1, second-stage problem, and even in the case when the second-stage problem is as simple as

$$\mathop{\mathrm{Min}}_{y}\langle q, y\rangle \text{ subject to } Tx + Wy \geq h, \tag{29}$$

with only the second-stage right hand side vector $h = h(\xi)$ being random.

To see that a generic problem of checking whether (29) is feasible for a given x is NP-hard, consider the case when the constraints $Tx + Wy \geq h(\xi)$ read $y \leq 0$, $y + x \geq h(\xi)$, where $x, y \in \mathbb{R}$,

$$h(\xi) := \sum_{i,j} Q_{ij}\xi_i\xi_j,$$

$Q = [Q_{ij}]$ is a given $d \times d$ symmetric matrix, and $\xi = (\xi_1, ..., \xi_d)$ is uniformly distributed in $[-1, 1]^d$. Here x results in feasible, with probability 1, second stage problem if and only if $x \geq \rho(Q)$, where

$$\rho(Q) := \max_{\xi} \left\{ \langle \xi, Q\xi \rangle : \xi \in [-1, 1]^d \right\}.$$

It is well-known that given x and Q, it is NP-hard to distinguish between the cases of $x \leq \rho(Q)$ and $x > 1.01\,\rho(Q)$. This NP-hard problem is, of course, not more difficult than to decide whether $x \geq \rho(Q)$. Note that replacing in the above example the uniform distribution on $[-1, 1]^d$ with the uniform distribution on the discrete set, of cardinality 2^d, of d-dimensional vectors with entries ± 1, we end up with an equally difficult problem.

Thus, if a two-stage (linear) problem has no relatively complete recourse (which in many applications is a rule rather than an exception), it is, in general, NP-hard just to find a feasible first-stage solution x (one which results in finite $f(x)$), not speaking about minimizing over these x's. As it was mentioned above, the standard way to avoid, to some extent, this difficulty is to pass to a penalized problem. For example, we can replace the second stage problem (4) with the penalized version:

$$\underset{y \geq 0,\, z \geq 0}{\text{Min}} \langle q, y \rangle + rz \text{ subject to } Tx + Wy \geq h - ze, \tag{30}$$

where e is vector of ones and $r \gg 1$ plays the role of the penalty coefficient. With this penalization, the second stage problem becomes always feasible. At the same time, one can hope that with large enough penalty coefficient r, the first-stage optimal solution will lead to "nearly always nearly feasible" second-stage problems, provided that the original problem is feasible. Unfortunately, in the situation where one cannot tolerate arising, with probability bigger than α, a second-stage infeasibility z bigger than τ (here α and τ are given thresholds), the penalty parameter r should be of order of $(\alpha\tau)^{-1}$. In the "high reliability" case $\alpha \ll 1$ we end up with problem (30) which contains large coefficients, which can lead to large value of the Lipschitz constant L_r of the optimal value function $F_r(\cdot, \xi)$ of the penalized second stage problem. As a result, quite moderate accuracy requirements (like ε being of order of 5% of the optimal value of the true problem) can result in the necessity to solve (30) within a pretty high relative accuracy $\nu = \varepsilon/(DL_r)$ like 10^{-6} or less, with all unpleasant consequences of this necessity.

3.1 What is difficult in the two-stage case?

We already know partial answer to this question: generically, under the premise of Theorem 2 it is difficult to solve problem (1) (even a convex one) to a high relative accuracy $\nu = \varepsilon/(DL)$. Note, however, that the statistical arguments demonstrating that this difficulty lies in the nature of the problem work *only for the black-box setting of* (1) *considered so far*, that is, only in the case when the distribution P of ξ is not known in advance, and all we have in our disposal is a black box generating realizations of ξ. With a "good description" of P available, the results could be quite different, as it is clear when looking at problems $(\text{SP}_{\pm 1})$ – with the underlying distributions given in advance, the problems become trivial. Note that in reality stochastic models are usually equipped with known in advance and easy-to-describe distributions, like Gaussian, or Bernoulli, or uniform on $[-1, 1]^d$. Thus, it might happen that our conclusion "it is difficult to solve (1) to high accuracy" is an artifact coming from the black-box model we used, and we could overcome this difficulty by using more advanced solution techniques based on utilizing a given in advance and "simple" description of P. *Unfortunately, this virtual possibility does not exist in reality.* Specifically, it is shown in [DS03] that indeed it is difficult to solve to high accuracy already two-stage linear stochastic programs with complete recourse and easy-to-describe discrete distributions.

Another difficulty, which we have already discussed, is the case of two-stage linear problems without complete recourse or, more generally, convex problems (1) with only partially defined integrand $F(x, \xi)$. As we have seen, this difficulty arises already when looking for feasible first-stage solutions with known in advance simple distribution P.

3.2 Complexity of multi-stage stochastic problems

In a multi-stage stochastic programming setting random data ξ is partitioned into $T \geq 2$ blocks ξ_t, $t = 1, ..., T$, i.e., ξ_t is viewed as a (discrete time) random process, and the decisions are made at time instants $0, 1, ..., T$. At time t the decision maker already knows the realizations ξ_τ, $\tau \leq t$, of the process up to time t, while realizations of the "future" blocks are still unknown. The goal is to find the first-stage decisions x (which should not depend on ξ) and decision rules $y_t = y_t(\xi_{[t]})$ which are functions of $\xi_{[t]} := (\xi_1, ..., \xi_t)$, $t = 1, ..., T$, which satisfy a given set of constraints

$$g_i(\xi, x, y_1, ..., y_T) \leq 0, \ i = 1, ..., I, \tag{31}$$

and minimize under these restrictions the expected value of a given cost function $f(x, y_1, ..., y_T)$. Note that even in the case when the functions g_i do not depend of ξ, the left hand sides of the constraints (31) are functions of ξ, since all y_t are so, and that the interpretation of (31) is that these functional constraints should be satisfied with probability one.

In the sequel, we focus on the case of linear multi-stage problems

$$\begin{aligned}
\underset{x, y(\cdot)}{\text{Min}} \ & \mathbb{E}_P \left\{ \langle c_0, x \rangle + \sum_{t=1}^{T} \langle c_t(\xi_{[t]}), y_t(\xi_{[t]}) \rangle \right\} \\
\text{s.t.} \ & A_0^0 x \geq b^0 & (C_0) \\
& A_0^1(\xi_{[1]})x + A_1^1(\xi_{[1]})y_1(\xi_{[1]}) \geq b^1(\xi_{[1]}) & (C_1) \\
& A_0^2(\xi_{[2]})x + A_1^2(\xi_{[2]})y_1(\xi_{[1]}) + A_2^2(\xi_{[2]})y_2(\xi_{[2]}) \geq b^2(\xi_{[2]}) & (C_2) \\
& \qquad\qquad \cdots\cdots \\
& A_0^T(\xi_{[T]})x + A_1^T(\xi_{[T]})y_1(\xi_{[1]}) + ... + A_T^T(\xi_{[T]})y_T(\xi_{[T]}) \geq b^T(\xi_{[T]}) & (C_T)
\end{aligned}$$
$$\tag{32}$$

where $y(\cdot) = (y_1(\cdot), ..., y_T(\cdot))$ and the constraints $(C_1), ..., (C_T)$ should be satisfied with probability one. Problems (32) are called problems with *complete recourse*, if for every instant t and whatever decisions x, $y_1, ... y_{t-1}$ made at preceding instants, the system of constraints (C_t) (treated as a system of linear inequalities in variable y_t) is feasible for almost all realizations of ξ. The major focus of theoretical research is on multi-stage problems even simpler than (32), specifically, on problems with *fixed recourse* where matrices $A_t^t = A_t^t(\xi_{[t]})$, $t = 1, ..., T$, are assumed to be deterministic (independent of ξ).

We argue that multi-stage problems, even linear of the form (32) with complete recourse, *generically* are *computationally intractable* already when medium-accuracy solutions are sought. (**Of course, this does not mean that some specific cases of multi-stage stochastic programming**

problems cannot be solved efficiently.) Note that this claim is rather a *belief* than a statement which we can rigorously prove. It is even not a formal statement which can be true or wrong since, in particular, we do not specify what does "medium accuracy" mean[14]. What we are trying to say is that we believe that in the multi-stage case (with T treated as varying parameter, and not as a once for ever fixed entity), even "moderately positive" results like the one stated in Theorem 2 are impossible. We are about to explain what are the reasons for our belief.

Often practitioners do not pay attention to a dramatic difference between two-stage and multi-stage case. It is argued that in both cases the problem of interest can be written in the form of (1), with appropriately defined integrand F. Specifically, in case of the linear two-stage problem, with relatively complete recourse, we have that $F(x,\xi) = \langle c, x \rangle + Q(x,\xi)$, where $Q(x,\xi)$ is the optimal value of the second stage problem (4). In the case of problem (32) with complete recourse, $F(x,\xi)$ is given by a recurrence as follows. We start with setting

$$F_T(x, y_1, ..., y_T, \xi_{[T]}) := \langle c_0, x \rangle + \langle c_1(\xi[1]), y_1 \rangle + ... + \langle c_{T-1}(\xi_{[T-1]}), y_{T-1} \rangle$$
$$+ \langle c_T(\xi_{[T]}), y_T \rangle$$

and specifying the conditional, given $\xi_{[T-1]}$, expected cost of the last-stage problem:

$$F_{T-1}(x, y_1, ..., y_{T-1}, \xi_{[T-1]}) := \mathbb{E}_{|\xi_{[T-1]}} \underset{y_T}{\text{Min}} \Big\{ F_T(x, y_1, ..., y_{T-1}, y_T, \xi_{[T]}) :$$
$$A_0^T(\xi_{[T]})x + A_1^T(\xi_{[T]})y_1 + ... + A_T^T(\xi_{[T]})y_T$$
$$\geq b^T(\xi_{[T]}) \Big\},$$

where $\mathbb{E}_{|\xi_{[T-1]}}$ is the conditional, given $\xi_{[T-1]}$, expectation. Observe that (32) is equivalent to the $(T-1)$-stage problem:

$$\underset{x, \{y_t(\cdot)\}_{t=1}^{T-1}}{\text{Min}} \mathbb{E}_{P^{T-1}} \Big\{ F_{T-1}(x, y_1, ..., y_{T-1}, \xi_{[T-1]}) \Big\}$$
$$\text{s.t.} \quad x, y_1(\cdot), ..., y_{T-1}(\cdot) \text{ satisfy } (C_0), (C_1), ..., (C_{T-1}) \text{ w.p.1,} \qquad (P_{T-1})$$

where P^{T-1} is the distribution of $\xi_{[T-1]}$. Now we can iterate this construction, ending up with the problem

$$\underset{x \in X}{\text{Min}} [F_0(x)].$$

It can be easily seen that under the assumption of complete recourse, plus mild boundedness assumptions, all functions $F_\ell(x, y_1, ..., y_\ell, \xi_{[\ell]})$ are Lipschitz continuous in the x, y-arguments.

---

[14] To the best of our knowledge, the complexity status of problem (32), even in the case of complete and fixed recourse and known in advance easy-to-describe distribution P, remains unknown (cf., [DS03]).

The "common wisdom" says that since both, two-stage and multi-stage, problems are of the same generic form (1), with the integrand convex in x, and both are processed numerically by generating a sample of scenarios and solving the resulting "scenario counterpart" of the problem of interest, there should be no much difference between the two and the multi-stage case, provided that in both cases one uses the same number of scenarios. This "reasoning", however, completely ignores a crucial point as follows: in order to solve generated SAA problems efficiently, the integrand F should be efficiently computable at every pair (x, ξ). This is indeed the case for a two-stage problem, since there $F(x, \xi)$ is the optimal value in an explicit Linear Programming problem and as such can be computed in polynomial time. In contrast to this, the integrand F produced by the outlined scheme, as applied to a multi-stage problem, is *not* easy to compute. For example, in 3-stage problem this integrand is the optimal value in a 2-stage stochastic problem, so that its computation at a point is a much more computationally involving task than similar task in the two-stage case. Moreover, in order to get just consistent estimates in an SAA type procedure (not talking about rate of convergence) one needs to employ a conditional sampling which typically results in an exponential growth of the number of generated scenarios with increase of the number T of stages (cf., [Sha03a]).

Analysis demonstrates that for an algorithm of the SAA type, the total number of scenarios needed to solve T-stage problem (32), with complete recourse, would grow, as ε diminishes, as ε^{-2T}, so that the computational effort blows up exponentially as the number of stages grows[15] (cf., [Sha05b]). Equivalently, *for a sampling-based algorithms with a given number of scenarios, existing theoretical quality guarantees deteriorate dramatically as the number of stages grows.* Of course, nobody told us that sampling-type algorithms are the only way to handle stochastic problems, so that the outlined reasoning does not pretend to justify "severe computational intractability" of multi-stage problems. Our goal is more modest, we only argue that the fact that when solving a particular stochastic program a sample of 10^7 scenarios was used does not say much about the quality of the resulting solution: in the two-stage case, there are good reasons to believe that this quality is reasonable, while in the 5-stage the quality may be disastrously bad.

We have described one source of severe difficulty arising when solving multi-stage stochastic problems – dramatic growth, with increase of the number of stages, in the complexity of evaluating the integrand F in representation (1) of the problem. We are about to demonstrate that even when this difficulty does not arise, a multi-stage problem still may be very difficult. To this end, consider the following story: at time $t = 0$, one has \$ 1, and should decide how to distribute this money between stocks and a bank account. When investing amount of money x into stocks, the value u_t of the portfolio at time t will be

---

[15] Note that in the considered framework, $T = 1$ corresponds to two-stage programming, $T = 2$ corresponds to 3-stage programming, and so on.

given by chain of t relations

$$u_1 = \rho_1(\xi_{[1]})x, u_2 = \rho_2(\xi_{[2]})u_1, ..., u_t = \rho_t(\xi_{[t]})u_{t-1},$$

where the returns $\rho_t(\xi_{[t]}) \equiv \rho_t(\xi_1, ..., \xi_t) \geq 0$ are known functions of the underlying random parameters. Amount of money $1 - x$ put to bank account reach at time t the value $v_t = \rho^t(1 - x)$, where $\rho > 0$ is a given constant. The goal is to maximize the total expected wealth $\mathbb{E}[u_T + v_T]$ at a given time T. The problem can be written as a simple-looking T-stage stochastic problem of the form (32):

$$
\begin{aligned}
\underset{x,y(\cdot)}{\text{Min}} \;\; & \mathbb{E}_P\left[u_T(\xi^T) + v_T(\xi^T)\right] \\
\text{s.t.} \;\; & 0 \leq x \leq 1 && (C_0) \\
& u_1(\xi_{[1]}) = \rho_1(\xi_{[1]})x, \;\; v_1(\xi_{[1]}) = \rho(1 - x) && (C_1) \\
& u_2(\xi_{[2]}) = \rho_2(\xi_{[2]})u_1(\xi_{[1]}), \;\; v_2(\xi_{[2]}) = \rho v_1(\xi_{[1]}) && (C_2) \\
& \qquad\qquad \cdots\cdots \\
& u_T(\xi_{[T]}) = \rho_{T-1}(\xi_{[T-1]})u_{T-1}(\xi_{[T-1]}), \;\; v_T(\xi_{[T]}) = \rho v_{T-1}(\xi_{[T-1]}) \; (C_T),
\end{aligned}
$$
(33)

where $y(\cdot) = (u_t(\cdot), v_t(\cdot))_{t=1}^T$. Now let us specify the structure and the distribution of ξ as follows: a realization of ξ is a permutation $\xi = (\xi_1, ..., \xi_T)$ of T elements $1, ..., T$, and P is the uniform distribution on the set of all $T!$ possible permutations. Further, let us specify the returns as follows: the returns are given by a $T \times T$ matrix A with 0-1 elements, and

$$\rho_t(\xi_1, ..., \xi_t) := \kappa A_{t,\xi_t}, \quad \kappa := (T!)^{1/T}$$

(Note that by Stirling's formula $\kappa = (T/e)(1 + o(1))$ as $T \to \infty$.) We end up with a simple-looking instance of (32) with complete recourse and given in advance "easy-to-describe" discrete distribution P; when represented in the form of (1), our problem becomes

$$\underset{x \in [0,1]}{\text{Min}} \left\{f(x) := \mathbb{E}_P F(x, \xi)\right\}, \quad F(x, \xi) = \rho^T(1 - x) + x \prod_{t=1}^{T}(\kappa A_{t\xi_t}), \quad (34)$$

so that F indeed is easy to compute. Thus, problem (33) looks nice – complete recourse, simple and known in advance distribution, no large data entries, easy-to-compute F in representation (1). At the same time the problem is disastrously difficult. Indeed, from (34) it is clear that $f(x) = \rho^T(1 - x) + x \operatorname{per}(A)$, where $\operatorname{per}(A)$ is the permanent of A:

$$\operatorname{per}(A) = \sum_{\xi} \prod_{t=1}^{T} A_{t\xi_t},$$

(the summation is taken over all permutations of T elements $1, ..., T$). Now, the solution to (34) is either $x = 1$ or $x = 0$, depending on whether or not per(A) $\geq \rho^T$. Thus, our simple-looking T-stage problem is, essentially, the problem of computing the permanent of a $T \times T$ matrix with 0-1 entries. The latter problem is known to be really difficult. First of all, it is NP-hard, [Val79]. Further, there are strong theoretical reasons to doubt that the permanent can be efficiently approximated within a given relative accuracy ε, provided that $\varepsilon > 0$ can be arbitrarily small, [DLMV88]. The best known to us algorithm capable to compute permanent of a $T \times T$ 0-1 matrix within relative accuracy ε has running time as large as $\varepsilon^{-2} \exp\{O(1)T^{1/2} \log^2(T)\}$ (cf., [JV96]), while the best known to us efficient algorithm for approximating permanent has relative error as large as c^T with certain fixed $c > 1$, see [LSW00]. Thus, simple-looking multi-stage stochastic problems can indeed be extremely difficult...

A reader could argue that in fact we deal with a two-stage problem (34) rather than with a multi-stage one, so that the outlined difficulties have nothing to do with our initial multi-stage setting. Our counter-argument is that the two-stage problem (34) honestly says about itself that it is very difficult: with moderate ρ and T, the data in (34) can be astronomically large (look at the coefficient ρ^T of $(1 - x)$ or at the products $\prod_{t=1}^{T}(\kappa A_{t\xi_t})$ which can be as large as $\kappa^T = T!$), and so is the Lipschitz constant of F. In contrast to this, the structure and the data in (33) look completely normal. Of course, it is immediate to recognize that this "nice image" is just a disguise, and in fact we are dealing with a disastrously difficult problem. Imagine, however, that we add to (33) a number of redundant variables and constraints; how could your favorite algorithm (or you, for that matter) recognize in the resulting messy problem that solving it numerically is, at least at the present level of our knowledge, a completely hopeless endeavor?

4 Some novel approaches

Here we outline some novel approaches to treating uncertainty which *perhaps* can cope, to some extent, with intrinsic difficulties arising in two-stage problems without complete recourse and in multi-stage problems.

4.1 Tractable approximations of chance constraints

As it was already mentioned, a natural way to handle two-stage stochastic problems without complete recourse is to impose *chance* constraints. That is, to require that a probability of insolvability of the second-stage problem is at most $\varepsilon \ll 1$ instead of being 0. The rationale behind this idea is twofold: first, from the practical viewpoint, "highly unlikely" events are not too dangerous: why should we bother about a marginal chance, like 10^{-6}, for the second stage to be infeasible, given that the level of various inaccuracies in our model, especially in its probabilistic data, usually is by orders of magnitude larger than 10^{-6}? Not speaking of the fact that 5 days a week we take worse chances in the morning traffic. Second, while it might be very difficult to check whether a given first-stage solution results in a feasible, with probability 1, second-stage problem, it seems to be possible to check whether this probability is at least $1 - \varepsilon$ by applying Monte-Carlo simulation. Note that chance constraints arise naturally not only in the context of two-stage problems without complete recourse, but in a much more general situation of solving a constrained optimization problem with the data affected by stochastic uncertainty. Thus, it makes sense to pose a question *how could one process numerically a chance constraint*

$$\phi(x) := \text{Prob}\{g(x, \xi) \leq 0\} \geq 1 - \varepsilon, \tag{35}$$

where x is the decision vector, ξ is the random disturbance with, say, known distribution, and $\varepsilon << 1$ is a given tolerance.

The concept of chance constraints originates from [CC59] and is one of the oldest concepts in Operations Research. Unfortunately, in its nearly 50 year old age, this concept still cannot be treated as practical. The first reason is that typically it is extremely difficult to verify *exactly* whether this constraint is satisfied at a given point. This problem is difficult already in the case of a single linear constraint $g(x, \xi) := \langle a_* + \xi, x \rangle$ with perturbations ξ uniformly distributed in a box. Another severe problem is that usually constraint (35), even with very simple, say bi-affine in x and in ξ, function $g(x, \xi)$ and simple-looking distribution of ξ (like uniform in a box) defines a nonconvex feasible set in the space of decision variables, which makes problematic subsequent optimization over this set of even pretty simple – just linear – objectives.

The difficulty we have just outlined rules out the idea to approximate (35) by a "sample version" of this constraint, that is, by

$$\widehat{\phi}_N(x) := \frac{1}{N} \sum_{j=1}^{N} \mathbb{1}_{\{g(x,\xi^j) \leq 0\}} \geq 1 - \theta\varepsilon, \tag{36}$$

where $\xi^1, ..., \xi^N$ is a sample of N independent realizations of ξ, $\mathbb{1}_{\{g(x,\xi^j) \leq 0\}}$ is the indicator function[16] of the event $\{g(x, \xi^j) \leq 0\}$, and $\theta < 1$ is fixed (say, $\theta = 0.99$). When $N >> \varepsilon^{-1}$, the validity of (36) at a point x implies, with probability close to 1, the validity of (35), so that (36) can be thought of as a "computable approximation" of (35). Unfortunately, the left hand side in (35) is, generically, a nonconvex (and even discontinuous) function of x, so that we have no way to optimize under this constraint.

To the best of our knowledge, the only generic case where both these severe difficulties disappear is the case of linear constraint $\langle a_* + \xi, x \rangle \leq 0$ with normally distributed data $\xi \sim \mathcal{N}(0, \Sigma)$. In this case, (35) is equivalent to the convex deterministic constraint

$$\langle a_*, x \rangle + \Omega(\varepsilon)\sqrt{\langle x, \Sigma x \rangle} \leq 0, \tag{37}$$

where the "safety parameter" $\Omega(\varepsilon) = \sqrt{2\log(1/\varepsilon)}(1 + o(1))$, $\varepsilon \to 0$, is readily given by ε (which we assume to be $\leq 1/2$).

---

[16] $\mathbb{1}_A = 1$ if the event A happens, and $\mathbb{1}_A = 0$ otherwise

There is another generic case when the feasible set given by a chance constraint is convex. This is the case when the constraint can be represented in the form $(x, \xi) \in Q$, where Q is a closed and convex set, and the distribution P of the random vector $\xi \in \mathbb{R}^d$ is *logarithmically quasi-concave*, meaning that

$$P(\lambda A + (1 - \lambda)B) \geq \max\left[P(A), P(B)\right]$$

for all closed and convex sets $A, B \subset \mathbb{R}^d$ (cf., Prekopa [Pre95]). Examples include uniform distributions on closed and bounded convex domains, normal distribution and every distribution on $\mathbb{R}^d$ with density $f(\xi)$ with respect to the Lebesgue measure such that the function $f^{-1/d}(\xi)$ is convex. The related result (due to Prekopa [Pre95]) is that in the situation in question, the set $\left\{x : P(\{\xi : (x, \xi) \in Q\}) \geq \alpha\right\}$ is closed and convex for every α. This result can be applied, e.g., to two-stage stochastic programs with chance constraints of the form

$$\underset{x \in X}{\text{Min}} \langle c, x \rangle \quad \text{s.t.} \quad \text{Prob}\{\exists y \in Y : Tx + Wy \geq \xi\} \geq 1 - \varepsilon,$$

where X, Y are closed convex sets and T, W are fixed matrices. Here the chance constraint indeed is of the form $\text{Prob}\{(x, \xi) \in Q\} \geq 1 - \varepsilon$, where

$$Q = \{(x, \xi) : \exists y \in Y : Tx + Wy \geq \xi\}.$$

The set Q clearly is convex; under mild additional assumptions, it is also closed. Thus, the feasible set of the chance constraint in question is convex, provided that the distribution of ξ is logarithmically quasi-concave.

Note that the outlined convexity results are applicable only to the chance constraints coming from scalar or vector inequalities where the only term affected by uncertainty is the right hand side, not the coefficients at the variables. For example, nothing similar is known for the chance constraint

$$\text{Prob}\left\{\langle a_* + \xi, x \rangle \leq 0\right\} \geq 1 - \varepsilon,$$

except for the already mentioned case of normally distributed vector ξ.

Aside of few special cases we have mentioned, chance constraint (35) "as it is" seems to be too difficult for efficient numerical processing, and what we can try to do is to replace it with its "tractable approximation". For the time being, there exist two approaches to building such an approximation: "deterministic" and "scenario".

Tractable deterministic approximations of chance constraints.

With this approach, one replaces (35) with a properly chosen deterministic constraint

$$\psi_\varepsilon(x) \leq 0, \tag{38}$$

which is a "safe computationally tractable" approximation of (35), with the latter notion defined as follows:

1. "Safety" means that the validity of (38) is a *sufficient* condition for the validity of (35);
2. "Tractability" means that (38) is an explicitly given convex constraint.

Just to give an example, consider a randomly perturbed linear constraint, that is, assume that

$$g(x, \xi) := \langle a_* + M\xi, x \rangle,$$

where the deterministic vector a_* is the "nominal data", M is a given deterministic matrix and $\xi = (\xi_1, ..., \xi_d)$ is a tuple of d independent scalar random variables with zero mean and "of order of 1":

$$\mathbb{E}\left[\exp(\xi_i^2)\right] \leq \exp\{1\}, \quad i = 1, ..., d,$$

e.g., ξ_i can have a distribution supported on the interval $[-1, 1]$, or ξ_i can have normal distribution $\mathcal{N}(0, 2^{-1/2})$, $= 1, ..., d$. In this case, applying standard results on probabilities of large deviations for sums of "light tail" independent random variables with zero means, one can easily verify that when $\varepsilon \in (0, 1)$ and $\Omega(\varepsilon) = O(1)\sqrt{\log(1/\varepsilon)}$ with properly chosen absolute constant $O(1)$, then the validity of the convex constraint

$$\langle a_*, x \rangle + \Omega(\varepsilon)\sqrt{\langle x, MM^T x \rangle} \leq 0 \tag{39}$$

is a sufficient condition for the validity of (35). (Note that under our assumptions MM^T is an upper bound on the covariance matrix of ξ, and compare with (37).)

The simple result we have just described is rather attractive. First, it does not require a detailed knowledge of the distribution of ξ. Second, the approximation, although being more complicated than a linear constraint we start with, still is pretty simple; modern convex optimization techniques can process routinely to high accuracy problems with thousands of decision variables and thousands of constraints of the form (39). Third, the approximation is "not too conservative" – the safety parameter $\Omega(\varepsilon)$ grows pretty slowly as $\varepsilon \to 0$ and is only by a moderate constant factor larger than the safety parameter in the case of Gaussian noise, where our approximation is not conservative at all.

Recently, "not too conservative" computationally tractable safe approximations were built (see [Nem03]) for chance versions of well-structured *non-linear* convex constraints with nice analytic structure, specifically, for affinely perturbed least squares constraints

$$\left\| \left[A_* + \sum_i A_i \xi_i\right] x - \left[b_* + \sum_i \xi_i b_i\right] \right\|_2 \leq \tau$$

and Linear Matrix Inequality constraints

$$\left[A_0^0 + \sum_i \xi_i A_i^0\right] + \sum_{j=1}^m x_j \left[A_0^j + \sum_i \xi_i A_i^j\right] \succeq 0$$

(A_q^p are symmetric matrices, $A \succeq 0$ means that A is symmetric positive semi-definite). In both cases, ξ_i are independent scalar disturbances with zero mean and "of order of 1". However, the outlined approach, whatever promising we believe it is, seemingly works for a very restricted family of "well-structured" functions $g(x, \xi)$, and even in these cases requires a lot of highly nontrivial "tailoring" to a particular structure in question. Consider, for example, the case of chance constraint associated with two-stage linear stochastic problem:

$$g(x, \xi) := \operatorname*{Min}_{z,y} \{z : T(\xi)x + W(\xi)y \geq h(\xi) - ze, z \geq 0\}, \qquad (40)$$

where e is vector of ones. Note that here $g(x, \xi)$ is convex in x, and $g(x, \xi) \leq 0$ if and only if the second-stage problem

$$\operatorname*{Min}_y \langle q(\xi), y \rangle \quad \text{s.t.} \quad T(\xi)x + W(\xi)y \geq h(\xi)$$

is feasible (cf., (30)). Thus, the chance constraint requires from x to result in a feasible, with probability at least $1 - \varepsilon$, second stage problem. Even in the case of simple recourse (T, W are independent of ξ) the chance constraint in question seems to be by far too difficult to admit a safe tractable deterministic approximation.

Scenario approximation.

In contrast to the "highly specialized and heavily restricted" approach we have just considered, the scenario-based approach is completely universal. We just generate a sample $\xi^1, ..., \xi^N$ of N "scenarios" – independent realizations of the random disturbance ξ – and approximate (35) by the random system of inequalities

$$g(x, \xi^j) \leq 0, \ j = 1, ..., N. \qquad (41)$$

Extremely nice features of this approach are its generality and computational tractability – whenever $g(x, \xi)$ is convex in x and efficiently computable (as it is the case, e.g., with the function (40)), (41) becomes a system of explicitly given convex constraints and as such can be efficiently processed numerically, *provided that the number of scenarios N is not prohibitively large.* The question, of course, is how large should be the sample in order to ensure, with reliability close to 1, that every feasible solution to (41) satisfies the chance constraint (35). This question is by far not easy, and we do not intend to discuss relevant nice and deep results known from the literature, since in fact we are more interested in a slightly different question, namely, as follows:

(Q)*Assume we are given a convex optimization problem*

$$\underset{x \in \mathbb{R}^n}{\text{Min}} \; f(x) \; \text{s.t.} \; g(x, \xi) \le 0 \tag{42}$$

(all f, g are convex in x) with ξ being a random vector with a known distribution, and, given tolerance $\varepsilon > 0$, replace this problem with its "scenario counterpart"

$$\underset{x \in \mathbb{R}^n}{\text{Min}} \; f(x) \; \text{s.t.} \; g(x, \xi^j) \le 0, j = 1, ..., N. \tag{43}$$

How large should be the sample size N in order for the optimal solution $\tilde{x}_N$ of (43) to be feasible for (42) with probability at least $1 - \varepsilon$?

The difference between the latter question and the former one is that now we do not require from *all* points feasible for (41) to satisfy (35), we require this property to be possessed by a specific point, $\tilde{x}_N$, we are interested in.

As it was discovered in [CC05, CC04], question (Q) admits a nice "universal" answer. Namely, under extremely mild assumptions it turns out that whenever $\varepsilon, \delta \in (0, 1/2)$ and

$$N \ge \frac{2n}{\varepsilon} \log\left(\frac{12}{\varepsilon}\right) + \frac{2}{\varepsilon} \log\left(\frac{2}{\delta}\right) + 2n, \tag{44}$$

the probability of "bad sampling" which results in $\tilde{x}_N$ not satisfying (35) is less than or equal to δ. Note that this result, which heavily utilizes the convexity of (42), is completely "distribution-free" – it is independent of any assumptions on the distribution of ξ and requires no knowledge of this distribution.

All this being said, there is a serious problem with the scenario approach as presented so far – it becomes impractical when the required value of ε is really small, like 10^{-6} or 10^{-8}. Indeed, for those ε relation (44) results in unrealistically large samples. Note that pretty small values of ε are completely reasonable when speaking about a "hard" constraint $g(x, \xi) \le 0$, that is, such that its violation has very severe or even catastrophic consequences, like heavy jam in a communication network, a blackout caused by malfunctioning of a power supply network, not speaking about exploding nuclear power plants or airliners falling from the sky. In a sense, in the context of chance constraints hard restrictions and implied pretty small values of ε seem to be a rule rather than exception. Indeed, "soft" constraints – those with ε like 1% or 0.1% – can be eliminated altogether by augmenting the objective with appropriate penalties[17].

---

[17] It should be added that the outlined "crude" scenario approach is not completely satisfactory even when ε is not too small. Indeed, assume that your problem has $n = 100$ variables and you are ready to take 10% chances ($\varepsilon = \delta = 0.1$). To this end, you use the scenario approach with the smallest N allowed by (44), that is, $N = 9835$. What should be the actual probability ε' for a fixed point $\bar{x}$ to violate

One could be surprised by the fact that we treat as acceptable the SAA method with the complexity proportional to ε^{-2}, ε being the required tolerance in terms of the objective, and are dissatisfied with the scenario approach where the sample size is merely inverse proportional to the tolerance ε. To explain our point, think whether you will agree (a) to use a portfolio management policy with the average profit by at most 0.5% less than the "ideal" – the optimal – one, and (b) to board an airliner which may crash during the flight with probability 0.5% (or 0.05%).

When handling hard chance constraints – those with really small ε, like 10^{-6} or less – we would like to have sample sizes polynomial in both $\log(1/\varepsilon)$ and $\log(1/\delta)$ rather than to be polynomial in $(1/\varepsilon)$ and $\log(1/\delta)$. We are about to explain that under favorable circumstances, such a possibility does exist; it is given by combining scenario approach with a kind of *importance sampling*. To proceed, assume that the constraint $g(x, \xi) \leq 0$ underlying (35) is of a specific structure as follows: there exists a closed convex set $K \subset \mathbb{R}^m$ and an affine mapping $x \mapsto A[\xi]x + b[\xi] : \mathbb{R}^n \to \mathbb{R}^m$ depending on ξ as on a parameter such that

$$g(x, \xi) \leq 0 \Leftrightarrow A[\xi]x + b[\xi] \in K. \tag{45}$$

Moreover, let us assume that the affine mapping in question is affinely parameterized by ξ, that is, both $A[\xi]$ and $b[\xi]$ depend affinely on ξ. Finally, we may assume without loss of generality that ξ has zero mean.

> As an instructive example, consider the feasibility constraint associated with the second-stage problem, that is, the constraint $g(x, \xi) \leq 0$ with $g(x, \xi)$ given by (40). Assuming fixed recourse, that is, $W(\xi) \equiv W$ being independent of ξ, let us set
>
> $$K := \{u : \exists y \text{ such that } u \leq Wy\}.$$
>
> Note that K is a convex polyhedral (and thus closed) set. Now, it is clear from (40) that $g(x, \xi) \leq 0$ if and only if $h(\xi) - T(\xi)x \in K$. It follows that when passing from uncertain parameter ξ to the new uncertain parameter $\bar{\xi} = [h(\xi), T(\xi)] - \mathbb{E}\{[h(\xi), T(\xi)]\}$ and updating accordingly the underlying distribution, we arrive at the situation described in (45).

Under our assumptions, the vector $A[\xi]x + b[\xi]$ is affine in ξ, and thus can be represented as $\alpha[x]\xi + \beta[x]$, where $\alpha[x]$, $\beta[x]$ are affine in x. It follows that

$$g(x, \xi) \leq 0 \Leftrightarrow A[\xi]x + b[\xi] \in K \Leftrightarrow \xi \in K_x := \{u : \alpha[x]u + \beta[x] \in K\}. \tag{46}$$

Note that the set K_x is closed and convex along with K. Now, numerous important distributions Π on $\mathbb{R}^p$ with zero mean (multivariate normal,

---

the constraint $g(x, \xi) \leq 0$ in order to be feasible for (43) with probability 0.9? The answer is: ε' should be as small as 10^{-5}. Thus, when applied with small ε, the crude scenario approach becomes impractical, while in the case of "large" ε it seems to be too conservative.

uniform on a multidimensional box, etc.) possess a kind of "concentration property" as follows: if Q is a closed convex set in $\mathbb{R}^p$ and $\Pi(Q) \geq c$, where $c < 1$ is a characteristic constant of Π, then the probability of the event $\Omega^{-1}\eta \in Q$, $\eta \sim P$, rapidly approaches 1 as $\Omega > 1$ grows, namely, $\Pi(\{\eta : \Omega^{-1}\eta \notin Q\}) \leq C^{-1}\exp\{-C\Omega^2\}$, where C is another characteristic constant of Π. For example, in the case of multivariate normal distribution Π with zero mean, then $\Pi(Q) \geq 0.8$ implies, for a closed convex set Q, that $\Pi(\{\eta : \eta/\Omega \notin Q\}) \leq \exp\{-\Omega^2/3\}$.

Now assume that we are in the situation of (46) and that the distribution of ξ possesses the outlined concentration property. Let us choose somehow a safety parameter $\Omega > 1$, and consider the scenario counterpart of (35), *where the disturbances are drawn from the distribution of $\Omega\xi$ rather than from the distribution of ξ*:

$$g(x, \Omega\xi^t) \leq 0, t = 1, ..., N$$
$$\Updownarrow \qquad\qquad (47)$$
$$\Omega\xi^t \in K_x, t = 1, ..., N$$

where $\xi^t \sim P$ are independent. Specifying N as

$$O(1)(1-c)^{-1}\log(1/\delta) \qquad\qquad (48)$$

with appropriate absolute constant $O(1)$, observe that if a *fixed* x satisfies (47), then it is "highly likely" that $\text{Prob}\{g(x, \Omega\xi) \leq 0\} \geq c$; specifically, in the case of $\text{Prob}\{g(x, \Omega\xi) \leq 0\} < c$, the probability to get a realization of N disturbances (with N given by (48)) which results in (47) is at most δ. Thus, when a given x turns out to satisfy (47), then, up to probability of "bad sampling" as small as δ, we have $\text{Prob}\{g(x, \Omega\xi) \leq 0\} \equiv \text{Prob}\{\Omega\xi \in K_x\} \geq c$. In the latter case, due to the concentration property of the distribution Π of $\eta = \Omega\xi$ (induced by similar property of the distribution P of ξ), we have $\text{Prob}\{g(x, \xi) > 0\} = \text{Prob}\{\xi \notin K_x\} \leq C^{-1}\exp\{-C\Omega^2\}$. When $\Omega = \sqrt{C^{-1}\log(C^{-1}\varepsilon^{-1})}$, the latter probability is $\leq \varepsilon$, that is, x satisfies the chance constraint (35). For example, in the case when P is a multivariate normal distribution with zero mean and ε in (35) is as small as 10^{-12}, the above rule results in $\Omega = 9.1$. Thus, when $\xi \sim \mathcal{N}(0, \Sigma)$, N is given by (48) and $\Omega = 9.1$, a fixed point x which satisfies (47) is, up to probability of "bad sampling" at most δ, feasible for the chance constraint (35) with $\varepsilon = 10^{-12}$.

The outlined idea – to apply the scenario approach with moderately amplified disturbances rather than with "true ones" – under favorable circumstances allows to approximate chance constraints via samples of size N which is polynomial in the "sizes" of the problem (the dimensions of x, ξ and K) and *logarithms* of $1/\varepsilon$, $1/\delta$, and thus allows to handle efficiently constraints (35) with really small tolerances ε. For detailed presentation and analysis of this approach, see [NS05].

4.2 Multistage Stochastic Programming in linear decision rules

Consider a linear multi-stage stochastic program

$$\operatorname*{Min}_{x,y(\cdot)} \mathbb{E}_P\left[\langle c_0(\xi), x\rangle + \sum_{t=1}^{T}\langle c_t, y_t(\xi_{[t]})\rangle\right] \text{ s.t. } A_0(\xi)x + \sum_{t=1}^{T} A_t y_t(\xi_{[t]}) \geq b(\xi) \quad (49)$$

with fixed recourse, where the cost coefficients c_t and the matrices A_t, $t \geq 1$, are not affected by uncertainty, as reflected in the notation. Besides this, in what follows we assume that the data affected by the uncertainty (that is, $c_0(\xi)$, $A_0(\xi)$, $b(\xi)$) are *affine* functions of ξ; as we remember from the previous section, this "assumption" is in fact a convention on how we use words: nobody forbids us to treat as the actual "random parameter" the collection $(c(\xi), A_0(\xi), b(\xi))$ rather than ξ itself.

As we have explained, a multistage problem (even much better structured than (49)) is, generically, "severely computationally intractable". We are about to propose a radical way to reduce the complexity of the problem, specifically, to pass from *arbitrary* decision rules $y_t(\cdot)$ to *affine* ones:

$$y_t(\xi) = x_t^0 + X_t Q_t \xi, \quad (50)$$

where x_t^0, X_t are our new – deterministic! – variables (a vector and a matrix of appropriate sizes), and $Q_t \xi$, Q_t being a given deterministic matrix, is the "portion" of uncertainty which is revealed at time t and thus can be used to make the decision y_t [18].

Now let us look at the problem we end up with. When substituting linear decision rules (50) into the constraint of (49), the constraint takes the form

$$\operatorname{Prob}\left\{ A_0(\xi)x + \sum_{t=1}^{T}\left[A_t x_t^0 + A_t X_t Q_t \xi\right] - b(\xi) \geq 0\right\} = 1.$$

The left hand side of the system of inequalities in the latter Prob$\{\cdot\}$ is affine in ξ, thus, the constraint in question says *exactly* that the system should be satisfied for all ξ from the support Ξ of the distribution P of ξ. Since the left hand side of the system is affine in ξ, the latter requirement is equivalent to the system to be valid for all $\xi \in \mathcal{Z}$, where $\mathcal{Z}$ is the closed convex hull of Ξ. Thus, the constraint of (49) is nothing but the semi-infinite system of linear inequalities

$$A_0(\xi)x + \sum_{t=1}^{T}\left[A_t x_t^0 + A_t X_t Q_t \xi\right] - b(\xi) \geq 0 \quad \forall \xi \in \mathcal{Z} \quad (51)$$

in variables $w = \{x, \{x_t^0, X_t\}_{t=1}^{T}\}$. Besides this, the coefficients of the semi-infinite inequalities in (51) depend affinely on ξ. Now let us use the following known fact (see [BN98]):

(!) *Assume that $\mathcal{Z}$ is a polyhedral set*

---

[18] In the notation of (32), $Q_t \xi = \xi_{[t]} = (\xi_1, ..., \xi_t)$.

$$\mathcal{Z} = \{\xi : \exists \eta \text{ such that } M\xi + N\eta + p \geq 0\},$$

given by the data M, N, p. Then the semi-infinite system (51) *is equivalent to a finite system $\mathcal{S}$ of linear inequalities:*

$$w \text{ satisfies } (51) \quad \Leftrightarrow \quad \exists u : \mathcal{A}w + \mathcal{B}u + q \geq 0.$$

The sizes of $\mathcal{S}$ (that is, the row and the column sizes of $\mathcal{A}, \mathcal{B}$) are polynomial in the sizes of the matrices A_0, $A_1,...,A_T$, M, N, and the data $\mathcal{A}, \mathcal{B}, q$ of $\mathcal{S}$ are readily given by the data of (51) *and M, N, p (that is, given the latter data, one can build $\mathcal{S}$ in polynomial time).*
In fact, [BN98] asserts much more than stated by (!), namely, that (51) is computationally tractable whenever $\mathcal{Z}$ is so. We, however, intend to stay within the grasp of Linear Programming, and to this end (!) is exactly what we need.
Example: interval uncertainty. Assume that $\mathcal{Z}$ is a box; without loss of generality, we may assume that $\mathcal{Z} = \{\xi : -1 \leq \xi_i \leq 1, \, i = 1,...,d\}$. Since $A_0(\xi)$, $b(\xi)$ are affine in ξ, (51) can be rewritten equivalently as the semi-infinite problem

$$s_0^j[X] + \sum_{i=1}^{d} s_i^j[X]\xi_i \leq 0 \; \forall \xi \in \mathcal{Z}, \, j = 1,...,J, \qquad (52)$$

where X stands for the collection $\{x, \{x_t^0, X_t\}_{t=1}^T\}$ of design variables in (51), and $s_i^j[X]$ are affine functions of X readily given by the data of (51). With our $\mathcal{Z}$, the semi-infinite system (52) is clearly equivalent to the system of constraints

$$s_0^j[X] + \sum_{i=1}^{d} |S_i^j[X]| \leq 0, \, j = 1,...,J,$$

that is, to an explicit system of convex constraints (which can be further straightforwardly converted to a system of linear inequalities).

By the outlined analysis, *when restricted to affine decision rules,* (49) *becomes an explicit deterministic linear program*

$$\text{Min}_{w=\{x,\{x_t^0,X_t\},u\}} \{\langle c, w \rangle : \mathcal{A}w + \mathcal{B}u + q \geq 0\},$$
$$\langle c, w \rangle \equiv \mathbb{E}\left\{\langle c_0(\xi), x \rangle + \sum_{t=1}^T \langle c_t, [x_t^0 + X_t P_t \xi]\rangle\right\}. \qquad (53)$$

in variables $w = \{x, \{x_t^0, X_t\}_{t=1}^T\}$.
Several remarks are in order.

Remark 2. The only reason for restricting ourselves with affine decision rules stems from the desire to end up with a computationally tractable problem. We do not pretend that affine decision rules approximate well the optimal ones – whether it is so or not, it depends on the problem, and we usually have no possibility to understand how good in this respect is a particular problem we should solve. The rationale behind restricting to affine decision rules is the belief that in actual applications it is better to pose a modest and achievable goal rather than an ambitious goal which we do not know how to achieve[19]

Remark 3. To some extent, what is affine and what is not is a matter of how we use words. Assume, e.g., that one wants to pass from affine decision rules to quadratic ones. This is exactly the same as to keep the rules affine and to add to the entries of ξ their pairwise products, and similarly for more complicated families of decision rules. Statement (!) explains what are the "limits of sophistication in the decision rules" we can achieve: representing a sophisticated decision rule as an affine one, the uncertainty vector ξ being properly extended, we need the convex hull of the support of this extended vector to be computationally tractable. In principle, this might be not the case already for "genuinely affine" decision rules; however, in typical applications distribution P of the "actual" uncertainty ξ is simple enough, so that $\mathrm{Conv}(\mathrm{supp}P)$ is computationally tractable. However, with P as simple as a uniform distribution on a box, the "quadratic extension" $\xi \mapsto (\xi, \{\xi_i \xi_j\}_{i,j})$ of ξ results in random vector with a distribution too complicated, as far as our needs are concerned. Thus, the limitations of affine decision rules are in fact limitations of our possibility to describe efficiently convex hulls of supports of nonlinear transformations of ξ.

Remark 4. One could bet that the idea of multi-stage decision making under uncertainty via linear decision rules is as old as the corresponding optimization model. It seems, however, that this idea remained completely forgotten for a long time; at least, we do not know who should be credited with it. Linear decision rules in optimization under uncertainty were recently "resurrected" in [BGGN04] in the framework of Robust Optimization. Our exposition follows the methodology developed in [BGGN04], with the only minor exception that in Robust Optimization one is aimed at minimizing the *worst-case* value of an uncertainty-affected objective under the restriction that a candidate solution remains feasible whatever be a realization of uncertainty-affected constraints, while here we intend to optimize, under the same restriction, the *expected* value of the objective.

Remark 5. We have assumed that (49) has a fixed recourse; the role of this assumption was to ensure affinity of the constraints in (51) in ξ, which in turn

---

[19] In this respect, it is very instructive to look at Control, where the idea of linear feedback dominates theoretical research, and, to some extent, applications. Aside of a handful of simple particular cases, there are no reasons to believe that "the abilities" of linear feedback are as good as those of a general nonlinear feedback. However, Control community realized long ago that a bird in the hand is worth two in the bush – it is much better to restrict ourselves with something which we indeed can analyze and process numerically. We believe this is an instructive example for the optimization community.

made it possible to use (!) in order to end up with tractable reformulation (53) of the problem of interest. In the case when the recourse is not fixed, that is, the matrices A_t, $t \geq 1$, in (49) depend affinely on ξ, the situation becomes much more complicated – the left hand sides of the inequalities in (51) become quadratic in ξ, which makes (!) inapplicable[20]. It turns out, however, that under not too restrictive assumptions the problem of optimizing under the constraints (51), although NP-hard, admits tractable approximations of reasonable quality [BGGN04].

Remark 6. Passing from arbitrary decision rules to affine ones seems to reduce dramatically the flexibility of our decision-making and thus – the expected results. Note, however, that the numerical results for inventory management models reported in [BGGN04, BGNV04] demonstrate that affinity may well be not as a severe restriction as one could expect it to be. In any case, we believe that when processing multi-stage problems, affine decision rules make a good and easy-to-implement starting point, and that it hardly makes sense to look for more sophisticated (and by far more computationally demanding) decision policies, unless there exists a clear indication of "severe non-optimality" of the affine rules.

References

[ADEH99] Artzner, P., Delbaen, F., Eber, J.-M., Heath, D.: Coherent measures of risk. Mathematical Finance, **9**, 203–228 (1999)

[ADEHK03] Artzner, P., Delbaen, F., Eber, J.-M., Heath, D. Ku, H.: Coherent multiperiod risk measurement, Manuscript, ETH Zürich (2003)

[BL97] Barmish, B.R., Lagoa, C.M.: The uniform distribution: a rigorous justification for the use in robustness analysis. Math. Control, Signals, Systems, **10**, 203–222 (1997)

[Bea55] Beale, E.M.L.: On minimizing a convex function subject to linear inequalities. Journal of the Royal Statistical Society, Series B, **17**, 173–184 (1955)

[BN98] Ben-Tal, A., Nemirovski, A.: Robust convex optimization. Mathematics of Operations Research, **23** (1998)

[BN01] Ben-Tal, A., Nemirovski, A.: Lectures on Modern Convex Optimization. SIAM, Philadelphia (2001)

[BGGN04] Ben-Tal, A., Goryashko, A., Guslitzer, E., Nemirovski, A.: Adjustable robust solutions of uncertain linear Programs. Mathematical Programming, **99**, 351–376 (2004)

[BGNV04] Ben-Tal, A., Golany, B., Nemirovski, A., Vial J.-Ph.: Retailer-supplier flexible commitments contracts: A robust optimization approach. Submitted to Manufacturing & Service Operations Management (2004)

[CC05] Calafiore G., Campi, M.C.: Uncertain convex programs: Randomized solutions and confidence levels. Mathematical Programming, **102**, 25–46 (2005)

---

[20] In fact, in this case the semi-infinite system (51) can become NP-hard already with $\mathcal{Z}$ as simple as a box [BGGN04].

[CC04]   Calafiore, G., Campi, M.C.: Decision making in an uncertain environment: the scenariobased optimization approach. Working paper (2004)

[CC59]   Charnes, A., Cooper, W.W.: Uncertain convex programs: randomized solutions and confidence levels. Management Science, **6**, 73–79 (1959)

[DLMV88]   Dagum, P., Luby, L., Mihail, M., Vazirani, U.: Polytopes, Permanents, and Graphs with Large Factors. Proc. 27th IEEE Symp. on Fondations of Comput. Sci. (1988)

[Dan55]   Dantzig, G.B.: Linear programming under uncertainty. Management Science, **1**, 197–206 (1955)

[DZ98]   Dembo, A., Zeitouni, O.: Large Deviations Techniques and Applications. Springer-Verlag, New York, NY (1998)

[Dup79]   Dupačová, J.: Minimax stochastic programs with nonseparable penalties. In: Optimization techniques (Proc. Ninth IFIP Conf., Warsaw, 1979), Part 1, **22** of Lecture Notes in Control and Information Sci., 157–163. Springer, Berlin (1980)

[Dup87]   Dupačová, J.: The minimax approach to stochastic programming and an illustrative application. Stochastics, **20**, 73–88 (1987)

[DS03]   Dyer, M., Stougie, L.: Computational complexity of stochastic programming problems. SPOR-Report 2003-20, Dept. of Mathematics and Computer Sci., Eindhoven Technical Univ., Eindhoven (2003)

[ER05]   Eichhorn, A., Römisch, W.: Polyhedral risk measures in stochastic programming. SIAM J. Optimization, to appear (2005)

[EGN85]   Ermoliev, Y., Gaivoronski, A., Nedeva, C.: Stochastic optimization problems with partially known distribution functions. SIAM Journal on Control and Optimization, **23**, 697–716 (1985)

[FS02]   Föllmer, H., Schied, A.: Convex measures of risk and trading constraints. Finance and Stochastics, **6**, 429–447 (2002)

[KSH01]   Kleywegt, A.J., Shapiro, A., Homem-De-Mello, T.: The sample average approximation method for stochastic discrete optimization. SIAM Journal of Optimization, **12**, 479–502 (2001)

[Gai91]   Gaivoronski, A.A.: A numerical method for solving stochastic programming problems with moment constraints on a distribution function. Annals of Operations Research, **31**, 347–370 (1991)

[JV96]   Jerrum, M., Vazirani, U.: A mildly exponential approximation algorithm for the permanent. Algorithmica, **16**, 392–401 (1996)

[LSW05]   Linderoth, J., Shapiro, A., Wright, S.: The empirical behavior of sampling methods for stochastic programming. Annals of Operations Research, to appear (2005)

[LSW00]   Linial, N., Samorodnitsky, A., Wigderson, A.: A deterministic strongly poilynomial algorithm for matrix scaling and approximate permanents. Combinatorica, **20**, 531–544 (2000)

[MMW99]   Mak, W.K., Morton, D.P., Wood, R.K.: Monte Carlo bounding techniques for determining solution quality in stochastic programs. Operations Research Letters, **24**, 47–56 (1999)

[Mar52]   Markowitz, H.M.: Portfolio selection. Journal of Finance, **7**, 77–91 (1952)

[Lan87]   H.J. Landau (ed): Moments in mathematics. Proc. Sympos. Appl. Math., **37**. Amer. Math. Soc., Providence, RI (1987)

[Nem03]   Nemirovski, A.: On tractable approximations of randomly perturbed convex constraints – Proceedings of the 42nd IEEE Conference on Decision and Control. Maui, Hawaii USA, December 2003, 2419-2422 (2003)

[NS05]   Nemirovski, A., Shapiro, A.: Scenario approximations of chance con-
           straints. In: Calafiore, G., Dabbene, F., (eds) Probabilistic and Random-
           ized Methods for Design under Uncertainty. Springer, Berlin (2005)
[Pre95]   Prékopa, A.: Stochastic Programming. Kluwer, Dordrecht, Boston (1995)
[Rie03]   Riedel, F.: Dynamic coherent risk measures. Working Paper 03004, De-
           partment of Economics, Stanford University (2003)
[RUZ02]   Rockafellar, R.T., Uryasev, S., Zabarankin, M.: Deviation measures in
           risk analysis and optimization, Research Report 2002-7, Department of
           Industrial and Systems Engineering, University of Florida (2002)
[RS04a]   Ruszczyński, A., Shapiro, A.: Optimization of convex risk functions. E-
           print available at: http://www.optimization-online.org (2004)
[RS04b]   Ruszczyński, A., Shapiro, A.: Conditional risk mappings. E-print available
           at: http://www.optimization-online.org (2004)
[SAGS05]   Santoso, T., Ahmed, S., Goetschalckx, M., Shapiro, A.: A stochastic pro-
           gramming approach for supply chain network design under uncertainty.
           European Journal of Operational Research, 167, 96–115 (2005)
[SH00]   Shapiro, A., Homem-de-Mello, T.: On rate of convergence of Monte Carlo
           approximations of stochastic programs. SIAM Journal on Optimization,
           11, 70–86 (2000)
[SK00]   Shapiro, A., Kleywegt, A.: Minimax analysis of stochastic programs. Op-
           timization Methods and Software, 17, 523–542 (2002)
[SHK02]   Shapiro, A., Homem de Mello, T., Kim, J.C.: Conditioning of stochastic
           programs. Mathematical Programming, 94, 1–19 (2002)
[Sha03a]   Shapiro, A.: Inference of statistical bounds for multistage stochastic pro-
           gramming problems. Mathematical Methods of Operations Research. 58,
           57–68 (2003)
[Sha03b]   Shapiro, A.: Monte Carlo sampling methods. In: Ruszczyński, A., Shapiro,
           A. (eds) Stochastic Programming, volume 10 of Handbooks in Operations
           Research and Management Science. North-Holland (2003)
[Sha04]   Shapiro, A.: Worst-case distribution analysis of stochastic programs. E-
           print available at: http://www.optimization-online.org (2004)
[Sha05a]   Shapiro, A.: Stochastic programming with equilibrium constraints. Jour-
           nal of Optimization Theory and Applications (to appear). E-print avail-
           able at: http://www.optimization-online.org (2005)
[Sha05b]   Shapiro, A.: On complexity of multistage stochastic programs.
           Operations Research Letters (to appear). E-print available at:
           http://www.optimization-online.org (2005)
[TA04]   Takriti, S., Ahmed, S.: On robust optimization of two-stage systems.
           Mathematical Programming, 99, 109–126 (2004)
[VAKNS03] Verweij, B., Ahmed, S., Kleywegt, A.J., Nemhauser, G., Shapiro, A.:
           The sample average approximation method applied to stochastic rout-
           ing problems: a computational study. Computational Optimization and
           Applications, 24, 289–333 (2003)
[Val79]   Valiant, L.G.: The complexity of computing the permanent. Theoretical
           Computer Science, 80, 189–201 (1979)
[Zac66]   Žáčková, J.: On minimax solutions of stochastic linear programming prob-
           lems. Čas. Pěst. Mat., 91, 423–430 (1966)

Nonlinear Optimization in Modeling Environments

Software Implementations for Compilers, Spreadsheets, Modeling Languages, and Integrated Computing Systems

János D. Pintér

Pintér Consulting Services, Inc.
129 Glenforest Drive, Halifax, NS, Canada B3M 1J2
jdpinter@hfx.eastlink.ca
http://www.pinterconsulting.com

Summary. We present a review of several professional software products that serve to analyze and solve nonlinear (global and local) optimization problems across a variety of hardware and software environments. The product versions discussed have been implemented for compiler platforms, spreadsheets, algebraic (optimization) modeling languages, and for integrated scientific-technical computing systems. The discussion highlights some of the key advantages of these implementations. Test examples, well-known numerical challenges and client applications illustrate the usage of the current software versions.

Key words: nonlinear (convex and global) optimization; LGO solver suite and its implementations; compiler platforms, spreadsheets, optimization modeling languages, scientific-technical computing systems; illustrative applications and case studies.

2000 MR Subject Classification. 65K30, 90C05, 90C31.

1 Introduction

Nonlinearity is literally ubiquitous in the development of natural objects, formations and processes, including also living organisms of all scales. Consequently, nonlinear descriptive models – and modeling paradigms even beyond a straightforward (analytical) function-based description – are of relevance in many areas of the sciences, engineering, and economics. For example, [BM68, Ric73, EW75, Man83, Mur83, Cas90, HJ91, Sch91, BSS93, Ste95, Gro96, PSX96, Pin96a, Ari99, Ber99, Ger99, Laf00, PW00, CZ01, EHL01, Jac01,

Sch02, TS02, Wol02, Diw03, Zab03, Neu04b, HL05, KP05, Pin05a, Pin05b] – as well as many other authors – present discussions and an extensive repertoire of examples to illustrate this point.

Decision-making (optimization) models that incorporate such a nonlinear system description frequently lead to complex models that (may or provably do) have multiple – local and global – optima. The objective of global optimization (GO) is to find the "absolutely best solution of nonlinear optimization (NLO) models under such circumstances.

The most important (currently available) GO model types and solution approaches are discussed in the *Handbook of Global Optimization* volumes, edited by Horst and Pardalos [HP95], and by Pardalos and Romeijn [PR02]. As of 2004, over a hundred textbooks and a growing number of informative web sites are devoted to this emerging subject.

We shall consider a general GO model form defined by the following ingredients:

- x decision vector, an element of the real Euclidean n-space R^n;
- $f(x)$ continuous objective function, $f : R^n \to R^1$;
- D non-empty set of admissible decisions, a proper subset of R^n.

The feasible set D is defined by

- l, u explicit, finite vector bounds of x (a "box" in R^n);
- $g(x)$ m-vector of continuous constraint functions, $g : R^n \to R^m$.

Applying the notation introduced above, the continuous global optimization (CGO) model is stated as

$$\min f(x) \quad \text{s.t.} \quad x \text{ belongs to} \tag{1}$$
$$D = \{x : l \leq x \leq u, g(x) \leq 0\}. \tag{2}$$

Note that in (2) all vector inequalities are meant component-wise (l, u, are n-vectors and the zero denotes an m-vector). Let us also remark that the set of the additional constraints g could be empty, thereby leading to – often much simpler, although still potentially multi-extremal – box-constrained models. Finally, note that formally more general optimization models (that include also = and $\geq$ constraint relations and/or explicit lower bounds on the constraint function values) can be simply reduced to the canonical model form (1)–(2). The canonical model itself is already very general: in fact, it trivially includes linear programming and convex nonlinear programming models (under corresponding additional specifications). Furthermore, it also includes the entire class of pure and mixed integer programming problems, since all (bounded) integer variables can be represented by a corresponding set of binary variables; and every binary variable $y \in \{0, 1\}$ can be equivalently represented by its continuous extension $y \in [0, 1]$ and the non-convex constraint $y(1 - y) \leq 0$. Of course, we do not claim that the above approach is best – or even suitable – for "all" optimization models: however, it certainly shows the generality of the CGO modeling framework.

Let us observe next that the above stated "minimal" analytical assumptions already guarantee that the optimal solution set X^* in the CGO model is non-empty. This key existence result directly follows by the classical theorem of Weierstrass (that states the existence of the minimizer point (set) of a continuous function over a non-empty compact set). For reasons of numerical tractability, the following additional requirements are also often postulated:

- D is a full-dimensional subset ("body") in R^n;
- the set of globally optimal solutions to (1)–(2) is at most countable;
- f and g (the latter component-wise) are Lipschitz-continuous functions on $[l, u]$.

Note that the first two of these requirements support the development and (easier) implementation of globally convergent algorithmic search procedures. Specifically, the first assumption – i.e., the fact that D is the closure of its non-empty interior – makes algorithmic search possible within the set D. This requirement also implies that e.g., nonlinear equality constraints need to be directly incorporated into the objective function as discussed in [Pin96a], Chapter 4.1.

With respect to the second assumption, let us note that in most well-posed practical problems the set of global optimizers consists only of a single point, or at most of several points. However, in full generality, GO models may have even manifold solution sets: in such cases, software implementations will typically find a single solution, or several of them. (There are theoretically straightforward iterative ways to provide a sequence of global solutions.)

The third assumption is a sufficient condition for estimating f^* on the basis of a finite set of feasible search points. (Recall that the real-valued function h is Lipschitz-continuous on its domain of definition $D \subset R^n$, if $|h(x_1) - h(x_2)| \leq L\|x_1 - x_2\|$ holds for all pairs $x_1 \in D, x_2 \in D$; here $L = L(D, h)$ is a suitable Lipschitz-constant of h on the set D: the inequality above directly supports lower bound estimates on sets of finite size.) We emphasize that the factual knowledge of the smallest suitable Lipschitz-constant – for each model function – is not required, and in practice such information is typically unavailable indeed.

Let us remark here that e.g., models defined by continuously differentiable functions f and g certainly belong to the CGO or even to the Lipschitz model class. In fact, even such "minimal" smooth structure is not essential: since e.g., "saw-tooth" like functions are also Lipschitz-continuous. This comment also implies that CGO indeed covers a very general class of optimization models. As a consequence of this generality, the CGO model class includes also many extremely difficult instances. To perceive this difficulty, one can think of model-instances that would require "the finding of the lowest valley across a range of islands" (since the feasible set may well be disconnected), based on an intelligent (adaptive, automatic), but otherwise completely "blind" sampling procedure...

For illustration, a merely one-dimensional, box-constrained model is shown in Fig. 1. This is a frequently used classical GO test problem, due to Shubert: it is defined as

$$\min \sum_{k=1,\ldots,5} k \sin(k + (k+1)x) \qquad 10 \leq x \leq 10.$$

Fig. 1. One-dimensional, box-constrained CGO model

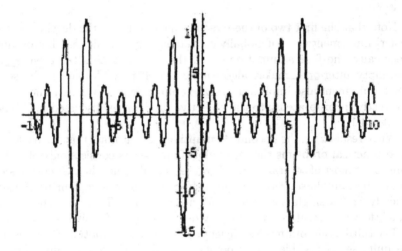

Model complexity may – and frequently will – increase dramatically, already in (very) low dimensions. For example, both the amplitude and the frequency of the trigonometric components in the model of Figure 1 could be increased arbitrarily, leading to more and more difficult problems.

Furthermore, increasing dimensionality *per se* can lead to a tremendous – theoretically exponential – increase of model complexity (e.g., in terms of the number of local/global solutions, for a given type of multi-extremal models). To illustrate this point, consider the – merely two-dimensional, box-constrained, yet visibly challenging – objective function shown in Fig. 2 below. The model is based on Problem 4 of the *Hundred-Dollar, Hundred-Digit Challenge Problems* [Tre02], and it is stated as

$$\min \frac{(x^2 + y^2)}{4} + \exp(\sin(50x)) - \sin(10(x+y)) + \sin(60\exp(y))$$
$$+ \sin(70\sin(x)) + \sin(\sin(80y))$$
$$-3 \leq x \leq 3 \quad -3 \leq y \leq 3.$$

Fig. 2. Two-dimensional, box-constrained CGO model

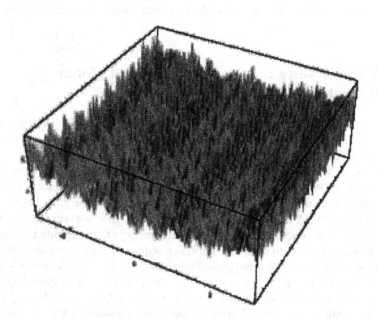

Needless to say, not all – and especially not all *practically motivated* – CGO models are as difficult as indicated by Figures 1 and 2. At the same time, we do not always have the possibility to directly inspect and estimate the difficulty of an optimization model, and perhaps unexpected complexity can be met under such circumstances. An important case in point is when the software user (client) has a confidential or otherwise visibly complex model that needs to be analyzed and solved. The model itself can be presented to the solver engine as an object code, dynamic link library (dll), or even as an executable program: in such situations, direct model inspection is simply not an option. In many other cases, the evaluation of the optimization model functions may require the numerical solution of a system of differential equations, the evaluation of special functions or integrals, the execution of a complex system of program code, stochastic simulation, even some physical experiments, and so on.

Traditional numerical optimization methods – discussed in most topical textbooks such as e.g. [BSS93, Ber99, CZ01] – search only for local optima. This approach is based on the tacit assumption that a "sufficiently good" initial solution (that is located in the region of attraction of the "true" solution) is immediately available. Both Fig. 1 and Fig. 2 suggest that this may not always be a realistic assumption ... Models with less "dramatic" difficulty, but in (perhaps much) higher dimensions also imply the need for global optimization. For instance, in advanced engineering design, models with hundreds or thousands of variables and constraints are analyzed. In similar cases to those mentioned above, even an approximately completed, but genuine global (exhaustive) search strategy may – and typically will – yield better results than the most sophisticated local search approach when "started from the wrong valley"...

2 A solver suite approach to practical global optimization

The general development philosophy followed by the software implementations discussed here is based on the seamless combination of rigorous (i.e., theoretically convergent) global and efficient local search strategies.

As it is well-known ([HT96, Pin96a]), the existence of valid overestimates of the actual (smallest possible) Lipschitz-constants, for f and for each component of g in the model (1)-(2), is sufficient to guarantee the global convergence of suitably defined adaptive partition algorithms. In other words, the application of a proper branch-and-bound search strategy (that exploits the Lipschitz information referred to above) generates a sequence of sample points that converges exactly to the (unique) global solution $x^* = \{X^*\}$ of the model instance considered. If the model has a finite or countable number of global solutions, then – theoretically, and under very general conditions – sub-sequences of search points are generated that respectively converge to the points of X^*. For further details related to the theoretical background, including also a detailed discussion of algorithm implementation aspects, consult [Pin96a] and references therein.

In numerical practice, deterministically guaranteed global convergence means that after a finite number of search steps – i.e., sample points and corresponding function evaluations – one has an incumbent solution (with a corresponding upper bound of the typically unknown optimum value), as well as a verified lower bound estimate. Furthermore, the "gap" between these estimates converges to zero, as the number of generated search points tends to infinity. For instance, interval arithmetic based approaches follow this avenue: consult, e.g., [RR95, Kea96, Neu04b]; [CK99] review a number of successful applications of rigorous search methods.

The essential difficulty of applying such rigorous approaches to "all" GO models is that their computational demand typically grows at an exponential pace with the size of the models considered. For example, the Lipschitz information referred to above is often not precise enough: "carefree" overestimates of the best possible (smallest) Lipschitz-constant lead to a search procedure that will, in effect, be close in efficiency to a passive uniform grid search. For this reason, in a practical GO context, other search strategies also need to be considered.

It is also well-known that properly constructed stochastic search algorithms also possess general theoretical global convergence properties (with probability 1): consult, for instance, the review of [BR95], or [Pin96a]. For a very simple example that illustrates this point, one can think of a pure random search mechanism applied in the interval $l \leq x \leq u$ to solve the CGO model: this will eventually converge, if the "basin of attraction" of the (say, unique) global optimizer x^* has a positive volume. In addition, stochastic sampling methods can also be directly combined with search steps of other – various global and efficient local – search strategies, and the overall global convergence of such strategies will be still maintained. The theoretical background of stochastic "hybrid" algorithms is discussed by [Pin96a]. The underlying general convergence theory of such combined methods allows for a broad range of implementations. In particular, a hybrid optimization program system supports the flexible usage of a selection of component solvers: one can execute a fully automatic global or local search based optimization run, can combine solvers, and can also design various interactive runs.

Obviously, there remains a significant issue regarding the (typically unforeseeable best) "switching point" from strategy to strategy: this is however, unavoidable, when choosing between theoretical rigor and numerical efficiency. (Even local nonlinear solvers would need, in theory, an infinite iterative procedure to converge, except in idealized special cases.) For example, in the stochastic search framework outlined above, it would suffice to find just one sample point in the "region of attraction" of the (unique) global solution x^*, and then that solution estimate could be refined by a suitably robust and efficient local solver. Of course, the region of attraction of x^* (e.g., its shape and relative size) is rarely known, and one needs to rely on computationally expensive estimates of the model structure (again, the reader is referred, e.g., to the review of [BR95]). Another important numerical aspect is that one loses the deterministic (lower) bound guarantees when applying a stochastic search procedure: instead, suitable statistical estimation methods can be applied, consult [Pin96a] and topical references therein. Again, the implementation of such methodology is far from trivial.

To summarize the discussion, there are good reasons to apply various search methods and heuristic global-to-local search "switching points" with a reasonable expectation of numerical success. Namely,

- one needs to apply proper global search methods to generate an initial good "coverage" of the search space;
- it is also advantageous to apply quality local search that enables the fast improvement of solution estimates generated by a preceding global search phase;
- using several – global or local – search methods based on different theoretical strategies, one has a better chance to find quality solutions in difficult models (or ideally, confirm the solution by comparing the results of several solver runs);
- one can always place more or less emphasis on rigorous search vs. efficiency, by selecting the appropriate solver combination, and by allocating search effort (time, function evaluations);

- we often have a priori knowledge regarding good quality solutions, based on practical, model-specific knowledge (for example, one can think of solving systems of equations: here a global solution that "nearly" satisfies the system can be deemed as a sufficiently good point from which local search can be directly started);
- practical circumstance and resource limitations may (will) dictate the use of additional numerical stopping and switching rules that can be flexibly built into the software implementation.

Based on the design philosophy outlined – that has been further confirmed and dictated by practical user demands – we have been developing for over a decade nonlinear optimization software implementations that are based on global and local solver combinations. The currently available software products will be briefly discussed below with illustrative examples; further related work is in progress.

3 Modeling systems and user demands

Due to advances in modeling, optimization methods and computer technology, there has been a rapidly growing interest towards modeling languages and environments. Consult, for example, the topical *Annals of Operations Research* volumes [MM95, MMS97, VMM00, CFO01], and the volume [Kal04]. Additional useful information can be found, for example, at the web sites [Fou04, MS04, Neu04a].

Prominent examples of widely used modeling systems that are focused on optimization include AIMMS ([PDT04]), AMPL ([FGK93]), GAMS ([BKM88]), the Excel Premium Solver Platform ([FS01]), ILOG ([I04]), the LINDO Solver Suite ([LS96]), MPL ([MS02]), and TOMLAB ([TO04]). (Please note that the literature references cited may not always reflect the current status of the modeling systems listed: for the latest information, contact the developers and/or visit their website.)

In addition, there exists also a large variety of core compiler platform-based solver systems with some built-in model development functionality: in principle, these all can be linked to the modeling languages listed above. At the other end of the spectrum, there is also significant development related to fully integrated scientific and technical computing (ISTC) systems such as Maple ([M04a]), *Mathematica* ([WR04]), and MATLAB ([TM04]). The ISTCs also incorporate a growing range of optimization-related functionality, supplemented by application products.

The modeling environments listed above are aimed at meeting the needs and demands of a broad range of clients. Major client groups include educational users (instructors and students); research scientists, engineers, economists, and consultants (possibly, but not necessarily equipped with an in-depth optimization related background); optimization experts, vertical application developers, and other "power users". Obviously, the user categories listed above are not necessarily disjoint: e.g., someone can be an expert researcher and software developer in a certain professional area, with a perhaps more modest optimization expertise. The pros and cons of the individual software products – in terms of ease of model prototyping, detailed code development and maintenance, optimization model processing tools, availability of solvers and other auxiliary tools, program execution speed, overall level of system integration, quality of related documentation and support – make such systems more or less attractive for the user groups listed.

It is also worth mentioning at this point that – especially in the context of nonlinear modeling and optimization – it can be a salient idea to tackle challenging problems by making use of several modeling systems and solver tools, if available. In general, dense NLO model formulations are far less easy to "standardize" than linear or even mixed integer linear models, since one typically needs an explicit, specific formula to describe a particular model function. Such formulae are relatively straightforward to transfer from one modeling system into another: some of the systems listed above even have such built-in converter capabilities, and their syntaxes are typically quite similar (whether it is $x^{**}2$ or x^2, $\sin(x)$ or $\mathrm{Sin}[x]$, bernoulli(n, x) or Bernoulli$B[n, x]$, and so on).

In subsequent sections we shall summarize the principal features of several current nonlinear optimization software implementations that have been developed with quite diverse user groups in mind. The range of products reviewed in this work includes the following:

- LGO Solver System with a Text I/O Interface
- LGO Integrated Development Environment
- LGO Solver Engine for Excel
- MathOptimizer Professional (LGO Solver Engine for *Mathematica*)
- Maple Global Optimization Toolbox (LGO Solver Engine for Maple).

We will also present relatively small, but non-trivial test problems to illustrate some of the key functionality of these implementations.

Note that all software products discussed are professionally developed and supported, and that they are commercially available. For this reason – and also in line with the objectives of this paper – some of the algorithmic technical details are only briefly mentioned. Additional technical information is available upon request; please consult also the publicly available references, including the software documentation and topical web sites.

In order to keep the length of this article within reasonable bounds, further product implementations not discussed here are

- LGO Solver Engine for GAMS
- LGO Solver Engine for MPL
- TOMLAB/LGO for MATLAB
- MathOptimizer for *Mathematica*.

With respect to these products, consult e.g. the references [Pin02a, PK03, KP04b, KP05, PHGE04, PK05].

4 Software implementation examples

4.1 LGO solver system with a text I/O interface

The Lipschitz Global Optimizer (LGO) software has been developed and used for more than a decade (as of 2004). Detailed technical descriptions and user documentation have appeared elsewhere: consult, for instance, [Pin96a, Pin97, Pin01a, Pin04], and the software review [BS00]. Let us also remark here that LGO was chosen to illustrate global optimization software (in connection with a demo version of the MPL modeling language) in the well-received textbook [HL05].

Since LGO serves as the core of most current implementations (with the exception of one product), we will provide its somewhat more detailed description, followed by concise summaries of the other platform-specific implementations.

In accordance with the approach advocated in Section 2, LGO is based on a seamless combination of a suite of global and local scope nonlinear solvers. Currently, LGO includes the following solver options:

- adaptive partition and search (branch-and-bound) based global search (BB)
- adaptive global random search (single-start) (GARS)
- adaptive global random search (multi-start) (MS)
- constrained local search (generalized reduced gradient method) (LS).

The global search methodology was discussed briefly in Section 2; the well-known GRG method is discussed in numerous textbooks, consult e.g. [EHL01]. Note that in all three global search modes the model functions are aggregated by an exact penalty function. By contrast, in the local search phase all model functions are considered and treated individually. Note also that the global search phases are equipped with stochastic sampling procedures that support the usage of statistical bound estimation methods.

All LGO search algorithms are derivative-free: specifically, in the local search phase central differences are used to approximate gradients. This choice reflects again our objective to handle (also) models with merely computable, continuous functions, including "black box" systems.

The compiler-based LGO solver suite is used as an option linked to various modeling environments. In its core text I/O based version, the application-specific LGO executable program (that includes a driver file and the model function file) reads an input text file that contains all remaining application information (model name, variable and constraint names, variable bounds and nominal values, and constraint types), as well as a few key solver options (global solver type, precision settings, resource and time limits). Upon completing the LGO run, a summary and a detailed report file are available. As can be expected, this LGO version has the lowest demands for hardware, it also runs fastest, and it can be directly embedded into vertical and proprietary user applications.

4.2 LGO integrated development environment

LGO can be also equipped – as a readily available implementation option – with a simple, but functional and user-friendly MS Windows interface. This enhanced version is referred to as the LGO Integrated Development Environment (IDE). The LGO IDE provides a menu that supports model development, compilation, linking, execution, and the inspection of results. To this end, a text editor is used that can be chosen optionally such as e.g. the freely downloadable ConTEXT and PFE editors, or others. Note here that even the simple Notebook Windows accessory – or the more sophisticated and still free Metapad text editor – would do. The IDE also includes external program call options and two concise help files: the latter discuss global optimization basics and the main application development steps when using LGO.

As already noted, this LGO implementation is compiler-based: user models can be connected to LGO using one of several programming languages on personal computers and workstations. Currently supported platforms include essentially all professional Fortran 77/90/95 and C compilers and some others: prominent examples are Borland C/C++ and Delphi, Compaq/Digital Visual Fortran; Lahey Fortran 77/90/95; Microsoft Visual Basic and C/C++; and Salford Fortran 77/95. Other customized versions can also be made available upon request, especially since the vendors of development environments often expand the list of compatible platforms.

This LGO software implementation (in both versions discussed above) fully supports communication with sophisticated user models, including entirely closed or confidential "black box" systems. These LGO versions are particularly advantageous in application areas, where program execution (solution) speed is a major concern: in the GO context, many projects fall into this category. The added features of the LGO IDE can also greatly assist in educational and research (prototyping) projects.

LGO deliveries are accompanied by an approximately 60-page User Guide. In addition to installation and technical notes, this document provides a brief introduction to GO; describes LGO and its solvers; discusses the model development procedure, including modeling and solution tips; and reviews a list of applications. The appendices provide examples of the user (main, model, and input parameter) files, as well as of the resulting output files; connectivity issues and workstation implementations are also discussed.

For a simple illustration, we display below the LGO model function file (in C format), and the input parameter file that correspond to a small, but not quite trivial GO model (this is a constrained extension of Shubert's model discussed earlier):

$$\min \sum_{k=1,\ldots,5} k \sin(k + (k+1)x)$$
$$\text{s.t. } x^2 + 3x + \sin(x) \le 6, \quad 10 \le x \le 10.$$

Both files are slightly edited for the present purposes. Note also that in the simplest usage mode, the driver file contains only a single statement that calls LGO: therefore we skip the display of that file. (Additional pre- and post-solver manipulations can also be inserted in the driver file: this can be useful in various customized applications.)

Model function file

```
#include <stdlib.h>
#include <stdio.h>
#include<math.h>

_stdcall USER _FCT( double x[], double fox[1], double gox[])
{
fox[0] = sin(1. + 2.*x[0]) + 2.*sin(2. + 3.*x[0]) + 3.*sin(3.
  + 4.*x[0]) + 4.*sin(4. + 5.*x[0]) + 5.*sin(5. + 6.*x[0]);
gox[0]=-6.+ pow(x[0],2.) + sin(x[0]) + 3.*x[0];
return 0;
}
```

Input parameter file

```
---Model Descriptors---
LGO Model       ! ModelName
1               !Number of Variables
1               ! Number of Constraints
Variable names ! Lower Bounds Nomimal Values Upper Bounds
x               -10. 0. 10.
ObjFct          ! Objective Function Name
Constraint Names and Constraint Types (0 for ==, -1 for <=)
Constraint1       -1


!--- SOLVER OPTIONS AND PARAMETERS ---
1               ! Operational modes 0: LS; 1: BB+LS; 2: GARS
                ! +LS; 3: MS+LS
2000            ! Maximal no. of fct evals in global search
                ! phase
400             ! Maximal no. of fct evals in global search
                ! w/o improvement
```

```
1.              ! Constraint penalty multiplier
-1000000.       ! Target objective function value in global
                ! search phase
-1000000.       ! Target objective function value in local
                ! search phase
0.000001        ! Merit function precision improvement
                ! threshold in local search phase
0.000001        ! Constraint violation tolerance in local
                ! search phase
0.000001        ! Kuhn-Tucker local optimality conditions
                ! tolerance in local search phase
0               ! Built-in random number generator seed value
300             ! Program execution time limit (seconds)
```

Summary result file

```
--- LGO Solver Results Summary ---
Model name: LGO Model
Total number of function evaluations      997
Objective function: ObjFct                -14.8379500257
Solution vector components
1                              x     -1.1140996879
Constraint function values at optimum estimate
1                    Constraint1    -8.9985950759
Solver status indicator value      4     TERMINATED BY
                                         SOLVER
Model status indicator value       1     GLOBALLY OPTIMAL
                                         SOLUTION FOUND
LGO solver system execution time (seconds) 0.01
For additional runtime information, please consult the
LGO.OUT file.
--- LGO application run completed. ---
```

4.3 LGO solver engine for Excel users

The LGO global solver engine for Microsoft Excel has been developed in
cooperation with Frontline Systems [FS01]. For details on the Excel Solver
and the currently available advanced engine options visit Frontline's web site
(www.solver.com). The site contains useful information, including for instance,
tutorial material, modeling tips, and various spreadsheet examples. The User
Guide provides a brief introduction to all current solver engines; discusses the
diagnosis of solver results, solver options and reports; and it also contains
a section on Solver VBA functions. Note that this information can also be
invoked through Excel's on-line help system. In this implementation, LGO
is a field-installable Solver Engine that seamlessly connects to the Premium

Solver Platform: the latter is fully compatible with the standard Excel Solver, but it has enhanced algorithmic capabilities and features.

LGO for Excel, in addition to continuous global and local capabilities, also provides basic support for handling integer variables: this feature has been implemented – as a generic option for all advanced solver engines – by Frontline Systems.

The LGO solver options available are essentially based on the stand-alone "silent" version of the software, with some modifications and added features. The LGO Solver Options dialog, shown by Fig. 3, allows the user to control solver choices and several other settings.

Fig. 3. Excel/ LGO solver engine: solver options and parameters dialog

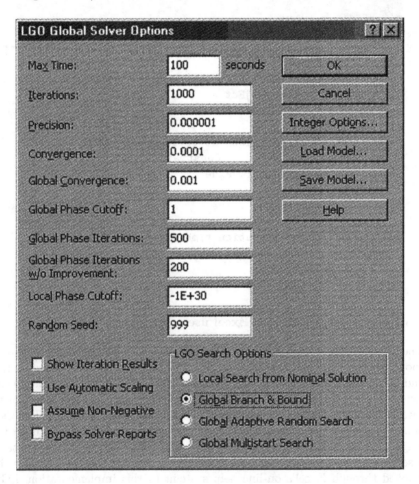

To illustrate the usage of the Excel/LGO implementation, we shall present and solve the *Electrical Circuit Design* (ECD) test problem. The ECD model has been extensively studied in the global optimization literature, as a well-known computational challenge: see, e.g., [RR93], with detailed historical notes and further references.

In the ECD problem, a bipolar transistor is modeled by an electrical circuit: this model leads to the following square system of nonlinear equations

$$a_k(x) = 0 \quad k = 1, \ldots, 4; \quad b_k(x) = 0 \quad k = 1, \ldots, 4; \quad c(x) = 0.$$

The individual equations are defined as follows:

$$a_k(x) = (1 - x_1 x_2) x_3 \{\exp[x_5(g_{1k} - g_{3k}x_7 - g_{5k}x_8)] - 1\} - g_{5k} + g_{4k}x_2,$$
$$k = 1, \ldots, 4;$$
$$b_k(x) = (1 - x_1 x_2) x_4 \{\exp[x_6(g_{1k} - g_{2k} - g_{3k}x_7 + g_{4k}x_9)] - 1\} - g_{5k}x_1 + g_{4k}$$
$$k = 1, \ldots, 4;$$
$$c(x) = x_1 x_3 - x_2 x_4.$$

By assumption, the vector variable x belongs to the box region $[0, 10]^9$. The numerical values of the constants $g_{ik}, i = 1, \ldots, 5, k = 1, \ldots, 4$ are listed in the paper of Ratschek and Rokne [RR93], and will not be repeated here. (Note that, in order to make the model functions more readable, several constants are simply aggregated in the above formulae, when compared to that paper.)

To solve the ECD model rigorously, Ratschek and Rokne applied a combination of interval arithmetic, subdivision and branch-and-bound strategies. They concluded that the rigorous solution was extremely costly (billions of model function evaluations were needed), in order to arrive at a *guaranteed* interval (i.e., embedding box) estimate that is component-wise within at least 10-4 precision of the postulated *approximate* solution:

$$x^* = (0.9, 0.45, 1.0, 2.0, 8.0, 8.0, 5.0, 1.0, 2.0).$$

Obviously, by taking e.g. the Euclidean norm of the overall error in the model equations, the problem of finding the solution can be formulated as a global optimization problem. This model has been set up in a demo spreadsheet, and then solved by the Excel LGO solver engine. The numerical solution found by LGO – directly imported from the answer report – is shown below:

```
Microsoft Excel 10.0 Answer Report
Worksheet: [CircuitDesign_9_9.XLS]Model
Report Created: 12/16/2004
12:39:29 AM
Result: Solver found a solution.  All constraints and
optimality conditions are satisfied.
```

Engine: LGO Global Solver

Target Cell (Min)

Cell	Name	Original Value	Final Value
B21	objective	767671534.2	9.02001E-11

Adjustable Cells

Cell	Name	Original Value	Final Value
D10	x_1	1	0.900000409
D11	x_2	2	0.450000021
D12	x_3	3	1.000000331
D13	x_4	4	2.000001476
D14	x_5	5	7.999999956
D15	x_6	6	7.999998226
D16	x_7	7	4.999999941
D17	x_8	8	1.000000001
D18	x_9	9	1.999999812

The error of the solution found is within 10^{-6} to the verified solution, for each component. The numerical solution of the ECD model in Excel takes less than 5 seconds on a personal computer (Intel Pentium 4, 2.4 GHz processor, 512 Mb RAM). Let us note that we have solved this model also using core LGO implementations with various C and Fortran compilers, with essentially identical success (in about a second or less). Although this finding should not lead per se to overly optimistic claims, it certainly shows the robustness and efficiency of LGO in solving this particular (non-trivial) example.

4.4 MathOptimizer Professional

Mathematica is an integrated environment for scientific and technical computing. This ISTC system supports functional, rule-based and procedural programming styles. *Mathematica* also offers advanced multimedia (graphics, image processing, animation, sound generation) tools, and it can be used to produce publication-quality documentation. For further information, consult the key reference [Wol03]; the website www.wolfram.com provides detailed information regarding also the range of other products and services related to *Mathematica*.

MathOptimizer Professional ([PK03]), combines the model development power of *Mathematica* with the robust performance and efficiency of the LGO solver suite. To this end, the general-purpose interface *MathLink* is used that supports communication between *Mathematica* and external programs. The functionality of *MathOptimizer Professional* is summarized by the following stages (note that all steps are fully automatic, except – obviously – the first one):

- model formulation in *Mathematica*

- translation of the *Mathematica* optimization model into C or Fortran code, to generate the LGO model function file
- generation of the LGO input parameter file
- compilation of the C or Fortran model code into object code or dynamic link library (dll): this step makes use of a corresponding compiler
- call to the LGO solver engine: the latter is typically provided as object code or an executable program that is now linked together with the model function object or dll file
- numerical solution and report generation by LGO
- report of LGO results back to the calling *Mathematica* notebook.

A "side-benefit" of using *MathOptimizer Professional* is that the *Mathematica* models formulated are automatically translated into C or Fortran format: this feature can be put to good use in a variety of contexts. (For example, the LGO model function and input parameter file examples shown in Section 4.2 were generated automatically.)

Let us also remark that the approach outlined supports "only" the solution of models defined in *Mathematica* that can be directly converted into C or Fortran program code. Of course, this model category still allows the handling of a broad range of optimization problems. The approximately 150-page *MathOptimizer Professional* manual is a "live" (notebook) document that can be directly invoked through *Mathematica*'s on-line help system. In addition to basic usage description, the User Guide also discusses a large number of simple and more challenging test problems, and several realistic application examples in detail.

As an illustrative example, we will present the solution of a new – and rather difficult – object packing model: we wish to find (numerically) the "best" non-overlapping arrangement of a set of non-uniform size circles in an embedding circle. Notice that this is not a standard model type (unlike uniform circle packings that have been studied for decades, yet still only special cases are solved to guaranteed optimality). Our approach can be directly generalized to find essentially arbitrary object arrangements.

The best packing is defined here by a combination of two criteria: the radius of the circumscribed circle, and the average pair-wise distance between the centers of the embedded circles. The relative weight of the two objective function components can be selected as a model-instance parameter.

Detailed numerical results are reported in [KP04a], for circles defined by the sequence of radii $ri = i^{-0.5}, i = 1, \ldots, N$, up to $N = 40$-circle configurations. Observe that the required (pair-wise) non-overlapping arrangement leads to $\frac{N(N-1)}{2}$ non-convex constraints, in addition to $2N + 1$ bound constraints on the circle center and circumscribed radius decision variables. Hence, in the 40-circle example, LGO solves this model with nearly 780 non-convex constraints: the corresponding runtime is about 3.5 hours on a P4 1.6 GHz personal computer.

As an illustration, the configuration found for the case of $N = 20$ circles is displayed in Fig. 4. In this example, equal consideration (weight) is given to minimizing the radius of the circumscribed circle and the average distance between the circle centers. As the picture shows, the circumscribed radius is about 2.2: in fact, the numerical value found is ~2.1874712123. Detailed results appeared and will appear in [KP04a] and [KP05], respectively.

Fig. 4. An illustrative non-uniform circle packing result for $N = 20$ circles with radii $r_i = i^{-0.5}, i = 1, \ldots, N$

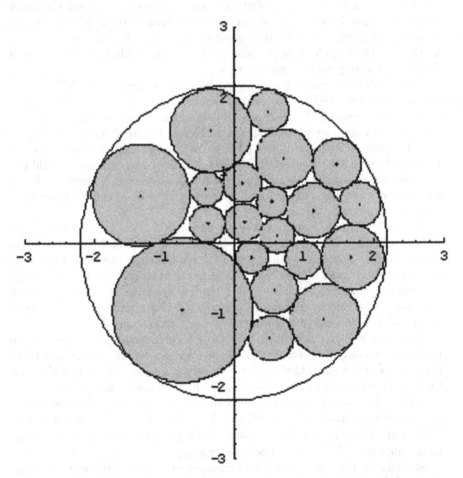

Let us also remark that we have attempted to solve instances of the same circle packing problem applying the built-in *Mathematica* function NMinimize for nonlinear (global) optimization, but – using it in all of its default solver modes – it could not find a solution of acceptable quality already for the case $N = 5$. Again, this is just a numerical observation, as opposed to an "all-purpose" conclusion, to illustrate the quality of the LGO solver suite. We have also conducted detailed numerical studies that provide a more systematic comparison of global solvers available for use with *Mathematica*: these results will appear in [KP05].

Finally, let us mention that *MathOptimizer Professional* is included in a recent peer review of optimization capabilities using *Mathematica* ([Cog03]).

4.5 Maple Global Optimization Toolbox

The integrated computing environment Maple [M04a] enables the development of sophisticated interactive documents that seamlessly combine technical description, calculations, simple and advanced computing, and visualization. Maple includes an extensive mathematical library: its more than 3, 500 built-in functions cover virtually all research areas in the scientific and technical disciplines. Maple also incorporates numerous supporting features and enhancements such as e.g. detailed on-line documentation, a built-in mathematical dictionary with definitions for more than 5000 mathematical terms, debugging tools, automated (ANSI C, Fortran 77, Java, Visual Basic and MATLAB) code generation, and document production (including HTML, MathML, TeX, and RTF converters). All these capabilities accelerate and expand the scope of optimization model development and solution.

To emphasize the key features pertaining to advanced systems modeling and optimization, a concise listing of these capabilities is provided below. Maple

- supports rapid prototyping and model development
- performance scales well to modeling large, complex problems
- offers context-specific "point and click" (essentially syntax-free) operations, including various "Assistants" (these are windows and dialogs that help to execute various tasks)
- has an extensive set of built-in mathematical and computational functions
- has comprehensive symbolic calculation capabilities
- supports advanced computations with arbitrary numeric precision
- is fully programmable, thus extendable by adding new functionality
- has sophisticated visualization and animation tools
- supports the development of GUIs (by using "Maplets")
- supports advanced technical documentation, desktop publishing, and presentation
- provides links to external software products.

Maple is portable across all major hardware platforms and operating systems (Windows, Macintosh, Linux, and Unix versions). Without going into further details that are outside of the scope of the present discussion, we refer to the web site www.maplesoft.com that provides a wealth of further topical information and product demos.

The core of the recently released Global Optimization Toolbox (GOT) is a customized implementation of the LGO solver suite for Maple [M04b]. To this end, LGO was auto-translated into C code, and then fully integrated with Maple. The advantage of this approach is that, in principle, the GOT can handle all (thousands) of functions that are defined in Maple, including their further extensions.

As an illustrative example, let us revisit Problem 4 posted by Trefethen [Tre02]; recall Fig. 2 from Section 1. We can easily set up this model in Maple:

```
> f := exp(sin(50*x1))+sin(60*exp(x2))+sin(70*sin(x1))
       +sin(sin(80*x2))-sin(10*(x1+x2))+(x1^2+x2^2)/4;
```

$$f := \exp(\sin(50x1)) + \sin(60\exp(x2)) + \sin(70\sin(x1)) + \sin(\sin(80x2))$$
$$- \sin(10x1 + 10x2) + \frac{1}{4}x1^2 + \frac{1}{4}x2^2$$

Now using the bounds $[-3, 3]$ for both variables, and applying the Global Optimization Toolbox we receive the numerical solution:

```
> GlobalSolve(f, x1=-3..3, x2=-3..3, evaluationlimit=100000,
      noimprovementlimit=100000);
```

$$[-3.30686864747523535, [x1 = -0.0244030794174338178,$$
$$x2 = 0.210612427162285371]]$$

We can compare the optimum estimate found to the corresponding 40-digit precision value as stated at the website http://web.comlab.ox.ac.uk/oucl/work /nick.trefethen/hundred.html (of Trefethen). The website provides the 40-digit numerical optimum value

$$-3.306868647\ 4752372800\ 7611377089\ 8515657166\ldots$$

Hence, the solution found by the Maple GOT (using default precision settings) is accurate to 15 digits.

It is probably just as noteworthy that one can find a reasonably good solution even in a much larger variable range, with the same solution effort:

```
> GlobalSolve(f, x1=-100..100, x2=-100..100, evaluationlimit
      =100000, noimprovementlimit=100000);
```

$$[-3.06433688275856530, [x1 = -0.233457978266705634e - 1,$$
$$x2 = .774154819772443825]]$$

A partial explanation is that the shape of the objective function f is close to quadratic, at least "from a distance". Note at the same time that the built-in Maple local solver produces much inferior results on the larger region (and it also misses the global optimum when using the variable bounds $[-3, 3]$, as can be expected):

> Minimize(f, x1=-100..100, x2=-100..100);

$$[-.713074709310511201, [x1 = -0.223022309405313465e - 1,$$
$$x2 = -0.472762143202519123e - 2]]$$

The corresponding GOT runtimes are a little more than one second in both cases. (Note that all such runtimes are approximate, and may vary a bit even between consecutive test runs, depending on the machine's actual runtime environment).

One of the advantages of using ISTCs that one can visualize models and verify their perceived difficulty. Fig. 5 is based on using the Maple Optimization Plotter dialog, a feature that can be used in conjunction with the GOT: it shows the box-constrained Trefethen model [Tre02] in the range $[-3, 3]^2$; observe also the location of the optimal solution (green dot).

Fig. 5. Problem 4 in [Tre02] solved and visualized using the Maple GOT

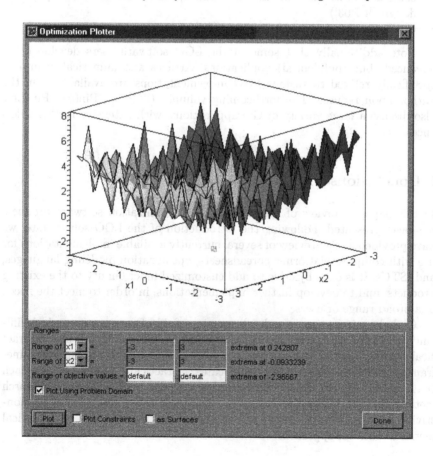

5 Further Applications

For over a decade, LGO has been applied in a variety of professional, as well as academic research and educational contexts (in some 20 countries, as of 2004). In recent years, LGO has been used to solve models in up to a few thousand variables and constraints. The software seems to be particularly well-suited to analyze and solve complex, sophisticated applications in advanced engineering, biotechnology, econometrics, financial modeling, process industries, medical studies, and in various other areas of scientific modeling.

Without aiming at completeness, let us refer to some recent (published) applications and case studies that are related to the following areas:

- model calibration ([Pin03a])
- potential energy models in computational chemistry ([Pin00, Pin01b]), ([SSP01])
- laser design ([IPC03])
- cancer therapy planning ([TKLPL03])
- combined finite element modeling and optimization in sonar equipment design ([PP03])
- Configuration analysis and design ([KP04b]).

Note additionally that some of the LGO software users develop other advanced (but confidential) applications. Articles and numerical examples, specifically related to various LGO implementations are available from the author upon request. The forthcoming volumes ([KP05]; [Pin05a, Pin05b]) also discuss a large variety of GO applications, with extensive further references.

6 Conclusions

In this paper, a review of several nonlinear optimization software products has been presented. Following the introduction of the LGO solver suite, we have provided a brief review of several currently available implementations for use with compiler platforms, spreadsheets, optimization modeling languages, and ISTCs. It is our objective to add customized functionality to the existing products, and to develop further implementations, in order to meet the needs of a broad range of users.

Global optimization is and will remain a field of extreme numerical difficulty, not only when considering "all possible" GO models, but also in practical attempts to handle complex, sizeable problems in an acceptable timeframe. Therefore the discussion advocates a practically motivated approach that combines rigorous global optimization strategies with efficient local search methodology, in integrated, flexible solver suites. The illustrative – yet nontrivial – application examples and the numerical results show the practical merits of such an approach.

We are interested to learn suggestions regarding future development directions. Test problems and challenges – as well as prospective application areas – are welcome.

Acknowledgements

First of all, I wish to thank my developer partners and colleagues for their cooperation and many useful discussions, quality software, documentation, and technical support. These partners include AMPL LLC, Frontline Systems, the GAMS Development Corporation, Dr. Frank J. Kampas, Lahey Computer Systems, LINDO Systems, Maplesoft, Maximal Software, Paragon Decision Technology, TOMLAB AB, and Wolfram Research.

Several application examples reviewed or cited in this paper are based on cooperation with colleagues: all such cooperation is gratefully acknowledged and is reflected by the references.

In addition to professional contributions and in-kind support offered by developer partners, the work summarized and reviewed in this paper has received financial support from the following organizations: DRDC Atlantic Region, Canada (Contract W7707-01-0746), the Dutch Technology Foundation (STW Grant CWI55.3638), the Hungarian Scientific Research Fund (OTKA Grant T 034350), Maplesoft, the National Research Council of Canada (NRC IRAP Project 362093), the University of Ballarat, Australia; the University of Kuopio, Finland; and the University of Tilburg, Netherlands.

References

[Ari99] Aris, R.: Mathematical Modeling: A Chemical Engineers Perspective. Academic Press, San Diego, CA (1999)

[BSS93] Bazaraa, M.S., Sherali, H.D., Shetty, C.M.: Nonlinear Programming: Theory and Algorithms. Wiley, New York (1993)

[BS00] Benson, H.P., Sun, E. LGO – Versatile tool for global optimization. In: OR/MS Today, **27**, 52-55 (2000)

[Ber99] Bertsekas, D.P.: Nonlinear Programming (2nd Edition). Athena Scientific, Cambridge, MA (1999)

[BR95] Boender, C.G.E., Romeijn, H.E. Stochastic methods. In: Horst and Pardalos (eds) Handbook of Global Optimization. Volume 1, pp. 829-869 (1995)

[BLWW04] Bornemann, F., Laurie, D., Wagon, S., Waldvogel, J.: The SIAM 100-Digit Challenge. A Study in High-Accuracy Numerical Computing. SIAM, Philadelphia, PA (2004)

[BM68] Bracken, J. and McCormick, G.P.: Selected Applications of Nonlinear Programming. Wiley, New York (1968)

[BKM88] Brooke, A., Kendrick, D. and Meeraus, A.: GAMS: A User's Guide. The Scientific Press, Redwood City, CA. (Revised versions are available from the GAMS Corporation.) See also http://www.gams.com (1988)

[Cas90] Casti, J.L.: Searching for Certainty. Morrow & Co., New York (1990)

[Cog03] Cogan, B. How to get the best out of optimization software. In: Scientific Computing World, **71**, 67-68 (2003)

[CK99] Corliss, G.F., Kearfott, R.B. Rigorous global search: industrial applications. In: Csendes, T. (ed) Developments in Reliable Computing, 1–16. Kluwer Academic Publishers, Boston/Dordrecht/London (1999)

[CFO01] Coullard, C., Fourer, R., Owen, J.H. (eds): Annals of Operations Research, **104**, Special Issue on Modeling Languages and Systems. Kluwer Academic Publishers, Boston/Dordrecht/London (2001)

[CZ01] Chong, E.K.P., Zak, S.H.: An Introduction to Optimization (2nd Edition). Wiley, New York (2001)

[Diw03] Diwekar, U.: Introduction to Applied Optimization. Kluwer Academic Publishers, Boston/Dordrecht/London (2003)

[EHL01] Edgar, T.F., Himmelblau, D.M., Lasdon, L.S. Optimization of Chemical Processes (2nd Edition). McGraw-Hill, New York (2001)

[EW75] Eigen, M. and Winkler, R.: Das Spiel. Piper & Co., München (1975)

[Fou04] Fourer, R.: Nonlinear Programming Frequently Asked Questions. Optimization Technology Center of Northwestern University and Argonne National Laboratory, http://www-unix.mcs.anl.gov/otc/Guide/faq/nonlinear-programming-faq.html (2004)

[FGK93] Fourer, R., Gay, D.M., Kernighan, B.W.: AMPL – A Modeling Language for Mathematical Programming. The Scientific Press, Redwood City, CA (Reprinted by Boyd and Fraser, Danvers, MA, 1996. See also http://www.ampl.com) (1993)

[FS01] Frontline Systems: Premium Solver Platform – Solver Engines. User Guide. Frontline Systems, Inc. Incline Village, NV (See http://www.solver.com, and http://www.solver.com/xlslgoeng.htm) (2001)

[Ger99] Gershenfeld, N.: The Nature of Mathematical Modeling. Cambridge University Press, Cambridge (1999)

[Gro96] Grossmann, I.E. (ed): Global Optimization in Engineering Design. Kluwer Academic Publishers, Boston/Dordrecht/London (1996)

[HJ91] Hansen, P.E. and Jørgensen, S.E. (eds): Introduction to Environmental Management. Elsevier, Amsterdam (1991)

[HL05] Hillier, F.J. and Lieberman, G.J. Introduction to Operations Research. (8th Edition.) McGraw-Hill, New York (2005)

[HP95] Horst, R., Pardalos, P.M. (eds): Handbook of Global Optimization (Volume 1). Kluwer Academic Publishers, Boston/Dordrecht/London (1995)

[HT96] Horst, R., Tuy, H.: Global Optimization – Determinsitic Approaches (3rd Edition). Springer-Verlag, Berlin / Heidelberg / New York (1996)

[I04] ILOG: ILOG OPL Studio and Solver Suite. http://www.ilog.com (2004)

[IPC03] Isenor, G., Pintér, J.D., Cada, M.: A global optimization approach to laser design. Optimization and Engineering 4, 177–196 (2003)

[Jac01] Jacob, C.: Illustrating Evolutionary Computation with Mathematica. Morgan Kaufmann Publishers, San Francisco (2001)

[Kal04] Kallrath, J. (ed): Modeling Languages in Mathematical Optimization. Kluwer Academic Publishers, Boston/Dordrecht/London (2004)

[KP04a] Kampas, F.J., Pintér, J.D.: Generalized circle packings: model formulations and numerical results. Proceedings of the International Mathematica Symposium (Banff, AB, Canada, August 2004)

[KP04b] Kampas, F.J., Pintér, J.D.: Configuration analysis and design by using optimization tools in Mathematica. The Mathematica Journal (to appear) (2004)

[KP05] Kampas, F.J., Pintér, J.D.: Advanced Optimization: Scientific, Engineering, and Economic Applications with Mathematica Examples. Elsevier, Amsterdam (to appear) (2005)

[Kea96] Kearfott, R.B.: Rigorous Global Search: Continuous Problems. Kluwer Academic Publishers, Boston/Dordrecht/London (1996)

[Laf00] Lafe, O.: Cellular Automata Transforms. Kluwer Academic Publishers, Boston / Dordrecht / London (2000)

[LCS02] Lahey Computer Systems. Fortran 90 User's Guide. Lahey Computer Systems, Inc., Incline Village, http://www.lahey.com (2002)

[LS96] LINDO Systems. Solver Suite. LINDO Systems, Inc., Chicago, IL. http://www.lindo.com (1996)

[Man83] Mandelbrot, B.B.: The Fractal Geometry of Nature. Freeman & Co., New York (1983)

[M04a] Maplesoft. Maple. (Current version: 9.5.) Maplesoft, Inc., Waterloo, ON. http://www.maplesoft.com (2004)

[M04b] Maplesoft. Global Optimization Toolbox. Maplesoft, Inc. Waterloo, ON. http://www.maplesoft.com (2004)

[MM95] Maros, I., Mitra, G. (eds): Annals of Operations Research, 58, Applied Mathematical Programming and Modeling II (APMOD 93) J.C. Baltzer AG, Science Publishers, Basel (1995)

[MMS97] Maros, I., Mitra, G., Sciomachen, A. (eds): Annals of Operations Research, 81, Applied Mathematical Programming and Modeling III (APMOD 95). J.C. Baltzer AG, Science Publishers, Basel (1997)

[MS04] Mittelmann, H.D., Spellucci, P. Decision Tree for Optimization Software. http://plato.la.asu.edu/guide.html (2004)

[MS02] Maximal Software. MPL Modeling System. Maximal Software, Inc. Arlington, VA. http://www.maximal-usa.com (2002)

[Mur83] Murray, J.D.: Mathematical Biology. Springer–Verlag, Berlin (1983)

[Neu04a] Neumaier, A.: Global Optimization. http://www.mat.univie.ac.at/ neum /glopt.html (2004)

[Neu04b] Neumaier, A.: Complete search in continuous global optimization and constraint satisfaction. In: Iserles, A. (ed) Acta Numerica 2004. Cambridge University Press, Cambridge (2004b)

[PW00] Papalambros, P.Y., Wilde, D.J.: Principles of Optimal Design. Cambridge University Press, Cambridge (2000)

[PDT04] Paragon Decision Technology: AIMMS (Current version 3.5). Paragon Decision Technology BV, Haarlem, The Netherlands. See http://www.aimms.com (2004)

[PSX96] Pardalos, P.M., Shalloway, D. and Xue, G.: Global minimization of nonconvex energy functions: molecular conformation and protein folding. In: DIMACS Series, 23, American Mathematical Society, Providence, RI (1996)

[PR02] Pardalos, P.M., Romeijn, H.E. (eds): Handbook of Global Optimization. Volume 2. Kluwer Academic Publishers, Boston/Dordrecht/London (2002)

[Pin96a] Pintér, J.D.: Global Optimization in Action. Kluwer Academic Publishers, Boston / Dordrecht / London (1996)

[Pin96b] Pintér, J.D.: Continuous global optimization software: A brief review. Optima, **52**, 1–8 (1996) (Web version is available at: http://plato.la.asu.edu/gom.html)

[Pin97] Pintér, J.D.: LGO – A Program System for Continuous and Lipschitz Optimization. In: Bomze, I.M., Csendes, T., Horst, R. and Pardalos, P.M. (eds) Developments in Global Optimization, 183-197. Kluwer Academic Publishers, Boston/Dordrecht/London (1997)

[Pin00] Pintér, J.D.: Extremal energy models and global optimization. In: Laguna, M., González-Velarde, J-L., (eds) Computing Tools for Modeling, Optimization and Simulation, 145–160. Kluwer Academic Publishers, Boston/Dordrecht/London (2000)

[Pin01a] Pintér, J.D.: Computational Global Optimization in Nonlinear Systems. Lionheart Publishing Inc., Atlanta, GA (2001)

[Pin01b] Pintér, J.D.: Globally optimized spherical point arrangements: model variants and illustrative results. Annals of Operations Research **104**, 213–230 (2001)

[Pin02a] Pintér, J.D.: MathOptimizer – An Advanced Modeling and Optimization System for Mathematica Users. User Guide. Pintér Consulting Services, Inc., Halifax, NS (2002a) (For a summary, see also http://www.wolfram.com/products/ applications/mathoptimizer/)

[Pin02b] Pintér, J.D.: Global optimization: software, test problems, and applications. In: Pardalos and Romeijn (eds) Handbook of Global Optimization. Volume 2, 515–569 (2002)

[Pin03a] Pintér, J.D.: (2003a) Globally optimized calibration of nonlinear models: techniques, software, and applications. Optimization Methods and Software, **18**, 335–355 (2003)

[Pin03b] Pintér, J.D.: GAMS /LGO nonlinear solver suite: key features, usage, and numerical performance. Submitted for publication. Downloadable at http://www.gams.com/solvers/lgo (2003)

[Pin04] Pintér, J.D.: LGO – A Model Development System for Continuous Global Optimization. Users Guide. (Current revision.) Pintér Consulting Services, Inc., Halifax, NS (2004) (For a summary, see http://www.pinterconsulting.com)

[Pin05a] Pintér, J.D.: Applied Nonlinear Optimization in Modeling Environments. CRC Press, Baton Rouge, FL (2005) (To appear)

[Pin05b] Pintér, J.D. (ed): Global Optimization – Selected Case Studies. Springer Science + Business Media, New York (2005) (To appear)

[PHGE04] Pintér, J.D., Holmström, K., Göran, A.O., Edvall, M.M.: User's Guide for TOMLAB /LGO. TOMLAB Optimization AB, Västerås, Sweden (2004) (See http://www.tomlab.biz)

[PK03] Pintér, J.D., Kampas, F.J.: MathOptimizer Professional – An Advanced Modeling and Optimization System for Mathematica Users with an External Solver Link. User Guide. Pintér Consulting Services, Inc., Halifax, NS, Canada (2003) (For a summary, see also http://www.wolfram.com/products/ applications/mathoptpro/)

[PK05] Pintér, J.D., Kampas, F.J.: Model development and optimization with Mathematica. In: Golden, B., Raghavan, S., Wasil, E. (eds) The Next Wave in Computing, Optimization, and Decision Technologies, 285–302. Springer Science + Business Media, New York (2005)

[PP03] Pintér, J.D., Purcell, C.J.: Optimization of finite element models with MathOptimizer and ModelMaker. Lecture presented at the 2003 Mathematica Developer Conference, Champaign, IL (2003) (Extended abstract is available upon request, and also from http://www.library.com)

[RR93] Ratschek, H., Rokne, J.: Experiments using interval analysis for solving a circuit design problem. Journal of Global Optimization 3, 501–518 (1993)

[RR95] Ratschek, H., Rokne, J.: Interval methods. In: Horst and Pardalos (eds) Handbook of Global Optimization. Volume 1, 751–828 (1995)

[Ric73] Rich, L.G.: Environmental Systems Engineering. McGraw-Hill, Tokyo (1973).

[Sch02] Schittkowski, K.: Numerical Data Fitting in Dynamical Systems. Kluwer Academic Publishers, Boston/Dordrecht/London (2002)

[Sch91] Schroeder, M.: Fractals, Chaos, Power Laws. Freeman & Co., New York (1991)

[Ste95] Stewart, I.: Nature's Numbers. Basic Books / Harper and Collins, New York (1995)

[SSP01] Stortelder, W.J.H., de Swart, J.J.B., Pintér, J.D.: Finding elliptic Fekete point sets: two numerical solution approaches. Journal of Computational and Applied Mathematics, 130, 205–216 (2001)

[TS02] Tawarmalani, M., Sahinidis, N.V.: Convexification and Global Optimization in Continuous and Mixed-integer Nonlinear Programming. Kluwer Academic Publishers, Boston/Dordrecht/London (2002)

[TKLPL03] Tervo, J., Kolmonen, P., Lyyra-Laitinen, T., Pintér, J.D., and Lahtinen, T. An optimization-based approach to the multiple static delivery technique in radiation therapy. Annals of Operations Research, 119, 205–227 (2003)

[TO04] TOMLAB Optimization. TOMLAB. TOMLAB Optimization AB, Västerås, Sweden (2004) (See http://www.tomlab.biz)

[Tre02] Trefethen, L.N.: The hundred-dollar, hundred-digit challenge problems. SIAM News, Issue 1, p. 3 (2002)

[TM04] The MathWorks: MATLAB. (Current version: 6.5) The MathWorks, Inc., Natick, MA (2004) (See http://www.mathworks.com)

[VMM00] Vladimirou, H., Maros, I., Mitra, G. (eds): Annals of Operations Research, 99, Applied Mathematical Programming and Modeling IV (APMOD 98) J.C. Baltzer AG, Science Publishers, Basel, Switzerland (2000)

[Wol02] Wolfram, S.: A New Kind of Science. Wolfram Media, Champaign, IL, and Cambridge University Press, Cambridge (2002)

[Wol03] Wolfram, S.: The Mathematica Book. (Fourth Edition) Wolfram Media, Champaign, IL, and Cambridge University Press, Cambridge (2003)

[WR04] Wolfram Research: Mathematica (Current version: 5.1). Wolfram Research, Inc., Champaign, IL (2004) (See http://www.wolfram.com)

[Zab03] Zabinsky, Z.B.: Stochastic Adaptive Search for Global Optimization. Kluwer Academic Publishers, Boston/Dordrecht/London (2003)

Supervised Data Classification via Max-min Separability

Adil M. Bagirov and Julien Ugon

CIAO, School of Information Technology and Mathematical Sciences
University of Ballarat
VIC 3353, Australia
a.bagirov@ballarat.edu.au, j.ugon@ballarat.edu.au

Summary. The problem of discriminating between the elements of two finite sets of points in n-dimensional space is a fundamental in supervised data classification. In practice, it is unlikely for the two sets to be linearly separable. In this paper we consider the problem of separating of two finite sets of points by means of piecewise linear functions. We prove that if these two sets are disjoint then they can be separated by a piecewise linear function and formulate the problem of finding the latter function as an optimization problem with an objective function containing max-min of linear functions. The differential properties of the objective function are studied and an algorithm for its minimization is developed. We present the results of numerical experiments with real world data sets. These results demonstrate the effectiveness of the proposed algorithm for separating two finite sets of points. They also demonstrate the effectiveness of an algorithm based on the concept of max-min separability for solving supervised data classification problems.

Key words: Supervised data classification, separability, nonconvex optimization, nonsmooth optimization.

1 Introduction

Supervised data classification is an important area in data mining. It has many applications in science, engineering, medicine etc. The aim of supervised data classification is to establish rules for the classification of some observations assuming that the classes of data are known. To find these rules, known training subsets of the given classes are used. During the last decades many algorithms have been proposed and studied to solve supervised data classification problems. One of the promising approaches to these problems is based on mathematical programming techniques. This approach has gained a great deal of attention over last years, see, for example, [AG02, Bag05, BRSY01, BRY00, BRY02, BB97, BB96, BM92, BM00, BFM99, Bur98, CM95, Man94, Man97, Tho02, Vap95].

There are different approaches for solving supervised data classification problems based on mathematical programming techniques. In one of them the use of mathematical programming techniques is carried out by reducing the classification problem to the problem of separation of two finite sets of points A and B in n-dimensional space. If $\operatorname{co} A \cap \operatorname{co} B = \emptyset$ then these two sets are linearly separable and there exists a hyperplane which separates these two sets. Linear programming techniques can be used to construct such a hyperplane. If the convex hulls of A and B intersect then linear programming techniques can be applied to obtain a hyperplane which minimizes some misclassification measure. Algorithms based on such an approach are developed in [BB96, BM92, CM95, Man94].

The paper [BM93] develops the concept of bilinear separability, where two sets are separated using two hyperplanes. The problem of finding of these hyperplanes is reduced to a certain bilinear programming problem. The paper [BM93] presents an algorithm for solving the latter problem.

In the paper [AG02] the concept of polyhedral separability was introduced. In this paper the case when $\operatorname{co} A \cap B = \emptyset$ was considered. The set A is approximated by a polyhedral set. It is proved that the sets A and B are h-polyhedrally separable for some $h \leq |B|$, where $|B|$ is the cardinality of the set B. Thus in this case the sets A and B can be separated by a certain piecewise linear function. The authors introduce an error function which is nonconvex piecewise linear function. An algorithm for minimizing this function is proposed. The problem of the calculation of the descent direction in this algorithm is reduced to a certain linear programming problem.

The paper [Bag05] introduces the notion of max-min separability where two sets are separated by a piecewise linear function. Since any piecewise linear function can be represented as a max-min of linear functions we call it max-min separability. This approach can be considered as a generalization of the linear, bilinear and polyhedral separabilities.

The problem of max-min separability is reduced to a certain nonsmooth, nonconvex optimization problem. The objective function in this problem is represented as a sum of functions containing max-min of linear functions and it is a locally Lipschitz continuous. However this function is not Clarke regular and the calculation of its subgradient is a difficult task. Therefore methods of nonsmooth optimization based on subgradient information are not appropriate for solving max-min separability problems.

In this paper we develop an algorithm for solving max-min separability problems which uses only values of the objective function. This algorithm calculates a descent direction by evaluating the so-called discrete gradient of the objective function. The form of the objective function allows to significantly reduce the number of its evaluations during the computation of the discrete gradient. This is very important because each evaluations of the objective function for large data sets is expensive.

We carried out some numerical experiments using large scale data sets. We present their results and discuss them.

The structure of this paper is as follows. Section 2 provides some preliminaries. In Section 3 the definition and some results related to the max-min separability are given. An algorithm for solving max-min separability problems is discussed in Section 4. Results of numerical experiments are presented in Section 5. Section 6 concludes the paper.

2 Preliminaries

In this section we present a brief review of the concepts of linear, bilinear and polyhedral separability.

2.1 Linear separability

Let A and B be given sets containing m and p n-dimensional vectors, respectively:

$$A = \{a^1, \ldots, a^m\}, \ a^i \in \mathbb{R}^n, \ i = 1, \ldots, m,$$

$$B = \{b^1, \ldots, b^p\}, \ b^j \in \mathbb{R}^n, \ j = 1, \ldots, p.$$

The sets A and B are linearly separable if there exists a hyperplane $\{x, y\}$, with $x \in \mathbb{R}^n$, $y \in \mathbb{R}^1$ such that

1) for any $j = 1, \ldots, m$

$$\langle x, a^j \rangle - y < 0,$$

2) for any $k = 1, \ldots, p$

$$\langle x, b^k \rangle - y > 0.$$

The sets A and B are linearly separable if and only if $\operatorname{co} A \cap \operatorname{co} B = \emptyset$.

In practice, it is unlikely for the two sets to be linearly separable. Therefore it is important to find a hyperplane which minimizes some misclassification cost. In the paper [BM92] the problem of finding this hyperplane is formulated as the following optimization problem:

$$\text{minimize } f(x, y) \text{ subject to } (x, y) \in \mathbb{R}^{n+1} \tag{1}$$

where

$$f(x, y) = \frac{1}{m} \sum_{i=1}^{m} \max\left(0, \langle x, a^i \rangle - y + 1\right) + \frac{1}{p} \sum_{j=1}^{p} \max\left(0, -\langle x, b^j \rangle + y + 1\right)$$

is an error function. Here $\langle \cdot, \cdot \rangle$ stands for the scalar product in $\mathbb{R}^n$. The authors describe an algorithm for solving problem (1). They show that the problem (1) is equivalent to the following linear program:

$$\text{minimize } \frac{1}{m} \sum_{i=1}^{m} t_i + \frac{1}{p} \sum_{j=1}^{p} z_j$$

subject to

$$t_i \geq \langle x, a^i \rangle - y + 1, \quad i = 1, \ldots, m,$$
$$z_j \geq -\langle x, b^j \rangle + y + 1, \quad j = 1, \ldots, p,$$
$$t \geq 0, \quad z \geq 0,$$

where t_i is nonnegative and represents the error for the point $a^i \in A$ and z_j is nonnegative and represents the error for the point $b^j \in B$.

The sets A and B are linearly separable if and only if $f^* = f(x^*, y_*) = 0$ where (x^*, y_*) is the solution to the problem (1). It is proved that the trivial solution $x = 0$ cannot occur.

2.2 Bilinear separability

The concept of bilinear separability was introduced in [BM93]. In this approach two sets are separated using two hyperplanes. We again assume that A and B are given sets containing m and p n-dimensional vectors, respectively.

Definition 1. *(see [BM93]). The sets A and B are bilinear separable if and only if there exist two hyperplanes (x^1, y_1) and (x^2, y_2) such that at least one of the following conditions holds:*

1. *For any $j = 1, \ldots, m$*

$$\langle x^l, a^j \rangle - y_l < 0, \quad l = 1, 2$$

   and for any $k = 1, \ldots, p$ there exists $l \in \{1, 2\}$ such that

$$\langle x^l, b^k \rangle - y_l > 0.$$

2. *For any $k = 1, \ldots, p$*

$$\langle x^l, b^k \rangle - y_l < 0, \quad l = 1, 2$$

   and for any $j = 1, \ldots, m$ there exists $l \in \{1, 2\}$ such that

$$\langle x^l, a^j \rangle - y_l > 0.$$

3. *For any $j = 1, \ldots, m$ either*

$$\langle x^l, a^j \rangle - y_l < 0, \quad l = 1, 2$$

   or

$$\langle -x^l, a^j \rangle + y_l < 0, \quad l = 1, 2$$

   and for any $k = 1, \ldots, p$ either

$$\langle x^1, b^k \rangle - y_1 < 0, \quad \langle -x^2, b^k \rangle + y_2 < 0$$

   or

$$\langle -x^1, b^k \rangle + y_1 < 0, \quad \langle x^2, b^k \rangle - y_2 > 0.$$

We reformulate Definition 1 using max and min statements.

Definition 2. *The sets A and B are bilinear separable if and only if there exist two hyperplanes (x^1, y_1) and (x^2, y_2) such that at least one of the following conditions holds:*

1. *For any $j = 1, \ldots, m$*

$$\max_{l=1,2}\{\langle x^l, a^j\rangle - y_l\} < 0$$

and for any $k = 1, \ldots, p$

$$\max_{l=1,2}\{\langle x^l, b^k\rangle - y_l\} > 0.$$

2. *For any $k = 1, \ldots, p$*

$$\max_{l=1,2}\{\langle x^l, b^k\rangle - y_l\} < 0$$

and for any $j = 1, \ldots, m$

$$\max_{l=1,2}\{\langle x^l, a^j\rangle - y_l\} > 0.$$

3. *For any $j = 1, \ldots, m$,*

$$\max\left[\min\{\langle x^1, a^j\rangle - y_1, -\langle x^2, a^j\rangle + y_2\}, \min\{-\langle x^1, a^j\rangle + y_1, \langle x^2, a^j\rangle - y_2\}\right] \\ < 0$$

and for any $k = 1, \ldots, p$,

$$\max\left[\min\{\langle x^1, b^k\rangle - y_1, -\langle x^2, b^k\rangle + y_2\}, \min\{-\langle x^1, b^k\rangle + y_1, \langle x^2, b^k\rangle - y_2\}\right] \\ > 0.$$

The problem of bilinear separability is reduced to a certain bilinear programming problem and the paper [BM93] presents an algorithm for its solution.

2.3 Polyhedral separability

The concept of h-polyhedral separability was developed in [AG02]. The sets A and B are h-polyhedrally separable if there exists a set of h hyperplanes $\{x^i, y_i\}$, with

$$x^i \in \mathbb{R}^n, \ y_i \in \mathbb{R}^1, \ i = 1, \ldots, h$$

such that

1) for any $j = 1, \ldots, m$ and $i = 1, \ldots, h$

$$\langle x^i, a^j\rangle - y_i < 0,$$

2) for any $k = 1, \ldots, p$ there exists at least one $i \in \{1, \ldots, h\}$ such that

$$\langle x^i, b^k \rangle - y_i > 0.$$

It is proved in [AG02] that the sets A and B are h-polyhedrally separable, for some $h \leq p$ if and only if

$$\mathrm{co}\, A \bigcap B = \emptyset.$$

Figure 1 presents one example of polyhedral separability.

The problem of polyhedral separability of the sets A and B is reduced to the following problem:

$$\text{minimize}\ \ f(x, y)\ \text{ subject to }\ (x, y) \in \mathbb{R}^{(n+1) \times h} \tag{2}$$

where

$$f(x, y) = \frac{1}{m} \sum_{j=1}^{m} \max \left[0, \max_{1 \leq i \leq h} \left\{ \langle x^i, a^j \rangle - y_i + 1 \right\} \right] +$$

$$\frac{1}{p} \sum_{k=1}^{p} \max \left[0, \min_{1 \leq i \leq h} \left\{ -\langle x^i, b^k \rangle + y_i + 1 \right\} \right]$$

is an error function. Note that this function is a nonconvex piecewise linear function. It is proved that $x^i = 0$, $i = 1, \ldots, h$ cannot be the optimal solution. Let $\{\bar{x}^i, \bar{y}_i\}, i = 1, \ldots, h$ be a global solution to the problem (2). The sets A and B are h-polyhedrally separable if and only if $f(\bar{x}, \bar{y}) = 0$. If there exists a nonempty set $\bar{I} \subset \{1, \ldots, h\}$ such that $x^i = 0$, $i \in \bar{I}$, then the sets A and B are $(h - |\bar{I}|)$-polyhedrally separable. In [AG02] an algorithm for solving problem (2) is developed. The calculation of the descent direction at each iteration of this algorithm is reduced to a certain linear programming problem.

The advantage of this technique is that it does not restrict the search to only a convex polyhedron, and thus allows both the sets A and B to be nonconvex. One disadvantage, however, is that it only considers the sets separately.

3 Max-min separability

In many practical applications two sets are not linearly, bilinearly or polyhedrally separable. Figure 2 presents one such case. In this case two sets are separable with more complicated piecewise linear function.

In this section we describe the concept of max-min separability and introduce an error function (see [Bag05]).

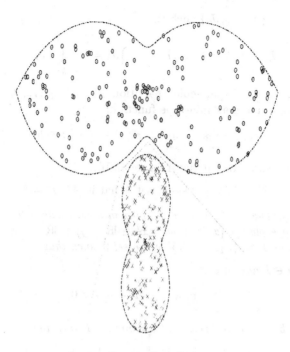

Fig. 1. Polyhedral separability.

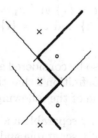

Fig. 2. The sets A and B are separated by a piecewise linear function.

3.1 Definition and properties

Let $H = \{h_1, \ldots, h_l\}$, where $h_j = \{x^j, y_j\}$, $j = 1, \ldots, l$ with $x^j \in \mathbb{R}^n$, $y_j \in \mathbb{R}^1$, be a finite set of hyperplanes. Let $J = \{1, \ldots, l\}$. Consider any partition of this set $J^r = \{J_1, \ldots, J_r\}$ such that

$$J_k \neq \emptyset, \ k = 1, \ldots, r, \ \ J_k \bigcap J_j = \emptyset, \ \ \bigcup_{k=1}^{r} J_k = J.$$

Let $I = \{1, \ldots, r\}$. A particular partition $J^r = \{J_1, \ldots, J_r\}$ of the set J defines the following max-min-type function:

$$\varphi(z) = \max_{i \in I} \ \min_{j \in J_i} \ \{\langle x^j, z \rangle - y_j\}, \ z \in \mathbb{R}^n. \tag{3}$$

In Figure 3 two sets are max-min separable.

Let $A, B \subset \mathbb{R}^n$ be given disjoint sets, that is $A \bigcap B = \emptyset$.

Definition 3. *The sets A and B are max-min separable if there exist a finite number of hyperplanes $\{x^j, y_j\}$ with $x^j \in \mathbb{R}^n$, $y_j \in \mathbb{R}^1$, $j \in J = \{1, \ldots, l\}$ and a partition $J^r = \{J_1, \ldots, J_r\}$ of the set J such that*

1) for all $i \in I$ and $a \in A$

$$\min_{j \in J_i} \{\langle x^j, a \rangle - y_j\} < 0;$$

2) for any $b \in B$ there exists at least one $i \in I$ such that

$$\min_{j \in J_i} \{\langle x^j, b \rangle - y_j\} > 0.$$

Remark 1. It follows from Definition 3 that if the sets A and B are max-min separable then $\varphi(a) < 0$ for any $a \in A$ and $\varphi(b) > 0$ for any $b \in B$, where the function φ is defined by (3). Thus the sets A and B can be separated by a function represented as a max-min of linear functions. Therefore this kind of separability is called a max-min separability.

Remark 2. Linear and polyhedral separability can be considered as particular cases of the max-min separability. If $I = \{1\}$ and $J_1 = \{1\}$ then we have the linear separability and if $I = \{1, \ldots, h\}$ and $J_i = \{i\}$, $i \in I$ we obtain the h-polyhedral separability.

Remark 3. Bilinear separability can also be considered as particular case of the max-min separability. It follows from Definition 2 that the bilinear separability of two sets A and B coincides with one of the following cases:

1. The sets A and B are 2-polyhedrally separable and $\text{co} \, A \bigcap B = \emptyset$;
2. The sets A and B are 2-polyhedrally separable and $\text{co} \, B \bigcap A = \emptyset$;
3. The sets A and B are max-min separable with the following hyperplanes:

$$\{(x^1, y_1), (-x^1, -y_1), (x^2, y_2), (-x^2, -y_2)\}.$$

In this case $I = \{1, 2\}$ and $J_1 = \{1, 4\}$, $J_2 = \{2, 3\}$. Thus the bilinear separable sets are also max-min separable.

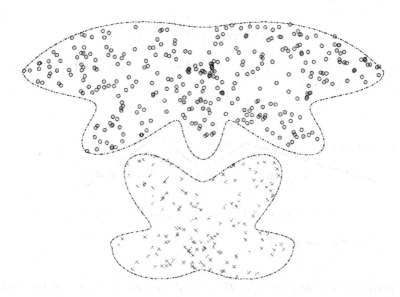

Fig. 3. Max-min separability.

Proposition 1. *(see [Bag05]). The sets A and B are max-min separable if and only if there exists a set of hyperplanes $\{x^j, y_j\}$ with $x^j \in \mathbb{R}^n$, $y_j \in \mathbb{R}^1$, $j \in J$ and a partition $J^r = \{J_1, \ldots, J_r\}$ of the set J such that*

1) for any $i \in I$ and $a \in A$

$$\min_{j \in J_i} \left\{ \langle x^j, a \rangle - y_j \right\} \leq -1;$$

2) for any $b \in B$ there exists at least one $i \in I$ such that

$$\min_{j \in J_i} \left\{ \langle x^j, b \rangle - y_j \right\} \geq 1.$$

Proof. Sufficiency is straightforward.
Necessity. Since A and B are max-min separable there exists a set of hyperplanes $\{\bar{x}^j, \bar{y}_j\}$ with $\bar{x}^j \in \mathbb{R}^n$, $\bar{y}_j \in \mathbb{R}^1$, $j \in J$, a partition J^r of the set J and numbers $\delta_1 > 0$, $\delta_2 > 0$ such that

$$\max_{a \in A} \max_{i \in I} \min_{j \in J_i} \left\{ \langle \bar{x}^j, a \rangle - \bar{y}_j \right\} = -\delta_1$$

and

$$\min_{b \in B} \max_{i \in I} \min_{j \in J_i} \left\{ \langle \bar{x}^j, b \rangle - \bar{y}_j \right\} = \delta_2.$$

We put $\delta = \min\{\delta_1, \delta_2\} > 0$. Then we have

$$\max_{i \in I} \min_{j \in J_i} \left\{ \langle \bar{x}^j, a \rangle - \bar{y}_j \right\} \leq -\delta, \ \forall a \in A, \tag{4}$$

$$\max_{i \in I} \min_{j \in J_i} \left\{ \langle \bar{x}^j, b \rangle - \bar{y}_j \right\} \geq \delta, \quad \forall b \in B. \tag{5}$$

We consider the new set of hyperplanes $\{x^j, y_j\}$ with $x^j \in \mathbb{R}^n$, $y_j \in \mathbb{R}^1$, $j \in J$, defined as follows:

$$x^j = \bar{x}^j / \delta, \; j \in J,$$

$$y^j = \bar{y}^j / \delta, \; j \in J.$$

Then it follows from (4) and (5) that

$$\max_{i \in I} \min_{j \in J_i} \left\{ \langle x^j, a \rangle - y_j \right\} \leq -1, \quad \forall a \in A,$$

$$\max_{i \in I} \min_{j \in J_i} \left\{ \langle x^j, b \rangle - y_j \right\} \geq 1, \quad \forall b \in B,$$

which completes the proof. $\square$

Proposition 2. (see [Bag05]). *The sets A and B are max-min separable if and only if there exists a piecewise linear function separating them.*

Proof. Since max-min of linear functions is piecewise linear function the necessity is straightforward.
Sufficiency. It is known that any piecewise linear function can be represented as a max-min of linear functions of the form (3) (see [BKS95]). Then we get that there exists max-min of linear functions that separates the sets A and B which in its turn means that these sets are max-min separable. $\square$

Remark 4. It follows from Proposition (2) that the notions of max-min and piecewise linear separability are equivalent.

Proposition 3. (see [Bag05]). *Assume that the set A can be represented as a union of sets A_i, $i = 1, \ldots, q$:*

$$A = \bigcup_{i=1}^{q} A_i$$

and for any $i = 1, \ldots, q$

$$B \bigcap \mathrm{co} A_i = \emptyset. \tag{6}$$

Then the sets A and B are max-min separable.

Proof. It follows from (6) that $b \notin \mathrm{co}\, A_i$ for all $b \in B$ and $i \in \{1, \ldots, q\}$. Then, for each $b \in B$ and $i \in \{1, \ldots, q\}$ there exists a hyperplane $\{x^i(b), y_i(b)\}$ separating b from the set $\mathrm{co}\, A_i$, that is

$$\langle x^i(b), b \rangle - y_i(b) > 0,$$

$$\langle x^i(b), a \rangle - y_i(b) < 0, \quad \forall a \in \text{co } A_i, \ i = 1, \ldots, q.$$

Then we have

$$\min_{i=1,\ldots,q} \left\{ \langle x^i(b), b \rangle - y_i(b) \right\} > 0$$

and

$$\min_{i=1,\ldots,q} \left\{ \langle x^i(b), a \rangle - y_i(b) \right\} < 0, \quad \forall a \in A.$$

Thus we obtain that for any $b^j \in B$, $j = 1, \ldots, p$ there exists a set of q hyperplanes $\{x^i(b^j), y_i(b^j)\}$, $i = 1, \ldots, q$ such that

$$\min_{i=1,\ldots,q} \left\{ \langle x^i(b^j), b^j \rangle - y_i(b^j) \right\} > 0 \tag{7}$$

and

$$\min_{i=1,\ldots,q} \left\{ \langle x^i(b^j), a \rangle - y_i(b^j) \right\} < 0, \quad \forall a \in A. \tag{8}$$

Consequently we have pq hyperplanes

$$\left\{ x^i(b^j), y_i(b^j) \right\}, \ i = 1, \ldots, q, \ j = 1, \ldots, p.$$

The set of these hyperplanes can be rewritten as follows:

$$H = \{h_1, \ldots, h_l\}, \quad h_{i+(j-1)q} = \left\{ x^i(b^j), y_i(b^j) \right\},$$
$$i = 1, \ldots, q, \ j = 1, \ldots, p, \ l = pq.$$

Let $J = \{1, \ldots, l\}$, $I = \{1, \ldots, p\}$ and

$$\bar{x}^{i+(j-1)q} = x^i(b^j), \ \bar{y}_{i+(j-1)q} = y_i(b^j), \ i = 1, \ldots, q, \ j = 1, \ldots, p.$$

Consider the following partition of the set J:

$$J^p = \{J_1, \ldots, J_p\}, \ J_k = \{(k-1)q + 1, \ldots, kq\}, \ k = 1, \ldots, p.$$

It follows from (7) and (8) that for all $k \in I$ and $a \in A$

$$\min_{j \in J_k} \left\{ \langle \bar{x}^j, a \rangle - \bar{y}_j \right\} < 0$$

and for any $b \in B$ there exists at least one $k \in I$ such that

$$\min_{j \in J_k} \left\{ \langle \bar{x}^j, b \rangle - \bar{y}_j \right\} > 0$$

which means that the sets A and B are max-min separable.   □

Corollary 1. *(see [Bag05]). The sets A and B are max-min separable if and only if they are disjoint: $A \cap B = \emptyset$.*

Proof. Necessity is straightforward.
Sufficiency. The set A can be represented as a union of its own points. Since the sets A and B are disjoint the condition (6) is satisfied. Then the proof of the corollary follows from Proposition 3.   □

In the next proposition we show that in most cases the number of hyper-lanes necessary for the max-min separation of the sets A and B is limited.

Proposition 4. *(see [Bag05]). Assume that the set A can be represented as a union of sets A_i, $i = 1, \ldots, q$ and the set B as a union of sets B_j, $j = 1, \ldots, d$ such that*

$$A = \bigcup_{i=1}^{q} A_i, \quad B = \bigcup_{j=1}^{d} B_j$$

and

$$co\, A_i \bigcap co\, B_j = \emptyset \quad \text{for all } i = 1, \ldots, q, \ j = 1, \ldots, d. \tag{9}$$

Then the number of hyperplanes necessary for the separation of the sets A and B is at most $q \cdot d$.

Proof. Let $i \in \{1, \ldots, q\}$ and $j \in \{1, \ldots, d\}$ be any fixed indices. Since $co\, A_i \bigcap co\, B_j = \emptyset$ there exists a hyperplane $\{x^{ij}, y_{ij}\}$ with $x^{ij} \in \mathbb{R}^n$, $y_{ij} \in \mathbb{R}^1$ such that

$$\langle x^{ij}, a \rangle - y_{ij} < 0 \quad \forall a \in co\, A_i$$

and

$$\langle x^{ij}, b \rangle - y_{ij} > 0 \quad \forall b \in co\, B_j.$$

Consequently for any $j \in \{1, \ldots, d\}$ there exists a set of hyperplanes $\{x^{ij}, y_{ij}\}$, $i = 1, \ldots, q$ such that

$$\min_{i=1,\ldots,q} \langle x^{ij}, b \rangle - y_{ij} > 0, \ \forall b \in B_j \tag{10}$$

and

$$\min_{i=1,\ldots,q} \langle x^{ij}, a \rangle - y_{ij} < 0, \ \forall a \in A. \tag{11}$$

Thus we get a system of $l = dq$ hyperplanes:

$$H = \{h_1, \ldots, h_l\}$$

where $h_{i+(j-1)q} = \{x^{ij}, y_{ij}\}$, $i = 1, \ldots, q$, $j = 1, \ldots, d$. Let $J = \{1, \ldots, l\}$, $I = \{1, \ldots, d\}$ and

$$\bar{x}^{i+(j-1)q} = x^{ij}, \ \bar{y}_{i+(j-1)q} = y_{ij}, \ i = 1, \ldots, q, \ j = 1, \ldots, d.$$

Consider the following partition of the set J:

$$J^d = \{J_1, \ldots, J_d\}, \ J_k = \{(k-1)q + 1, \ldots, kq\}, \ k = 1, \ldots, d.$$

It follows from (10) and (11) that for all $k \in I$ and $a \in A$

$$\min_{j \in J_k} \{\langle \bar{x}^j, a \rangle - \bar{y}_j\} < 0$$

and for any $b \in B$ there exists at least one $k \in I$ such that

$$\min_{j \in J_k} \left\{ \langle \bar{x}^j, b \rangle - \bar{y}_j \right\} > 0,$$

that is the sets A and B are max-min separable with at most $q \cdot d$ hyperplanes.

$\square$

Remark 5. The only cases where the number of hyperplanes necessary is large are when the sets A_i and B_j contain a very small number of points. This situation appears only in the particular case where the distribution of the points is like a "chessboard".

3.2 Error function

Given any set of hyperplanes $\{x^j, y_j\}$, $j \in J = \{1, \ldots, l\}$ with $x^j \in \mathbb{R}^n$, $y_j \in \mathbb{R}^1$ and a partition $J^r = \{J_1, \ldots, J_r\}$ of the set J, we say that a point $a \in A$ is well separated from the set B if the following condition is satisfied:

$$\max_{i \in I} \min_{j \in J_i} \left\{ \langle x^j, a \rangle - y_j \right\} + 1 \leq 0.$$

Then we can define the separation error for a point $a \in A$ as follows:

$$\max \left[0, \max_{i \in I} \min_{j \in J_i} \left\{ \langle x^j, a \rangle - y_j + 1 \right\} \right]. \tag{12}$$

Analogously, a point $b \in B$ is said to be well separated from the set A if the following condition is satisfied:

$$\min_{i \in I} \max_{j \in J_i} \left\{ -\langle x^j, b \rangle + y_j \right\} + 1 \leq 0.$$

Then the separation error for a point $b \in B$ can be written as

$$\max \left[0, \min_{i \in I} \max_{j \in J_i} \left\{ -\langle x^j, b \rangle + y_j + 1 \right\} \right]. \tag{13}$$

Thus, an averaged error function can be defined as

$$f(x, y) = (1/m) \sum_{k=1}^{m} \max \left[0, \max_{i \in I} \min_{j \in J_i} \left\{ \langle x^j, a^k \rangle - y_j + 1 \right\} \right]$$

$$+ (1/p) \sum_{t=1}^{p} \max \left[0, \min_{i \in I} \max_{j \in J_i} \left\{ -\langle x^j, b^t \rangle + y_j + 1 \right\} \right] \tag{14}$$

where $x = (x^1, \ldots, x^l) \in \mathbb{R}^{l \times n}$, $y = (y_1, \ldots, y_l) \in \mathbb{R}^l$. It is clear that $f(x, y) \geq 0$ for all $x \in \mathbb{R}^{l \times n}$ and $y \in \mathbb{R}^l$.

Proposition 5. *(see [Bag05]). The sets A and B are max-min separable if and only if there exists a set of hyperplanes $\{x^j, y_j\}, j \in J = \{1, \ldots, l\}$ and a partition $J^r = \{J_1, \ldots, J_r\}$ of the set J such that $f(x, y) = 0$.*

Proof. Necessity. Assume that the sets A and B are max-min separable. Then it follows from Proposition 1 that there exists a set of hyperplanes $\{x^j, y_j\}, j \in J$ and a partition $J^r = \{J_1, \ldots, J_r\}$ of the set J such that

$$\min_{j \in J_i}\{\langle x^j, a\rangle - y_j\} \leq -1, \ \forall a \in A, \ i \in I = \{1, \ldots, r\} \tag{15}$$

and for any $b \in B$ there exists at least one $t \in I$ such that

$$\min_{j \in J_t}\{\langle x^j, b\rangle - y_j\} \geq 1. \tag{16}$$

Consequently we have

$$\max_{i \in I} \min_{j \in J_i}\{\langle x^j, a\rangle - y_j + 1\} \leq 0, \ \ \forall a \in A,$$

$$\min_{i \in I} \max_{j \in J_i}\{-\langle x^j, b\rangle + y_j + 1\} \leq 0, \ \ \forall b \in B.$$

Then from the definition of the error function we obtain that $f(x, y) = 0$.

Sufficiency. Assume that there exist a set of hyperplanes $\{x^j, y_j\}, j \in J = \{1, \ldots, l\}$ and a partition $J^r = \{J_1, \ldots, J_r\}$ of the set J such that $f(x, y) = 0$. Then from the definition of the error function f we immediately get that the inequalities (15) and (16) are satisfied, that is the sets A and B are max-min separable.   □

Proposition 6. *(see [Bag05]). Assume that the sets A and B are max-min separable with a set of hyperplanes $\{x^j, y_j\}, j \in J = \{1, \ldots, l\}$ and a partition $J^r = \{J_1, \ldots, J_r\}$ of the set J. Then*

1) $x^j = 0, \ j \in J$ cannot be an optimal solution;
2) if
   (a) for any $t \in I$ there exists at least one $b \in B$ such that

$$\max_{j \in J_t}\left\{-\langle x^j, b\rangle + y_j + 1\right\} = \min_{i \in I} \max_{j \in J_i}\left\{-\langle x^j, b\rangle + y_j + 1\right\}, \tag{17}$$

(b) there exists $\tilde{J} = \{\tilde{J}_1, \ldots, \tilde{J}_r\}$ such that $\tilde{J}_t \subset J_t, \ \forall t \in I, \ \tilde{J}_t$ is nonempty at least for one $t \in I$ and $x^j = 0$ for any $j \in \tilde{J}_t, \ t \in I$.
Then the sets A and B are max-min separable with a set of hyperplanes $\{x^j, y_j\}, j \in J^0$ and a partition $\bar{J} = \{\bar{J}_1, \ldots, \bar{J}_r\}$ of the set J^0 where

$$\bar{J}_t = J_t \setminus \tilde{J}_t, \ t \in I \ \ and \ \ J^0 = \bigcup_{i=1}^{r} \bar{J}_i.$$

Proof. 1) Since the sets A and B are max-min separable we get from Proposition 5 that $f(x, y) = 0$. If $x^j = 0$, $j \in J$ then it follows from (14) that for any $y \in \mathbb{R}^l$

$$f(0, y) = (1/m) \sum_{k=1}^{m} \max \left[0, \max_{i \in I} \min_{j \in J_i} \{-y_j + 1\} \right]$$

$$+ (1/p) \sum_{t=1}^{p} \max \left[0, \min_{i \in I} \max_{j \in J_i} \{y_j + 1\} \right].$$

We denote

$$R = \max_{i \in I} \min_{j \in J_i} \{-y_j\}.$$

Then we have

$$\min_{i \in I} \max_{j \in J_i} y_j = - \max_{i \in I} \min_{j \in J_i} \{-y_j\} = -R.$$

Thus

$$f(0, y) = \max[0, R + 1] + \max[0, -R + 1].$$

It is clear that

$$\max[0, R + 1] + \max[0, -R + 1] = \begin{cases} -R + 1 & \text{if } R \le -1, \\ 2 & \text{if } -1 < R < 1, \\ R + 1 & \text{if } R \ge 1. \end{cases}$$

Thus for any $y \in \mathbb{R}^l$

$$f(0, y) \ge 2.$$

On the other side $f(x, y) = 0$ for the optimal solution (x, y), that is $x^j = 0$, $j \in J$ cannot be the optimal solution.

2) Consider the following sets:

$$I^1 = \{i \in I : \bar{J}_i \ne \emptyset\},$$

$$I^2 = \{i \in I : \tilde{J}_i \ne \emptyset\}, \quad I^3 = I^1 \bigcap I^2.$$

It is clear that $\tilde{J}_i = \emptyset$ for any $i \in I^1 \setminus I^3$ and $\bar{J}_i = \emptyset$ for any $i \in I^2 \setminus I^3$.
It follows from the definition of the error function that

$$0 = f(x, y) = \frac{1}{m} \sum_{k=1}^{m} \max \left[0, \max_{i \in I} \min_{j \in J_i} \{\langle x^j, a^k \rangle - y_j + 1\} \right]$$

$$+ \frac{1}{p} \sum_{t=1}^{p} \max \left[0, \min_{i \in I} \max_{j \in J_i} \{-\langle x^j, b^t \rangle + y_j + 1\} \right].$$

Since the function f is nonnegative we obtain

$$\max_{i \in I} \min_{j \in J_i} \left\{ \langle x^j, a \rangle - y_j + 1 \right\} \le 0, \ \forall a \in A, \tag{18}$$

$$\min_{i \in I} \max_{j \in J_i} \left\{ -\langle x^j, b \rangle + y_j + 1 \right\} \le 0, \ \forall b \in B. \tag{19}$$

It follows from (17) and (19) that for any $i \in I^2$ there exists a point $b \in B$ such that

$$\max_{j \in J_i} \left\{ -\langle x^j, b \rangle + y_j + 1 \right\} \le 0. \tag{20}$$

If $i \in I^3 \subset I^2$ then we have

$$0 \ge \max_{j \in J_i} \left\{ -\langle x^j, b \rangle + y_j + 1 \right\} = \max \left\{ \max_{j \in \bar{J}_i} \left\{ -\langle x^j, b \rangle + y_j + 1 \right\}, \max_{i \in \bar{J}_i} \left\{ y_j + 1 \right\} \right\}$$

which means that

$$\max_{j \in \bar{J}_i} \left\{ -\langle x^j, b \rangle + y_j + 1 \right\} \le 0 \tag{21}$$

and

$$\max_{j \in \bar{J}_i} \left\{ y_j + 1 \right\} \le 0. \tag{22}$$

If $i \in I^2 \setminus I^3$ then from (20) we obtain

$$0 \ge \max_{j \in J_i} \left\{ -\langle x^j, b \rangle + y_j + 1 \right\} = \max_{i \in \bar{J}_i} \left\{ y_j + 1 \right\}.$$

Thus we get that for all $i \in I^2$ the inequality (22) is true. (22) can be rewritten as follows:

$$\max_{j \in \bar{J}_i} y_j \le -1, \ \forall i \in I^2. \tag{23}$$

Consequently for any $i \in I^2$

$$\min_{j \in \bar{J}_i} \left\{ -y_j + 1 \right\} = -\max_{j \in \bar{J}_i} y_j + 1 \ge 2. \tag{24}$$

It follows from (18) that for any $i \in I$ and $a \in A$

$$\min_{j \in J_i} \left\{ \langle x^j, a \rangle - y_j + 1 \right\} \le 0. \tag{25}$$

Then for any $i \in I^3$ we have

$$0 \ge \min_{j \in J_i} \left\{ \langle x^j, a \rangle - y_j + 1 \right\} = \min \left\{ \min_{j \in \bar{J}_i} \left\{ \langle x^j, a \rangle - y_j + 1 \right\}, \min_{j \in \bar{J}_i} \left\{ -y_j + 1 \right\} \right\}.$$

Taking into account (24) we get that for any $i \in I^3$ and $a \in A$

$$\min_{j \in \bar{J}_i} \left\{ \langle x^j, a \rangle - y_j + 1 \right\} \le 0. \tag{26}$$

If $i \in I^2 \setminus I^3$ then it follows from (25) that

$$\min_{j\in \bar{J}_i}\{-y_j + 1\} \leq 0$$

which contradicts (24). Thus we obtain that $I^2 \setminus I^3 \neq \emptyset$ cannot occur, $I^2 \subset I^1$ and $I^3 = I^2$. It is clear that $\bar{J}_i = J_i$ for any $i \in I^1 \setminus I^2$. Then it follows from (18) that for any $i \in I^1 \setminus I^2$ and $a \in A$

$$\min_{j\in \bar{J}_i}\{\langle x^j, a\rangle - y_j + 1\} \leq 0. \tag{27}$$

From (26) and (27) we can conclude that for any $i \in I$ and $a \in A$

$$\min_{j\in \bar{J}_i}\{\langle x^j, a\rangle - y_j + 1\} \leq 0. \tag{28}$$

It follows from (19) that for any $b \in B$ there exists at least one $i \in I$

$$\max_{j\in J_i}\{-\langle x^j, b\rangle + y_j + 1\} \leq 0.$$

Then from expression

$$\max_{j\in J_i}\{-\langle x^j, b\rangle + y_j + 1\} = \max\left\{\max_{j\in \bar{J}_i}\{-\langle x^j, b\rangle + y_j + 1\}, \max_{i\in \bar{J}_i}\{y_j + 1\}\right\}$$

we get that for any $b \in B$ there exists at least one $i \in I$ such that

$$\max_{j\in \bar{J}_i}\{-\langle x^j, b\rangle + y_j + 1\} \leq 0. \tag{29}$$

Thus it follows from (28) and (29) that the sets A and B are max-min separable with the set of hyperplanes $\{x^j, y_j\}$, $j \in J^0$ and a partition $\bar{J}$ of the set J^0.   □

Remark 6. In most cases, if a given set of hyperplanes with a particular partition separates the sets A and B, then there are other sets of hyperplanes with the same partition which will also separate the sets A and B (see Figure 4). The error function (14) is nonconvex and if the sets A and B are max-min separable, then the global minimum of this function $f(x^*, y_*) = 0$ and the global minimizer is not unique.

4 Minimization of the error function

In this section we discuss an algorithm for minimization of the error function.

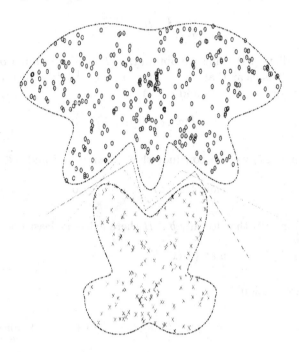

Fig. 4. Max-min separability.

4.1 Statement of problem

The problem of the max-min separability is reduced to the following mathematical programming problem:

$$\text{minimize } f(x,y) \text{ subject to } (x,y) \in \mathbb{R}^{(n+1)\times l} \tag{30}$$

where the objective function f has the following form:

$$f(x,y) = f_1(x,y) + f_2(x,y)$$

and

$$f_1(x,y) = \frac{1}{m} \sum_{k=1}^{m} \max\left[0, \max_{i\in I} \min_{j\in J_i}\left\{\langle x^j, a^k\rangle - y_j + 1\right\}\right], \tag{31}$$

$$f_2(x,y) = \frac{1}{p} \sum_{t=1}^{p} \max\left[0, \min_{i\in I} \max_{j\in J_i}\left\{-\langle x^j, b^t\rangle + y_j + 1\right\}\right]. \tag{32}$$

The problem (30) is a global optimization problem. However, the number of variables in this problem is large and the global optimization methods cannot be directly applied to solve it. Therefore we will discuss algorithms for finding local minima of the function f.

The function f_1 contains the following max-min functions:

$$\varphi_{1k}(x,y) = \max_{i \in I} \min_{j \in J_i} \left\{ \langle x^j, a^k \rangle - y_j + 1 \right\}, \quad k = 1, \ldots, m$$

and the function f_2 contains the following min-max functions:

$$\varphi_{2t}(x,y) = \min_{i \in I} \max_{j \in J_i} \left\{ -\langle x^j, b^t \rangle + y_j + 1 \right\}, \quad t = 1, \ldots, p.$$

4.2 Differential properties of the objective function

Both functions f_1 and f_2 are nonsmooth, nonconvex piecewise linear. These functions contain some max-min-type functions. The functions f_1 and f_2 and consequently, the function f are locally Lipschitz continuous. We will recall some definitions from nonsmooth analysis.

We consider a locally Lipschitz function φ defined on $\mathbb{R}^n$. This function is differentiable almost everywhere and one can define for it a Clarke subdifferential (see [Cla83]), by

$$\partial\varphi(x)$$
$$= \text{co} \left\{ v \in \mathbb{R}^n : \exists (x^k \in D(\varphi), x^k \longrightarrow x, k \longrightarrow +\infty) : v = \lim_{k \longrightarrow +\infty} \nabla\varphi(x^k) \right\},$$

here $D(\varphi)$ denotes the set where φ is differentiable, co denotes the convex hull of a set.

The function φ is differentiable at the point $x \in \mathbb{R}^n$ with respect to the direction $g \in \mathbb{R}^n$ if the limit

$$\varphi'(x,g) = \lim_{\alpha \to +0} \frac{\varphi(x + \alpha g) - \varphi(x)}{\alpha}$$

exists. The number $\varphi'(x,g)$ is said to be the derivative of the function φ with respect to the direction $g \in \mathbb{R}^n$ at the point x.

The Clarke upper derivative $\varphi^0(x,g)$ of the function φ at the point x with respect to the direction $g \in \mathbb{R}^n$ is defined as follows:

$$\varphi^0(x,g) = \limsup_{\alpha \to +0, y \to x} \frac{\varphi(y + \alpha g) - \varphi(y)}{\alpha}.$$

The following is true (see [Cla83])

$$\varphi^0(x,g) = \max\{\langle v, g \rangle : v \in \partial\varphi(x)\}.$$

It should be noted that the Clarke upper derivative always exists for locally Lipschitz continuous functions. The function φ is said to be Clarke regular at the point $x \in \mathbb{R}^n$ if

$$\varphi'(x,g) = \varphi^0(x,g)$$

for all $g \in \mathbb{R}^n$. For Clarke regular functions there exists a calculus (see [Cla83, DR95]). However in general for non-regular functions such a calculus does not exist.

The function φ is called semismooth at $x \in \mathbb{R}^n$, if it is locally Lipschitz continuous at x and for every $g \in \mathbb{R}^n$, the limit

$$\lim_{v \in \partial\varphi(x+tg'),g'\to g,t\to+0} \langle v, g \rangle$$

exists (see [Mif77]).

Let us return to the objective function f of problem (30). Since this function is locally Lipschitz continuous it is Clarke subdifferentiable.

Proposition 7. *The function f is semismooth.*

Proof. The sum, the maximum and the minimum of semismooth functions are semismooth (see [Mif77]). A linear function, as a smooth function, is semismooth. Thus the function f which is the sum of functions represented as the maximum of 0 and max-min of linear functions, is semismooth.   □

The properties of max-min type functions were studied, for example, in [DDM02, Pol97]. Max-min-type functions in general are not Clarke regular.

Example 1. Consider the function

$$\varphi(x) = \max\{\min\{3x_1 + x_2, \ 2x_1 + 3x_2\}, \ \min\{x_1 + 2x_2, \ 4x_1 + 4x_2\}\}.$$

The Clarke subdifferential of this function at the point $x = (0,0)$ is

$$\partial\varphi(x) = \mathrm{co}\{(3,1),(2,3),(1,2),(4,4)\}.$$

Then the Clarke upper derivative $\varphi^0(x, g^0)$ of the function φ at the point $x = (0,0)$ with respect to the direction $g^0 = (0,1)$ is

$$\varphi^0(x, g^0) = \max\{\langle v, g^0 \rangle : \ v \in \partial\varphi(x)\} = 4.$$

However, the directional derivative of this function with respect to the direction $g^0 = (0,1)$ is $\varphi'(x, g) = 2$ that is $\varphi'(x, g^0) < \varphi^0(x, g^0)$. Thus the function φ is not Clarke regular.

Since the function f contains max-min of linear functions this function is not Clarke regular apart from linear separability. Therefore, subgradients of the function f cannot be calculated using subgradients of the involved max-min-type functions. We can conclude that the calculation of the subgradients of the function f is a very difficult task and therefore the application of methods of nonsmooth optimization requiring a subgradient evaluation at each iteration, including bundle method and its variations([HL93, Kiw85, MN92]), cannot be effective.

In the paper [KP98] optimization problems with twice continuously differentiable objective functions and max-min constraints were considered and these problems were converted to problems with smooth objective and constraint functions. However, this approach cannot be applied to the problem (30), because the function f contains not only max-min-type functions but also min-max-type functions.

Since the evaluation of subgradients of the function f is difficult, direct search methods of optimization seem to be the best option for solving problem (30). Among such methods we mention here two widely used methods: Powell's method (see [Pow02]) which is based on a quadratic approximation of the objective function and Nelder-Mead's simplex method [NM65]. As was mentioned in [Pow02] Powell's method performs well when the number of variables is less than 20. For the simplex method this number is even smaller. Moreover, both methods are effective when the objective function is smooth. However, in the max-min separability problem the number of variables is $n_v = (n+1) \times l$ where n is the dimension of the sets A and B (ranging from 5 to thousands in real world datasets), and l is the number of separating hyperplanes. In many cases the number n_v is greater than 20. Furthermore, the objective function in this problem is a quite complicated nonsmooth function.

In this paper we use the discrete gradient method to solve the problem (30). The description of this method can be found in [Bag99a, Bag99b] (see, also, [Bag02]). The discrete gradient method can be considered as a version of the bundle method ([HL93, Kiw85, MN92]), where subgradients of the objective function are replaced by its discrete gradients.

The discrete gradient method uses only values of the objective function. It should be noted that the calculation of the objective function in the problem (30) can be expensive. We will show that the use of the discrete gradient method allow to significantly reduce the number of objective function evaluations.

4.3 Discrete gradient method

In this subsection we will briefly describe the discrete gradient method. We start with the definition of the discrete gradient.

Definition of the discrete gradient

Let f be a locally Lipschitz continuous function defined on $\mathbb{R}^n$. Let

$$S_1 = \{g \in \mathbb{R}^n : \|g\| = 1\},$$
$$G = \{e \in \mathbb{R}^n : e = (e_1, \ldots, e_n), |e_j| = 1, j = 1, \ldots, n\},$$
$$P = \{z(\lambda) : z(\lambda) \in \mathbb{R}^1, z(\lambda) > 0, \lambda > 0, \lambda^{-1}z(\lambda) \to 0, \lambda \to 0\},$$
$$I(g, \alpha) = \{i \in \{1, \ldots, n\} : |g_i| \geq \alpha\},$$

where $\alpha \in (0, n^{-1/2}]$ is a fixed number.

Here S_1 is the unit sphere, G is the set of vertices of the unit hypercube in $\mathbb{R}^n$ and P is the set of univariate positive infinitesimal functions.

We define operators $H_i^j : \mathbb{R}^n \to \mathbb{R}^n$ for $i = 1, \ldots, n$, $j = 0, \ldots, n$ by the formula

$$H_i^j g = \begin{cases} (g_1, \ldots, g_j, 0, \ldots, 0) & \text{if } j < i, \\ (g_1, \ldots, g_{i-1}, 0, g_{i+1}, \ldots, g_j, 0, \ldots, 0) & \text{if } j \geq i. \end{cases} \tag{33}$$

We can see that

$$H_i^j g - H_i^{j-1} g = \begin{cases} (0, \ldots, 0, g_j, 0, \ldots, 0) & \text{if } j = 1, \ldots, n, j \neq i, \\ 0 & \text{if } j = i. \end{cases} \tag{34}$$

Let $e(\beta) = (\beta e_1, \beta^2 e_2, \ldots, \beta^n e_n)$, where $\beta \in (0, 1]$. For $x \in \mathbb{R}^n$ we consider vectors

$$x_i^j \equiv x_i^j(g, e, z, \lambda, \beta) = x + \lambda g - z(\lambda) H_i^j e(\beta), \tag{35}$$

where $g \in S_1$, $e \in G$, $i \in I(g, \alpha)$, $z \in P$, $\lambda > 0$, $j = 0, \ldots, n$, $j \neq i$.

It follows from (34) that

$$x_i^{j-1} - x_i^j = \begin{cases} (0, \ldots, 0, z(\lambda) e_j(\beta), 0, \ldots, 0) & \text{if } j = 1, \ldots, n, j \neq i, \\ 0 & \text{if } j = i. \end{cases} \tag{36}$$

It is clear that $H_i^0 g = 0$ and $x_i^0(g, e, z, \lambda, \beta) = x + \lambda g$ for all $i \in I(g, \alpha)$.

Definition 4. *(see [BG95]) The discrete gradient of the function f at the point $x \in \mathbb{R}^n$ is the vector $\Gamma^i(x, g, e, z, \lambda, \beta) = (\Gamma_1^i, \ldots, \Gamma_n^i) \in \mathbb{R}^n, g \in S_1, i \in I(g, \alpha)$, with the following coordinates:*

$$\Gamma_j^i = [z(\lambda) e_j(\beta)]^{-1} \left[f(x_i^{j-1}(g, e, z, \lambda, \beta)) - f(x_i^j(g, e, z, \lambda, \beta)) \right],$$

$$j = 1, \ldots, n, j \neq i,$$

$$\Gamma_i^i = (\lambda g_i)^{-1} \left[f(x_i^n(g, e, z, \lambda, \beta)) - f(x) - \sum_{j=1, j \neq i}^{n} \Gamma_j^i(\lambda g_j - z(\lambda) e_j(\beta)) \right].$$

A more detailed description of the discrete gradient and examples can be found in [Bag99b].

Remark 7. It follows from Definition 4 that for the calculation of the discrete gradient $\Gamma^i(x, g, e, z, \lambda, \beta), i \in I(g, \alpha)$ we define a sequence of points

$$x_i^0, \ldots, x_i^{i-1}, x_i^{i+1}, \ldots, x_i^n.$$

For the calculation of the discrete gradient it is sufficient to evaluate the function f at each point of this sequence.

Remark 8. The discrete gradient is defined with respect to a given direction $g \in S_1$. We can see that for the calculation of one discrete gradient we have to calculate $(n+1)$ values of the function f: at the point x and at the points $x_i^j(g, e, z, \lambda, \beta)$, $j = 0, \ldots, n$, $j \neq i$. For the calculation of the next discrete gradient at the same point with respect to any other direction $g^1 \in S_1$ we have to calculate this function n times, because we have already calculated f at the point x.

Calculation of the discrete gradients of the objective function (30)

Now let us return to the objective function f of the problem (30). This function depends on $(n+1)l$ variables where l is the number of hyperplanes. The function f_1 contains max-min functions φ_{1k}

$$\varphi_{1k}(x, y) = \max_{i \in I} \min_{j \in J_i} \psi_{1jk}(x, y), \quad k = 1, \ldots, m$$

where

$$\psi_{1jk}(x, y) = \langle x^j, a^k \rangle - y_j + 1, \quad j \in J_i, \ i \in I.$$

We can see that for every $k = 1, \ldots, m$, each pair of variables $\{x^j, y_j\}$ appears in only one function ψ_{1jk}.

For a given $i = 1, \ldots, (n+1)l$ we set

$$q_i = \left\lfloor \frac{i-1}{n+1} \right\rfloor + 1, \quad d_i = i - (q_i - 1)(n+1)$$

where $\lfloor u \rfloor$ stands for the floor of a number u. We define by X the vector of all variables $\{x^j, y_j\}, j = 1, \ldots, l$:

$$X = (X_1, X_2, \ldots, X_{(n+1)l})$$

where

$$X_i = \begin{cases} x_{d_i}^{q_i} & \text{if } 1 \leq d_i \leq n, \\ y_{q_i} & \text{if } d_i = n+1. \end{cases}$$

We use the vector of variables X to define a sequence

$$X_t^0, \ldots, X_t^{t-1}, X_t^{t+1}, \ldots, X_t^{(n+1)l}, \ t \in I(g, \alpha), \ g \in \mathbb{R}^{(n+1)l}$$

as in Remark 7. It follows from (36) that the points X_t^{i-1} and X_t^i differ by one coordinate only. This coordinate appears in only one linear function ψ_{1q_ik}. It follows from the definition of the operator H_i^j that $X_t^t = X_t^{t-1}$ and thus this observation is also true for X_t^{t+1}. Then we get

$$\psi_{1jk}(X_t^i) = \psi_{1jk}(X_t^{i-1}) \ \forall j \neq q_i.$$

Moreover the function ψ_{1q_ik} can be calculated at the point X_t^i using the value of this function at the point X_t^{i-1}, $i \geq 1$:

$$\psi_{1q_i k}(X_t^i) = \begin{cases} \psi_{1q_i k}(X_t^{i-1}) - z(\lambda)a_{d_i}^k e_i(\beta) & \text{if } 1 \le d_i \le n, \\ \psi_{1q_i k}(X_t^{i-1}) + z(\lambda)e_i(\beta) & \text{if } d_i = n+1 \end{cases} \tag{37}$$

In order to calculate the function f_1 at the point X_t^i, $i \ge 1$ first we have to calculate the values of the functions $\psi_{1q_i k}$ for all $a^k \in A, k = 1, \ldots, m$ using (37). Then we update f_1 using these values and the values of all other linear functions at the point X_t^{i-1} according to (31). Thus we have to apply a full calculation of the function f_1 using the formula(31) only at the point $X_t^0 = X + \lambda g$.

Since the function f_2 has a similar structure as f_1 we can calculate it in the same manner using a formula similar to (37).

Thus for the calculation of each discrete gradient we have to apply a full calculation of the objective function f only at the point $X_t^0 = X + \lambda g$ and this function can be updated at the points X_t^i, $i \ge 1$ using a simplified scheme.

We can conclude that for the calculation of the discrete gradient at a point X with respect to the direction $g^0 \in S_1$ we calculate the function f at two points: X and $X_t^0 = X + \lambda g^0$. For the calculation of another discrete gradient at the same point X with respect to any other direction $g^1 \in S_1$ we calculate the function f only at the point: $X + \lambda g^1$.

Since the number of variables $(n + 1)l$ in the problem (30) can be large this algorithm allows to significantly reduce the number of objective function evaluations during the calculation of a discrete gradient.

On the other hand the function f_1 contains max-min-type functions and their computation can be simplified using an algorithm proposed in [Evt72]. The function f_2 contains min-max-type functions and a similar algorithm can be used for their calculation.

Results of numerical experiments show that the use of these algorithms allows one to significantly accelerate the computation of the objective function f and its discrete gradients.

Discrete gradient method

We consider the following unconstrained minimization problem:

$$\text{minimize } \varphi(x) \text{ subject to } x \in \mathbb{R}^n \tag{38}$$

where the function φ is assumed to be semismooth. We consider the following algorithm for solving this problem. An important step in this algorithm is the calculation of a descent direction of the objective function φ. So first, we describe an algorithm for the computation of this descent direction.

Let $z \in P, \lambda > 0, \beta \in (0, 1]$, the number $c \in (0, 1)$ and a small enough number $\delta > 0$ be given.

Algorithm 1. An algorithm for the computation of the descent direction.

Step 1. Choose any $g^1 \in S_1, e \in G, i \in I(g^1, \alpha)$ and compute a discrete gradient $v^1 = \Gamma^i(x, g^1, e, z, \lambda, \beta)$. Set $\overline{D}_1(x) = \{v^1\}$ and $k = 1$.

Step 2. Calculate the vector $\|w^k\| = \min\{\|w\| : w \in \overline{D}_k(x)\}$. If

$$\|w^k\| \le \delta, \tag{39}$$

then stop. Otherwise go to Step 3.

Step 3. Calculate the search direction by $g^{k+1} = -\|w^k\|^{-1}w^k$.
Step 4. If

$$\varphi(x + \lambda g^{k+1}) - \varphi(x) \le -c\lambda\|w^k\|, \tag{40}$$

then stop. Otherwise go to Step 5.

Step 5. Calculate a discrete gradient

$$v^{k+1} = \Gamma^i(x, g^{k+1}, e, z, \lambda, \beta), \ i \in I(g^{k+1}, \alpha),$$

construct the set $\overline{D}_{k+1}(x) = \text{co}\,\{\overline{D}_k(x) \bigcup \{v^{k+1}\}\}$, set $k = k + 1$ and go to Step 2.

Algorithm 1 contains some steps which deserve some explanations. In Step 1 we calculate the first discrete gradient. The distance between the convex hull of all calculated discrete gradients and the origin is calculated in Step 2. If this distance is less than the tolerance $\delta > 0$ then we accept the point x as an approximate stationary point (Step 2), otherwise we calculate another search direction in Step 3. In Step 4 we check whether this direction is a descent direction. If it is we stop and the descent direction has been calculated, otherwise we calculate another discrete gradient with respect to this direction in Step 5 and add it to the set $\overline{D}_k$.

It is proved that Algorithm 1 is terminating (see [Bag99a, Bag99b]).

Let numbers $c_1 \in (0, 1), c_2 \in (0, c_1]$ be given.

Algorithm 2. Discrete gradient method

Step 1. Choose any starting point $x^0 \in \mathbb{R}^n$ and set $k = 0$.
Step 2. Set $s = 0$ and $x_s^k = x^k$.

Step 3. Apply Algorithm 1 for the calculation of the descent direction at $x = x_s^k, \delta = \delta_k, z = z_k, \lambda = \lambda_k, \beta = \beta_k, c = c_1$. This algorithm terminates after a finite number of iterations $m > 0$. As a result we get the set $D_m(x_s^k)$ and an element v_s^k such that

$$\|v_s^k\| = \min\{\|v\| : v \in \overline{D}_m(x_s^k)\}.$$

Furthermore either $\|v_s^k\| \le \delta_k$ or for the search direction $g_s^k = -\|v_s^k\|^{-1}v_s^k$

$$\varphi(x_s^k + \lambda_k g_s^k) - \varphi(x_s^k) \leq -c_1 \lambda_k \|v_s^k\|. \tag{41}$$

Step 4. If
$$\|v_s^k\| \leq \delta_k \tag{42}$$
then set $x^{k+1} = x_s^k, k = k + 1$ and go to Step 2. Otherwise go to Step 5.

Step 5. Construct the following iteration $x_{s+1}^k = x_s^k + \sigma_s g_s^k$, where σ_s is defined as follows

$$\sigma_s = \arg\max\{\sigma \geq 0 : \varphi(x_s^k + \sigma g_s^k) - \varphi(x_s^k) \leq -c_2 \sigma \|v_s^k\|\}.$$

Step 6. Set $s = s + 1$ and go to Step 3.

For the point $x^0 \in \mathbb{R}^n$ we consider the set $M(x^0) = \{x \in \mathbb{R}^n : \varphi(x) \leq \varphi(x^0)\}$.

Theorem 1. *Assume that the set $M(x^0)$ is bounded for starting points $x^0 \in \mathbb{R}^n$. Then every accumulation point of $\{x^k\}$ belongs to the set $X^0 = \{x \in \mathbb{R}^n : 0 \in \partial\varphi(x)\}$.*

Since the objective function in problem (30) is semismooth the discrete gradient method can be applied to solve it. Discrete gradients in Step 5 of Algorithm 1 can be calculated using the simplified scheme described above.

5 Results of numerical experiments

We applied the max-min separation to solve supervised data classification problems on some real-world datasets. In this section we present results of numerical experiments. Our algorithm has been implemented in Lahey Fortran 95 on a Pentium 4 1.7 GHz.

5.1 Supervised data classification via max-min separability

We are given a dataset A containing a finite number of points in $\mathbb{R}^n$. This dataset contains d disjoint subsets $A_1, \ldots, A_d$ where A_i represents a training set for the class i. The aim of supervised data classification is to establish rules for the classification of some new observations using these training subsets of the classes. This problem is reduced to d set separation problems.

Each of these problems consists in separating one class from the rest of the dataset. To separate the class i from all others, we separate sets A_i and $\bigcup_{j \neq i} A_j$, with a piecewise linear function by solving problem (30).

One of the important question in supervised data classification is the estimation of performance measure. Different performance measures are discussed in [Tho02]. When the dataset contains two classes the classification problem can be reduced to only one separation problem, therefore the classification rules are straightforward. We consider that the separation function obtained from the training set, separates the two classes.

When the dataset contains more than two classes we have more than one separation function. In our case for each class i of the dataset A we have one piecewise linear function φ_i separating the training set A_i from all other training points $\bigcup_{j \neq i} A_j$. We approximate the training set A_i using the following set

$$\bar{A}_i = \{a \in \mathbb{R}^n : \varphi_i(a) < 0\}.$$

Thus we get the sets $\bar{A}_1, \ldots, \bar{A}_d$ which approximate the training sets $A_1, \ldots, A_d$, respectively. Then for each $i \in \{1, \ldots, d\}$ we can consider the following two sets:

$$A_i^0 = \bar{A}_i, \quad \bar{A}_i^0 = \bigcup_{j=1, j \neq i}^{d} \bar{A}_j$$

These two sets define the following four sets (see Figure 5):

1. $A_i^0 \bigcap (\mathbb{R}^n \backslash \bar{A}_i^0)$
2. $(\mathbb{R}^n \backslash A_i^0) \bigcap \bar{A}_i^0$
3. $A_i^0 \bigcap \bar{A}_i^0$
4. $(\mathbb{R}^n \backslash A_i^0) \bigcap (\mathbb{R}^n \backslash \bar{A}_i^0)$

If a new observation a belongs to the first set we classify it in class i, if it belongs to the second set we classify it not to be in class i. If this point belongs to the third or fourth set in this case if $\varphi_i(a) < \min_{j=1,\ldots,d, j \neq i} \varphi_j(a)$ then we classify it in class i, otherwise we classify it not to be in class i.

In order to evaluate the classification algorithm we use two performance measures. First we present the average accuracy (a_{2c} in Tables 3 and 4) for well-classified points in two classes classification (when one particular class is separated from all others) and the multi-class classification accuracy (a_{mc} in Tables 3 and 4) as described above. First accuracy is an indication of separation quality and the second one is an indication of multi-class classification quality.

5.2 Results on small and middle size datasets

In this subsection we present results of numerical experiments with some small and middle size datasets in order to demonstrate the separation ability of the proposed algorithm. The datasets used are the Wisconsin Breast Cancer Diagnosis (WBCD), the Wisconsin Breast Cancer Prognosis (WBCP), the Cleveland Heart Disease (Heart), the Pima Indians Diabetes (Diabetes), the BUPA Liver Disorders (Liver), the United States Congressional Voting Records (Votes) and the Ionosphere. All datasets contain 2 classes. The description of these datasets can be found in [MA92].

We take entire datasets and check their polyhedral or max-min separability considering various number of hyperplanes. Results of numerical experiments

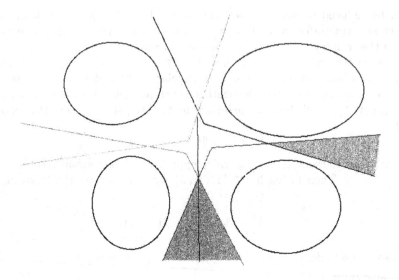

Fig. 5. Multi-class classification by a max-min separation

are presented in Table 1. We use the following notation: m - is the number of instances in the first class, p - is the number of instances in the second class, n - number of attributes, h number of hyperplanes used for polyhedral separability, r is the cardinality of the set I and j is the cardinality of the sets J_i, $i \in I$ in the max-min separability. The sets J_i contain the same number of indices for all $i \in I$. In our experiments we restrict r to 15 and j to 5. The accuracy is defined as the ratio between the number of well-classified points of both A and B and the total number of points in the dataset.

Table 1. Results of numerical experiments with small and middle size datasets

Database	$m/p/n$	Linear	Polyhedral		Max-min	
			h	accuracy	$r \times j$	accuracy
WBCD	239/444/9	97.36	7	98.98	5×2	100
Heart	137/160/13	84.19	10	100	2×5	100
Ionosphere	126/225/34	93.73	4	97.44	2×2	100
Votes	168/267/16	96.80	5	100	2×3	100
WBCP	46/148/32	76.80	4	100	3×2	100
Diabetes	268/500/8	76.95	12	80.60	15×2	90.10
Liver	145/200/6	68.41	12	74.20	6×5	89.86

From the results presented in Table 1 we can conclude that in none of the datasets classes are linearly separable. Classes in heart, votes and WBCP are polyhedrally separable and in WBCD they are "almost" polyhedrally separable. We considered different values for h in diabetes and liver datasets and present best results. These results show that classes in these datasets are unlikely to be polyhedrally separable. Classes in WBCD, heart, ionosphere, votes and WBCP are max-min separable with a presented number of hyperplanes whereas classes in diabetes and liver datasets are likely to be max-min separable with quite large number of hyperplanes. On the other side results for these datasets show that the use of max-min separability allows one to achieve significantly better separation.

5.3 Results on larger datasets

Datasets

The datasets used are the Shuttle control , the Letter recognition, the Landsat satellite image, the Pen-based recognition of handwritten and the Page blocks classification databases. Table 2 presents some characteristics of these databases. More detailed information can be found in [MA92]. It should be noted that all attributes in these datasets are continuous.

Table 2. Large datasets

Database	(train,test)	No. of attributes	No. of classes
Shuttle control	(43500,14500)	9	7
Letter recognition	(15000,5000)	16	26
Landsat satellite image	(4435,2000)	36	6
Pen-based recognition of handwritten	(7494,3498)	16	10
Page blocks	(4000,1473)	10	5

Results and discussion

We took $X^0 = 0 \in \mathbb{R}^{(n+1)l}$ as a starting point for solving each separation problem (30). At each iteration of the discrete gradient method the line search is carried out by approximation of the objective function using univariate piecewise linear function (see [Bag99a]). In each separation problem (30) all J_i, $i \in I$ have the same cardinality.

Results of numerical experiments are presented in Tables 3 and 4. In these tables *fct eval*, *DG eval* and *CPU time* show respectively the average number of objective function evaluations, discrete gradient evaluations and CPU time required to solve an optimization problem. CPU time is presented in seconds.

From the results presented in these tables we can see that the use of the max-min separability algorithm allows to achieve a high classification accuracy for both training and test phases. Results on training sets show that this algorithm provides a high quality of separation between two sets. In our experiments we used only large-scale datasets. Results on these datasets show that a few hyperplanes are sufficient to separate efficiently sets with large numbers of points. Since we use a derivative-free method to solve problem (30) the number of objective function evaluations is a significant characteristic for estimation of the complexity of the max-min separability algorithm. Results presented in Tables 3 and 4 confirm that the proposed algorithm is effective for solving classification problems on large-scale databases.

Table 3. Results of numerical experiments with Shuttle control, Letter recognition and Landsat satellite image datasets

		Training		Test				
$|I|$	$|J_i|$	a_{2c}	a_{mc}	a_{2c}	a_{mc}	fct eval	DG eval	CPU time
			Shuttle control dataset					
1	1	94.63	87.84	94.66	87.86	265	268	54.44
2	1	97.26	97.58	97.08	97.49	396	399	145.12
3	1	97.04	99.36	96.87	99.21	379	379	211.23
4	1	97.35	99.50	97.19	99.35	402	405	310.54
2	2	99.86	99.57	99.86	99.39	391	394	281.92
3	2	99.48	99.92	99.43	99.86	636	639	825.99
4	2	99.84	99.76	99.82	99.70	447	450	810.58
			Letter recognition dataset					
1	1	92.51	66.89	92.32	66.00	280	284	17.57
2	1	96.83	79.86	95.24	79.36	568	572	60.98
3	1	98.34	85.73	95.94	84.82	573	575	93.72
4	1	99.08	89.32	96.36	86.86	665	667	158.29
2	2	98.12	86.89	96.20	84.56	683	686	143.07
3	2	98.97	91.46	96.32	89.12	634	635	366.16
3	3	99.52	93.73	96.16	90.32	511	511	436.37
			Landsat satellite image dataset					
1	1	93.12	86.00	91.30	83.45	298	301	4.62
2	1	96.73	88.12	94.40	85.65	549	552	19.12
3	1	97.54	89.80	94.80	87.00	618	621	37.37
4	1	97.81	91.14	94.35	87.45	656	659	61.64
2	2	97.56	90.85	94.25	87.10	606	609	48.83
3	2	98.02	90.98	94.60	86.70	712	715	116.86
4	2	98.47	93.33	94.80	86.70	533	536	137.07

6 Conclusions and further work

In this paper we have developed the concept of the max-min separability. If finite point sets A and B are disjoint then they can be separated by a certain

Table 4. Results of numerical experiments with Pen-based recognition of handwritten and Page blocks datasets

		Training		Test				
$|I|$	$|J_i|$	a_{2c}	a_{mc}	a_{2c}	a_{mc}	fct eval.	DG. eval.	CPU time
		Pen-based recognition of handwritten dataset						
1	1	97.54	94.93	93.68	89.94	385	388	6.97
2	1	99.45	98.91	96.05	95.37	582	585	19.71
3	1	99.91	99.65	96.51	96.54	865	868	48.19
4	1	99.97	99.79	96.23	97.11	841	844	70.21
2	2	99.91	99.69	96.68	96.31	888	890	63.94
3	2	99.97	99.88	97.37	97.40	727	730	124.91
4	2	99.99	99.89	97.06	97.28	733	736	191.71
		Page blocks dataset						
1	1	93.48	92.60	81.87	82.48	623	626	2.93
2	1	93.88	93.48	80.52	85.61	369	372	3.59
3	1	95.38	94.20	87.24	86.69	550	553	9.65
4	1	95.68	94.88	85.81	87.44	822	825	22.09
2	2	95.55	94.33	88.53	86.97	505	508	11.51
3	2	96.55	95.68	89.34	88.46	779	782	40.71
4	2	96.45	95.40	87.71	86.08	682	685	54.60

piecewise linear function presented as a max-min of linear functions. We have proposed an algorithm to find this piecewise linear function by minimizing an error function.

This algorithm has been applied to solve data classification problems in some large-scale datasets. Results from numerical experiment show the effectiveness of this algorithm.

However the number of hyperplanes needed to separate the two sets has to be known. In further research some methods to find automatically this number will be introduced. Problem (30) is a global optimization problem on which we use a local optimization method. Therefore it is very crucial to find a good initial point in order to reduce computational cost and to improve the solution. These questions are the subject of our further research.

Acknowledgements

This research was supported by the Australian Research Council.

References

[AG02]   Astorino, A., Gaudioso, M.: Polyhedral separability through successive LP. Journal of Optimization Theory and Applications, **112**, 265–293 (2002)

[BG95]   Bagirov, A.M., Gasanov, A.A.: A method of approximating a quasidifferential. Russian Journal of Computational Mathematics and Mathematical Physics, **35**, 403–409 (1995)

[Bag99a]   Bagirov, A.M.: Derivative-free methods for unconstrained nonsmooth optimization and its numerical analysis. Investigacao Operacional, **19**, 75-93 (1999)

[Bag99b]   Bagirov, A.M.: Minimization methods for one class of nonsmooth functions and calculation of semi-equilibrium prices. In: Eberhard. A. et al (eds) Progress in Optimization: Contribution from Australasia, 147–175. Kluwer Academic Publishers (1999)

[Bag02]   A.M. Bagirov, A method for minimization of quasidifferentiable functions. Optimization Methods and Software, **17**, 31–60 (2002)

[Bag05]   Bagirov, A.M.: Max-min separability. Optimization Methods and Software, **20**, 271-290 (2005)

[BRSY01]   Bagirov, A.M., Rubinov, A., Soukhoroukova, N., Yearwood, J.: Unsupervised and supervised data classification via nonsmooth and global optimization. TOP, **11**, 1–93, Sociedad de Estadistica Operativa, Madrid, Spain, June 2003 (2003)

[BRY00]   Bagirov, A.M., Rubinov, A.M., Yearwood, J.: Using global optimization to improve classification for medical diagnosis and prognosis. Topics in Health Information Management, **22**, 65–74 (2001)

[BRY02]   Bagirov, A.M., Rubinov, A.M., Yearwood, J.: A global optimization approach to classification. Optimization and Engineering, **3**, 129–155 (2002)

[BKS95]   Bartels, S.G., Kuntz, L., Sholtes, S.: Continuous selections of linear functions and nonsmooth critical point theory. Nonlinear Analysis, TMA, **24**, 385–407 (1995)

[BB97]   Bennet, K.P., Blue, J.: A support vector machine approach to decision trees. Mathematics Report 97-100, Rensselaer Polytechnic Institute, Troy, New York (1997)

[BB96]   Bennet, K.P., Bredersteiner, E.J.: A parametric optimization method for machine learning. INFORMS Journal on Computing, **9**, 311-318, (1997)

[BM92]   Bennett, K.P., Mangasarian, O.L.: Robust linear programming discrimination of two linearly inseparable sets. Optimization Methods and Software, **1**, 23–34 (1992)

[BM93]   Bennett, K.P., Mangasarian, O.L.: Bilinear separation of two sets in n-space. Computational Optimization and Applications, **2**, 207-227 (1993)

[BM00]   Bradley, P.S., Mangasarian, O.L.: Massive data discrimination via linear support vector machines. Optimization Methods and Software, **13**, 1–10 (2000)

[BFM99]   Bradley, P.S., Fayyad, U.M., O.L. Mangasarian: Data mining: overview and optimization opportunities. INFORMS Journal on Computing, **11**, 217–238 (1999)

[Bur98]   Burges, C.J.C.: A tutorial on support vector machines for pattern recognition. Data Mining and Knowledge Discovery, **2**, 121–167 (1998)

[CM95]   C. Chen, Mangasarian, O.L.: Hybrid misclassification minimization. Mathematical Programming Technical Report, 95-05, University of Wisconsin (1995)

[Cla83]   Clarke, F.H.: Optimization and Nonsmooth Analysis, Wiley-Interscience, New York (1983)

[DDM02] Demyanov, A.V., Demyanov, V.F., Malozemov, V.N.: Minmaxmin problems revisited. Optimization Methods and Software, **17**, 783–804 (2002)

[DR95] Demyanov, V.F., Rubinov, A.M., Constructive Nonsmooth Analysis. Peter Lang, Frankfurt am Main (1995)

[Evt72] Evtushenko, Yu.G.: A numerical method for finding best guaranteed estimates. USSR Journal of Computational Mathematics and Mathematical Physics, **12**, 109–128 (1972)

[HL93] Hiriart-Urruty, J.-B., Lemarechal, C.: Convex Analysis and Minimization Algorithms, Vol. 1 and Vol. 2. Springer Verlag, Berlin, Heidelberg, New York (1993)

[Kiw85] Kiwiel, K.C.: Methods of Descent for Nondifferentiable Optimization. Lecture Notes in Mathematics, **1133**, Springer Verlag, Berlin (1985)

[Pol97] Polak, E.: Optimization: Algorithms and Consistent Approximations. Springer Verlag, New York (1997)

[KP98] Kirjner-Neto, C., Polak, E.: On the conversion of optimization problems with max-min constraints to standard optimization problems. SIAM J. Optimization, **8**, 887-915 (1998)

[MN92] Makela, M.M., Neittaanmaki, P.: Nonsmooth Optimization. World Scientific, Singapore (1992)

[Man94] Mangasarian, O.L.: Misclassification minimization. Journal of Global Optimization, **5**, 309-323 (1994)

[Man97] Mangasarian, O.L.: Mathematical programming in data mining. Data Mining and Knowledge Discovery, **1**, 183–201 (1997)

[Mif77] Mifflin, R.: Semismooth and semiconvex functions in constrained optimization. SIAM Journal on Control and Optimization, **15**, 957-972 (1977)

[MA92] Murphy, P.M., Aha, D.W.: UCI repository of machine learning databases. Technical report, Department of Information and Computer science, University of California, Irvine (1992) (www.ics.uci.edu/mlearn/MLRepository.html)

[NM65] Nelder, J.A., Mead, R.: A simplex method for function minimization. Comput. J., **7**, 308-313 (1965)

[Pow02] Powell, M.J.D.: UOBYQA: unconstrained optimization by quadratic approximation. Mathematical Programming, Series B, **92**, 555–582 (2002)

[Tho02] Thorsten, J.: Learning to Classify Text Using Support Vector Machines. Kluwer Academic Publishers, Dordrecht (2002)

[Vap95] Vapnik, V.N.: The Nature of Statistical Learning Theory. Springer, New York (1995)

A Review of Applications of the Cutting Angle Methods

Gleb Beliakov

School of Information Technology
Deakin University
221 Burwood Hwy, Burwood, 3125, Australia
gleb@deakin.edu.au

Summary. The theory of abstract convexity provides us with the necessary tools for building accurate one-sided approximations of functions. Cutting angle methods have recently emerged as a tool for global optimization of families of abstract convex functions. Their applicability have been subsequently extended to other problems, such as scattered data interpolation. This paper reviews three different applications of cutting angle methods, namely global optimization, generation of nonuniform random variates and multivatiate interpolation.

Key words: Global optimization, Abstract convexity, Cutting angle method, Random variate generation, Uniform approximation.

1 Introduction

The theory of abstract convexity [Rub00] provides the necessary tools for building accurate lower and upper approximations of various classes of functions. Such approximations arise from a generalization of the following classical result: each convex function is the upper envelop of its affine minorants [Roc70]. In abstract convex analysis the requirement of linearity of the minorants is dropped, and abstract convex functions are represented as the upper envelops of some simple minorants, or support functions, which are not necessarily affine. Depending on the choice of the support functions, one obtains different flavours of abstract convex analysis.

By using a subset of support functions, one obtains an approximation of an abstract convex function from below. Such one-sided approximation, or underestimate, can be very useful in various applications. For instance, in optimization, the global minimum of the underestimate provides a lower bound on the global minimum of the objective function. One can find the global minimum of the objective function as the limiting point of the sequence of global minima of underestimates. This is the principle of the cutting angle method of global optimization [ARG99, BR00, Rub00], reviewed in section 3.

This paper discusses two other applications of one-sided approximations. The second application is generation of random variates from a given distribution using acceptance/rejection approach. Non-uniform random variates generation is an important task in statistical sumulation. The method of acceptance/ rejection consists in approximating the required probability density from above, using a simpler function, called the hat function. Then the random variates are generated using a multiple of the hat function as the density, and these random variates are either accepted or rejected based on the value of an independent uniform random number. In section 4 we discuss this approach in detail, and show how one-sided approximation (from above) can be used to build suitable hat functions.

The last application comes from the field of scattered data interpolation. Here we combine the upper and lower approximations of the function known to us through a set of its values, and obtain an accurate interpolant, which as we show, solves the best uniform approximation problem.

2 Support functions and lower approximations

2.1 Basic definitions

We will use the following notations.

- R^n_+ denotes the cone of vectors with non-negative components $\{x \in R^n : x_i \geq 0, i = 1, \ldots, n\}$;
- R^n_{++} denotes the cone of vectors with strictly positive components $\{x \in R^n : x_i > 0, i = 1, \ldots, n\}$;
- $R_{+\infty}$ denotes $(-\infty, +\infty]$;
- S denotes the unit simplex $S = \{x \in R^n : x_i \geq 0, \sum_{i=1}^n x_i = 1\}$;
- riS is the relative interior of S, $riS = \{x \in R^n : x_i > 0, \sum_{i=1}^n x_i = 1\}$;
- Index set $I = \{1, 2, \ldots, n\}$;
- $x = (x_1, x_2, \ldots, x_n) \in R^n$;
- $x^k \in S$ denotes the k-th vector of some sequence $\{x^k\}_{k=1}^K$;
- Vector inequality $x \geq y$ denotes dominance $x_i \geq y_i, \forall i \in I$.

Definition 1. *The function $f : X \to R$ is called Lipschitz-continuous in X, if there exists a number $M: \forall x, y \in X : |f(x) - f(y)| \leq M\|x - y\|$. The smallest such number is called the Lipschitz constant of f in the norm $\| \cdot \|$.*

Definition 2. *A function $f : R^n_+ \to R$ is called IPH (Increasing positively homogeneous functions of degree one) if*
$\forall x, y \in R^n_+, \ x \geq y \Rightarrow f(x) \geq f(y); \forall x \in R^n_+, \forall \lambda \in R_{++} : f(\lambda x) = \lambda f(x).$

Let X be some set, and let H be a nonempty set of functions $h : X \to V \subset [-\infty, +\infty]$. We have the following definitions [Rub00].

Definition 3. *A function f is abstract convex with respect to the set of functions H (or H-convex) if there exists $U \subset H$:*

$$f(x) = \sup\{h(x) : h \in U\}, \forall x \in X.$$

Definition 4. *The set U of H-minorants of f is called the support set of f with respect to the set of functions H:*

$$supp(f, H) = \{h \in H, h(x) \le f(x) \, \forall x \in X\}.$$

Definition 5. *H-subgradient of f at x is a function*

$$h \in H : f(y) \ge h(y) - (h(x) - f(x)), \forall y \in X.$$

The set of all H-subgradients of f at x is called H-subdifferential

$$\partial_H f(x) = \{h \in H : \forall y \in X, f(y) \ge h(y) - (h(x) - f(x))\}.$$

Definition 6. *The set $\partial_H^* f(x)$ at x is defined as*

$$\partial_H^* f(x) = \{h \in supp(f, H) : h(x) = f(x)\}.$$

Proposition 1. *[Rub00], p.10. If the set H is closed under vertical shifts, i.e., $(h \in H, c \in R)$ implies $h - c \in H$, then $\partial_H^* f(x) = \partial_H f(x)$.*

Definition 7. *Polyhedral distance.*

Let P be a finite convex polyhedron in R^n defined by the intersection of r halfspaces, containing the origin in its interior (example 7.2 from [DR95])

$$P = \bigcap_{i=1}^{r} \{x : x \cdot h_i \le 1\}, \tag{1}$$

where $h_i \in R^n$ are the directional vectors. The polyhedral distance is

$$d_P(x, y) = \max\{(x - y) \cdot h_i : 1 \le i \le r\}.$$

As a special case consider the distance defined by a simplex centered at 0.

Definition 8. *Simplicial distance.*

Let P be a simplex defined as the intersection of $n + 1$ halfspaces (1), defined by the vectors

$$h_1 = (-v_1, 0, 0, \ldots),$$
$$h_2 = (0, -v_2, 0, \ldots),$$
$$\vdots$$
$$h_{m+1} = (v_{m+1}, \ldots, v_{m+1}), \tag{2}$$

$v_i > 0$. The simplicial distance is

$$d_P(x, y) = \max\{\max_{i=1,\ldots,m} v_i(y_i - x_i), v_{n+1} \sum_{i=1}^{n}(x_i - y_i)\}. \tag{3}$$

Let us now for the purposes of convenience introduce a slack variable $x_{n+1} = 1 - \sum_{i=1}^{n} x_i$. With the help of the new coordinate, and using $\sum_{i=1}^{n}(x_i - y_i) = 1 - \sum_{i=1}^{n} y_i - (1 - \sum_{i=1}^{n} x_i) = y_{n+1} - x_{n+1}$, we can write (3) in a more symmetric form

$$d_P(x,y) = \max_{i=1,\ldots,n+1} v_i(y_i - x_i). \tag{4}$$

2.2 Choices of support functions

We start with the classical case of affine support functions [Roc70, Rub00].

Example 1. Let the set H denote the set of all affine functions

$$H = \{h : h(x) = a \cdot x + b, \ x, a \in R^n, b \in R\}.$$

A function $f : R^n \to R_{+\infty}$ is H-convex if and only if f a is lower semicontinuous convex function.

As a consequence of this result, we can approximate convex lower semicontinuous functions from below using a finite subset of functions from $supp(f, H)$. For instance, suppose know a number of values of function f at points $x^k, k = 1, \ldots, K$. Then the pointwise maximum of the support functions h^k

$$H^K(x) = \max_{k=1,\ldots,K} h^k(x) = \max_{k=1,\ldots,K} (f(x^k) + A_f^k(x - x^k)) \tag{5}$$

is a lower approximation, or underestimate of f. A_f^k denotes a subgradient of f at x^k. The function H^K is a piecewise linear convex function, illustrated on Fig.1.

Example 2. [Rub00]. Let the set H be the set of min-type functions

$$H = \{h : h(x) = \min_{i \in I} a_i x_i, a \in R_+^n, x \in R_+^n\}.$$

A function $f : R_+^n \to R_+$ is H-convex if and only if f is IPH.

As a consequence, we can approximate IPH functions from below using pointwise maxima of subsets of its support functions,

$$H^K(x) = \max_{k=1,\ldots,K} h^k(x) = \max_{k=1,\ldots,K} \min_{i \in I} a_i^k x_i, \tag{6}$$

where $a_i^k = \frac{f(x^k)}{x_i^k}$ if $x_i^k > 0$ and 0 otherwise.

Further, it is shown in [Rub00] that IHP functions are closely related to Lipschitz functions, in the sense that every Lipschitz function g defined on the unit simplex S can be transformed to a restriction of an IPH f to S using an additive constant: $f = g + C$, where $C \geq -\min g(x) + 2M$, where M is

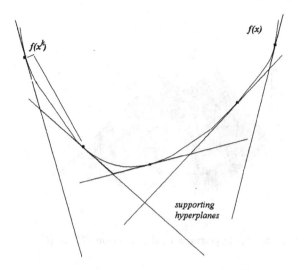

Fig. 1. The graph of the function H^K in (5).

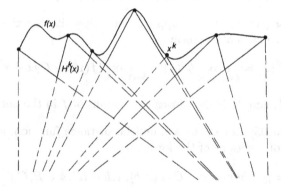

Fig. 2. Saw-tooth underestimate of f in CAM using functions (6).

the Lipschitz constant of g in l_1-norm. Thus the underestimate (6) can also be used to approximate Lipschitz functions on the unit simplex.

Function (6) has a very irregular shape illustrated on Figs. 2,3, the reason why it is often called the saw-tooth underestimate (or saw-tooth cover) of f.

Example 3. [Rub00]. Let the set H be the set of functions of the form

$$H = \{h : h(x) = a - C\|x - b\|, x, b \in R^n, a \in R, C \in R_+\}$$

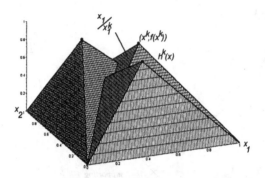

Fig. 3. The hypograph of the function H^K in (6).

Then $f : R^n \to R_{+\infty}$ is H-convex if and only if f is a lower semicontinuous function. The H-subdifferential of f is not empty if f is Lipschitz.

As a consequence, we can approximate Lipschitz functions from below using underestimates of the form

$$H^K(x) = \max_{k=1,\dots,K} h^k(x) = \max_{k=1,\dots,K}(f(x^k) - C\|x - x^k\|), \qquad (7)$$

where $C \geq M$, and M is the Lipschitz constant of f in the norm $\|\cdot\|$.

Example 4. [Bel05]. Let d_P be a simplicial distance function, and let the set H be the set of functions of the form

$$H = \{h : h(x) = a - Cd_P(x, b), x, b \in R^n, a \in R, C \in R_+\}$$

Then $f : R^n \to R_{+\infty}$ is H-convex if and only if f is a lower semicontinuous function. The H-subdifferential of f is not empty if f is Lipschitz.

Since d_P can also be written as (4), we can use the following underestimate of a Lipschitz f

$$H^K(x) = \max_{k=1,\dots,K}(f(x^k) - Cd_P(x, x^k)) = \max_{k=1,\dots,K} \min_{i \in I}(f(x^k) - C_i(x_i^k - x_i)), \qquad (8)$$

where $C_i = Cv_i$, and C satisfies $Cd_P(x, y) \geq M\|x - y\|$, where M is the Lipschitz constant of f in the norm $\|\cdot\|$ [Bel05]. We remind that here we use a slack variable, as in (4), and the components of $x \in R^{n+1}$ are restricted by $\sum x_i = 1$. The shape of H^K is illustrated on Figs. 4,5, and it is also called the saw-tooth underestimate.

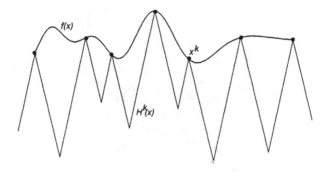

Fig. 4. Univariate saw-tooth underestimate of f using functions (8).

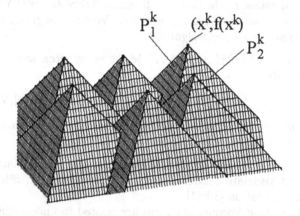

Fig. 5. The hypograph of the function H^K in (8) in the case of two variables.

2.3 Relation to Voronoi diagrams

Consider a set of points $\{x^k\}_{k=1}^K, x^k \in R^n$, called *sites*.

Definition 9. *The set*

$$Vor(x^k) = \{x \in R^n : \|x - x^k\| \le \|x - x^j\|, \forall j \ne k\}$$

is called the Voronoi cell of x^k.

One can choose any norm, or in fact any distance function d_P in this definition. The collection of Voronoi cells for all sites $x^k, k = 1, \ldots, K$ is called the Voronoi diagram of the data set. Voronoi diagram is one of the most fundamental data structures of a data set with a long history [Aur91, OBSC00, BSTY98]. An example is presented on Fig.6.

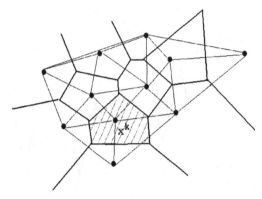

Fig. 6. The Voronoi diagram of a set of sites, and its dual Delaunay triangulation.

There are multiple extensions of the Voronoi diagram, notably those based on the generalization of the distance function [OBSC00, BSTY98]. One such generalization is called additively weighted Voronoi diagram, in which case each site has an associated weight w_k.

Definition 10. *Let* $\{x^k\}_{k=1}^K, x^k \in R^n$ *be the set of sites, and* $w \in R^K$ *be the vector of weights. The set*

$$Vor(x^k, w) = \{x \in R^n : w_k + \|x - x^k\| \leq w_j + \|x - x^j\|, \forall j \neq k\}.$$

is called Additively Weighted Voronoi cell. The collection of such cells is called Additively Weighted Voronoi diagram.

Voronoi diagrams and their duals, Delaunay (pre-)triangulations, are very popular in multivariate scattered data interpolation, e.g., Sibson's natural neighbour interpolation [Sib81].

Let us show how Voronoi diagrams are related to underestimates (7),(8). First consider the special case $f = 1$. For the function H^K in (7), and for each $k = 1, \ldots, K$ define the set

$$S^k = \{x \in R^n : h^k(x) \geq h^j(x), \forall j \neq k\}.$$

It is easy to show that sets S^k coincide with Voronoi cells $Vor(x^k)$. Indeed, $h^k(x) \geq h^j(x)$ implies $1 - C\|x - x^k\| \geq 1 - C\|x - x^j\|$, and then $\|x - x^k\| \leq \|x - x^j\|$. Furthermore, if we now take H^K in (8), the sets S^k coincide with Voronoi cells in distance d_P.

Let us now take an arbitrary Lipschitz f and (7). Consider an additively weighted Voronoi diagram with weights w_k given as $w_k = -\frac{f(x^k)}{C}$. It is not difficult to show that Voronoi cells $Vor(x^k, w)$ can be written as

$$Vor(x^k, w) = \{x \in R^n : h^k(x) \geq h^j(x), \forall j \neq k\}.$$

The last equation is also valid for other distance functions, and in particular d_P and h^k in (8).

This interesting relation of saw-tooth underestimates and Voronoi diagrams has two implications. Firstly, we can use existing results on computational complexity of Voronoi diagrams to estimate the number of "teeth" of the saw-tooth underestimate, i.e., the number of local minimizers of H^K. These miminizers correspond to the vertices of the Voronoi diagram. It is known that the number of vertices of Voronoi diagram grows as $O(K^{\lceil \frac{n}{2} \rceil})$ in any simplicial distance function or l_∞-metric [BSTY98]. $\lceil a \rceil$ denotes the smallest integer greater or equal to a. Thus we obtain an estimate on the number of local minimizers of H^K.

Secondly, we can apply methods of enumerating local minima of H^K discussed in the next section as a tool for building Voronoi diagrams, and in particular weighted Voronoi diagrams, as well as their dual Delaunay triangulations.

3 Optimization: the Cutting Angle method

3.1 Problem formulation

We consider the following global optimization problem. Let f be an H-convex function on some compact set $D \subset R^n$. We solve

$$\min f(x) \qquad (9)$$
$$\text{s.t. } x \in D.$$

Depending on the set H we obtain different classes of abstract convex functions. Consider the following instances of Problem (9). In the case of H being the set of affine functions, f is convex and possesses the unique local minimum. While there are many alternative efficient methods of local minimization, we consider below the cutting plane method of Kelley [Kel60], as other instances of Problem (9) essentially rely on the same approach.

If H is the set of min-functions as in Example 2, f is IPH. The class of IPH functions is quite broad, and includes the following functions on R_+^n or R_{++}^n

1. $f(x) = a^t x, a_i \geq 0$;
2. $f(x) = \|x\|_p, p > 0$;
3. $f(x) = \prod_{j \in J} x_j^{t_j}, J \subset I = \{1, \ldots, n\}, t_j > 0, \sum_{j \in J} t_j = 1$;
4. $f(x) = \sqrt{[Ax, x]}$, where A is a matrix with nonnegative entries

and $[\cdot, \cdot]$ is the usual inner product in R^n.

In addition, since Lipschitz functions on S, modified with a suitable constant, can be seen as restrictions of IPH functions, we can effectively solve Lipschitz optimization problems on S or its subsets.

If the set H is chosen as in Examples 3 and 4, f is lower semicontinuous, and if we require the subdifferential to be non-empty, then f is Lipschitz. Lipschitz functions appear very frequently in applications [HP95, HPT00, HJ95, HPT00, Neu97, Pin96]. The difficulty of the optimization problem in this case is that the objective function f may possess a huge number of local minimizers (in some instances $10^{20} - 10^{60}$ [Flo00, LS02], which are impossible to enumerate (and hence find the global minimum) using local optimization methods.

Lipschitz properties of f allow one to put accurate bounds on the value of the global minimum on D and also on parts of D. Those parts of the domain on which the lower bound is too high are automatically excluded, the technique known as fathoming. This way a largely reduced subset of D will eventually be searched for the global minimum, and the majority of local minima of f can be avoided.

3.2 The Cutting Angle algorithm

Below we present the generalized cutting plane method, of which cutting angle method (CAM) is a particular instance, following [Rub00, ARG99, BR00]. The principle of this method is to replace the original global optimization problem with a sequence of relaxed problems

$$\min H^K(x) \tag{10}$$
$$\text{s.t. } x \in D,$$

$K = 1, 2, \ldots$. The sequence of solutions to the relaxed problems converges to the global minimum of f under very mild assumptions [Rub00].

Generalized Cutting Plane Algorithm
Step 0. (Initialisation)

0.1 Set $K = 0$.
0.2 Choose an arbitrary initial point $x^0 \in D$.

Step 1. (Calculate H-subdifferential)

1.1 Calculate $h^K \in \partial_H^* f(x^K)$.
1.2 Define $H^K(x) = \max_{k=0,\ldots,K} h^k(x)$, for all $x \in D$.

Step 2. (Minimize H^K)

2.1 Solve Problem (10). Let x^* be its solution.
2.2 Set $K = K + 1, x^K = x^*$.

Step 3. (Stopping criterion)

3.1 If $K < K_{max}$ and $f_{best} - H^K(x^*) > \epsilon$ go to Step 1.

The relaxed problems (10) are required at every iteration of the algorithm, and as such their solution must be efficient. In the case of convex f we obtain Kelley's cutting plane method. In this case the relaxed problem can be solved using linear programming techniques.

For Lipschitz and IPH functions, the relaxed problems are very challenging. In the univariate case, the above algorithm is known as Pijavski-Shubert method [HJ95, Pij72, Shu72, SS00], and many its variations are available. However its multivariate generalizations, like Mladineo's method [Mla86], did not succeed for more than 2-3 variables because of significant computational challenges [HP95, HPT00].

To solve the relaxed Problem (10) with H^K given by (6),(7) or (8), one has to enumerate all local minimizers of the saw-tooth underestimate. The number of these minimizers grows exponentially with the dimension n, and until recently this task was impractical. Below we review a new method for enumerating local minimizers of H^K, as published in the series of papers [BR00, BR01, BB02, Bel03].

3.3 Enumeration of local minima

We are concerned with enumerating all local minimizers of the function H^K (6) on S or $D \subset S$, where D is a polytope. This function is illustrated on Figs.2,3. For convenience, let us introduce the support vectors $l^k \in R_+^n \cup \infty$

$$l_i^k = \frac{f(x^k)}{x_i^k}, \text{ if } x_i^k > 0, \text{ or } \infty \text{ otherwise.} \tag{11}$$

At the K-th iteration of the algoritm we have K support vectors. Consider ordered combinations of n support vectors, $L = \{l^{k_1}, l^{k_2}, \ldots, l^{k_n}\}$, which we can visualize as $n \times n$ matrix whose rows are given by the participating support vectors

$$L = \begin{pmatrix} l_1^{k_1} & l_2^{k_1} & \ldots & l_n^{k_1} \\ l_1^{k_2} & l_2^{k_2} & \ldots & l_n^{k_2} \\ \vdots & \vdots & \ddots & \vdots \\ l_1^{k_n} & l_2^{k_n} & \ldots & l_n^{k_n} \end{pmatrix}. \tag{12}$$

The following result is proven in [BR00]: every local minimizer x^* of H^K in $ri\, S$ corresponds to a combination L satisfying two conditions

(I) $\forall i, j \in I, i \neq j : l_i^{k_i} > l_i^{k_j}$

(II) $\forall v \in \mathcal{K} \setminus L, \exists i \in I : l_i^{k_i} \leq v_i$

where $\mathcal{K} = \{l^1, l^2, \ldots, l^K\}$ is the set of all support vectors. Further, the actual local minima are found from L using

$$d = H^K(x^*) = Trace(L)^{-1}, \tag{13}$$
$$x^*(L) = d\, diag(L).$$

Condition (I) implies that the diagonal elements dominate their respective columns, and condition (II) implies that the diagonal of L does not dominate any other support vector v. Thus we obtain a combinatorial problem of enumerating all combinations L that satisfy conditions (I) and (II).

It is infeasible to enumerate all such combinations directly for large K. Fortunately there is no need to do so. It was shown in [BB02, Bel03, Bel04] that the required combinations can be put into a tree structure. The leaves of the tree correspond to the local minimizers of H^K, whereas the intermediate nodes correspond to the minimizers of $H^n, H^{n+1}, \ldots, H^{K-1}$. Such a tree is illustrated on Fig.7. The use of the tree structure makes the algorithm very efficient numerically (as processing of queries using trees requires logarithmic time of the number of nodes).

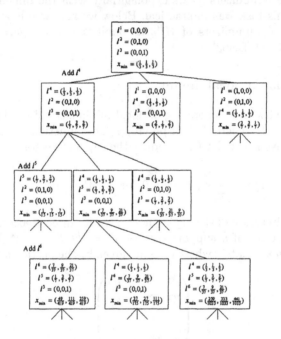

Fig. 7. The tree of combinations of support vectors L that satisfy conditions (I) and (II) and define local minima of H^K.

To enumerate local minimizers in a polytope $D \subset S$ one proceeds as follows. Using the enumeration technique from [BB02, Bel03], find all local minimizers on $ri\, S$. Each such minimizer has an associated set $A(L)$ on which it is unique. The set $A(L)$ is characterized by [Bel03, Bel04]

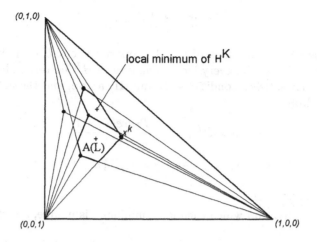

Fig. 8. Sets $A(L)$ on which the saw-tooth underestimate has unique local minimum. Two such sets are shown. Black circles denote points x^k.

$$x_j x_i^{k_j} \leq x_i x_j^{k_j}, \; i, j \in I, i > j,$$
$$x_i x_j^{k_j} > x_j x_i^{k_j}, \; i, j \in I, i < j. \tag{14}$$

The sets $A(L)$ form a nonintersecting partition of S. They are illustrated on Fig.8.

For each local minimizer x^* on $ri\, S$ we can have three situations: a) $x^* \in D$, in which case we just record it, b) $x^* \notin D$ and $A(L) \cap D = \emptyset$, in which case we discard x^*, and c) $A(L) \cap D \neq \emptyset$, in which case we look for a constrained minimum on the boundary of D. This can be done by solving an optimization problem

$$\min \max_{i \in I} l_i^{k_i} x_i \tag{15}$$
$$\text{s.t. } x \in \bar{A}(L) \cap D,$$

which is subsequently transformed into a linear programming problem. To do this, introduce an auxiliary variable $a = \max_{i \in I} l_i^{k_i} x_i$, and write (15) as

$$\min a \tag{16}$$
$$\text{s.t. } \forall i \in I : a - l_i^{k_i} x_i \geq 0,$$
$$x \in \bar{A}(L) \cap D,$$

and recall that the set $A(L)$ is an intersection of halfspaces (14) and D is a polytope. The details are given in [Bel04].

Consider now functions (8), illustrated on Figs.4,5. In this case we can use a similar enumeration technique. Define the support vectors

$$l_i^k = \frac{f(x^k)}{C_i} - x_i^k. \tag{17}$$

Form ordered combinations of $n + 1$ support vectors L (12). We have the following result [Bel05]: every local minimizer of H^K corresponds to a combination L that satisfies conditions (I) and (II) above, and the actual minima are found from

$$d = H^K(x^*) = \frac{Trace(L) + 1}{C^-}, \tag{18}$$

$$x_i^* = \frac{d}{C_i} - l_i^{k_i}, i = 1, \ldots, n + 1,$$

where $C^- = \sum_{i \in I} \frac{1}{C_i}$.

The sets $A(L)$, on which each local minimum is unique, are characterized by

$$\forall i, j \in \{1, \ldots, n + 1\}, i \neq j : C_j(x_j^* - x_j^{k_j}) < C_i(x_i^* - x_i^{k_j}). \tag{19}$$

3.4 Numerical experiments

We performed extensive testing of various versions of CAM on test and real life problems [BB02, Bel03, Bel04, BTMRB03, LBB03, LBB03]. In this section, to indicate the performance of the algorithm, we present a selection of results of numerical experiments. We took the following test optimization problems.

Test Problem 1 (Six-hump camel back function)

$$f(x) = \left(4 - 2.1x_1^2 + \frac{x_1^4}{3}\right)x_1^2 + x_1x_2 + 4(x_2^2 - 1)x_2^2$$

$$-2 \leq x_i \leq 2, i = 1, 2.$$

Test Problem 2 ([HPT00],p.261)

$$f(x) = -\sum_{i=1}^{10} \frac{1}{\|x - a^i\|^2 + c^i}$$

$$0 \leq x_i \leq 10, i = 1, 2.$$

Parameters a^i and c^i are given in [HPT00],p.262.

Test Problem 3 [HJ95]

$$f(x) = \sin(x_1)\sin(x_1x_2)\sin(x_1x_2x_3)$$

$$0 \leq x_i \leq 4.$$

Test Problem 4 (Griewanks function)

$$f(x) = \frac{1}{d}\sum_{i=1}^{n} x_i^2 - \prod_{i=1}^{n} \cos\left(\frac{x_i}{\sqrt{i}}\right) + 1, d = 4000,$$

$$-50 \leq x_i \leq 50.$$

In Table 1 we compare the performance of CAM which uses underestimate (6) and Extended CAM (ECAM), which uses underestimate (8). For functions 1-3, ECAM was able to compute the same lower bound on the global minimum using less function evaluations (and significantly less time) than CAM. For function 4, we ran both CAM and ECAM algorithms the same number of iterations (function evaluations), and compared the values of the lower bound on the global minimum. It appears from Table 1 that ECAM consistently produces better results than CAM. This is not surprising, as all test problems involve Lipschitz functions. Approximation (6) used in CAM is more suitable for IPH functions, and the conversion of Lipschitz objective functions to IPH functions resulted in somewhat less efficient algorithm than ECAM.

3.5 Applications

Various versions of CAM have been applied to solving real life practical problems. In [BRY01, BRY02] the authors successfully used CAM in problems of supervised classification. In particular they applied CAM for automatic classification of medical diagnosis. In [BRS03] the same authors extended the use of CAM for unsupervised classification problems.

CAM has been applied as a tool to find parameters of a function in univariate and multivariate nonlinear approximation. [Bel03] applies CAM to optimize position of knots in univariate spline approximation, whereas in [Bel02] CAM was used to fit aggregation operators to empirical data.

Recently we applied CAM to the molecular structure prediction problem [Neu97, Flo00, LBB03]. This is a very challenging problem in computational chemistry, which consists in predicting the geometry of a molecule by minimizing its potential energy as a function of atomic coordinates. We chose the benchmark problem of unsolvated met-enkephalin [Flo00, LBB03]. As independent variables we used the 24 dihedral angles of this pentapeptide, and following [Flo00], 10 of the dihedral angles (the backbone) were used as global variables in ECAM, while the rest were treated as local variables (i.e., each function evaluation involved a local optimization problem with respect to the dihedral angles treated as local variables). This objective function (the potential energy) involves in the order of 10^{11} local minima. The problem is very challenging because of the existence of several strong local minima which trap local descent algorithms. For instance all reported multistart local search algorithms failed to identify the global minimum [Flo00].

Previously we reported that a combination of CAM with local search algorithms allowed us to locate the global minimum of the potential energy function in 120,000 iterations of CAM, which took 4740 seconds (79 min) on a cluster of 36 DEC Alpha workstations (1 MHz processors) [LBB03, LBB03]. Using ECAM and the same hardware and software configuration the global minimum was found in 80,000 iterations, which took 50 min on the cluster of 36 DEC Alpha workstations.

It is worth noting that CAM can be efficiently parallelized to take advantage of the distributed memory architecture of computer clusters. Various branches of the tree of local minima are stored on different processors, and are processed independently of each other. It allows one to use the combined RAM of many processors. Our experiments with parallelization of CAM are described in [BTMRB03, BTM01].

Table 1. Comparison of performance of CAM and Extended CAM on a set of test problems. CPU is measured on Pentium 4 1.2GHz PC with 512 MB RAM, under Windows XP. The algorithms were implemented in C++ language (Visual C++ 6 compiler). The values in the last column are the global minima of the functions, found by a local descent algorithm starting from the approximate minimum found by CAM/ECAM.

Problem	m	Iterations	CPU (sec)	upper bound f_{best}	lower bound	Solution improved by local method
1 (CAM)	2	30000	3.12	-1.0302	-1.07	-1.03163
1 (ECAM)	2	10000	1.31	-1.0316	-1.07	-1.03163
2 (CAM)	2	10000	1.10	-2.1452	-2.152	-2.14520
2 (ECAM)	2	10000	1.03	-2.1452	-2.148	-2.14520
3 (CAM)	3	40000	21.5	-0.999	-1.09	-1
3 (ECAM)	3	10000	2.7	-0.9998	-1.10	-1
4 (CAM)	2	10000	0.99	0.0022	-0.61	0
4 (ECAM)	2	10000	1.30	0.000012	-0.06	0
4 (CAM)	3	40000	21.1	0.0071	-0.41	0
4 (ECAM)	3	40000	17.2	0.0058	-0.138	0
4 (CAM)	4	60000	380	0.00	-1.02	0
4 (ECAM)	4	60000	231	0.00	-0.91	0
4 (CAM)	5	90000	523	0.00	-1.18	0
4 (ECAM)	5	90000	460	0.00	-0.51	0

4 Random variate generation: acceptance/ rejection

4.1 Problem formulation

Efficient non-uniform random number generators are important in many applications, such as Markov Chain sampling. Many specialized algorithms for a variety of standard distributions are available; however more recently so-called black box methods have attracted substantial attention [HLD04]. These methods are applicable to a large class of distributions, but require a setup stage and are generally less efficient than the specialized methods. The monographs [Dag88, Dev86, HLD04] present a wide range of methods used in this area.

There are two main approaches for generating random numbers from arbitrary distributions. The *inversion method* relies on knowledge of the inverse of the cumulative distribution $P(x)$, $P^{-1}(y)$. If this inverse is given explicitly, then one generates uniformly distributed random numbers Z and transforms them to X using $X = P^{-1}(Z)$. This approach is very useful when distributions are simple enough to find P^{-1} analytically, however, in case of more complicated distributions, P^{-1} may not available, and one has to invert P numerically by solving the equation $Z = P(X)$ for X, e.g., using bisection or Newton's method. Given the slowness of numerical solution, this method becomes very inefficient. This method cannot be used for multivariate densities.

The second approach, so-called *acceptance/ rejection method*, relies on efficient generation of random numbers from another distribution, whose density $h(x)$ multiplied by a suitable positive constant, dominates the density $\rho(x)$ of the required distribution, $\forall x \in Dom[\rho] : \rho(x) \leq g(x) = ch(x)$. The function $g(x)$ is often called the *hat function* of the distribution with density ρ. In this case we need two independent random variates, a random number X with density $h(x)$ and a uniform random number Z on $[0,1]$. If $Zg(X) \leq \rho(X)$, then X is accepted (and returned by the generator), otherwise X is rejected, and the process repeats until some X is accepted.

The acceptance/rejection approach does not rely on the analytic form of the distribution or its inversion. However, its effectiveness depends on how accurate ρ is approximated from above by the hat function. The less accurate is the approximation, the greater is the chance of rejection (and hence inefficiency of the algorithm). A number of important inequalities relating densities of various distributions are presented in [Dev86]. These inequalities allow one to choose an appropriate hat function for a given ρ.

The acceptance/rejection approach generalizes well for multivariate distributions. In fact, this method does not change at all if X is a random vector rather than a random number. The challenge lies in efficient construction of the hat function for a multivariate density $\rho(x)$, and finding an efficient way to sample from the distribution defined by this hat function. With the increasing dimension, the need for tight upper approximation to ρ becomes more important, as the number of wasted calculations in case of X rejected increases.

Subdivision of the domain of ρ is frequently used in universal random number generators [Hor95, LH98, LH01]. If little information about ρ is available (i.e., no analytical form), a piecewise constant (or piecewise linear) hat function can be used. It is constructed by taking values of ρ at a number of points (Fig.9). For instance, some methods use concavity of ρ to guarantee that such an approximation overestimates ρ, whereas in [Hor95, LH98, LH01] the log-concavity or T-concavity is exploited. A function is called log-concave (or T-concave for a monotone continuous function T), if the transformed density $\hat{\rho} = ln(\rho)$ (or $\hat{\rho} = T(\rho)$) is concave. In [ES98] the authors rely on detecting the inflection points of ρ in their construction of the hat function.

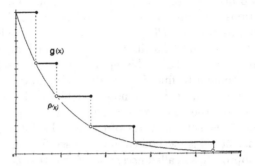

Fig. 9. A piecewise constant upper semicontinuous hat function (thick solid line) that approximates a monotone density ρ.

However, regardless of the way the hat function is obtained at this preprocessing step, the random numbers are always generated in a similar fashion. First the interval is chosen using a universal discrete generator (e.g., using alias method [Dev86, Wal74]). Then a random variate X is generated that has a multiple of the hat function on this interval as its density. Then X is either accepted or rejected (according to whether $Zg(X) \le \rho(X)$ for a uniform random variate Z on $[0,1]$). In case of rejection we have to restart from the first step. The intervals are chosen with probabilities proportional to the area under g on each interval.

It is clear that the form of the hat function g on each interval of the subdivision needs not be the same. While constant or linear functions can be used for some intervals, on intervals where ρ is has a vertical asymptote, or on infinite intervals (for the tails of the distributions) other forms are more appropriate (e.g., multiples of Pareto or Cauchy tails). It is also clear that the multivariate case can be treated in exactly the same way, by partitioning the domain into small regions. For T-concave distributions such method is described in [LH98].

Hence, efficient universal generators of non-uniform random numbers or random vectors can be built in a standardized fashion, by partitioning the domain of ρ, and constructing a piecewise continuous hat function. The problem is how to build an accurate upper approximation that can serve as a hat function. In this section we review the methods of building the hat functions based on one-sided approximations discussed earlier in this paper.

4.2 Log-concave densities

The use of envelop representation of convex functions, and one-sided approximation of type (5) has been used to construct the hat function of univariate log-concave densities for some time [Dev86, HLD04]. In [LH98] the authors described the transformed density rejection approach applicable to multivariate T-concave distributions. Consider a continuous strictly increasing function T. A density ρ is called T-concave if the transformed density $\hat{\rho} = T(\rho)$ is concave. A typical example is $T = \ln$, in which case ρ is called log-concave.

Let us define a convex function $f = -T(\rho)$. We shall build an underestimate of f using Eq.(5), and then change its sign to obtain the overestimate $g^K = -H^K$. After this, the hat function of ρ is computed as $g = T^{-1}(g^K)$.

In the univariate case, generation of random numbers using a multiple of the hat function $g = T^{-1}(g^K)$ as the density is quite simple. Firstly, one calculates the intersections of linear segments of functions $H^K = \max_k(f(x^k) + A_f^k(x - x^k))$, which gives a partition of the domain of ρ into subintervals. H^K is linear on each subinterval, and since T is given, generation of random numbers on each subinterval using g as the hat function is easily achieved by inversion [Dev86, HLD04]. The choice of the subinterval is performed using a discrete randon mumber generator.

The multivariate case proceeds in a similar fasion, but with a more complicated generation step. The authors of [LH98] use piecewise linear function H^K (5) to build the hat function $g = T^{-1}(-H^K)$. Then they determine the partition of the domain of ρ into the set of convex polyhedra (bounded or unbounded), so than on each polyhedron H^K is linear. Then the authors use the sweep-plane algorithm to generate random vectors on each polyhedron. As earlier, the choice of the polyhedron is performed using a discrete randon mumber generator. The programming library UNURAN implements several universal random variate generation algorithms for T-concave densities [HLD04].

4.3 Univariate Lipschitz densities

In this section we consider univariate Lipschitz-continuous densities ρ on a compact set. As we mentioned earlier, the infinite domains can be treated by splitting them into a compact and semi-infinite interval (say, $[0, a], [a, \infty)$). The hat function of the tail of the distribution on $[a, \infty)$ can be the multiple of Pareto heavy tail distribution $g(x) = c/x^{1+\alpha}$, and will not be treated here. We are interested in the compact subdomain $[0, a]$ (or $[a, b]$ for generality).

Let us subdivide the interval $[a, b]$ into a finite number of subintervals $[x^k, x^{k+1}]$:

$$[a, b] = \bigcup_{k=1,\ldots,K-1} [x^k, x^{k+1}],$$

whose interiors do not intersect $(x^k, x^{k+1}) \cap (x^j, x^{j+1}) = \emptyset$, if $j \neq k$.

Lipschitz continuity can be exploited in order to put upper and lower bounds on the values of ρ on any subinterval $[x^k, x^{k+1}]$, given its values at the ends $\rho_k = \rho(x^k), \rho_{k+1} = \rho(x^{k+1})$, namely

$$\max \left\{ \rho_k - M|x^k - x|, \rho_{k+1} - M|x^{k+1} - x| \right\} \leq \rho(x)$$
$$\leq \min \left\{ \rho_k + M|x^k - x|, \rho_{k+1} + M|x^{k+1} - x| \right\},$$
$$x \in [x^k, x^{k+1}].$$

As earlier M denotes the Lipschitz constant of ρ. By having K values of ρ on $[a, b]$ we can build the saw-tooth overestimate of ρ, which we can use as the hat function

$$g^K(x) = \min_{k=1,\ldots,K} (\rho_k + M|x^k - x|). \qquad (20)$$

One can recognize Eq.(7), in which we use $f = -\rho$ and $H^K = -g^K$, as we are interested in the upper, rather than lower approximation.

The use of saw-tooth overestimates as hat functions in the acceptance/ rejection approach was described in [Dev86], p.348. The process of building the saw-tooth overestimate of ρ can be organized very efficiently (in $O(K \log K)$ operations), and the points x^k can be chosen either randomly on $[a, b]$, or, which is more efficient, by choosing one of the schemata described in [HJ95, SS00]. For example, in Pijavski-Shubert algorithm [Pij72], given a set of K function values ρ_k, $k = 1, \ldots, K$, one chooses the $K + 1$-st value at the global maximum of the function $g^K(x)$ in (20). The global maximum of (20) is found by sorting out all local maxima (the teeth of the saw-tooth cover). This way the saw-tooth overestimate tends to be closer to ρ, which reduces the chance of rejection.

There are two ways to proceed with building the hat function after the saw-tooth overestimate is built. Firstly, we can use a constant hat function $g(x) = \max g^K(x), x \in [x^k, x^{k+1}]$ on every subinterval $[x^k, x^{k+1}], k = 1, \ldots, K - 1$. Secondly, we can use the saw-tooth overestimate itself as the hat function, $g(x) = g^K(x)$, in which case we need to divide $[a, b]$ into as twice as many subintervals $[x^k, \xi^k], [\xi^k, x^{k+1}], k = 1, \ldots, K - 1$, where ξ^k is the local maximizer of g^K, $\xi^k = \arg\max_{x \in [x^k, x^{k+1}]} g^K(x)$. On each subinterval the hat function is linear, and the random variate X with (a multiple of) such density, as required by the acceptance/ rejection method, is generated using inversion (Fig.10).

It is worth noting that the described approach is applicable to multimodal distributions (as opposed to T-concave distributions in [Hor95, LH01]). However, this method requires knowledge of the Lipschitz constant of ρ, M, which is a crucial piece of information. If unknown, the Lipschitz constant can be safely overestimated, at the price of less accurate upper approximation. In references [WZ96, SL97, SS00, Ser03] various methods of estimating Lipschitz constants are developed. These methods are based only on the ability to compute the values of ρ, not on its analytic formula. On the other hand, the value of M can sometimes follow from theoretical considerations.

Using saw-tooth overestimates as hat functions requires more function values K than methods applicable to T-concave distributions, which translates into a longer pre-processing step (building saw-tooth overestimate and tables

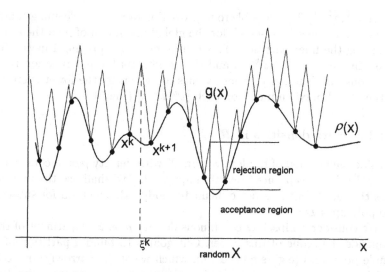

Fig. 10. A piecewise linear hat function g built using the saw-tooth overestimate in the univariate case.

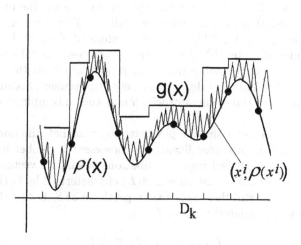

Fig. 11. A piecewise constant hat function g built using the saw-tooth overestimate in the univariate case. The value g_k is chosen as the absolute maximum of the saw-tooth overestimate on each D_k.

for the alias method) and longer tables in the alias method, but not in longer generation time once preprocessing has been finalized.

One variation of this method is to use shorter tables (i.e., less subintervals), but to improve the lower overestimate of the maximum of ρ on each subinterval. Previously we assumed that such lower overestimate is the maximum of

g^K on $[x^k, x^{k+1}]$. It is possible to improve this value by performing subdivision of these subintervals in search for the global maximum of ρ on them, without recording the finer partition. This can be done by applying Pijavski-Shubert algorithm on each $[x^k, x^{k+1}]$, and then taking as the hat function the piecewise constant function g, whose values are given by the lower overestimates of the global maximum of ρ on each $[x^k, x^{k+1}]$ (Fig.11).

4.4 Lipschitz densities in R^n

Consider generation of random vectors X with density ρ on a compact subset $A \in R^n$ using acceptance/ rejection approach. We shall use the unit simplex S as the set A, but it is not difficult to modify this method for subsets of S, like polytopes $D \subset S$.

We consider a Lipschitz continuous density ρ on S. Treatment of the tails is outside the scope of this paper. Our goal is to build a partition of S into simple polytopes (e.g., simplices) on which we shall (narrowly) overestimate ρ with a constant function. This piecewise constant upper approximation will be our hat function in the acceptance/ rejection approach.

Because Lipschitz functions on S can be seen as restrictions of a suitable IPH function (see discussion after Example 2), we will use the underestimate (6) in our computations. Let us define an auxilirary IPH function $f = -\rho + C$, with $C \geq \max_{x \in S} \rho(x) + 2M$. Using the values of f at $x^k, k = 1, \ldots, K$, build the underestimate H^K (6). At this stage, we can take the function $g^K = -H^K + C$ as the overestimate of ρ, and use it in the acceptance/ rejection algorithm. However this is extremely inconvenient, because it is hard to build a random variate generator which uses such a complicated g^K as the density.

Instead, we will use a simpler piecewise constant hat function. We know that function H^K is piecewise linear, and possesses a number of local minima, which can be identified from combinations of support vectors (12) using Eq.(13). We further know that on sets $A(L)$ characterized by (14), each local minimum is unique, and these sets form a partition of S. Define the following piecewise constant underestimate of f

$$H(x) = d(L), \text{ if } x \in A(L),$$

where L is the combination of support vectors which identifies the minimizer x^*, and $A(L)$ is the set (14) on which it is unique. Now we take $g = -H + C$ as the hat function.

We now need an efficient method of generating random variates with a multiple of the hat function as the density. In our case the hat function is piecewise constant, which means that we can generate random variates in two steps: 1) randomly choose an element of the partition $A(L)$, with probability proportional to the volume of $A(L)$ times $d(L)$; 2) generate X uniformly distributed on $A(L)$. The first step requires an efficient discrete random variate generator. We can use the *alias method* [Dev86, Wal74] for this purpose. The second step requires additional processing, as generation of random variates on a polytope requires its triangulation.

Generation of random variates uniformly distributed in a simplex is relatively easy using sorting or uniform spacings [Dev86],p.214. The way to generate uniform random variates on a general polytope $A(L)$ is to subdivide it into simplices, the procedure known as triangulation. Further, it is easy to compute the volume of a polytope given its triangulation. Hence we will triangulate every polytope $A(L)$ as part of the preprocessing.

For our purposes any triangulation of the polytope is suitable, and we used the revised Cohen and Hickey triangulation as described in [BEF00]. This triangulation method requires the vertex representation of the polytope $A(L)$, whereas it is given as the set of inequalities (14). The calculation of vertex representation of $A(L)$ can be done using the Double Description method [FP96, MRT53]. The software package CDD, which implements the Double Description method is available from [Fuk05]. The software package Vinci, available from [Eng05] can be used for the revised Cohen and Hickey triangulation.

Once the triangulation of the sets $A(L)$ is done, the volume of each simplex needs to be computed and multiplied by the value of the hat function g on it. The volume computation is performed by taking the determinant of an $n \times n$-matrix of vertex coordinates [BEF00]. The vertices and volumes (times the value of g) of the simplices that partition the domain of ρ are stored for the random vector generator.

Summarizing this section, given an arbitrary Lipschitz density ρ on S, we can find an underestimate H^K of an auxiliary function $f = -\rho + C$, and a partition of S into polytopes $A(L)$, such that on each $A(L)$, the local minimum of H^K, $d(L)$ in (13), is the greatest lower bound on f. This lower bound is tight, i.e., one can find such a Lipschitz function, that $\min_{x \in A(L)} f(x) = d$, for instance $f = H^K$ itself. Based on H^K, we define the hat function as $g = -H + C$, where $H(x) = d$, if $x \in A(L)$. Then we subdivide each polytope $A(L)$ into simplices to facilitate generation of random variates, and compute the volume of each simplex for the discrete random variate generator.

4.5 Description of the algorithm

Let us now detail some of the steps required to build a universal random vector generator using the hat function described in the previous section. The algorithm consists of two parts, preprocessing and generation. First, given the set of values $\rho(x^k), k = 1, \ldots, K$, we build the saw-tooth underestimate of an IPH function $f = -\rho + C$. Points x^k can be given a priori, or can be determined by the algorithm itself, for instance each $x^k, k = n + 1, \ldots$ can be chosen as a global minimizer of the function H^{k-1}, i.e., at the teeth of the saw-tooth underestimate at the current iteration. The first n points are always chosen as the vertices of S.

We build the saw-tooth underestimate (6) by enumerating its local minimizers using the combinatorial technique presented in section 3.3. Based on these local minimizers, we partition the domain S into polytopes $A(L)$, and then further into simplices. On each $A(L)$ the hat function is defined by $g = -d + C$. We complete the preprocessing part by computing the volumes of each simplex S_i in the partition.

The generation part now works as usual: 1) randomly choose a simplex S_i of the partition of S according to the probability, which is proportional to the volume of S_i times the value of g on it; for this we use the alias method. 2) generate a random vector X uniformly distributed in the chosen simplex, see [Dev86]. 3) generate an independent random number Z, uniformly distributed in $[0, 1]$; if $Zg(X) \leq \rho(X)$ then accept X, otherwise reject X and return to step 1).

The overall algorithm to generate random vectors with density ρ follows.

Acceptance/rejection Algorithm for Lipschitz densities
Requires: density ρ (not necessarily an analytic expression), its Lipschitz constant M in l_1-norm (or its overestimate) and $\rho_{max} = \max_{x \in S} \rho(x)$.
The number of points K as a control parameter.

Preprocessing

1 Choose constant $C \geq \rho_{max} + 2M$
2 Build the saw-tooth underestimate H^K of the function $f = -\rho + C$ using K points x^k within the domain of ρ, by using the algorithm from [BB02, Bel03]. Except for the first n points, x^k are chosen automatically by the algorithm.
3 For each local minimum of H^K compute the polytope $A(L)$ using (14).
4 Convert each $A(L)$ to the vertex representation using the Double Description method from [FP96, MRT53] and find its triangulation.
5 For each simplex S_i from the triangulation of $A(L)$ find its volume and multiply it by $P(S_i) = C - d(L)$.
6 Store the list of all simplices as the list of vertices and computed values P and $VP(S_i) = Volume(S_i) \times P(S_i)$.
7 Create two tables for the alias method using the values VP as the vector of probabilities.

Random vector generation

1 Using the alias method randomly choose simplex S_i.
2 Generate random vector $\mathbf{R}$ uniformly distributed in the unit simplex S ([Dev86], p.214, via either sorting or uniform spacings).
3 Compute vector $\mathbf{X} = \sum_{j=1}^{n} R_j \mathbf{S}_i^j$, where $\mathbf{S}_i^j$ is the j-th vertex of the chosen simplex S_i ([Dev86], p.568).
4 Generate an independent uniform random number Z in $[0, 1]$
5 If $ZP(S) \leq \rho(\mathbf{X})$ then return $\mathbf{X}$ otherwise go to Step 1.

Generation step clearly requires $n + 1$ random numbers (either uniform or exponential, see [Dev86], p.214), and calculations take $O(n^2)$ operations, because of computing the n components of X in the sum. Bucket sort is assumed to take on average $O(n)$ operations ([Dev86], p.216). Probability of rejection depends on how accurate is the computed upper approximation to ρ, which in turn depends on its Lipschitz constant and the number of points K. The latter value is the control parameter for the algorithm: the more points are used, the better is the approximation, but the longer is preprocessing step, dominated by building the saw-tooth underestimate and triangulation.

The number of simplices in the partition of the domain of ρ is difficult to calculate a priori, but Table 2 provides some indicative values.

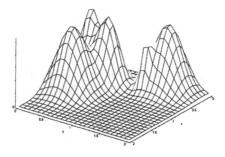

Fig. 12. Multimodal density ρ used to generate random vectors in R^2. ρ in this example is a mixture of five normal distributions. The algorithm uses exclusively numerical values of ρ and its Lipschitz constant.

4.6 Numerical experiments

We tested the acceptance/rejection method for Lipschitz densities on some multivariate multimodal distributions, such as a mixture of several normal distributions with different weights a_i, μ and covariance matrices. One such distribution is plotted on Fig.12 for the case of two variables. Of course, one can easily generate random variates from such a mixture using alternative methods (e.g., composition method, if the parameters a_i, μ, Σ are known). However, none of this information was available to the algorithm, which relies only on the ability to compute the value of ρ at a given point (plus its Lipschitz properties). Figs. 13,14 depict graphs of other densities used for testing. Sampling from these non-standard densities is a much more challenging problem than sampling from a mixture of normal distributions, yet the described algorithm easily accomplishes this task with the same efficiency.

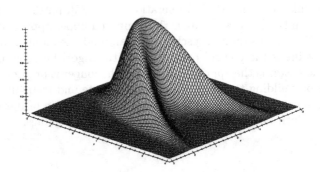

Fig. 13. Density ρ used to generate random vectors in R^2, given by $\rho(x,y) = k \exp(-(y - x^2)^2 - \frac{x^2 + y^2}{2})$. This density is not log-concave.

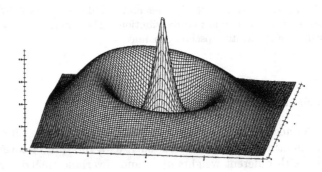

Fig. 14. Density ρ used to generate random vectors in R^2 is given by $\rho(r) = (|r| - 1)^2 \times \exp(-\frac{|r + 0.2|^2}{3})$, where $r = (x, y)$.

Table 3 presents timing of preprocessing and generation steps for various n and K for one such ρ, taken as

$$\rho = \sum_{i=1}^{5} a_i Norm(\mu_i, \Sigma_i).$$

Covariance matrices were all diagonal. For the reference, the time to generate one uniform random number was 0.271×10^{-6} sec. The Ranlux lagged Fibonacci generator with the period 10^{171} was used for uniform random numbers [Lue94].

Table 2. The number of local minima of H^K. Function $f = 1$ was used in the calculations.

K	$n=1$	$n=3$	$n=5$	$n=7$	$n=9$
1000	999	4699	13495	24810	31217
2000	1999	9631	28210	50526	74132
4000	3999	20435	104117	177358	187973
8000	7999	42031	270328	527995	886249
15000	14999	81301	532387	1093040	1956075
20000	19999	109587	738888	1605995	2661807
25000	24999	137770	993812	3861070	6175083
30000	29999	167251	1234810	6340898	10521070

Table 3 clearly shows that as K increases, the upper approximation becomes more tight, and the acceptance ratio improves. However, this is at the cost of a rapidly growing number of simplices in the subdivision of the domain of ρ, and thus at the cost of increased preprocessing time, especially for $n > 3$.

5 Scattered data interpolation: Lipschitz approximation

5.1 Problem formulation

Multivariate data interpolation and approximation is a very common problem in many branches of science. Sometimes this problem is referred to as regression, estimation, data fitting, learning of functions and other names. There is a great number of techniques developed for various instances of this problem, such as polynomial regression, spline interpolation and smoothing, wavelets, nearest neighbour search, Sibson interpolation, MARS (multivariate adaptive regression splines), machine learning techniques (e.g., decision trees), neural networks, radial basis functions, etc. For an overview the reader is referred to [Alf89].

Shape preserving approximation refers to the approximation problem in which in addition to the data, other information about the function in question

Table 3. Performance of the acceptance/rejection method as a function of dimension n and the number of points K. Preprocessing step includes building the saw-tooth underestimate and triangulation. Generation time is the average time to generate one random n-vector. Acceptance ratio is the criterion of efficiency.

n	K	Number of simplices	Time to build saw-tooth underestimate (s)	Time for triangulation (s)	Generation time ($s \times 10^{-6}$)	Acceptance ratio
2	300	1276	0.05	0.27	9.28	0.24
	1000	4424	0.18	0.73	6.11	0.36
	2000	8972	0.33	1.39	5.24	0.44
	4000	18078	0.56	2.82	4.82	0.53
	8000	36369	1.11	5.78	4.31	0.61
	16000	73208	2.29	11.70	4.08	0.69
3	300	16166	0.39	2.20	23.8	0.13
	1000	60080	1.08	8.38	18.0	0.18
	2000	124300	1.98	17.32	15.3	0.21
	4000	259428	3.74	35.22	12.9	0.26
	8000	530237	7.24	69.28	11.2	0.31
4	300	333522	3.18	30.58	63.5	0.06
	1000	1399372	11.06	116.6	58.2	0.09
	2000	3087003	22.51	268.6	50.1	0.11
5	50	509560	1.39	41.12	29950	0.00012
	100	1904996	4.80	102.3	27130	0.0002
	200	5378880	14.4	370.8	21411	0.00028

is available. For instance, it may be known a priori that the function must be monotone, convex, positive, symmetric, unimodal, etc. These conditions determine additional constraints on the approximant, which may find explicit representation in terms of the parameters that are fitted to the data. In spline approximation, this problem has been thoroughly studied (see [Die95, KM97, Kva00, Bel00]), and such constraints as monotonicity or convexity usually translate into restrictions on spline coefficients.

More recently, the concept of shape preserving interpolation and approximation has been extended to include other known a priori restrictions on the approximant, such as generalized convexity, unimodality, possessing peaks or discontinuities, Lipschitz property, associativity [KM97, Bel03]. These restrictions require new problem formulations leading to new specific methods of approximation.

In this section we consider interpolation of scattered multivariate data which restricts the Lipschitz constant of the interpolant. Lipschitz condition ensures reasonable bounds on the interpolated values of the function, which is sometimes hard to achieve in nonlinear interpolation. As we shall see, preservation of the Lipschitz condition implies strict bounds on the difference between the interpolant and the function it models in the Chebyshev max-norm, so that Lipschitz interpolation *guarantees* the performance of the interpolant in the *worst case* scenario, whereas other methods target the average performance. In this sense, Lipschitz approximation translates into *reliable* learning of functions [Coo95].

As the interpolant, we will use a combination of the lower and upper approximations of Lipschitz f defined by (7) or (8). We will show that such an interpolant is not a matter of arbitrary choice, but arises as the solution to the best uniform approximation problem, as formulated in the next section. On the other hand, the obtained solution is a piecewise continuous function (piecewise linear in case of (8), i.e., a linear spline). Splines possess many desirable features, such as stability and speed of evaluation, local behaviour, ability to model functions of virtually any shape, and so on [Die95]. We also obtain continuous dependence of the interpolant on the data, which is frequently hard to achieve [Alf89].

5.2 Best uniform approximation

Assume that we are given a data set $\{(x^k, y^k)\}_{k=1}^{K}$, $x^k \in R^n, y^k \in R$. We also assume that y^k are the values of some function $f(x^k) = y^k$, which is unknown to us and which we want to approximate with g, $g \approx f$. Thus we look for an interpolant $g : R^n \to R$, such that

$$g(x^k) = y^k, k = 1, \ldots, K.$$

It is known (e.g., see [GW59]) that it is impossible to give finite bounds on the values $f(x)$, $x \neq x^k, k = 1, \ldots, K$ in terms of the data set, if the only additional information is that f is the element of a linear space $\mathcal{V}$, no matter how restricted the space $\mathcal{V}$ is in terms of conditions of continuity, smoothness, analyticity, etc. Therefore it is meaningless to speak about the goodness of approximation without a reference to some nonlinear constraint on $\mathcal{V}$.

We shall work in the space of continuous functions with the supremum norm, i.e., $\mathcal{V} = C(X), X \subset R^n$. We shall assume that f is bounded and Lipschitz continuous, with the Lipschitz constant M in the norm $\| \cdot \|$. We denote the class of functions whose smallest Lipschitz constant is equal or smaller than M by $Lip(M)$. We can use any norm, or any distance function d_P. Our goal is to find an interpolant g that approximates f well at the points x distinct from the data, given that $f \in Lip(M)$. That is, we solve the following problem.

Given the data set as above, find an optimal interpolating function $g_M : R^n \to R$,

$$g_M = \arg \inf_{g \in C(X), f \in Lip(M)} \{\|f - g\|_{C(X)}\} \qquad (21)$$

such that

$$g(x^k) = f(x^k) = y^k, k = 1, \ldots, K.$$

Golomb and Weinberger [GW59] have considered the problem of approximation in linear spaces subject to finite bounds on some nonlinear functional in a very general setting. Let $\mathcal{V}$ be a linear vector space, u is an element of $\mathcal{V}$, and $F(u), F_1(u), \ldots, F_K(u)$ are linear functionals on this space. Given the values of functionals $F_k(u), k = 1, \ldots, K$, the goal is to approximate $F(u)$, subject to u being restricted to some subset $\mathcal{S} \subset \mathcal{V}$. The subset $\mathcal{S}$ is defined by means of a non-negative nonlinear positively homogeneous and continuous functional $\rho(u)$: $\mathcal{S} = \{u \in \mathcal{V} : \rho(u) \leq r\}$. Thus the unknown function u is known to lie in the intersection of the set $\mathcal{S}$ and the plane $v \in \mathcal{V}$, defined by $F_k(v) = f_k, k = 1, \ldots, K$.

Consider the set σ of values the functional $F(v)$ assumes as v ranges over this intersection. Under certain conditions on ρ (namely, the triangular inequality), σ is a closed interval, and the best approximation problem has a solution $\bar{u}$ that corresponds to the midpoint of this interval, while the error bounds on $F(u)$ are easily computed as half-length of σ.

In our case, $\mathcal{V}$ is the space of continuous functions $C(X)$, the functionals $F, F_k, k = 1, \ldots, K$ are defined as the values $u(x), u(x^k)$, and $\rho(u)$ is the Lipschitz seminorm (i.e.,

$$\forall v \in \mathcal{V} : \rho(v) = \inf\{M : |v(x) - v(z)| \leq M\|x - z\|, \forall x, z\}).$$

For every $x \in X$, denote the interval σ by $[H^{lower}(x), H^{upper}(x)]$, where H^{lower}, H^{upper} are respectively the lower and the upper bounds on u. Then the solution to the best uniform approximation problem (21) is given by

$$\bar{u}(x) = \frac{1}{2}[H^{lower}(x) + H^{upper}(x)].$$

To build a constructive interpolation algorithm, we need a suitable representation for functions H^{lower}, H^{upper}. This representation is given by Eqs. (7) or (8), and involves only the values of $f(x^k)$ and its Lipschitz constant M. We already used this representation to build both the lower and the upper approximations of f. We now combine the two approximations.

5.3 Description of the algorithm

First we describe Lipschitz approximation algorithm in the univariate case for the purposes of illustration, and then we proceed to the general multivariate case. Given the dataset $\{(x^k, y^k)\}_{k=1}^K$, $x^k, y^k \in R$, and the Lipschitz constant of f, M, define the lower and upper approximations

$$H^{lower}(x) = \max_k(y^k - M|x - x^k|), \quad H^{upper}(x) = \min_k(y^k + M|x - x^k|).$$

The lower approximation directly follows from (7), whereas the upper approximation is built from the lower approximation of an auxiliary function $\hat{f} = -f$, cf. Eq. (20).

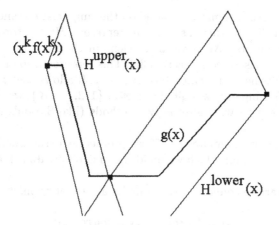

Fig. 15. The lower and upper approximations of a Lipschitz function f, and the best uniform approximation g.

Both approximations are piecewise linear functions, illustrated on Fig. 15. Their calculation for any x can be performed very efficiently in $O(\log K)$ operations by locating the interval $x \in [x^k, x^{k+1})$, assuming that x^k are sorted in increasing order. Under our assumption that $f \in Lip(M)$,

$$\forall x \in X : H^{lower}(x) \leq f(x) \leq H^{upper}(x),$$

and the bounds are tight. The optimal interpolant is then

$$g(x) = \frac{1}{2} \left(H^{lower}(x) + H^{upper}(x) \right).$$

Such an interpolant was considered in [Coo95, ZKS02]; the authors used it as a tool for reliable learning of Lipschitz functions. It possesses a number of desirable features listed below.

(1) g is a piecewise linear continuous function.
(2) g has Lipschitz constant M, i.e., $g \in Lip(M)$.
(3) g reproduces constant and linear functions.
(4) g preserves the range of the data $\min\{y^k\} \leq g \leq \max\{y^k\}$.
(5) g preserves monotonicity of the data, if for all k: $x^k \leq x^{k+1}$ implies $y^k \leq y^{k+1}$, then $g(x) \leq g(z) \; \forall x, z : x \leq z$.
(6) g continuously depends on x^k and y^k.
(7) The tight bound on the largest error of approximation is computed as $C = M \max_x \min_{x^k} |x - x^k|$. That is $\forall f \in Lip(M), f(x^k) = y^k$, $\max_x |f(x) - g(x)| \leq C$, and this bound is achieved, e.g., when $f = H^{lower}$ or $f = H^{upper}$.
(8) g is a minimum of the functional $F(g) = \int_X |g'(x)| dx$.

Now consider the multivariate case. We use the underestimate (7) (and the respective overestimate) as the functions H^{lower}, H^{upper}. We can use any norm in (7). However the method based on the simplicial distance (8) is very efficient numerically. In this case we can represent H^{lower} through the list of its local minimizers. We have an efficient method of enumerating local minimizers of (8), described in Section 3.3. This representation is useful when a value of H^{lower} is needed for an arbitrary $x \in X$. It allows one to compute the maximum in (8) using only a limited subset of $\{1, 2, \ldots, K\}$, which makes the algorithm competitive with alternative methods (like Sibson's interpolation [Sib81]).

To obtain the overestimate H^{upper} we proceed as earlier: define an auxiliary function $\hat{f} = -f$, for which we build the underestimate (8), then we take $H^{upper} = -H^K$.

Like its univariate counterpart, the multivariate interpolant

$$g(x) = \frac{1}{2} \left(H^{lower}(x) + H^{upper}(x) \right)$$

also possesses a number of desirable features. It provides uniform approximation to f, preserves its range, preserves the Lipschitz constant of f, and provides local approximation scheme (i.e., values of g depend only on the nearest data points). Furthermore, g depends continuously on the data. The latter property is very desirable [Alf89], but only a few multivariate interpolants possess this property. For instance, none of the schemata based on triangulation of the domain of f has this property.

However, the most important feature of the interpolant g is that it provides the best approximation of f in the worst case scenario: no matter how "bad" was the Lipschitz function f that generated the input data, or how inconveniently these data are distributed, g is the best approximation of f based on the available data. Thus our method translates into reliable approximation of f: even in the worst case the error bounds are guaranteed.

5.4 Numerical experiments

To illustrate the performance of the interpolant g we approximate the following Lipschitz functions.

Test function 1

$$f(x) = \sin x_1 \sin x_2 + 0.05(\sin 5x_1 \sin 5x_2)^3, x \in [0, 1]^2.$$

Test function 2

$$f(x) = \sin 5x_1 \sin 2x_2 + 0.2(\sin 20x_1 \sin 20x_2)^3, x \in [0, 3]^2.$$

Test function 3

$$f(x) = \prod_{i=1}^{n} \sin 2x_i, x \in [0, 3]^n.$$

Table 4. Performance of the algorithm for test function 1 as a function of the number of data points.

K	preprocessing time(s)	evaluation time (s $\times 10^{-3}$)	max error	root mean squared error
10000	1.021	0.293	0.173	0.025
20000	2.393	0.28	0.116	0.018
40000	5.578	0.39	0.0923	0.013
80000	12.878	0.45	0.069	0.0090

Table 5. Performance of the algorithm for test function 2 as a function of the number of data points.

K	preprocessing time(s)	evaluation time (s $\times 10^{-3}$)	max error	root mean squared error
10000	1.021	0.25	0.34	0.045
20000	2.43	0.28	0.18	0.031
40000	5.83	0.34	0.15	0.021
80000	12.9	0.40	0.021	0.013

Table 6. Performance of the algorithm for test function 3 as a function of the number of data points and dimension.

n	K	preprocessing time(s)	evaluation time (s $\times 10^{-3}$)	max error	root mean squared error
3	1000	0.17	0.72	0.63	0.14
	10000	2.8	1.43	0.32	0.063
	20000	6.67	1.66	0.27	0.050
	40000	15.69	1.85	0.18	0.038
	80000	35.57	2.09	0.17	0.031
4	1000	0.78	4.42	1.01	0.19
	5000	7.29	8.91	0.72	0.13
	10000	18.2	11.0	0.61	0.113
	20000	45.3	20.8	0.33	0.08
	40000	110.0	15.7	0.29	0.076
5	1000	4.66	29.84	1.04	0.19
	5000	54.08	69.06	0.83	0.14
	10000	211.8	98.40	0.62	0.12

The approximations of test functions 1 and 2 are plotted on Figs.16-19. Tables 4-6 provide quantitative information about the quality of fit and the speed of evaluation.

There are two steps of the algorithm that need benchmarking. The first step of building the interpolant g is called preprocessing, and the second step is the evaluation of g for an arbitrary x. Evaluation step was performed $N =$

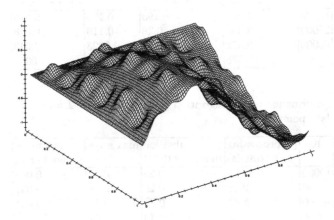

Fig. 16. Test function 1

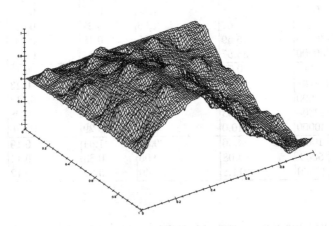

Fig. 17. Uniform approximation of the test function 1 using 20000 data points.

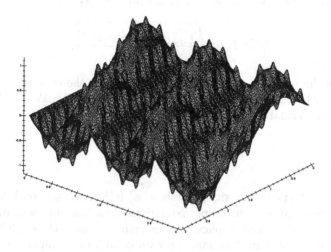

Fig. 18. Test function 2

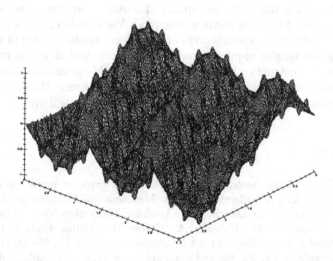

Fig. 19. Uniform approximation of the test function 2 using 80000 data points.

100000 times at random points to gather statistics, and the average time is reported. Further, the maximum and mean errors of approximation are reported. The root mean squared error is computed as

$$RMSE = \sqrt{\frac{\sum_{i=1}^{N}(f(x^i) - g(x^i))^2}{N}},$$

where N is the number of test points x^i not used in the construction of the interpolant. All computations were performed on a Pentium-IV PC, 1.2 GHz, 512 MB Ram, Visual C++ (version 6) compiler.

6 Conclusion

The theory of abstract convexity provides us with the necessary tools for building guaranteed tight one-sided approximations of various classes of functions. Such approximations find applications in many areas, such as global optimization, statistical simulation and approximation. In this paper we reviewed methods of building lower (upper) approximations of convex, log-convex, IPH and Lipschitz functions, which commonly arise in practice.

We presented an overview of three important applications of one-sided approximations: global optimization, random variate generation and scattered data interpolation. In all three applications we used essentially the same construction, in which the lower (or upper) approximation was represented by means of the list of its local minima (maxima). We also described a fast combinatorial algorithm for identification of these local minima. Each of the presented applications also requires a number of specific techniques to make use of this general construction. This paper addresses this issue and presents the details of the algorithms used in each case, and also illustrates the performance of the algorithms using numerical experiments, and practical applications.

References

[Alf89]   Alfeld, P.: Scattered data interpolation in three or more variables. In Schumaker, L.L., Lyche, T. (eds) Mathematical Methods in Computer Aided Geometric Design, 1–34. Academic Press, New York (1989)

[ARG99]   Andramonov, M., Rubinov, A., Glover, B.: Cutting angle methods in global optimization. Applied Mathematics Letters, **12**, 95–100 (1999)

[Aur91]   Aurenhammer, F.: Voronoi diagrams - a survey of a fundamental data structure. ACM Computing Surveys, **23**, 345–405 (1991)

[BR00]   Bagirov, A., Rubinov, A.: Global minimization of increasing positively homogeneous function over the unit simplex. Annals of Operations Research, **98**, 171–187 (2000)

[BR01]   Bagirov, A., Rubinov, A.: Modified versions of the cutting angle method. In: Hadjisavvas, N., Pardalos, P.M., (eds) Convex Analysis and Global optimization, Nonconvex optimization and its applications, **54**, 245–268. Kluwer, Dordrecht (2001)

[BRS03]   Bagirov, A., Rubinov, A.M., Soukhoroukova, N.V., Yearwood, J.L.: Unsupervised and supervised data classification via nonsmooth and global optimization. TOP (Formerly Trabajos Investigacin Operativa), **11**, 1–93 (2003)

[BRY01]   Bagirov, A., Rubinov, A.M., Yearwood, J.L.: Using global optimization to improve classification for medical diagnosis. Topics in Health Information Management, **22**, 65–74 (2001)

[BRY02]   Bagirov, A., Rubinov, A.M., Yearwood, J.L.: A global optimization approach to classification. Optimization and Engineering, **3**, 129–155 (2002)

[BB02]   Batten, L.M., Beliakov, G.: Fast algorithm for the cutting angle method of global optimization. Journal of Global Optimization, **24**, 149–161 (2002)

[Bel00]   Beliakov, G.: Shape preserving approximation using least squares splines. Approximation theory and applications, **16**, 80–98 (2000)

[Bel02]   Beliakov, G.: Approximation of membership functions and aggregation operators using splines. In Bouchon-Meunier, B., Gutierrez-Rios, Magdalena, L., and Yager, R. (eds) Technologies for Constructing Intelligent Systems, **2**, 159–172. Springer, Berlin (2002)

[Bel03]   Beliakov, G.: Geometry and combinatorics of the cutting angle method. Optimization, **52**, 379–394 (2003)

[Bel03]   Beliakov, G: How to build aggregation operators from data? Int. J. Intelligent Systems, **18**, 903–923, (2003)

[Bel04]   Beliakov, G.: The cutting angle method – a tool for constrained global optimization. Optimization Methods and Software, **19**, 137–151 (2004)

[Bel03]   Beliakov, G.: Least squares splines with free knots: global optimization approach. Applied Mathematics and Computation, **149**, 783–798 (2004)

[Bel05]   Beliakov, G: Extended cutting angle method of constrained global optimization. In: Caccetta, L. (eds) Optimization in Industry (in press). Kluwer, Dordrecht (2005)

[BTM01]   Beliakov, G., Ting, K.-M., Murshed, M.: Efficient serial and parallel implementation of the cutting angle global optimization technique. In: 5th International Conference on Optimization: Techniques and Applications, **1**, 80–87, Hong Kong (2001)

[BTMRB03] Beliakov, G., Ting, K.M., Murshed, M., Rubinov, A., Bertoli, M.: Efficient serial and parallel implementations of the cutting angle method. In: Di Pillo, G. (ed) High Performance Algorithms and Software for Nonlinear Optimization, 57–74. Kluwer Academic Publishers (2003)

[BSTY98] Boissonnat, J.-D., Sharir, M., Tagansky, B., Yvinec, M.: Voronoi diagrams in higher dimensions under certain polyhedral distance functions. Discrete and Comput. Geometry, **19**, 485–519 (1998)

[BEF00]   Büeler, B., Enge, A., Fukuda, K.: Exact volume computation for convex polytopes: a practical study. In: Kalai, G., Ziegler, G.M. (eds) Polytopes – Combinatorics and Computation, 131–154. Birkhäuser, Basel (2000)

[Coo95]   Cooper, D.A.: Learning Lipschitz functions. Int. J. of Computer Mathematics, **59**, 15–26 (1995)

[Dag88]   Dagpunar, J.: Principles of Random Variate Generation. Clarendon Press, Oxford (1988)

246   G. Beliakov

[DR95]   Demyanov, V.F., Rubinov, A.M.: Constructive Nonsmooth Analysis. Pe-
           ter Lang, Frankfurt am Main (1995)
[Dev86]   Devroye, L.: Non-uniform Random Variate Generation. Springer Verlag,
           New York (1986)
[Die95]   Dierckx, P.: Curve and Surface Fitting with Splines. Clarendon press,
           Oxford (1995)
[Eng05]   Enge, A.: http://www.lix.polytechnique.fr/labo/andreas.enge/volumen.html
           (2005)
[ES98]   Evans, M., Swartz, T.: Random variable generation using concavity prop-
           erties of the transformed densities. J. of Computational and Graphical
           Statistics, 7, 514–528 (1998)
[Flo00]   Floudas, C.A.: Deterministic Global Optimization: Theory, Methods, and
           Applications. Nonconvex optimization and its applications, 37. Kluwer
           Academic Publishers, Dordrecht/London, (2000)
[Fuk05]   Fukuda, K.: http://www.cs.mcgill.ca/~fukuda/soft/cdd_home/cdd.html
           (2005)
[FP96]   Fukuda, K., Prodon, A.: Double description method revisited. In: Deza,
           M., Euler, R., Manoussakis, I. (eds) Combinatorics and Computer Science,
           91–111. Springer-Verlag, Heidelberg (1996)
[GW59]   Golomb, M., Weinberger, H.F.: Optimal approximation and error bounds.
           In: Langer, R.E. (ed) On Numerical Approximation, 117–190. The Univ.
           of Wisconsin Press, Madison (1959)
[HJ95]   Hansen, P., Jaumard, B.: Lipschitz optimization. In: Horst, R, Pardalos,
           P. (eds) Handbook of Global Optimization, 407–493. Kluwer, Dordrecht
           (1995)
[Hor95]   Hörmann, W.: A rejection technique for sampling from t-concave distrib-
           utions. ACM Transactions on Mathematical Software, 21, 182–193 (1995)
[HLD04]   Hörmann, W., Leydold, J., Derflinger, G.: Automatic Nonuniform Ran-
           dom Variate Generation. Springer, Berlin (2004)
[HPT00]   Horst, R., Pardalos, P., Thoai, N.: Introduction to Global Optimization
           (2nd edition). Kluwer Academic Publishers, Dordrecht (2000)
[HP95]   Horst, R., Pardalos, P.M.: Handbook of Global Optimization. Noncon-
           vex optimization and its applications, 2. Kluwer Academic Publishers,
           Dordrecht/Boston (1995)
[Kel60]   Kelley, J.E.: The cutting-plane method for solving convex programs. J. of
           SIAM, 8, 703–712 (1960)
[KM97]   Kocic, L.M., Milovanovic, G.V.: Shape-preserving approximations by
           polynomials and splines. Computer and Mathematics with Applications,
           33, 59–97 (1997)
[Kva00]   Kvasov, B.: Methods of Shape Preserving Spline Approximation. World
           Scientific, Singapore (2000)
[LH98]   Leydold, J., Hörmann, W.: A sweep-plane algorithm for generating ran-
           dom tuples in simple polytopes. Mathematics of Computation, 67, 1617–
           1635 (1998)
[LH01]   Leydold, J., Hörmann, W.: Universal algorithms as an alternative for
           generating non-uniform continuous random variates. In Schüeler, G.I.,
           Spanos, P.D. (eds) Monte Carlo Simulation, 177–183. A. A. Balkema,
           Rotterdam (2001)

[LBB03]   Lim, K.F., Beliakov, G., Batten, L.M.: A new method for locating the global optimum: Application of the cutting angle method to molecular structure prediction. In: Proceedings of the 3rd International Conference on Computational Science, 4, 1040–1049. Springer-Verlag, Heidelberg (2003)

[LBB03]   Lim, K.F., Beliakov, G., Batten, L.M.: Predicting molecular structures: Application of the cutting angle method. Physical Chemistry Chemical Physics, 5, 3884–3890 (2003)

[LS02]   Locatelli, M., Schoen, F.: Fast global optimization of difficult lennard-Jones clusters, Computational Optimization and Applications, 21, 55–70 (2002)

[Lue94]   Luescher, M.: A portable high-quality random number generator for lattice field theory calculations. Computer Physics Communications, 79, 100–110 (1994)

[Mla86]   Mladineo, R.: An algorithm for finding the global maximum of a multimodal, multivariate function. Math. Prog., 34, 188–200 (1986)

[MRT53]   Motzkin, T.S., Raiffa, H., Thompson, G.L., Thrall, R.M.: The double description method. In: Kuhn, H.W., Tucker, A.W. (eds) Contribution to Theory of Games, 2. Princeton University Press, Princeton, RI (1953)

[Neu97]   Neumaier, A.: Molecular modeling of proteins and mathematical prediction of protein structure. SIAM Review, 39, 407–460 (1997)

[OBSC00] Okabe, A., Boots, B., Sugihara, K., Chiu, S.N.: Spatial Tessellations: Concepts and Applications of Voronoi Diagrams (2nd edition). John Wiley, Chichester (2000)

[Pij72]   Pijavski, S.A.: An algorithm for finding the absolute extremum of a function. USSR Comput. Math. and Math. Phys., 2, 57–67 (1972)

[Pin96]   Pinter, J.: Global Optimization in Action: Continuous and Lipschitz Optimization–algorithms, implementations, and applications. Nonconvex optimization and its applications, 6. Kluwer Academic Publishers, Dordrecht/Boston (1996)

[Roc70]   Rockafellar, R.T.: Convex Analysis. Princeton University Press, Princeton (1970)

[Rub00]   Rubinov, A.M.: Abstract Convexity and Global Optimization. Nonconvex optimization and its applications, 44. Kluwer Academic Publishers, Dordrecht/Boston (2000)

[Ser03]   Sergeyev, Y.D.: Finding the minimal root of an equation: applications and algorithms based on Lipschitz condition. In Pintér, J. (ed) Global Optimization – Selected Case Studies. Kluwer Academic Publishers (2003)

[Shu72]   Shubert, B.: A sequential method seeking the global maximum of a function. SIAM J. Numer. Anal., 9, 379–388 (1972)

[Sib81]   Sibson, R.: A brief description of natural neighbor interpolation. In: Barnett, V. (ed) Interpreting Multivariate Data, 21–36. John Wiley, Chichester (1981)

[SL97]   Sio, K.C., Lee, C.K.: Estimation of the Lipschitz norm with neural networks. Neural Processing Letters, 6, 99–108 (1997)

[SS00]   Strongin, R.G., Sergeyev, Y.D.: Global Optimization with Non-convex Constraints: Sequential and Parallel Algorithms. Nonconvex optimization and its applications, 45. Kluwer Academic, Dordrecht/London (2000)

[Wal74]   Walker, A.J.: New fast method for generating discrete random numbers with arbitrary frequency distributions. Electron. Lett., 10, 127–128 (1974)

[WZ96]   Wood, G.R., Zhang, B.P.: Estimation of the Lipschitz constant of a func-
          tion. J. Global Optim., **8**, 91–103 (1996)
[ZKS02]   Zabinsky, Z.B., Kristinsdottir, B.P., Smith, R.L.: Optimal estimation
          of univariate black box Lipschitz functions with upper and lower error
          bounds. Int. J. of Computers and Operations Research (2002)

Part II

Theory and Numerical Methods

A Numerical Method for Concave Programming Problems

Altannar Chinchuluun[1], Enkhbat Rentsen[2], and Panos M. Pardalos[1]

[1] Department of Industrial and Systems Engineering
University of Florida
303 Weil Hall, Gainesville, FL, 32611, USA
altannar@ufl.edu, pardalos@ufl.edu

[2] Department of Mathematical Modeling
School of Mathematics and Computer Science
National University of Mongolia
Ulaanbaatar, Mongolia
renkhbat@ses.edu.mn

Summary. Concave programming problems constitute one of the most important and fundamental classes of problems in global optimization. Concave minimization problems have a diverse range of direct and indirect applications. Moreover, concave minimization problems are well known to be NP-hard. In this paper, we present three algorithms which are similar to each other for concave minimization problems. In each iteration of the algorithms, linear programming problems with the same constraints as the initial problem are required to solve and a local search method is required to use. Furthermore, the convergence result is given. From the result, we see that the local search method is not necessarily required but we require that some conditions must hold on the constraint.

Key words: Approximation set; Trivial Approximation Set; Improved Approximation Set; General Orthogonal Approximation Set; Level Set; Concave Programming; Quasiconcave Function; Global Optimization

1 Introduction

Concave minimization techniques play an important role in other fields of global optimization. Large classes of optimization problems can be transformed into equivalent concave minimization problems. Concave minimization can be applied in the large number of fields. For instance, many problems from such fields as economics, telecommunications, transportation, computer design and finance can be formulated as concave minimization problems. More applications of concave minimization can be found in [HT93, PR87]. Concave minimization problems are NP-hard, even in most special cases. For instance,

[PS88] has shown that minimizing a concave quadratic function over a very simple polyhedron such as a hypercube is an NP-hard problem. More complete surveys of the complexity of these and other problems can be found in [Par93]. General concave minimization problem can be written as follows:

$$\min \quad f(x)$$
$$\text{s.t.} \quad x \in D,$$

where f is a concave function and D is a convex set. Concave minimization problems generally possess many local solutions that are not global. Moreover, we know that the global minimum of the above problem is attained at a vertex of D when D is a polytope. Many deterministic and stochastic approaches have been proposed for the local and global solutions to the concave minimization problem. There are three fundamental algorithmic approaches. The first approach is the enumerative method and it can be used only when D is a polyhedron. The other two approaches are the successive approximation approach and the branch and bound approach. These approaches can be found in most global optimization books [HT93, HP95, HPT01].

In this paper, we present a numerical method to solve the concave minimization problem with specific constraints. Basic idea of the method is to find an approximate solution to the problem solving linear programming problems with the same constraints as the initial problem. The paper is organized as follows: In Section 2, an optimality condition for the quasiconcave minimization problem is presented. In Section 3, the concept of approximation techniques and an approximation set, which are helpful to construct the algorithms, are introduced. In Section 4, three global optimization algorithms, which are based on the global optimality condition for the concave quadratic problem, are presented and their convergence properties are established.

2 Global Optimality Condition

Consider the quasiconcave minimization problem

$$\min \quad f(x) \tag{1}$$
$$\text{s.t.} \quad x \in D,$$

where $f : \mathbb{R}^n \rightarrow \mathbb{R}$ is a quasiconcave and differentiable function and D is a convex set in $\mathbb{R}^n$. Then the following theorem generalizes the result in Strekalovsky [Str98, SE90] .

Theorem 1. *Let z is a solution of Problem (1), and let*

$$E_c(f) = \{y \in \mathbb{R}^n \mid f(y) = c\}.$$

Then

$$(x - y)^T \nabla f(y) \geq 0 \text{ for all } y \in E_{f(z)}(f) \text{ and } x \in D, \tag{2}$$

If, in addition, $\nabla f(y) \neq 0$ holds for all $y \in E_{f(z)}(f)$, then condition (2) is sufficient for $z \in D$ being a solution to Problem (1).

Proof. Necessity. Suppose that z is a global minimizer of problem (1) and let $y \in E_{f(z)}(f)$ and $x \in D$. Then we have $f(x) \geq f(y)$. Since the function f is a quasiconcave, it follows that

$$f(\alpha x + (1 - \alpha)y) \geq \min\{f(x), f(y)\} = f(y) \quad \text{for all } \alpha \in [0, 1].$$

By Taylor's formula, there is a neighborhood of the point y on which:

$$f(y + \alpha(x - y)) - f(y) = \alpha \left((x - y)^T \nabla f(y) + \frac{o(\alpha\|x - y\|)}{\alpha} \right) \geq 0, \ \alpha > 0.$$

Note that $\lim_{\alpha \to 0} \frac{o(\alpha\|x-y\|)}{\alpha} = 0$. This implies that $(x - y)^T \nabla f(y) \geq 0$.

Sufficiency. Conversely, suppose that z is not a solution to problem (1); i.e., there exists an $u \in D$ such that $f(u) < f(z)$. By the definition of quasiconcave function, $U_{f(z)}(f) = \{x \in \mathbb{R}^n \mid f(x) \geq f(z)\}$ is a closed and convex set. Note that int $U_{f(z)}(f) \neq \emptyset$ according to the assumption in the theorem. Denote the projection of u on $U_{f(z)}(f)$ by y. It satisfies

$$\|y - u\| = \min_{x \in U_{f(z)}(f)} \|x - u\|.$$

Clearly,

$$\|y - u\| > 0 \tag{3}$$

holds because $u \notin U_{f(z)}(f)$. Moreover, this y can be considered as a solution of the following convex minimization problem:

$$\min \ g(x) = \frac{1}{2}\|x - u\|^2$$
$$\text{s.t.} \ x \in U_{f(z)}(f).$$

Since $U_{f(z)}(f) \neq \emptyset$ and this set is convex, the Slater's constraint qualification condition holds. Under this condition, y is a solution to the above problem if and only if there exists Lagrange multiplier λ such that (y, λ) is a solution to the following mixed nonlinear complementary problem :

$$\begin{cases} \nabla g(y) - \lambda \nabla f(y) = 0 \\ \lambda(f(z) - f(y)) = 0 \\ f(z) - f(y) \leq 0, \ \lambda \geq 0 \end{cases} \tag{4}$$

If $\lambda = 0$, then we have $\nabla g(y) = y - u = 0$, which contradicts (3). Thus, $\lambda > 0$ in (4). Then we obtain

$$y - u - \lambda \nabla f(y) = 0, \quad \lambda > 0,$$
$$f(y) = f(z).$$

From this we conclude that $(u-y)^T \nabla f(y) < 0$, which contradicts (2). This last contradiction implies that the assumption that z is not a solution of Problem (1) must be false. This completes the proof.   □

This theorem tells us that we need to find a pair $x, y \in \mathbb{R}^n$ such that

$$(x - y)^T \nabla f(y) < 0 , \ f(y) = f(z') , \ x \in D$$

in order to conclude that the point $z' \in D$ is not a solution to Problem (1). The following example illustrate the use of this property.

Example 1.

$$\min \ f(x) = \frac{x_1^2 + x_2^2}{1 - x_1 - x_2}$$
$$\text{s.t.} \quad 0.6 \le x_1 \le 7,$$
$$0.6 \le x_2 \le 2.$$

We can easily show that f is a quasiconcave function over the constraint set. The gradient of the function is found as follows.

$$\nabla f(x) = \left(\frac{x_2^2 - 2x_1 x_2 - x_1^2 + 2x_1}{(1 - x_1 - x_2)^2}, \frac{x_1^2 - 2x_1 x_2 - x_2^2 + 2x_2}{(1 - x_1 - x_2)^2} \right)^T .$$

Now we want to check whether a feasible point $x^0 = (0.6, 0.6)^T$, which is clearly local minimizer to the problem, is global minimizer or not. Then consider a pair $u = (5, 2)^T$ and $y = (3, 3)^T$ satisfying $f(y) = f(x^0) = -3.6$. We have $(u - y)^T \nabla f(y) = -\frac{12}{15} < 0$ and it follows that x^0 is not a global solution. In fact, we can show that the global solution is $x^* = (7, 0.6)^T$.

3 Approximation Techniques of the Level Set

For further discussion, we will consider only the concave case of the Problem (1), which is

$$\min \ f(x) \qquad\qquad (5)$$
$$\text{s.t.} \quad x \in D,$$

where f is a concave and differentiable function and D is a convex compact set in $\mathbb{R}^n$.

Definition 1. *The set $E_{f(z)}(f)$ defined by*

$$E_{f(z)}(f) = \{y \in \mathbb{R}^n \mid f(y) = f(z)\}$$

is called the level set of f at z.

Note that the optimality condition (2) for Problem (5) requires to check the linear programming problem

$$\min \quad (x - y)^T \nabla f(y)$$
$$\text{s.t.} \quad x \in D$$

for every $y \in E_{f(z)}(f)$. This is a hard problem. Thus, we need to find an appropriate approximation set so that one could check the optimality condition at a finite number of points.

The following lemmas show that finding a point at the level set of $f(x)$ is theoretically possible.

Lemma 1. *Let $h \in \mathbb{R}^n$, $z \in D$ which is not a global maximizer of $f(x)$ over $\mathbb{R}^n$ and let x^* be an optimal solution of the problem*

$$\max \quad f(x)$$
$$\text{s.t.} \quad x \in \mathbb{R}^n,$$

and let the set of all optimal solutions of this problem be bounded. Then there exists a unique positive number α such that $x^* + \alpha h \in E_{f(z)}(f)$.

Proof. We will prove that there exists a positive number α such that $x^* + \alpha h \in E_{f(z)}(f)$ at first. Suppose conversely that there is no number which satisfies the above condition; i.e., $f(x^* + \alpha h) > f(z)$ holds for all $\alpha \geq 0$. Note that $hyp(f) = \{(x, r) \in \mathbb{R}^{n+1} : r \leq f(x)\}$ is a convex set since f is a concave function. For $\alpha \geq 0$, we obtain $(x^* + \alpha h, f(z)) \in hyp(f)$. Next we show that $(h, 0)$ is a direction of $hyp(f)$. Suppose conversely that there exist a vector $y \in hyp(f)$ and a positive scalar β such that $y + \beta(h, 0) \in \mathbb{R}^{n+1} \setminus hyp(f)$. Since $\mathbb{R}^{n+1} \setminus hyp(f)$ is an open set, there exists a scalar μ that satisfies the following conditions:

$$\mu(x^*, f(z)) + (1 - \mu)(y + \beta(h, 0)) \in \mathbb{R}^{n+1} \setminus hyp(f) , \quad 0 < \mu < 1 \qquad (6)$$

On the other hand, we can show that $\mu(x^*, f(z)) + (1 - \mu)(y + \beta(h, 0))$ lies on the line segment joining some two points of $hyp(f)$. For the points $(x^*, f(z)) + \frac{(1-\mu)\beta}{\mu}(h, 0)$ and y, the following equation holds.

$$\mu((x^*, f(z)) + \frac{(1 - \mu)\beta}{\mu}(h, 0)) + (1 - \mu)y = \mu(x^*, f(z)) + (1 - \mu)(y + \beta(h, 0))$$

By convexity of $hyp(f)$, we have $\mu(x^*, f(z)) + (1 - \mu)(y + \beta(h, 0)) \in hyp(f)$. This contradicts (6), hence, $(h, 0)$ is a direction of $hyp(f)$. Since $(x^*, f(x^*)) \in hyp(f)$, the following statement is true.

$$(x^*, f(x^*)) + \alpha(h, 0) \in hyp(f) \text{ for all } \alpha \geq 0$$

We can conclude that $x^* + \alpha h$ is also a global maximizer of f for all $\alpha \geq 0$ because x^* is a global maximizer of f over $\mathbb{R}^n$. This contradicts the assumption in the lemma. Now, we prove the uniqueness property. Assume that there are two positive scalars α_1 and α_2 such that $x^* + \alpha_i h \in E_{f(z)}(f)$, $i = 1, 2$. Without loss of generality, we can assume that $0 < \alpha_1 \leq \alpha_2$. By concavity of f, we have

$$f(z) = f(x^* + \alpha_1 h) = f\left(\left(1 - \frac{\alpha_1}{\alpha_2}\right)x^* + \frac{\alpha_1}{\alpha_2}(x^* + \alpha_2 h)\right)$$

$$\geq \left(1 - \frac{\alpha_1}{\alpha_2}\right)f(x^*) + \frac{\alpha_1}{\alpha_2}f(x^* + \alpha_2 h)$$

$$= \left(1 - \frac{\alpha_1}{\alpha_2}\right)f(x^*) + \frac{\alpha_1}{\alpha_2}f(z) \geq f(z)$$

This inequality is valid only if $\alpha_1 = \alpha_2$ □

Under some condition, it is possible to compute a point on the level set. This is shown by the following statement.

Lemma 2. *Let a point* $z \in D$ *and a vector* $h \in \mathbb{R}^n$ *satisfy* $h^T \nabla f(z) < 0$ *and let* x^* *be global maximizer of* f *over* D. *Then there exists a unique positive number* α *such that* $x^* + \alpha h \in E_{f(z)}(f)$.

Proof. Suppose conversely that

$$f(z) < f(x^* + \alpha h) \text{ for all } \alpha \geq 0.$$

Note that $f(x^*) \geq f(z)$ for any $z \in D$. By convexity of f, we have

$$(x^* - z)^T \nabla f(z) \geq 0.$$

From the last inequality and assumption $h^T \nabla f(z) < 0$, we can conclude that

$$\frac{(x^* - z)^T \nabla f(z)}{h^T \nabla f(z)} \leq 0$$

Since f is a concave function, for all $\alpha \geq 0$, we have

$$0 < f(x^* + \alpha h) - f(z) \leq (x^* + \alpha h - z)^T \nabla f(z)$$
$$= (x^* - z)^T \nabla f(z) + \alpha h^T \nabla f(z).$$

For $\alpha = -\frac{(x^* - z)^T \nabla f(z)}{h^T \nabla f(z)} \geq 0$, we get

$$0 < f(x^* + \alpha h) - f(z) \leq 0.$$

This gives contradiction.   □

Example 2. Consider the quadratic concave minimization problem

$$\min \quad f(x) = \frac{1}{2}x^T C x + d^T x \tag{7}$$

$$\text{s.t.} \quad x \in D$$

where D is a convex set in $\mathbb{R}^n$, $d \in \mathbb{R}^n$ and C is a symmetric negative definite $n \times n$ matrix.

Since C is negative definite, we have

$$h^T Ch < 0$$

for all $h \neq 0$. Let us solve the equation $f(x^* + \alpha h) = f(z)$ with respect to α.

$$\frac{1}{2}(x^* + \alpha h)^T C(x^* + \alpha h)^T + d^T(x^* + \alpha h) = f(z)$$

or

$$f(x^*) + \alpha h^T(Cx^* + d) + \frac{1}{2}\alpha^2 h^T Ch = f(z)$$

Note that x^* satisfies $Cx^* + d = 0$. Using this fact, we have

$$\alpha = \left(\frac{2(f(z) - f(x^*))}{h^T Ch}\right)^{\frac{1}{2}}.$$

Constructing Points on the Level Set

As we have seen in Example 2, the number α can be found analytically for the quadratic case. In a general case, this analytical formula is not always available but Lemmas 1 and 2 give us an opportunity to find a point on the level set using numerical methods. For this purpose, let us introduce the following function of one variable in $\mathbb{R}^+$.

$$\psi(t) = f(x^* + th) - f(z). \tag{8}$$

The above lemmas state that this function has a unique root in $\mathbb{R}^+$. Our goal is to find the root of the function and , now, we can use numerical methods for this problem such as the Fixed point method, the Newton's method, the Bracketing methods and so on . We could use the following method to find initial guesses a and b such that $\psi(a) > 0$ and $\psi(b) \leq 0$ for the Bracketing methods as follows:

1. Choose a step size $\rho > 0$.
2. Determine $\psi(q\rho)$, $q = 1, 2, \ldots, q_0$
3. $a = (q_0 - 1)\rho$, $b = q_0\rho$

where q_0 is the smallest positive integer number such that $\psi(q_0\rho) \leq 0$. Moreover the Bisection method can be stated in the following form.

1. Determine ψ at the midpoint $\frac{a+b}{2}$ of the interval $[a, b]$.
2. If $\psi(\frac{a+b}{2}) > 0$, then the root is in the interval $\left[\frac{a+b}{2}, b\right]$. Otherwise, the root is in the interval $\left[a, \frac{a+b}{2}\right]$. The length of the interval containing the root is reduced by a factor of $\frac{1}{2}$.
3. Repeat the procedure until a prescribed precision is attained.

Finally, we choose a number α as an approximate root of the function $\psi(t)$ such that

$$\psi(\alpha) \leq 0 \tag{9}$$

i.e., α lies on the right hand side of the exact root. We will see that this selection helps us when we construct algorithms for Problem (1) in the next section.

In order to check the optimality condition at a finite number of points of the level set, it is necessary to introduce a notion of an approximation set.

Definition 2. *The set A_z^m defined for a given integer m by*

$$A_z^m = \{y^1, y^2, \ldots, y^m \mid y^i \in E_{f(z)}(f), \quad i = 1, 2, \ldots, m\} \tag{10}$$

is called an approximation set to the level set $E_{f(z)}(f)$ at z.

Since we can construct a point on the level set, an approximation set can be constructed in same way. Assume that A_z^m is given. Then for each $y^i \in A_z^m$, $i = 1, 2, \ldots, m$, solve the auxiliary problem

$$\min \quad x^T \nabla f(y^i) \tag{11}$$
$$\text{s.t.} \quad x \in D.$$

Let u^i, $i = 1, 2, \ldots, m$, be the solutions of those problems, which always exist due to the compactness of D:

$$u^{i^T} \nabla f(y^i) = \min_{x \in D} x^T \nabla f(y^i) \tag{12}$$

Let us define θ_m as follows:

$$\theta_m = \min_{i=1,2,\ldots,m} (u^i - y^i)^T \nabla f(y^i) \tag{13}$$

There are some properties of A_z^m and θ_m.

Lemma 3. *If there is a point $y^i \in A_z^m$ for $z \in D$ such that $(u^i - y^i)^T \nabla f(y^i) < 0$, where $u^i \in D$ satisfies $u^{i^T} \nabla f(y^i) = \min_{x \in D} x^T \nabla f(y^i)$, then*

$$f(u^i) < f(z)$$

holds.

Proof. By the definition of u^i, we have

$$\min_{x \in D} (x - y^i)^T \nabla f(y^i) = (u^i - y^i)^T \nabla f(y^i)$$

Since f is concave,

$$f(u) - f(v) \leq (u - v)^T \nabla f(v)$$

holds for all $u, v \in \mathbb{R}^n$. Therefore, the assumption in the lemma implies that

$$f(u^i) - f(z) = f(u^i) - f(y^i) \leq (u^i - y^i)^T \nabla f(y^i) < 0.$$

$\square$

Trivial Approximation Set

Consider the following set of vectors :

$$A_z^{2n} = \{y^1, y^2, \ldots, y^{2n} \mid y^j = x^* + \alpha_j l^j \in E_{f(z)}(f), \quad j = 1, 2, \ldots, 2n\}, \quad (14)$$

where α_j's are positive numbers, l^j's are orthogonal vectors such that $l^j = -l^{n+j}$ for $j = 1, \ldots, n$ and x^* is a solution to the problem

$$\max f(x)$$
$$\text{s.t. } x \in \mathbb{R}^n.$$

Without loss of generality we can assume that l^j is the jth unit (coordinate) vector and there exists some α_j such that $y^j \in E_{f(z)}(f)$ (if this number does not exist, we just eliminate y^j from the set), therefore A_z^{2n} is an approximation set to the level set $E_{f(z)}(f)$ at the point z. When z is not a global maximizer of f over R^n, clearly, A_z^{2n} is a nonempty set and it contains at least n points.

Definition 3. *The approximation set constructed according to (14) is called the trivial approximation set.*

Second order Approximation Set

In order to improve the approximation set, it is helpful to define another approximation set based on the previous approximation set. Assume that we have an approximation set A_z^m. We can construct another approximation set B_z^m based on the approximation set as follows.

$$B_z^m = \{\bar{y}^1, \bar{y}^2, \ldots, \bar{y}^m \mid \bar{y}^i = x^* + \bar{\alpha}_i(u^i - x^*) \in E_{f(z)}(f), \quad i = 1, 2, \ldots, m\}, \quad (15)$$

where x^* is a solution to the problem

$$\max f(x)$$
$$\text{s.t. } x \in \mathbb{R}^n.$$

and u^i is a solution to the problem

$$\min x^T \nabla f(y^i)$$
$$\text{s.t. } x \in D.$$

The use of x^* is justified by the relationship between A_z^m and B_z^m in the following lemma.

Lemma 4. *Let $f(z) \neq f(x^*)$. If $\theta_m < 0$ then there exists a $j \in \{1, 2, \ldots, m\}$ and $v \in D$ such that $\bar{y}^j \in B_z^m$ satisfies $(v - \bar{y}^j)^T \nabla f(\bar{y}^j) < 0$.*

Proof. According to (13), there exists a $j \in \{1, 2, \ldots, m\}$ such that

$$\theta_m = (u^j - y^j)^T \nabla f(y^j) = \min_{i=1,2,\ldots,m} (u^i - y^i)^T \nabla f(y^i) < 0,$$

where u^i satisfies $u^{i^T} \nabla f(y^i) = \min_{x \in D} x^T \nabla f(y^i)$.

$$0 > (u^j - y^j) \nabla f(y^j) = (u^j - y^j + x^* - x^*) \nabla f(y^j)$$
$$= (u^j - x^*) \nabla f(y^j) + (x^* - y^j) \nabla f(y^j)$$

or

$$(u^j - x^*) \nabla f(y^j) < (y^j - x^*) \nabla f(y^j).$$

Using the concavity of f, we can show that the right hand side of the last inequality is negative as follows

$$(y^j - x^*) \nabla f(y^j) \leq f(y^j) - f(x^*) < 0.$$

Since $(u^j - x^*) \nabla f(y^j) < 0$ from the last two inequalities, according to Lemma 2, there exists a unique positive number $\bar{\alpha}_j$ such that $\bar{y}^j = x^* + \bar{\alpha}_j (u^j - x^*) \in E_{f(y^j)}(f) = E_{f(z)}(f)$. Clearly $\bar{y}^j \in B_z^m$. Now, we will show that $\bar{\alpha}^j < 1$. Conversely, suppose that $\bar{\alpha}^j \geq 1$ or $0 < \frac{1}{\bar{\alpha}^j} \leq 1$. As we have seen in Lemma 3, we can write

$$f(u^j) < f(z).$$

Since x^* is the global maximizer of f over $\mathbb{R}^n$, we have

$$f(u^j) < f(z) = f(x^* + \bar{\alpha}_j(u^j - x^*)) < f(x^*).$$

On the other hand, by concavity of f

$$\frac{1}{\bar{\alpha}_j} f\left(x^* + \bar{\alpha}_j(u^j - x^*)\right) + \left(1 - \frac{1}{\bar{\alpha}_j}\right) f(x^*)$$
$$\leq f\left(\frac{1}{\bar{\alpha}_j}(x^* + \bar{\alpha}_j(u^j - x^*)) + \left(1 - \frac{1}{\bar{\alpha}_j}\right) x^*\right) = f(u^j).$$

This contradicts the previous inequality. Thus $0 < \bar{\alpha}^j < 1$. Now we are ready to prove the lemma.

Consider the point $\bar{y}^j = x^* + \bar{\alpha}_j(u^j - x^*)$ in B_z^m. From the concavity of f and the above observations, it follows that

$$(u^j - \bar{y}^j)^T \nabla f(\bar{y}^j) = (1 - \bar{\alpha}_j)(u^j - x^*)^T \nabla f(x^* + \bar{\alpha}_j(u^j - x^*))$$
$$= \frac{1 - \bar{\alpha}_j}{\bar{\alpha}_j} \bar{\alpha}_j (u^j - x^*)^T \nabla f(x^* + \bar{\alpha}_j(u^j - x^*))$$
$$= \frac{1 - \bar{\alpha}_j}{\bar{\alpha}_j} (x^* + \bar{\alpha}_j(u^j - x^*) - x^*)^T \nabla f(x^* + \bar{\alpha}_j(u^j - x^*))$$
$$\leq \frac{1 - \bar{\alpha}_j}{\bar{\alpha}_j} (f(x^* + \bar{\alpha}_j(u^j - x^*)) - f(x^*)) < 0.$$

Now, if we take a point $v = u^j \in D$, then we have $(v - \bar{y}^j)^T \nabla f(\bar{y}^j) < 0$, and the assertion is proven. $\qquad\qquad\qquad\qquad\qquad\qquad\qquad\qquad\qquad\qquad\qquad\qquad$ $\square$

Remark 1. Note that $\theta_m \geq 0$ does not always imply

$$\min_{i=1,2,\ldots,m} \; \min_{\bar{y}^i \in B_z^m} (u^i - \bar{y}^i)^T \nabla f(\bar{y}^i) \geq 0.$$

Remark 2. If we use Selection (14), it is easy to see that the lemma is still true when α_j and $\bar{\alpha}_j$ are approximate roots to the functions $\psi_1(t) = f(x^* + tl^j) - f(z)$ and $\psi_2(t) = f(x^* + t(u^j - x^*)) - f(z)$, respectively.

In analogy with θ_m for A_z^m, introduce $\bar{\theta}_m$ for the set B_z^m as follows.

$$\bar{\theta}_m = \min_{i=1,2,\ldots,m} (v^i - \bar{y}^i)^T \nabla f(\bar{y}^i),$$

where v^i is defined by $v^{i^T} \nabla f(\bar{y}^i) = \min_{x \in D} x^T \nabla f(\bar{y}^i)$

Definition 4. *The approximation set constructed according to (15) is called the second order approximation set to the level set $E_{f(z)}(f)$ at point z.*

Orthogonal Approximation Set

Another way to construct an approximation set is extracting an approximation set from the trivial approximation set using the rotation. Consider the coordinate vectors l^j, $j = 1, \ldots, n$, l^{n+j} such that $l^{n+j} = -l^j$, $j = 1, \ldots, n$ and a rotation matrix R. Let us define the following vectors and a set of vectors.

$$C_z^{2n} = \{\hat{y}^1, \hat{y}^2, \ldots, \hat{y}^{2n} \mid \hat{y}^j = x^* + \beta_j q^j, \; i = 1, 2, \ldots, 2n\}, \qquad (16)$$

where x^* is a solution to the problem

$$\max f(x)$$
$$\text{s.t. } x \in \mathbb{R}^n.$$

and $q^j = Rl^j$, $j = 1, \ldots, 2n$.

Without loss of generality, we can assume that there exist positive numbers β_j such that $x^* + \beta_j q^j \in E_{f(z)}f$ and, therefore, C_z^{2n} is an approximation set to the level set $E_{f(z)}(f)$ at point z. Also we can introduce $\hat{\theta}_m$ as follows:

$$\hat{\theta}_m = \min_{i=1,2,\ldots,m} (w^i - \hat{y}^i)^T \nabla f(\hat{y}^i),$$

where w^i is defined by $w^{i^T} \nabla f(\bar{y}^i) = \min_{x \in D} x^T \nabla f(\bar{y}^i)$.

Definition 5. *The approximation set constructed according to (16) is called the orthogonal approximation set to the level set $E_{f(z)}(f)$.*

4 Algorithms and their Convergence

In this section, we discuss three algorithms based on observations discussed in Section 3 to solve Problem (5). We begin by explaining the main idea of our methods for the problem. The idea of the algorithms is to check whether θ_m, which is defined in (13), is negative or nonnegative solving linear programming problems. Therefore, if it is negative, a new improved solution can be found according to Lemma 3; otherwise, terminate the algorithm. When the new improved solution is found, we will use one of the existing local search methods to get faster convergence. Also, we assume here that Problem (5) has finite stationary points on the constraint set D in order to ensure convergence of the algorithms. The first algorithm uses only the trivial approximation set and the second algorithm uses a combination of the trivial and the second order approximation sets, finally, the last algorithm uses a combination of the three approximation sets defined in Section 3. The basic algorithm can be summarized as follows:

Algorithm 1. INPUT : A concave differentiable function f, a convex compact set D and x^*, a global maximizer of f.
OUTPUT : A global solution x to Problem (5).
Step 1. Choose a point $x^0 \in D$. Set $k = 0$.
Step 2. Find a local minimizer $z^k \in D$ using one of the existing methods starting with an initial approximation point x^k.
Step 3. Construct the trivial approximation set $A_{z^k}^{2n}$ at z^k.
Step 4. For each $y^i \in A_{z^k}^{2n}$, $i = 1, 2, \ldots, 2n$, solve the problem

$$\min \quad x^T \nabla f(y^i)$$
$$\text{s.t.} \quad x \in D$$

to obtain a solution u^i, i.e.,

$$u^{i^T} \nabla f(y^i) = \min_{x \in D} x^T \nabla f(y^i)$$

Step 5. Find the number $j \in \{1, 2, \ldots, 2n\}$ such that

$$\theta_{2n}^k = (u^j - y^j)^T \nabla f(y^j) = \min_{i=1,2,\ldots,2n} (u^i - y^i)^T \nabla f(y^i)$$

Step 6. If $\theta_{2n}^k < 0$ then $x^{k+1} := u^j$, $k := k + 1$ and return to Step 2. Otherwise, z^k is an approximate global minimizer and terminate.

The convergence of Algorithm 1 is given by the following statement.

Theorem 2. *The sequence $\{z^k,\ k = 0, 1, \ldots\}$ generated by Algorithm 1 converges to a solution of Problem (5) in a finite number of steps or finds an approximate solution as a local solution to the problem.*

Proof. We show that if $\theta_{2n}^k < 0$ holds for all k, then z^k converges to a global minimizer of Problem (5) in a finite number of steps. In fact, take a $j \in \{1, 2, \ldots, 2n\}$ such that $y^j \in A_{z^k}^{2n}$ and $u^j \in D$ satisfy

$$\theta_{2n}^k = (u^j - y^j)^T \nabla f(y^j) < 0.$$

According to Lemma 3, we have

$$f(u^j) < f(z^k)$$

We show that this inequality holds even when α_j is an approximate root of the function $\psi(t) = f(x^* + tl^j) - f(z^k)$. From Selection (9), we can conclude that

$$f(y^j) \leq f(z^k).$$

Then it follows that

$$f(u^j) - f(z^k) \leq f(u^j) - f(y^j) \leq (u^j - y^j)\nabla f(y^j) < 0.$$

Since u^j is a starting point for finding a local solution z^{k+1}, finally, it can be deduced that

$$f(z^{k+1}) < f(z^k) \text{ for all } k = 0, 1, 2, \ldots.$$

By the assumption, the number of local minimizers z^k is finite, and this sequence reaches a global minimizer in a finite number of steps or stops at an approximate local solution. This completes the proof. $\qquad\square$

Remark 3. When D is a polytope, we can use Algorithm 1 without a local search method since every auxiliary problem finds a vertex of the set D and number of the vertices is finite.

Example 3. [HPT01]. To illustrate Algorithm 1, let us consider the following example.

$$\min f(x) = \frac{1}{2} x^T C x + d^T x$$

$$\text{s.t. } Ax \leq b$$

$$x \geq 0,$$

where

$$C = \begin{pmatrix} -2 & 4 \\ 4 & -8 \end{pmatrix}, \quad d = \begin{pmatrix} 2 \\ 4 \end{pmatrix}, \quad A^T = \begin{pmatrix} -4 & 0 & 1 & 1 & 1 \\ 2 & 1 & 1 & 0 & -4 \end{pmatrix}, \quad b^T = (1, 2, 4, 3, 1).$$

Iteration 1.
An initial feasible solution is $x_0^1 = (0, 0)^T$. Note that this vertex is a local solution to the problem. In this case, a local search method cannot affect the current approximate solution. The current best objective function value

is 0. There does not exist a global maximizer of the function $f(x)$ over $\mathbb{R}^2$. Thus, we consider a global maximizer of the function over the constraint set; therefore, it can be used for constructing an approximation set. The maximizer is $x_1^* = (2.555, 1.444)^T$. The trivial approximation set can be constructed easily solving quadratic equations.

$$y_1^1 = (7.432, 1.444)^T, \ y_1^2 = (2.555, 3.452)^T,$$

$$y_1^3 = (0.345, 1.444)^T, \ y_1^4 = (2.555, 0.102)^T.$$

Solving linear programming problems, we find the following vectors.

$$u_1^1 = (3.0, 0.5)^T, \ u_1^2 = (0.75, 2.0)^T, \ u_1^3 = (0.75, 2.0)^T, \ u_1^4 = (1.0, 0.0)^T$$

Moreover, $\theta_4^1 = -0.563$ and the initial feasible point to the next iteration is $u_1^3 = (0.75, 2.0)^T$.

Iteration 2.
New feasible solution is $x_0^2 = (0.75, 2.0)^T$. The local search method cannot improve this solution. The current objective function value is -1.0625. The trivial approximation set to the level set $E_{-1.0625}(f)$ is

$$y_2^1 = (7.579, 1.444)^T, \ y_2^2 = (2.555, 3.530)^T,$$

$$y_2^3 = (0.199, 1.444)^T, \ y_2^4 = (2.555, 0.025)^T.$$

Solving linear programming problems, we have

$$u_2^1 = (3.0, 0.5)^T, \ u_2^2 = (0.75, 2.0)^T,$$

$$u_2^3 = (0.75, 2.0)^T, \ u_2^4 = (1.0, 0.0)^T$$

which is the same as we find at Iteration 1. Therefore, $\theta_4^2 = 0.313$ at the vertex $u_2^3 = (0.75, 2.0)^T$. The algorithm terminates at this iteration and the global approximate solution is $(0.75, 2.0)^T$. Note that this is a global solution to the problem.

Unfortunately, Algorithm 1 cannot always guarantee for a global optimal solution. In this case, we can extend Algorithm 1 using the improved approximation set and present an outline of the next algorithms as follows :

Algorithm 2. INPUT : A concave differentiable function f, a convex compact set D and x^*, a global maximizer of f.
OUTPUT : A global solution x to Problem (5).
Step 1. Choose a point $x^0 \in D$. Set $k = 0$.
Step 2. Find a local minimizer $z^k \in D$ using one of the existing methods starting with an initial approximation point x^k.
Step 3. Construct a trivial approximation set $A_{z^k}^{2n}$ at z^k.

Step 4. For each $y^i \in A_{z^k}^{2n}$, $i = 1, 2, \ldots, 2n$, solve the problem

$$\min \quad x^T \nabla f(y^i)$$
$$\text{s.t.} \quad x \in D$$

to obtain a solution u^i, i.e.,

$$u^{i^T} \nabla f(y^i) = \min_{x \in D} x^T \nabla f(y^i)$$

Step 5. Find the number $j \in \{1, 2, \ldots, 2n\}$ such that

$$\theta_{2n}^k = (u^j - y^j)^T \nabla f(y^j) = \min_{i=1,2,\ldots,2n} (u^i - y^i)^T \nabla f(y^i)$$

Step 6. If $\theta_{2n}^k < 0$ then $x^{k+1} := u^j$, $k := k+1$ and return to Step 2. Otherwise go to the next step.

Step 7. Construct a second order approximation set $B_{z^k}^{2n}$ at z^k by (15).

Step 8. For each $\bar{y}^i \in B_{z^k}^{2n}$, $i = 1, 2, \ldots, 2n$, solve the problem

$$\min \quad x^T \nabla f(\bar{y}^i)$$
$$\text{s.t.} \quad x \in D$$

to obtain a solution v^i, i.e.,

$$v^{i^T} \nabla f(\bar{y}^i) = \min_{x \in D} x^T \nabla f(\bar{y}^i)$$

Step 9. Find the number $s \in \{1, 2, \ldots, 2n\}$ such that

$$\bar{\theta}_{2n}^k = (v^s - \bar{y}^s)^T \nabla f(\bar{y}^s) = \min_{i=1,2,\ldots,2n} (v^i - \bar{y}^i)^T \nabla f(\bar{y}^i)$$

Step 10. If $\bar{\theta}_{2n}^k < 0$ then $x^{k+1} := v^s$, $k := k+1$ and return to Step 2. Otherwise, z^k is an approximate global minimizer and terminate the algorithm.

The convergence of this algorithm is the same as in Theorem 2.

Theorem 3. *The sequence $\{z^k, \ k = 0, 1, \ldots\}$ generated by Algorithm 2 converges to a solution of problem (5) in a finite number of steps or finds an approximate solution as a local solution to the problem.*

Remark 4. When D is a polytope, Algorithm 2 can be used without a local search method.

Example 4. To illustrate Algorithm 2, let us consider the following concave quadratic programming problem.

$$\min f(x) = \frac{1}{2} x^T C x$$
$$\text{s.t.} \quad Ax \leq b$$
$$x \geq 0,$$

where

$$C = \begin{pmatrix} -4 & -0.5 \\ -0.5 & -4 \end{pmatrix}, \quad A^T = \begin{pmatrix} 5 & 4 & 0 \\ 1 & 2 & 1 \end{pmatrix}, \quad b^T = (20, 19, 3).$$

Since the constraint set D is a polytope, we can use the algorithm without a local search method.

Iteration 1.
Let us choose $x^1 = (1.0, 0.0)^T$ as an initial feasible solution. The current objective function value is -2.0. The global maximizer of the function $f(x)$ over $\mathbb{R}^2$ is $x^* = (0.0, 0.0)^T$. The trivial approximation set can be computed as we have seen in Example 2, and the vectors are

$$y_1^1 = (1.0, 0.0)^T, \quad y_1^2 = (0.0, 1.0)^T, \quad y_1^3 = (-1.0, 0.0)^T, \quad y_1^4 = (0.0, -1.0)^T.$$

Solving linear programming problems, the following vertices of the polytope are found.

$$u_1^1 = (4.0, 0.0)^T, \quad u_1^2 = (3.25, 3.0)^T, \quad u_1^3 = (0.0, 0.0)^T, \quad u_1^4 = (0.0, 0.0)^T$$

Therefore, $\theta_4^1 = -12$ at the vertex $u_1^1 = (4.0, 0.0)^T$.

Iteration 2.
The current best feasible solution is $x^2 = (4.0, 0.0)^T$ and the objective function value is -32. The trivial approximation set to the level set $E_{-32}(f)$ is

$$y_2^1 = (4.0, 0.0)^T, \quad y_2^2 = (0.0, 4.0)^T, \quad y_2^3 = (-4.0, 0.0)^T, \quad y_2^4 = (0.0, -4.0)^T.$$

The u_2^j vectors are

$$u_2^1 = (4.0, 0.0)^T, \quad u_2^2 = (3.25, 3.0)^T, \quad u_2^3 = (0.0, 0.0)^T, \quad u_2^4 = (0.0, 0.0)^T$$

which is same as we find at Iteration 1. Nevertheless, $\theta_4^2 = 0$ at the vertex $u_1^1 = (4.0, 0.0)^T$; i.e., in this case, the trivial approximation set does not work. Next, we construct the improved approximation set, which is derived from the last two sets of vectors according to (15).

$$\bar{y}_2^1 = (4.0, 0.0)^T, \quad \bar{y}_2^2 = (2.772, 2.558)^T, \quad \bar{y}_2^3 = (-4.0, 0.0)^T, \quad \bar{y}_2^4 = (-2.772, -2.558)^T.$$

Solving linear programming problems, we have

$$v_2^1 = (4.0, 0.0)^T, \quad v_2^2 = (3.25, 3.0)^T, \quad v_2^3 = (0.0, 0.0)^T, \quad v_2^4 = (0.0, 0.0)^T.$$

Therefore, $\bar{\theta}_4^2 = -11.047$ at the vertex $v_2^2 = (3.25, 3.0)^T$.

Iteration 3.
The current approximate feasible solution is $x^2 = (3.25, 3.0)^T$ and the objective function value at this point is -44. The trivial approximation set to the level set $E_{-44}(f)$ is

$y_3^1 = (4.69, 0.0)^T$, $y_3^2 = (0.0, 4.69)^T$, $y_3^3 = (-4.69, 0.0)^T$, $y_3^4 = (0.0, -4.69)^T$.

The vectors u_3^j's are same as the vectors u_2^j's. Therefore, $\theta_2^3 = 12.953$ at the vertex $u_3^1 = (4.0, 0.0)^T$. Thus, the current approximate solution did not change. Also, for the improved approximation set to the level set $E_{-44}(f)$, the following sets can be found:

$\bar{y}_3^1 = (4.69, 0.0)^T$, $\bar{y}_3^2 = (3.25, 3.0)^T$, $\bar{y}_3^3 = (-4.69, 0.0)^T$, $\bar{y}_3^4 = (-3.25, -3.0)^T$.

and

$v_3^1 = (4.0, 0.0)^T$, $v_3^2 = (3.25, 3.0)^T$, $v_3^3 = (0.0, 0.0)^T$, $v_3^4 = (0.0, 0.0)^T$.

Therefore, $\bar{\theta}_4^3 = 0.0$ at the vertex $v_2^2 = (3.25, 3.0)^T$. The algorithm terminates at this iteration. Hence, the algorithm terminates at this iteration, and the global approximate solution is $(3.25, 3.0)^T$.

Note that this approximate solution is the global optimal solution and $(4.0, 0.0)^T$ is the local solution to the problem.

Algorithm 3. INPUT : A concave differentiable function f, a convex compact set D and x^*, a global maximizer of f.
OUTPUT : A global solution x to Problem (5).
Step 1. Choose a point $x^0 \in D$. Set $k = 0$.
Step 2. Find a local minimizer $z^k \in D$ using one of the existing methods starting with an initial approximation point x^k.
Step 3. Construct a trivial approximation set $A_{z^k}^{2n}$ at z^k.
Step 4. For each $y^i \in A_{z^k}^{2n}$, $i = 1, 2, \ldots, 2n$, solve the problem

$$\min \quad x^T \nabla f(y^i)$$
$$\text{s.t.} \quad x \in D$$

to obtain a solution u^i, i.e.,

$$u^{i^T} \nabla f(y^i) = \min_{x \in D} x^T \nabla f(y^i)$$

Step 5. Find the number $j \in \{1, 2, \ldots, 2n\}$ such that

$$\theta_{2n}^k = (u^j - y^j)^T \nabla f(y^j) = \min_{i=1,2,\ldots,2n} (u^i - y^i)^T \nabla f(y^i)$$

Step 6. If $\theta_{2n}^k < 0$ then $x^{k+1} := u^j$, $k := k+1$ and return to Step 2. Otherwise go to the next step.
Step 7. Construct a second order approximation set $B_{z^k}^{2n}$ at z^k by (15).
Step 8. For each $\bar{y}^i \in B_{z^k}^{2n}$, $i = 1, 2, \ldots, 2n$, solve the problem

$$\min \quad x^T \nabla f(\bar{y}^i)$$
$$\text{s.t.} \quad x \in D$$

to obtain a solution v^i, i.e.,

$$v^{i^T} \nabla f(\bar{y}^i) = \min_{x \in D} x^T \nabla f(\bar{y}^i)$$

Step 9. Find the number $s \in \{1, 2, \ldots, 2n\}$ such that

$$\bar{\theta}_{2n}^k = (v^s - \bar{y}^s)^T \nabla f(\bar{y}^s) = \min_{i=1,2,\ldots,2n} (v^i - \bar{y}^i)^T \nabla f(\bar{y}^i)$$

Step 10. If $\bar{\theta}_{2n}^k < 0$ then $x^{k+1} := v^s$, $k := k+1$ and return to Step 2. Otherwise go to the next step.

Step 11. Construct an orthogonal approximation set $C_{z^k}^{2n}$ at z^k by (16).

Step 12. For each $\hat{y}^i \in C_{z^k}^{2n}$, $i = 1, 2, \ldots, 2n$, solve the problem

$$\min \quad x^T \nabla f(\hat{y}^i)$$
$$\text{s.t.} \quad x \in D$$

to obtain a solution w^i, i.e.,

$$w^{i^T} \nabla f(\hat{y}^i) = \min_{x \in D} x^T \nabla f(\hat{y}^i)$$

Step 12. Find the number $s \in \{1, 2, \ldots, 2n\}$ such that

$$\hat{\theta}_{2n}^k = (w^s - \bar{y}^s)^T \nabla f(\hat{y}^s) = \min_{i=1,2,\ldots,2n} (w^i - \hat{y}^i)^T \nabla f(\hat{y}^i)$$

Step 13. If $\hat{\theta}_{2n}^k < 0$ then $x^{k+1} := w^s$, $k := k+1$ and return to Step 2. Otherwise z^k is an approximate global minimizer and terminate the algorithm.

The convergence of the algorithm is given by the following theorem and the proof is similar to the proof of Theorem 2.

Theorem 4. *The sequence $\{z^k, \ k = 0, 1, \ldots\}$ generated by Algorithm 3 converges to a solution of Problem (5) in a finite number of steps or finds an approximate solution as a local solution to the problem.*

Remark 5. When D is a polytope, we can use Algorithm 3 without a local search method.

Example 5. Consider the following problem to illustrate Algorithm 3.

$$\min f(x) = -\|x\|^2$$
$$\text{s.t.} \ Ax \le b,$$

where

$$A^T = \begin{pmatrix} -1 & -17 & 7 & 6 & 8 & 2 \\ -1 & 5 & -19 & 0.5 & 4.5 & 15 \end{pmatrix}, \quad b^T = (4, 90, 102, 121, 192, 270).$$

Since the constraint set D is a polytope, we can use the algorithm without a local search method.

Iteration 1.

An initial feasible solution is $x^1 = (1.0, 0.0)^T$. The current objective function value is -1.0. The global maximizer of the function $f(x)$ over $\mathbb{R}^2$ is $x^* = (0.0, 0.0)^T$. $\theta_4^1 = -38$ at the vertex $u_1^1 = (20.0, 2.0)^T$. Therefore, this vertex is the initial point of the next iteration.

Iteration 2.
$x^2 = (20.0, 2.0)^T$. The current objective function value is -404.0. In this case, the approach of the trivial approximation set cannot improve the current approximate solution, i.e., $\theta_4^2 = 4.01$ at the vertex $u_2^1 = (20.0, 2.0)^T$. Introducing the improved approximation set, we get $\bar{\theta}_4^2 = -4.00$ at the vertex $v_2^1 = (19.5, 8.0)^T$.

Iteration 3.
The current objective function value is -444.25 at $x^3 = (19.5, 8.0)^T$. Constructing the trivial and the improved approximation sets cannot improve the current approximate solution, i.e., $\theta_4^3 = 45.41$ at the vertex $u_3^1 = (20.0, 2.0)^T$ and $\bar{\theta}_4^3 = 37.01$ at the vertex $v_3^1 = (19.5, 8.0)^T$. Next, we introduce the rotation matrix

$$R = \begin{pmatrix} \frac{1}{\sqrt{2}} & \frac{1}{\sqrt{2}} \\ -\frac{1}{\sqrt{2}} & \frac{1}{\sqrt{2}} \end{pmatrix}.$$

Using this rotation matrix, the following new orthogonal approximation set is found.

$$\hat{y}_3^1 = (15.508, -15.508)^T, \ \hat{y}_3^2 = (15.508, 15.508)^T,$$
$$\hat{y}_3^3 = (-15.508, 15.508)^T, \ \hat{y}_3^4 = (-15.508, -15.508)^T.$$

The solutions of the corresponding linear programming problems are

$$w_3^1 = (20.0, 2.0)^T, \ w_3^2 = (15.0, 16.0)^T,$$
$$w_3^3 = (0.0, 18.0)^T, \ w_3^4 = (-2.0, -2.0)^T$$

Since $\hat{\theta}_4^3 = -35.54$ at the vertex $w_3^2 = (15.0, 16.0)^T$, according to Lemma 3, new approximate solution is $(15.0, 16.0)^T$.

Iteration 4.
The current approximate solution is $x^4 = (15.0, 16.0)^T$. The objective function value at this point is -481. We can check that the algorithm stops at this iteration. Thus, $(15.0, 16.0)^T$ is the approximate global optimal solution to the problem.

Note that this solution is the global optimal solution to the problem. The points $x^2 = (20.0, 2.0)^T$ and $x^3 = (19.5, 8.0)^T$ which we found during the algorithm are local solutions to the problem.

5 Numerical Examples

In this section, we present four examples which are implemented by the proposed algorithms.

Problem 1.

$$\min \quad f(x) = -\sum_{i=1}^{n}(x_i + i)^2 \qquad (17)$$

$$\text{s.t.} \quad 1 - i \le x_i \le 2i \ , \ i = 1, 2, \ldots, n$$

The global solutions to these problems are obtained by Algorithm 1 and the computational results are shown in Table 1.

Table 1. Computational results for Problem (17) and Problem (20).

Problems	Dimension of the problems	Initial Value	Global Value	Computational time (sec.)
(17)	10	-10	-3465	0.090
(17)	50	-50	-386325	1.542
(17)	100	-100	-3045150	3.565
(17)	200	-200	-24180300	25.447
(17)	500	-500	-376125750	158.989
(17)	1000	-1000	-3004501500	694.889
(20)	10	-10	-1540	0.731
(20)	50	-50	-171700	17.315
(20)	100	-100	-1353400	45.470
(20)	200	-200	-10746800	143.707
(20)	500	-500	-167167000	826.769
(20)	1000	-1000	-1335334000	3253.549

Next, we consider the two test problems given in [Tho94].

Problem 2. Consider the following concave minimization problem

$$\min \quad f(x) = -(a^T x)^{\frac{3}{2}} \qquad (18)$$

$$\text{s.t.} \quad Ax \le b$$

$$x_i \ge -1 \ , \ i = 1, 2, \ldots, n$$

where A is an $n \times n$ matrix with positive entries, a and b are n vectors with positive entries. Let u^1 and u^2 be the optimal solutions to the linear programming problems $\min\{a^T x \ : \ Ax \le b \ , \ x_i \ge -1, \ i = 1, 2, \ldots, n\}$ and

$\max\{a^T x : Ax \leq b , x_i \geq -1, i = 1, 2, \ldots, n\}$, respectively. Then the above problem has a global solution $u \in \{u^1, u^2\}$ [Tho94]. Algorithm 1 finds a global solution to the problem for various dimensions without a local search method, and results are shown in Table 2.

Problem 3.

$$\min \quad f(x) = \frac{\alpha - (a^T x)^2}{1 + \alpha - (a^T x)^2} + \ln(1 + \alpha - (a^T x)^2) \qquad (19)$$

$$\text{s.t.} \quad Ax \leq b$$

$$x_i \geq -1 , \ i = 1, 2, \ldots, n$$

where A is an $n \times n$ matrix with positive entries, a and b are n vectors with positive entries, and α is a real number such that $\alpha - (a^T x)^2 > 0$ for the all feasible points of the problem. Consider the following concave quadratic programming problem.

$$\min \quad g(x) = -(a^T x)^2$$

$$\text{s.t.} \quad Ax \leq b$$

$$x_i \geq -1 , \ i = 1, 2, \ldots, n$$

Let v be an n vector which its all entries are equal to -1. Then, whenever the linear programming problem $\max\{a^T x : Ax \leq b , x_i \geq -1, i = 1, 2, \ldots, n\}$ has an optimal solution w, the concave quadratic minimization problem has a global solution $u \in \{v, w\}$. Moreover, u is also a global solution of Problem (19) [Tho94]. This solution can be found using Algorithm 1 without a local search method for Problem (19) and the computational results are shown in Table 2.

Problem 4.

$$\min \quad f(x) = -\sum_{i=1}^{n} (x_i - i)^2 \qquad (20)$$

$$\text{s.t.} \quad -i \leq x_i \leq i - 1 , \ i = 1, 2, \ldots, n$$

Algorithm 2 can be used for the above problem without a local search method. Table 1 shows the computational results for the problem.

The numerical experiments were conducted using MATLAB 6.1 on a PC with an Intel Pentium 4 CPU 2.20GHz processor and memory equal to 512 MB. The primal-dual interior-point method [Meh92] and the active set method [Dan55], which is a variation of the simplex method, were implemented by calling subroutines linprog.m from Matlab 6.1 regarding the size of the problem. For Problem (17), the subspace trust region-method [CL96] based on the interior-reflective Newton method and the active set method [GMW81], which is a projection method, were implemented as local search methods by calling the subroutine quadprog.m from Matlab 6.1.

Table 2. Computational results for Problem (18) and Problem (19).

Problems	Constraint type	Dimension of the problems	Computational time (sec.)
(18)	random generated	10	0.551
(18)	random generated	50	16.283
(18)	random generated	100	126.502
(18)	random generated	200	1571.760
(19)	random generated	10	0.432
(19)	random generated	50	17.185
(19)	random generated	100	140.642
(19)	random generated	200	1906.512

6 Conclusions

In this paper, we developed three algorithms for concave programming problems based on a global optimality condition. Under some condition, the convergence of the algorithms have been established. For the implementation purpose, three kinds of approximation sets are introduced and it is shown that some numerical methods are available to construct the approximation sets. At each iteration, it is required to solve $2n$ linear programming problems with the same constraints as the initial problem. Some existing test problems were solved by the proposed algorithms and the computational results have shown that the algorithms are efficient and easy in computing a solution.

References

[Ber95]   Bertsekas, D.P.: Nonlinear programming. Athena Scientific, Belmont, Mass. (1995)

[CL96]   Coleman, T.F., Li, Y.: A reflective Newton method for minimizing a quadratic function subject to bounds on some of the variables. SIAM Journal on Optimization, 6, 1040–1058 (1996)

[Dan55]   Dantzig, G.B., Orden, A., Wolfe, P.: The generalized Simplex Method for minimizing a linear form under linear inequality constraints. Pacific Journal Math., **5**, 183–195.

[Die94]   Dietrich, H.: Global optimization conditions for certain nonconvex minimization problems. Journal of Global Optimization, **5**, (359-370) (1994)

[Enk96]   Enkhbat, R.: An algorithm for maximizing a convex function over a simple set. Journal of Global Optimization, **8**, 379–391 (1996)

[GMW81]   Gill, P.E., Murray, W., Wright, M.H.: Practical Optimization. Academic Press, London, UK (1981)

[Hir89]   Hiriart-Urruty, J.B.: From convex optimization to nonconvex optimization. In: Nonsmooth Optimization and Related Topics, 219–239. Plenum (1989)

[HT93]   Horst, R., Tuy, H.: Global Optimization: Deterministic Approaches (second edition). Springer Verlag, Heidelberg (1993)

[HP95]   Horst, R., Pardalos, P.M. (eds): Handbook of Global Optimization. Kluwer Academic, Netherlands (1995)

[HPT01]   Horst, R., Pardalos, P.M., Thoai, N.V.: Introduction to Global Optimization (second edition). Kluwer Academic, Netherlands (2001)

[Kha79]   Khachiyan, L.: A polynomial algorithm in linear programming. Math. Doklady, 20, 191–194 (1979)

[Meh92]   Mehrotra, S.: On the implementation of a Primal-Dual Interior Point Method, SIAM Journal on Optimization, 2, 575–601 (1992)

[Par93]   Pardalos, P.M.: Complexity in Numerical Optimization. World Scientific Publishing, River Edge, New Jersey (1993)

[PR87]   Pardalos, P.M., Rosen, J.B.: Constrained Global Optimization: Algorithms and Applications. Lecture Notes in Computer Science 268, Springer-Verlag (1987)

[PR86]   Pardalos, P.M., Rosen, J.B.: Methods for Global Concave Minimization: A Bibliographic Survey. SIAM Review, 28, 367–379 (1986)

[PS88]   Pardalos, P.M., Schnitger, G.: Checking local optimality in constrained quadratic programming is NP-hard. Operations Research Letters, 7, 33–35 (1988)

[Roc70]   Rockafellar, R.T.: Convex Analysis. Princeton University Press, Princeton (1970)

[Str98]   Strekalovsky, A.S.: Global optimality conditions for nonconvex optimization. Journal of Global Optimization, 12, 415–434 (1998)

[SE90]   Strekalovsky, A.S., Enkhbat, R.: Global maximum of convex functions on an arbitrary set. Dep.in VINITI, Irkutsk, No. 1063, 1–27 (1990)

[Tho94]   Thoai, N.V.: On the construction of test problems for concave minimization problem. Journal of Global Optimization, 5, 399–402 (1994)

Convexification and Monotone Optimization

Xiaoling Sun[1], Jianling Li[12], and Duan Li[3]

[1] Department of Mathematics
Shanghai University
Shanghai 200444, P.R. China
xlsun@staff.shu.edu.cn
[2] College of Mathematics and Information Science
Guangxi University
Nanning, Guangxi 530004, P.R. China
ljl123@gxu.edu.cn
[3] Department of Systems Engineering and Engineering Management
The Chinese University of Hong Kong
Shatin, N.T., Hong Kong, P.R. China
dli@se.cuhk.edu.hk

Summary. Monotone maximization is a global optimization problem that maximizes an increasing function subject to increasing constraints. Due to the often existence of multiple local optimal solutions, finding a global optimal solution of such a problem is computationally difficult. In this survey paper, we summarize global solution methods for the monotone optimization problem. In particular, we propose a unified framework for the recent progress on convexification methods for the monotone optimization problem. Suggestions for further research are also presented.

1 Introduction

Global optimization has been one of the important yet challenging research areas in optimization. It appears very difficult, if not impossible, to design an efficient method in finding global optimal solutions for general global optimization problems. Over the last four decades, much attention has been drawn to the investigation of specially structured global optimization problems. In particular, concave minimization problems have been studied extensively. Various algorithms including extreme point ranking methods, cutting plane methods and outer approximation methods have been developed for concave minimization problems (see e.g. [Ben96, HT93, RP87] and a bibliographical survey in [PR86]). Monotone optimization problems, as an important class of specially structured global optimization problems, have been also studied in recent years by many researchers (see e.g. [LSBG01, RTM01, SML01, Tuy00, TL00]).

The monotone optimization problem can be posted in the following form:

$$(P) \qquad \max \ f(x)$$
$$\text{s.t.} \ \ g_i(x) \le b_i, \ i = 1, \ldots, m,$$
$$x \in X = \{x \mid l_j \le x_j \le u_j, \ j = 1, \ldots, n\},$$

where f and all g_is are increasing functions on $[l, u]$ with $l = (l_1, l_2, \ldots, l_n)^T$ and $u = (u_1, u_2, \ldots, u_n)^T$. Note that functions f and g_is are not necessarily convex or separable. Due to the monotonicity of f and g_is, optimal solutions of (P) always locate on the boundary of the feasible region. It is easy to see that the problem of maximizing a decreasing function subject to decreasing constraints can be reduced to problem (P). Since there may exist multiple local optimal solutions on the boundary, problem (P) is of a specially structured global optimization problem. In real-world applications, the monotonicity often arises naturally from certain inherent structure of the problem under consideration. For example, in resource allocation problems ([IK88]), the profit or return is increasing as the assigned amount of resource increases. In reliability networks, the overall reliability of the system and the weight, volume and cost are increasing as the reliability in subsystems increases ([Tza80]). Partial or total monotone properties are also encountered in globally optimal design problems ([HJL89]).

The purpose of this survey paper is to summarize the recent progress on convexification methods for monotone optimization problems. In Section 2, we discuss the convexification schemes for monotone functions. In Section 3 we first establish the equivalence between problem (P) and its convexified problem. Outer approximation method for the transformed convex maximization problem is then described. Polyblock outer approximation method is presented in Section 4. In Section 5, a hybrid method that combines partition, convexification and local search is described. Finally, concluding remarks with some suggestions for further studies are given in Section 6.

2 Monotonicity and convexity

Monotonicity and convexity are two closely related yet different properties of a real function in classical convex analysis. One of the interesting questions is whether or not a nonconvex monotone function can be transformed into a convex function via certain variable transformations. Since linear transformation does not change the convexity of a real function, we have to appeal to nonlinear transformation for converting a monotone function into a convex function.

To motivate the convexification method for general monotone functions, let us consider a univariate function $f(x)$. Suppose that $f(x)$ is a strictly increasing function and $t(x)$ is a strictly monotone and convex function. Define a composite function $g(x) = f(t(x))$. If f is twice differentiable, then $g'(x) = f'(t(x))t'(x)$ and

$$g''(x) = f''(t(x))[t'(x)]^2 + f'(t(x))t''(x).$$

Thus, $g(x)$ is a convex function if and only if

$$\frac{t''(x)}{[t'(x)]^2} \geq -\frac{f''(t(x))}{f'(t(x))}, \qquad \text{for all } x. \tag{1}$$

Inequality (1) characterizes the condition for a nonlinear transformation t to convexify a univariate twice differentiable increasing function via a variable transformation or domain transformation. We now turn to derive conditions for convexifying a multivariate monotone function. A function $f : D \rightarrow \mathbb{R}$ is said to be increasing (decreasing) on $D \subseteq \mathbb{R}^n$ if $f(x) \geq f(y)$ ($f(x) \leq f(y)$) for any two vectors $x, y \in D$ whenever $x \geq y$. If the strict inequality holds, then f is said to be strictly increasing (decreasing) on $D \subset \mathbb{R}^n$.

Let $\alpha, \beta \in \mathbb{R}^n$ with $0 < \alpha < \beta$. Denote $[\alpha, \beta] = \{x \in \mathbb{R}^n \mid \alpha \leq x \leq \beta\}$. Let $t : \mathbb{R}^n \mapsto \mathbb{R}^n$ be a one-to-one mapping. Define

$$f_t(y) = f(t(y)). \tag{2}$$

The domain of f_t is $Y^t = t^{-1}(X)$. Define

$$\sigma = \min\{d^T \nabla^2 f(x)d \mid x \in [\alpha, \beta],\ \|d\|_2 = 1\}, \tag{3}$$

$$\mu = \min\{\frac{\partial f}{\partial x_j} \mid x \in [\alpha, \beta],\ j = 1, \ldots, n\}. \tag{4}$$

We have the following theorem on convexification.

Theorem 1. ([LSM05]) *Assume that f is a twice differentiable and strictly increasing function on $[\alpha, \beta]$ and $\mu > 0$. Suppose that $t(y) = (t_1(y_1), \ldots, t_n(y_n))$ is a separable mapping, and each t_i is twice differentiable and strictly monotonic. If t satisfies the following condition:*

$$\frac{t_j''(y_j)}{[t_j'(y_j)]^2} \geq -\frac{\sigma}{\mu}, \qquad \text{for } y_j \in Y_j^t = t_j^{-1}([\alpha_j, \beta_j]),\ j = 1, \ldots, n, \tag{5}$$

then $f_t(y)$ is a convex function on any convex subset of Y^t.

Similarly, a strictly decreasing function can also be converted into a convex function via a variable transformation satisfying:

$$\frac{t_j''(y_j)}{[t_j'(y_j)]^2} \leq -\frac{\sigma}{\mu}, \qquad \text{for } y_j \in Y_j^t,\ j = 1, \ldots, n.$$

There are many specific mappings that satisfy condition (5). In particular, consider the following two functions:

$$t_j(y_j) = \frac{1}{p} \ln(1 - \frac{1}{y_j}), \qquad p > 0,\ j = 1, \ldots, n, \tag{6}$$

$$t_j(y_j) = y_j^{-\frac{1}{p}}, \qquad p > 0,\ j = 1, \ldots, n. \tag{7}$$

Corollary 1. *Let $p_1 = \max\{0, -\sigma/\mu\}$ and $p_2 = \max\{0, -(\bar{\beta}\sigma)/\mu - 1\}$, where $\bar{\beta} = \min_{1 \le j \le n} \beta_j$. Then, the mapping t with t_j defined by (6) satisfies condition (5) when $p > p_1$, and the mapping t with t_j defined by (7) satisfies condition (5) when $p > p_2$.*

For illustration, let us consider a one-dimensional function:

$$f(x) = (x-2)^3 + 2x, \quad x \in X = [1,3].$$

Note that $f(x)$ is a nonconvex and strictly increasing function. The plot of $f(x)$ is shown in Figure 1. We have $f'(x) = 3(x-2)^2 + 2 \ge 2$ and $f''(x) = 6(x-2) \ge -6$ for $x \in [1,3]$. Take $t(y) = (1/p)\ln(1 - \frac{1}{y})$ in (2). By Corollary 1, $p_1 = -(-6)/2 = 3$. So, any $p \ge 3$ guarantees the convexity of $f_t(y)$ on $Y^t = [-1/(e^p - 1), -1/(e^{3p} - 1)]$. Figure 2 shows the convexified function $f_t(y)$ with $p = 3$. In practice, p can be chosen much smaller than the bound defined in Corollary 1.

Fig. 1. The plot of $f(x)$.

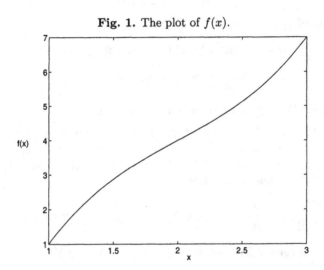

Range transformation can be also incorporated into the convexification formulation (2) to enhance the convexification effect. Let T be a strictly increasing and convex function on $\mathbb{R}$. Define

$$f_{T,t}(y) = T(f(t(y))). \tag{8}$$

Certain conditions [SML01] can be derived for $f_{T,t}(y)$ to be a convex function on Y^t. Typical range transformation functions are $T(z) = e^{rz}$ and $T(z) = z^p$, where $r > 0$ and $p > 0$ are positive parameters. One advantage to use both range and domain transformations is to reduce possible ill conditions caused by the convexification process.

Fig. 2. The plot of $f_t(y)$.

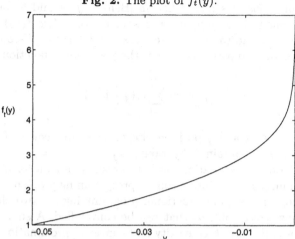

Theorem 1 was generalized in [SLL04] to convexify a class of nonsmooth functions. Let $\partial f(x)$ denote the set of Clarke's generalized gradient of f at x and $\partial_w f(x)$ the set of Clarke's generalized gradient in direction w. Denote by $f''_-(x, v, w)$ Chaney's second-order derivative of f at x (cf. [Cha85]).

Theorem 2. ([SLL04]) *Assume in (2) that*
   (i) f *is semismooth and regular on* X.
   (ii) f *is a strictly increasing function on* X *and*

$$\inf_{} \min_{i=1,\dots,n} \{\xi_i \mid \xi = (\xi_1, \dots, \xi_n)^T \in \partial f(x), \ x \in X\} \geq \epsilon > 0, \qquad (9)$$

$$\sigma = \inf\{f''_-(x, v, w) \mid x \in X, \ \|w\|_2 = 1, \ v \in \partial_w f(x)\} > -\infty. \qquad (10)$$

   (iii) $t(y) = (t_1(y_1), \dots, t_n(y_n))$ *and* t_j, $j = 1, \dots, n$, *are twice differentiable and strictly monotone convex functions satisfying*

$$\frac{t''_j(y_j)}{[t'_j(y_j)]^2} \geq -\frac{\sigma}{\epsilon}, \quad y_j \in Y^t_j, \ j = 1, \dots, n. \qquad (11)$$

Then $f_t(y)$ *defined in (2) is a convex function on any convex subset of* Y^t.

   Note that convex functions, C^1 functions, and pointwise maximum or minimum of C^1 functions are semismooth. Furthermore, certain composite semismooth functions are also semismooth (see [Muf77]).
   The idea of convexifying a nonconvex function via both domain transformation and range transformation can be traced back to 1950s. Convex (or concave) transformable functions were introduced in [Fen51]. Let f be defined on a convex subset $C \subseteq \mathbb{R}^n$. f is said to be convex range transformable or F-convex if there exists a continuous strictly increasing function F such

that $F(f(x))$ is convex. The concept of domain transformation for general nonconvex functions was introduced in [Ben77]. f is said to be h-convex if a continuous one-to-one mapping h exists such that $f(h^{-1}(y))$ is convex on domain $h(C)$. f is said to be (h, F)-convex if $f(h^{-1}(y))$ is F-convex on $h(C)$. A special class of h-convex functions is the posynomial function defined as

$$p(x) = \sum_{i=1}^{q} c_i (\prod_{j=1}^{n} x_j^{a_{ij}})$$

with $c_i > 0$, $a_{ij} \in \mathbb{R}$ and q positive integer. A simple convexification transformation is readily obtained by taking $x_j = e^{y_j}$, $j = 1, \ldots, n$. Previous research work on convexification has led to results on transforming a nonconvex programming problem into a convex programming problem. In particular, geometric programming and fractional programming are two classes of nonconvex optimization problems that can be convexified. A survey of applications of F-convexity and h-convexity and convex approximation in nonlinear programming was given in [Hor84].

Convexifying monotone functions was inspired by a success of convexifying a perturbation function in nonconvex optimization. Li [Li95] first introduced a p-th power method for convexifying the perturbation function of a nonconvex optimization problem (see also [LS01]). In [Li96], the p-th power method was applied to convexify the noninferior frontier in multi-objective programming. Two p-th power transformation schemes were proposed in [LSBG01] for convexifying monotone functions:

$$f_p^1(y) = -[f(y^{1/p})]^{-p}, \quad p > 0, \tag{12}$$

$$f_p^2(y) = [f(y^{1/p})]^p, \quad p > 0. \tag{13}$$

It is shown in [LSBG01] that if $f(x)$ is a strictly increasing function, then $f_p^1(y)$ is a concave function for sufficiently large p, and if $f(x)$ is a strictly decreasing function, then $f_p^2(y)$ is a convex function for sufficiently large p. Another class of convexification transformations is defined as follows:

$$f_p(y) = T(pf(\frac{1}{p}t(y))), \quad p > 0, \tag{14}$$

where t is a one-to-one mapping without parameter. Conditions for convexifying f via transformation (14) were derived in [SML01]. Obviously, (14) is a special case of the general formulation (8). A more general transformation that includes (12), (13) and (14) as special cases was proposed in [WBZ05]. Convexification method was used in [LWLYZ05] to identify a class of hidden convex optimization problems.

3 Monotone optimization and concave minimization

3.1 Equivalence to concave minimization

Given a mapping $t: \mathbb{R}^n \to \mathbb{R}^n$. We now consider the following transformed problem of (P):

$$\max \ \phi(y) = f(t(y))$$
$$\text{s.t.} \ \psi_i(y) = g_i(t(y)) \leq b_i, \ i = 1, \ldots, m, \tag{15}$$
$$y \in Y^t,$$

where $t : Y^t \to X$ is an onto mapping with $X = t(Y^t)$. Denote by S and S_t the feasible region of problem (P) and problem (15), respectively, i.e.

$$S = \{x \in X \mid g_i(x) \leq b_i, \ i = 1, \ldots, m\}, \tag{16}$$
$$S_t = \{y \in Y^t \mid \psi_i(y) \leq b_i, \ i = 1, \ldots, m\}. \tag{17}$$

The following theorem establishes the equivalence between the monotone optimization problem (P) and the transformed problem (15).

Theorem 3. ([SML01])
  (i) $y^* \in Y^t$ is a global optimal solution to problem (15) if and only if $x^* = t(y^*)$ is a global optimal solution to problem (P).
  (ii) If t^{-1} exists and both t and t^{-1} are continuous mappings, then $y^* \in Y^t$ is a local optimal solution to problem (15) if and only if $x^* = t(y^*)$ is a local optimal solution to problem (P).

Combining Theorem 3 with Theorems 1-2 implies that if t in (15) is a one-to-one mapping satisfying the conditions in Theorems 1 or 2, then the monotone optimization (P) is equivalent to the convex maximization (or concave minimization) problem (15). Especially, when t_i takes the form of (6) or (7) and the parameter p is greater than certain threshold value, then the transformed problem (15) is a concave minimization problem.

3.2 Outer approximation algorithm for concave minimization problems

Concave minimization is a class of global optimization problems studied intensively in the literature. It is well-known that a convex function always achieves its maximum over a bounded polyhedron at one of its vertices. Ranking the function values at all vertices of the polyhedron gives an optimal solution. For a convex maximization (or concave minimization) problem with a general convex feasible set, Hoffman [Hof81] proposed an outer approximation algorithm. The convex objective function is successively maximized on a sequence of polyhedra that enclose the feasible region. At each iteration the current enclosing polyhedron is refined by adding a cutting plane tangential to the feasible region at a boundary point. The algorithm generates a nonincreasing sequence of upper bounds for the optimal value of problem (15) and terminates when the current feasible solution is within a given tolerance of the optimal solution.

An outer approximation procedure for problem (15) can be described briefly as follows:

Algorithm 1 (Polyhedral Outer Approximation Method).

Step 1. Choose an initial polyhedron P_0 that contains S_t with vertex set V_0 and set $k = 0$.

Step 2. Compute v^k and ϕ^k such that $\phi^k = \phi(v^k) = \max_{v \in V_k} \phi(v)$, i.e., v^k is the best vertex in the current enclosing polyhedron.

Step 3. Find a feasible point y^k on the boundary of S_t. Let i be such that $\psi_i(y^k) = b_i$. Form a new polyhedron P_{k+1} by adding a cutting plane inequality: $\xi_k^T(y - y^k) \leq 0$, where ξ_k is a subgradient of the binding constraint ψ_i at y^k.

Step 4. Calculate the vertex set V_{k+1} of P_{k+1}. Set $k := k + 1$, return to Step 2.

It was shown in [Hof81] that the above method converges to a global optimal solution to problem (15). In implementation, the above procedure can be terminated when $\phi^k - \phi(y^k) \leq \epsilon$, where $\epsilon > 0$ is a given tolerance. There are many ways to generate the feasible point y^k in Step 3. A simple method proposed in [Hof81] is to find the (relative) boundary point of S_t on the line connecting v^k and a fixed (relative) interior point of S_t. Horst and Tuy [HT93] suggested projecting v^k onto the boundary of S_t and choosing y^k to be the projected point. Finding vertices of P_{k+1} is the major computational burden in the outer approximation method. After adding a cutting plane $\{y \mid \xi_k^T(y - y^k) = 0\}$, the new vertices can be generated by computing the intersection point of each edge of P_k with the cutting plane. Techniques of computing new vertices resulted from an intersection of a polyhedron with a cutting plane are discussed in [CHJ91, HV88].

Let us consider a small example to illustrate the convexification and outer approximation method.

Example 1.

$$\max \ f(x) = 4.5(1 - 0.40^{x_1-1})(1 - 0.40^{x_2-1}) + 0.2\exp(x_1 + x_2 - 7)$$
$$\text{s.t. } g_1(x) = 5x_1x_2 - 4x_1 - 4.5x_2 \leq 32,$$
$$x \in X = \{x \mid 2 \leq x_1 \leq 6.2, \ 2 \leq x_2 \leq 6\}.$$

It is clear that f and g_1 are strictly increasing functions on X. The problem has three local optimal solutions: $x_{loc}^1 = (2.2692, 6)^T$ with $f(x_{loc}^1) = 3.7735$, $x_{loc}^2 = (3.4528, 3.5890)^T$ with $f(x_{loc}^2) = 3.857736$ and $x_{loc}^3 = (6.2, 2.1434)^T$ with $f(x_{loc}^3) = 3.6631$. Figure 3 shows the feasible region of the example.

It is clear that the global optimal solution x_{loc}^2 is not on the boundary of the convex hull of the nonconvex feasible region S. Take t to be the convexification transformation (6) with $p = 2$. The convexified feasible region is shown in Figure 4. Set $\epsilon = 10^{-4}$. The outer approximation procedure finds an approximate global optimal solution $y^* = (-0.21642, -0.19934)$ of (15) after 17 iterations and generating 36 vertices. The point y^* corresponds to $x^* = (3.45290, 3.58899)$, an approximate optimal solution to Example 1 with $f(x^*) = 3.857736887$.

Fig. 3. Feasible region of Example 1.

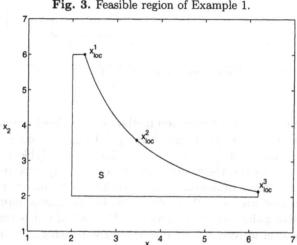

4 Polyblock outer approximation method

Polyblock approximation methods for monotone optimization were proposed in [RTM01, Tuy00, TL00]. A polyblock is a union of a finite number of boxes $[a, z]$, where point a is called the lower corner point and point $z \in V$ is called a vertex of the polyblock, with V being a finite set in $\mathbb{R}^n$. The polyblock outer approximation method is based on the following two key observations: (i) the feasible region S of (P) can be approximated from outside by a polyblock, no matter S is convex or nonconvex, and (ii) any increasing function achieves its maximum on a polyblock at one of its vertices. These two properties are analogous to those of the polyhedral outer approximation in concave minimization. Recall that a convex set can be approximated from outside by a polyhedron and any convex function achieves its maximum on a polyhedron at one of its vertices.

Fig. 4. Convexified feasible region with $p = 2$.

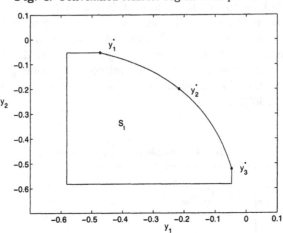

A polyblock outer approximation method can be developed for monotone optimization by successively constructing polyblock that covers the feasible region S. The algorithm first uses $[l, u]$ as the initial polyblock. At the k-th iteration, let z^k be the vertex with the maximum objective function value among all the vertices of the enclosing polyblock. A boundary point x^k of S on the line connecting l and z^k is calculated. The polyblock approximation is refined by cutting the box $(x^k, z^k]$ from $[l, z^k]$. A set of n new vertices is then generated by alternatively setting one of the components equal to that of x^k and the other components equal to those of z^k. The iteration process repeats until the difference between the upper bound (the maximum objective value of the vertices) and the lower bound (the objective value of the current best boundary point) is within a given tolerance. A vertex z is called improper if there exists another vertex w of the polyblock such that $z \leq w$ with at least one component satisfying $z_i < w_i$. By the monotonicity of the problem, any improper vertex generated during the polyblock approximation process can be deleted.

Let $S_\epsilon = S \cap [l + \epsilon e, u]$, where $\epsilon > 0$ and $e = (1, \ldots, 1)^T$. A feasible solution x^* is said to be an ϵ-optimal solution to (P) if $x^* \in \arg\max\{f(x) \mid x \in S_\epsilon\}$. A feasible solution x^* is said to be an (ϵ, η)-optimal solution to (P) if $f(x^*) \geq f^* - \eta$, where f^* is the global maximum of $f(x)$ over S_ϵ. It is easy to see that both ϵ-optimal and (ϵ, η)-optimal solutions can be regarded as approximate optimal solutions to (P).

Algorithm 2 (Polyblock Approximation Algorithm).

Step 0 (Initialization). Choose tolerance parameters $\epsilon > 0$ and $\eta > 0$. If l is infeasible then (P) has no feasible solution. If u is feasible then u is the optimal solution to (P), stop. Otherwise, set $\tilde{x}^0 = l$, $V_1 = \{u\}$ and $k = 1$.

Step 1. Compute

$$z^k \in \arg\max\{f(z) \mid z \in V_k, \ z \geq l + \epsilon e\}.$$

If $z^k \in S$, stop and z^k is an ϵ-optimal solution to (P).

Step 2. Compute a boundary point x^k of S on the line linking l and z^k. Set $\tilde{x}^k = \arg\max\{f(\tilde{x}^{k-1}), f(x^k)\}$. If $f(\tilde{x}^k) \geq f(z^k) - \eta$, stop and $\tilde{x}^k$ is an (ϵ, η)-approximate solution to problem (P).

Step 3. Compute the n new vertices of the box $[x^k, z^k]$ that are adjacent to z^k:

$$z^{k,i} = z^k - (z_i^k - x_i^k)e^i, \quad i = 1, \ldots, n,$$

where e^i is the i-th unit vector of $\mathbb{R}^n$. Set

$$\widetilde{V}_{k+1} = (V_k \setminus \{z^k\}) \cup \{z^{k,1}, \ldots, z^{k,n}\}.$$

Let V_{k+1} be the set of the remaining vertices after removing all improper vertices in $\widetilde{V}_{k+1}$.

Step 4. Set $k := k + 1$, return to Step 1.

Remark 1. We note that in Algorithm 2 at most n new vertices are added to the vertex set V_k at each iteration. However, the number of vertices accumulated during the iteration process could be so large such that storing all vertices is prohibitive from the computational point of view. In order to avoid such a storage problem, restarting strategy can be adopted. Specifically, Step 4 can be replaced by the following two steps:

Step 4. If $|V_k| \leq N$ (N is the critical size of the vertex set), then set $k := k+1$ and return to Step 1. Otherwise go to Step 5.

Step 5. Redefine $V_{k+1} = \{u - (u_i - x_i^k)e^i, \ i = 1, \ldots, n\}$. Set $k := k + 1$ and return to Step 1.

It was shown that Algorithm 2 either stops at an ϵ-optimal solution in Step 1 or stops at an (ϵ, η)-approximate solution after finite number of iterations (see [RTM01]).

Figure 5 illustrates the first 3 iterations of the polyblock approximation for Example 1. Using the same accuracy $\epsilon = 10^{-4}$ as in Example 1, the method finds an ϵ-approximate global solution $x_{best} = (3.4526, 3.5890)^T$ with $f(x_{best}) = 3.857736$ after 359 iterations and generating 718 vertices.

Algorithm 2 can be extended to deal with monotone optimization problems with an additional reverse monotone constraint $h(x) \geq c$, where $h(x)$ is an increasing function (see [Tuy00, TL00]). Various applications of monotone optimization can be found in [Tuy00].

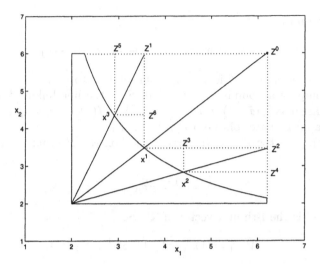

Fig. 5. Polyblock approximation for Example 1.

5 A hybrid method

Despite its relatively easy implementation, the polyblock approximation method may suffer from its slow convergence due to the poor quality of upper bounds, as witnessed from illustrative examples and computational experiments. The convexification method discussed in Sections 2 and 3, on the other hand, is essentially a polyhedral approximation method for solving the transformed concave minimization problem. Therefore, it may suffer from the rapid (exponentially) increase of the number of polyhedral vertices generated by the outer approximation, as is the case for any polyhedral outer approximation methods for concave minimization (see [Ben96, HT93]). Moreover, it is difficult to determine a suitable convexification parameter that controls the degree of the convexity of the functions on a large size domain. A large parameter may cause an ill-conditional transformed problem.

To overcome the computational difficulties of the convexification method, a hybrid method was developed in [SL04] to incorporate three basic strategies: partition, convexification and local search, into a branch-and-bound framework. The partition scheme is used to decompose the domain X into a union of subboxes. The union of these subboxes forms a *generalized* polyblock that covers the boundary of the feasible region. Figure 6 illustrates the partition process for Example 1. To obtain a better upper bound on each subbox, convexification method is used to construct polyhedral outer approximation, thus enabling more efficient node fathoming and speeding up the convergence of the branch-and-bound process. A local search procedure is employed to improve the lower bound of the optimal value. Since only an approximate solution is needed in the upper bounding procedure, the number of polyhedral vertices can be limited and controlled. Moreover, as the domain shrinks during the branch-and-bound process, the convexity can be achieved with a smaller parameter, thus avoiding the ill-conditional effect for the transformed subproblems.

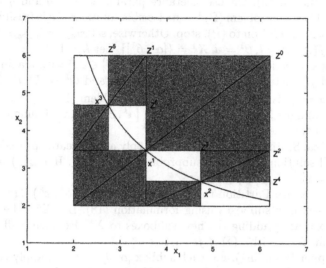

Fig. 6. Partition process for Example 1.

Consider a subproblem of (P) by replacing $X = [l, u]$ with a subbox $[\alpha, \beta] \subseteq X$:

$$(SP) \quad \max \ f(x)$$
$$\text{s.t.} \ \ g_i(x) \le b_i, \ i = 1, \ldots, m,$$
$$x \in [\alpha, \beta].$$

Let x_b be the boundary point of S on the line connecting α and β. By the monotonicity of f and g_is, there are no better feasible points than x_b in $[\alpha, x_b)$ and there are no feasible points in $(x_b, \beta]$. Therefore, the two boxes $[\alpha, x_b)$ and $(x_b, \beta]$ can be removed from $[\alpha, \beta]$ without missing any optimal solution of (SP) (cf. Figure 6). The following lemma shows that the set of the points left in $[\alpha, \beta]$ after removing $[\alpha, x_b)$ and $(x_b, \beta]$ can be partitioned into at most $2n - 2$ subboxes.

Lemma 1. ([SL04]) *Let $\alpha < \beta$. Denote $A = [\alpha, \beta]$, $B = [\alpha, \gamma)$ and $C = (\gamma, \beta]$. Then $A \setminus (B \cup C)$ can be partitioned into $2n - 2$ subboxes.*

$$A \setminus (B \cup C) = \{\cup_{i=2}^n \left(\Pi_{k=1}^{i-1}[\alpha_k, \gamma_k] \times [\gamma_i, \beta_i] \times \Pi_{k=i+1}^n[\alpha_k, \beta_k]\right)\}$$
$$\cup \{\cup_{i=2}^n \left(\Pi_{k=1}^{i-1}[\gamma_k, \beta_k] \times [\alpha_i, \gamma_i] \times \Pi_{k=i+1}^n[\alpha_k, \beta_k]\right)\}. \tag{18}$$

Let $J = \{1, \ldots, n\}$. The hybrid algorithm can be formally described as follows.

Algorithm 3 (Hybrid Algorithm for Monotone Optimization).

Step 0 (Initialization). Choose tolerance parameters $\epsilon > 0$ and $\eta > 0$. If l is infeasible then problem (P) has no feasible solution. If u is feasible then u is the optimal solution to (P), stop. Otherwise, set $x_{best} = l$, $f_{best} = f(x_{best})$, $f_u^1 = f(u)$, $\alpha^1 = l$, $\beta^1 = u$, $X^1 = \{[\alpha^1, \beta^1]\}$. Set $k = 1$.

Step 1 (Box Selection). Select the subbox $[\alpha^k, \beta^k] \in X^k$ with maximum upper bound f_u^k. Let $I^k = \{j \in J \mid \beta_j^k - \alpha_j^k \leq \eta\}$ and $Q^k = J \setminus I^k$. If $Q^k = \emptyset$, stop, $x = \alpha^k$ is an η-optimal solution to problem (P).

Step 2 (Boundary Point). Set $X^k := X^k \setminus [\alpha^k, \beta^k]$. Find a boundary point x^k of S on the line connecting α^k and β^k.

Step 3 (Local Search). Starting from x^k, apply a local search procedure to find a local solution x_{loc}^k of the subproblem on $[\alpha^k, \beta^k]$. If $f(x_{loc}^k) > f_{best}$, set $x_{best} = x_{loc}^k$, $f_{best} = f(x_{loc}^k)$.

Step 4 (Partition). Partition the set $\Omega^k = [\alpha^k, \beta^k] \setminus ([\alpha^k, x^k) \cup (x^k, \beta^k])$ into $2|Q^k| - 2$ new subboxes using formulation (18). Let X^{k+1} be the set of subboxes after adding the new subboxes to X^k. Removing all sub-boxes $[\gamma, \delta]$ in X^{k+1} with $f(\delta) \leq f_{best}$.

Step 5 (Upper Bounding). For each subbox $[\alpha, \beta]$ in X^{k+1}, apply the convexification and outer approximation (Algorithm 1) to find an upper bound $UB_{[\alpha,\beta]}$ of the objective function.

Step 6 (Fathoming). Removing all subboxes $[\alpha, \beta]$ in X^{k+1} with $UB_{[\alpha,\beta]} \leq f_{best}$. Let f_u^{k+1} be the maximum upper bound of all the subboxes. If $f_u^{k+1} - f_{best} < \epsilon$, then stop, x_{best} is an ϵ-optimal solution to (P). Otherwise, set $k := k + 1$, goto Step 1.

In the implementation of Step 5, the outer approximation iteration can be terminated upon finding a new vertex whose objective function value is less than or equal to the lower bound f_{best}. It was shown that Algorithm 3 either stops at an η-optimal solution in Step 1 or stops at an ϵ-optimal solution in Step 6 within finite number of iterations (see [SL04]). Details of computational considerations and extensive numerical results of Algorithm 3 were given in [SL04].

6 Conclusions

We have summarized in this paper basic ideas and results on convexification methods for monotone optimization. Applying convexification to a monotone optimization problem results in a concave minimization problem that can be solved by the polyhedral outer approximation method. The polyblock approximation method can also be viewed as an outer approximation method where polyblocks are used to approximate the feasible region and upper bounds are computed by ranking the extreme points of the polyblock. Integrating the promising features of convexification schemes and the polyblock approximation method, the newly proposed branch-and-bound framework that combines partition, convexification and local search is promising from the computational point of view.

Among many interesting topics for the future research, we mention the following three areas:

(i) D.I. functions (difference of two increasing functions) constitute a large class of nonconvex functions in global optimization (e.g., polynomials). Direct application of the convexification method to problem with D.I. functions involved gives rise to a D.C. (difference of convex functions) optimization problem ([HT99]). It is of a great interest to study efficient convexification methods for different types of D.I. programming problems and develop efficient global optimization methods for the transformed D.C. problems.

(ii) Many real-world optimization models may only have partial monotonicity. For example, the function is monotone with respect to some variables and nonmonotone with respect to other variables, or the function is a sum of a monotone function and a nonmonotone function. In global optimal design problems ([HJL89]), partial monotonicity properties are often inherent in objective and constraint functions. How to exploit the partial monotonicity by certain convexification scheme is an interesting topic for future study.

(iii) Many computational issues of the outer approximation method still need to be further investigated. The major computation burden in the outer approximation method is the computation and storage of the vertices of the polyhedron containing the feasible region. Vertex elimination technique could be a possible remedy for preventing a rapid increase of the number of vertices of the outer approximation polyhedron.

7 Acknowledgement

This research was supported by the National Natural Science Foundation of China under Grants 10271073 and 10261001, and the Research Grants Council of Hong Kong under Grant CUHK 4214/01E.

References

[Ben77] Ben-Tal, A.: On generalized means and generalized convexity. Journal of Optimization Theory and Applications, **21**, 1–13 (1977)

[Ben96] Benson, H.P.: Deterministic algorithm for constrained concave minimization: A unified critical survey. Naval Research Logistics, **43**, 765–795 (1996)

[Cha85] Chaney, R.W.: On second derivatives for nonsmooth functions. Nonlinear Analysis: Theory and Methods and Application, **9**, 1189–1209 (1985)

[CHJ91]   Chen, P.C., Hansen, P., Jaumard, B.: On-line and off-line vertex enumeration by adjacency lists. Operations Research Letters, **10**, 403–409 (1991)

[Fen51]   Fenchel, W.: Convex cones, sets and functions, mimeographed lecture notes. Technical report, Princeton University, NJ, 1951

[HJL89]   Hansen, P., Jaumard, B., Lu, S.H.: Some further results on monotonicity in globally optimal design. Journal of Mechanisms, Transmissions, and Automation Design, **111**, 345–352 (1989)

[Hof81]   Hoffman, K.L.A.: A method for globally minimizing concave functions over convex set. Mathematical Programming, **20**, 22–32 (1981)

[Hor84]   Horst, R.: On the convexification of nonlinear programming problems: An applications-oriented survey. European Journal of Operational Research, **15**, 382–392, (1984)

[HT99]   Horst, R., Thoai, N.V.: D.C. programming: Overview. Journal of Optimization Theory and Applications, **103**, 1–43 (1999)

[HT93]   Horst, R., Tuy, H.: Global Optimization: Deterministic Approaches. Springer-Verlag, Heidelberg (1993)

[HV88]   Horst, R., Vries, J.D.: On finding new vertices and redundant constraints in cutting plane algorithms for global optimization. Operations Research Letters, **7**, 85–90, (1988)

[IK88]   Ibaraki, T., Katoh, N.: Resource Allocation Problems: Algorithmic Approaches. MIT Press, Cambridge, Mass. (1988)

[Li95]   Li, D.: Zero duality gap for a class of nonconvex optimization problems. Journal of Optimization Theory and Applications, **85**, 309–324 (1995)

[Li96]   Li, D.: Convexification of noninferior frontier. Journal of Optimization Theory and Applications, **88**, 177–196 (1996)

[LS01]   Li, D., Sun X.L.: Convexification and existence of saddle point in a p-th power reformulation for nonconvex constrained optimization. Journal of Nonlinear Analysis: Theory and Methods (Series A), **47**, 5611–5622 (2001)

[LSBG01]   Li, D., Sun, X.L., Biswal, M.P., Gao, F.: Convexification, concavification and monotonization in global optimization. Annals of Operations Research, **105**, 213–226 (2001)

[LSM05]   Li, D., Sun, X.L., McKinnon, K.: An exact solution method for reliability optimization in complex systems. Annals of Operations Research, **133**, 129–148 (2005)

[LWLYZ05]   Li, D., Wu, Z.Y., Lee, H.W.J., Yang, X.M., Zhang, L.S.: Hidden convex minimization. Journal of Global Optimization, **31**, 211–233 (2005)

[Muf77]   Mufflin, R.: Semismooth and semiconvex functions in constrained optimization. SIAM Journal on Control and Optimization, **15**, 959–972 (1977)

[PR86]   Pardalos, P.M., Rosen, J.B.: Methods for global concave minimization: a bibliogrphic survey. SIAM Review, **28**, 367–379 (1986)

[RP87]   Rosen, J.B., Pardalos, P.M.: Constrained Global Optimization: Algorithms and Applications. Springer-Verlag (1987)

[RTM01]   Rubinov, A., Tuy, H., Mays, H.: An algorithm for monotonic global optimization problems. Optimization, **49**, 205–221 (2001)

[SL04]   Sun, X.L., Li, J.L.: A new branch-and-bound method for monotone optimization problems. Technical report, Department of Mathematics, Shanghai University (2004)

[SLL04]   Sun, X.L., Luo, H.Z., Li, D.: Convexification of nonsmooth monotone functions. Technical report, Department of Mathematics, Shanghai University, (2004)

[SML01] Sun, X.L., McKinnon, K.I.M., Li, D.: A convexification method for a class of global optimization problems with applications to reliability optimization. Journal of Global Optimization, **21**, 185–199 (2001)

[Tuy00] Tuy, H.: Monotonic optimization: problems and solution approaches. SIAM Journal on Optimization, **11**, 464–494 (2000)

[TL00] Tuy, H., Luc, L.T.: A new approach to optimization under monotonic constraint. Journal of Global Optimization, **18**, 1–15 (2000)

[Tza80] Tzafestas, S.G. Optimization of system reliability: A survey of problems and techniques. International Journal of Systems Science, **11**, 455–486 (1980)

[WBZ05] Wu, Z.Y., Bai, F.S., Zhang, L.S.: Convexification and concavification for a general class of global optimization problems. Journal of Global Optimization, **31**, 45–60 (2005)

Generalized Lagrange Multipliers for Nonconvex Directionally Differentiable Programs

Nguyen Dinh[1], Gue Myung Lee[2], and Le Anh Tuan[3]

[1] Department of Mathematics-Informatics
   Ho Chi Minh City University of Pedagogy
   280 An Duong Vuong St., District 5, HCM city, Vietnam
   ndinh@hcmup.edu.vn
[2] Division of Mathematical Sciences
   Pukyong National University
   599 - 1, Daeyeon-3Dong, Nam-Gu, Pusan 608 - 737, Korea
   gmlee@pknu.ac.kr
[3] Ninh Thuan College of Pedagogy
   Ninh Thuan, Vietnam
   latuan02@yahoo.com

Summary. A class of nonconvex optimization problems in which all the functions involved are directionally differentiable is considered. Necessary optimality conditions of Kuhn-Tucker type based on the directional derivatives are proved. Here the Lagrange multipliers generally depend on the directions. It is shown that for various concrete classes of problems (including classes convex problems, locally Lipschitz problems, composite nonsmooth problems), generalized Lagrange multipliers collapse to the standard ones (i.e., Lagrange multipliers are constants as usual). Optimality conditions for quasidifferentiable problems are derived from the main results. Optimality conditions for a class of problems in which all the functions possess upper DSL-approximates are also derived from the framework.

2000 MR Subject Classification. Primary: 90C30; Secondary: 90C46; 49K27

Key words: Directional Kuhn-Tucker condition, quasidifferentiable functions, regularity conditions, upper approximates, invexity, composite problems, optimality conditions.

1 Introduction and Preliminaries

We consider the following mathematical programming problem (P):

$$\min \quad f(x)$$

subject to $x \in C, \quad g_i(x) \leq 0, \quad i = 1, 2, \cdots, m.$

where $f, g_i : X \longrightarrow \mathbb{R} \cup \{\infty\}$, $i \in I := \{1, 2, \cdots, m\}$, X is a real Banach space and C is a closed convex subset of X.

In the case where the directional derivatives of the functions f and g_i, $i = 1, 2, \cdots, m$ exist but are not convex functions of the directions, the standard necessary condition for a feasible point x_0 to be a solution of (P) stating that there exist $\lambda_i \geq 0$, $i = 0, 1, 2, \ldots, m$ such that

$$\lambda_0 f'(x_0, r) + \sum_{i=1}^{m} \lambda_i g_i'(x_0, r) \geq 0, \ \forall r \in \text{cone}(C - x_0), \tag{1}$$

$$\lambda_i g_i(x_0) = 0, \ \text{for all } i = 1, 2, \ldots, m \tag{2}$$

fails to hold (see [Cra00, DT03], and Examples 1 and 3). For this class of problems, an optimality condition based on directional derivatives (an extended version of (1)-(2)) with Lagrange multipliers are functions of directions was introduced recently by B.D. Craven in [Cra00]. Such type of conditions were also established in [DT03] for quasidifferentiable problems.

In this paper we introduce a more general approach which can apply to larger classes of directionally differentiable problems. Concretely, we are dealing with a class of problems where the functions involved are directionally differentiable and possess upper approximates in each direction. A necessary condition for optimality of Kuhn-Tucker form where Lagrange multipliers are functions of directions is established. This condition is also sufficient under invex type hypothesis. It is shown that for various concrete classes of problems (including classes of convex problems, locally Lipschitz problems, composite nonsmooth problems), generalized Lagrange multipliers collapse to the standard ones (i.e., Lagrange multipliers are constants). As an application, optimality conditions for quasidifferentiable problems are derived from the main results. Optimality conditions for a class of problems in which all the functions possess upper DSL-approximates (see [Sha86, MW90]) are also derived from the framework.

In Section 2, a necessary condition of Kuhn-Tucker type (called "directional Kuhn-Tucker condition") based on directional derivatives is established. Here the Lagrange multiplier λ is a map of the directions, $\lambda : \text{cone}(C - x_0) \longrightarrow \mathbb{R}_+^m$. It is called "generalized Lagrange multiplier". Under some generalized convexity condition, the condition is also sufficient for optimality. Also, necessary optimality conditions (Fritz-John and Kuhn-Tucker type conditions) associated with upper approximates of the objective and the constrained functions are given. In Section 3 we examine some special cases where the generalized Lagrange multipliers collapse to constants as usual. As applications, in this section optimality conditions for a class of composite nonsmooth problems with Gateaux differentiability and also for quasidifferentiable problems are given. For the class of quasidifferentiable problems, it is shown that the "directional Kuhn-Tucker condition" is weaker than the well-known Kuhn-Tucker condition in the set inclusion form established earlier in [War91, LRW91]. It should be noted that for this class of problems the optimality conditions obtained in this section are based on the directional derivatives only and hence, do not depend on any specific choice of quasidifferentials of the functions involved. This is one of the special interest aspects of this class of problems (see [LRW91, War91]). An example is given at the end of this section to show that for quasidifferentiable problems, in general, the Lagrange multipliers can not be a constants. Furthermore, it is shown by this example that the candidates for minimizers can be sorted out by using the directional Kuhn-Tucker condition. In the last section, Section 4, we show the ability of applying the framework introduced in Section 2 to some larger class of problems. Concretely, it is shown that the framework is applicable to the class of problems for which the functions involved possess upper DSL-approximates in the sense of [Sha86, MW90] (that is, upper approximates can be represented as a difference of two sublinear functions). A necessary condition parallel to those given in [Sha86, MW90] and a sufficient condition are proved. A relation between these conditions and the one in [MW90, Sha86] is also established.

We close this section by recalling the notions of directional differentiability and recession functions of extended real-valued functions.

Let X be a real Banach space and $f : X \longrightarrow \mathbb{R} \cup \{+\infty\}$. For $x_0, h \in X$, if the limit

$$f'(x_0, h) := \lim_{\lambda \to 0^+} \frac{f(x_0 + \lambda h) - f(x_0)}{\lambda}$$

exists and is finite then $f'(x_0, h)$ is called the directional derivative of f at x_0 in the direction h. The function f is called directionally differentiable at x_0 if $f'(x_0, h)$ exists and is finite for any direction $h \in X$.

Note that if f is directionally differentiable at x_0 then the directional derivative $f'(x_0, \cdot)$ is positive homogeneous but in general it is not convex.

Let $g : X \longrightarrow \mathbb{R} \cup \{+\infty\}$ be directionally differentiable at x_0. The *recession function* of g' at x_0 is defined by

$$(g')^{\infty}(x_0, y) := \sup_{d \in X} [g'(x_0, d + y) - g'(x_0, d)].$$

The notion of recession function was widely used (see [War91, MW90] and the references therein). It is worth noting that $(g')^{\infty}(x_0, \cdot)$ is a sublinear function and $g'(x_0, \cdot) \leq (g')^{\infty}(x_0, \cdot)$. Concerning the recession function, the following lemma [MW90, Corollary 3.5] will be used in the next section.

Lemma 1. *[MW90] Suppose that g is directionally differentiable at x_0 and $g'(x_0, .)$ is lower semicontinuous (l.s.c.) on X. If $p(.)$ is an upper approximate of g at x_0, then there exists an upper approximate h of g at x_0 such that*

$$h(x) \leq \min\{p(x), (g')^\infty(x_0, x)\} \text{ for all } x \in X.$$

It is worth noting that the conclusion of Lemma 1 still holds without the assumption on the lower semicontinuity of $g'(x_0, \cdot)$ if X is finite dimensional. This was established in [War91, Lemma 2.6].

2 Generalized Lagrange Multipliers

In this section we will concern the Problem (P) where $f, g_i : X \longrightarrow \mathbb{R} \cup \{+\infty\}$, $i \in I := \{1, 2, \cdots, m\}$. Let S be the feasible set of (P), that is, $S := C \cap \{x \in X \mid g_i(x) \leq 0, \ i = 1, 2, \cdots, m\}$. Let also $x_0 \in S$ and $I(x_0) := \{i \in I \mid g_i(x_0) = 0\}$. We assume in the following that all the functions f and g_i, $i \in I$ are directionally differentiable at x_0. It is not assumed that the functions $f'(x_0, \cdot)$ and $g_i'(x_0, \cdot)$, $i \in I(x_0)$, are convex.

2.1 Necessary conditions for optimality

We begin with the necessary condition of Fritz-John type whose the proof is quite elementary. Note that no extra assumptions are needed here but the directional differentiability of f, g_i, and the continuity of g_i (at x_0) with $i \notin I(x_0)$. The same condition (holds for feasible directions from x_0 and $X = \mathbb{R}^n$) was recently proved in [Cra00].

Theorem 1. *Suppose that f and g_i, $i \in I(x_0)$ are directionally differentiable at x_0 and g_i with $i \notin I(x_0)$ are continuous at x_0. If $x_0 \in S$ is a local minimizer of (P) then for each $r \in cone(C - x_0)$, there exists $\lambda = (\lambda_0, \lambda_1, \cdots, \lambda_m) \in \mathbb{R}_+^{m+1}$, $\lambda \neq 0$ such that the following conditions hold:*

$$\lambda_0 f'(x_0, r) + \sum_{i=1}^m \lambda_i g_i'(x_0, r) \geq 0, \tag{3}$$

$$\lambda_i g_i(x_0) = 0, \text{ for all } i = 1, 2, \cdots, m. \tag{4}$$

Proof. We first note that the conditions (3)-(4) are equivalent to

$$\lambda_0 f'(x_0, r) + \sum_{i \in I(x_0)} \lambda_i g_i'(x_0, r) \geq 0$$

and $\lambda_i = 0$ for $i \notin I(x_0)$.

Suppose that $x_0 \in S$ is a local minimizer of (P). Assume on the contrary that there exists $\bar{r} \in cone(C - x_0)$ such that for any $\lambda \in \mathbb{R}_+^{|I(x_0)|+1}$, $\lambda \neq 0$ one has

$$\lambda_0 f'(x_0, \bar{r}) + \sum_{i \in I(x_0)} \lambda_i g_i'(x_0, \bar{r}) < 0. \tag{5}$$

Then by the arbitrariness of $\lambda \in \mathbb{R}_+^{|I(x_0)|+1}$ we get from (5) that

$$f'(x_0, \bar{r}) < 0, \quad g_i'(x_0, \bar{r}) < 0, \quad i \in I(x_0).$$

It follows from the definition of directional derivatives and the continuity of g_i, $i \notin I(x_0)$ that for sufficiently small $\mu > 0$,

$$x_0 + \mu\bar{r} \in C, \quad f(x_0 + \mu\bar{r}) < f(x_0), \quad g_i(x_0 + \mu\bar{r}) < 0, \quad \forall i \in I.$$

This contradicts the fact that x_0 is a minimizer of (P). □

It is worth noting that the multiplier $\lambda = (\lambda_0, \lambda_1, \cdots, \lambda_m) \in \mathbb{R}_+^{m+1}$, $\lambda \neq 0$ that exists in Theorem 1 depends on the direction $r \in \text{cone}(C - x_0)$. Precisely, the Lagrange multiplier λ is a map of direction $r \in \text{cone}(C - x_0)$. The conclusion of Theorem 1 can be expressed as follows:

There exists a map $\lambda(.) : cone(C - x_0) \longrightarrow \mathbb{R}_+^{m+1}$, $\lambda = (\lambda_0, \lambda_1, \cdots, \lambda_m)$ *with nonzero values, satisfying*

$$\lambda_0(r) f'(x_0, r) + \sum_{i=1}^{m} \lambda_i(r) g_i'(x_0, r) \geq 0, \quad \forall r \in \text{cone}(C - x_0),$$

$$\lambda_i(r) g_i(x_0) = 0, \quad \text{for all } i = 1, 2, \ldots, m \text{ and } r \in \text{cone}(C - x_0).$$

The map $\lambda(.) : \text{cone}(C - x_0) \longrightarrow \mathbb{R}_+^{m+1}$ is then called a *generalized Lagrange multiplier* (of Fritz-John type). It is shown in the next section that in many cases (e.g., differentiable, convex, Lipschitz problems) λ can be taken to be a constant (constant map) as usual.

It is easy to see that if for some $r \in \text{cone}(C - x_0)$, $g_i'(x_0, r) < 0$ for all $i \in I(x_0)$ then $\lambda_0(r) \neq 0$. So if we want to have a condition of Kuhn-Tucker type (which is of the most interest), such condition have to be satisfied for all $r \in X$. But it seems that this is quite strong in comparision with the well-known ones (see [Man94]). In the following we will search for some weaker ones that imply $\lambda_0(r) \neq 0$ for all $r \in \text{cone}(C - x_0)$. Such conditions are often known as regularity conditions .

Let $g : X \longrightarrow \mathbb{R} \cup \{+\infty\}$ be directionally differentiable at $x_0 \in X$.

Definition 1. *A lower semicontinuous sublinear function* $\phi : X \longrightarrow \mathbb{R}$ *is called an upper approximate of g at x_0 if*

$$g'(x_0, x) \leq \phi(x), \quad \text{for all } x \in X.$$

If this condition satisfies for all $x \in D$ *where D is a cone in X then we say that* ϕ *is an upper approximate of g at x_0 on D.*

An upper approximate of a function g, if it exists, may not be unique. So in general it may not give "good enough" information about the function g near x_0. We introduce another kind of upper approximates.

Definition 2. *Let ξ be a point of X. A function* $\phi : X \longrightarrow \mathbb{R}$ *is called an upper approximate of g at x_0 in the direction $\xi \in X$ if ϕ is an upper approximate of g at x_0 and*

$$g'(x_0, \xi) = \phi(\xi).$$

Note that if g is a proper convex function on X then $g'(x_0, .)$ is an upper approximate of g at x_0. Moreover, in this case $g'(x_0, .)$ is also an upper approximate of g at x_0 in any direction $\xi \in X$. If further, g is locally Lipschitz at x_0 then $g^0(x_0, .)$ (the Clarke generalized derivative at x_0) is an upper approximate of g at x_0. $g^0(x_0, .)$ is an upper approximate of g at x_0 in the direction $\xi \in X$ if and only if $g^0(x_0, \xi) = g'(x_0, \xi)$.

A function g which possesses upper approximates at x_0 in every direction $\xi \in X$ means that there exists a family of upper approximates $(\phi^\xi(.))_{\xi \in X}$ of g at x_0 such that $\phi^\xi(\xi) = g'(x_0, \xi)$, for every $\xi \in X$. Such classes of functions contain, for example, the class of convex functions, differentiable functions, locally Lipschitz and regular functions (in the sense of Clarke), and the class of quasidifferentiable functions in the sense of Demyanov and Rubinov (see Section 3).

We now introduce a regularity condition for (P), which is of Slater type constraint qualifications and involves upper approximates. Suppose that g_i, $i \in I(x_0)$, possesses upper approximates at x_0 in any direction $\xi \in X$.

Definition 3. *The Problem (P) is called (CQ1) regular at x_0 if there exists* $\bar{x} \in cone(C - x_0)$ *such that for any direction $\xi \in X$ there are upper approximates* $\Phi_i^\xi(.)$ *of g_i, $i \in I(x_0)$, at x_0 in the direction ξ satisfying*

$$\Phi_i^\xi(\bar{x}) < 0 \quad \text{for all } i \in I(x_0).$$

Definition 4. *[MW90] The Problem (P) is called (CQ2) regular at x_0 if there exists $\bar{x} \in cone(C - x_0)$ such that*

$$(g_i')^\infty(x_0, \bar{x}) < 0 \quad \text{for all } i \in I(x_0).$$

We are now able to establish a necessary optimality condition of Kuhn-Tucker type for (P).

Theorem 2. *(Directional Kuhn-Tucker condition) Suppose that $f, g_i, i \in I$, are directionally differentiable at x_0 and possess upper approximates at x_0 in any direction $\xi \in X$; g_i is continuous at x_0 for all $i \notin I(x_0)$. If x_0 is a local minimizer of (P) and if one of the following holds*
    (a) (P) is (CQ1) regular at x_0,
    (b) dim $X < +\infty$ and (P) is (CQ2) regular at x_0,
    (c) (P) is (CQ2) regular at x_0 and $g_i'(x_0, \cdot)$ is l.s.c. for all $i \in I(x_0)$
then the following directional Kuhn-Tucker condition (DKT) holds
    (DKT) For each $r \in cone(C - x_0)$, there exists $\lambda = (\lambda_i)_{i \in I} \in \mathbb{R}_+^m$ such that

$$f'(x_0, r) + \sum_{i \in I} \lambda_i g_i'(x_0, r) \geq 0, \quad \lambda_i g_i(x_0) = 0, \quad \forall i \in I.$$

A point $x_0 \in S$ that satisfies (DKT) is called a *directional Kuhn-Tucker point* of (P).

Proof. Suppose that x_0 is a minimizer of (P). It follows from Theorem 1 that for each $r \in cone(C - x_0)$, there exists $\lambda(r) = (\lambda_0(r), \lambda_1(r), \cdots, \lambda_m(r)) \neq 0$, $\lambda_i(r) \geq 0$ for all $i \in I$ such that

$$\lambda_0(r) f'(x_0, r) + \sum_{i \in I} \lambda_i(r) g_i'(x_0, r) \geq 0, \quad \lambda_i(r).g_i(x_0) = 0, \quad \forall i \in I. \quad (6)$$

It suffices to prove that for each $r \in cone(C - x_0)$, $\lambda_0(r) \neq 0$. Assume on the contrary that there is $\bar{r} \in cone(C - x_0)$ with $\lambda_0(\bar{r}) = 0$. We will prove that in this case it is possible to replace the multiplier $\lambda(\bar{r})$ by some other $\bar{\lambda}(\bar{r})$ with $\bar{\lambda}_0(\bar{r}) \neq 0$ such that (6) holds at $r = \bar{r}$ with $\bar{\lambda}(\bar{r})$ instead of $\lambda(\bar{r})$.

Since x_0 is a local minimizer of (P), the following system of variable $\xi \in X$ is inconsistent:

$$\xi \in cone(C - x_0), \quad f'(x_0, \xi) < 0, \quad g_i'(x_0, \xi) < 0, \quad \forall i \in I(x_0). \quad (7)$$

(i) Suppose that (c) holds, i.e., (P) is (CQ2) regular and $g'(x_0, .)$ is l.s.c. for all $i \in I(x_0)$. Let $\Phi^{\bar{r}}(.), \Phi_i^{\bar{r}}(.)$ be upper approximates of f and $g_i, i \in I(x_0)$ at x_0 in the direction $\bar{r}$, respectively. By Lemma 1 there exist h, h_i, upper approximates of $f, g_i, i \in I(x_0)$ at x_0 (respectively), satisfying for all $x \in X$,

$$\begin{cases} h(x) \leq \min\{\Phi^{\bar{r}}(x), (f')^{\infty}(x_0, x)\}, \\ h_i(x) \leq \min\{\Phi_i^{\bar{r}}(x), (g_i')^{\infty}(x_0, x)\}, \quad \forall i \in I(x_0). \end{cases}$$

Since (7) is inconsistent, the following system of convex functions is inconsistent:

$$\xi \in cone(C - x_0), \quad h(x) < 0, \quad h_i(x) < 0, \quad i \in I(x_0).$$

By Gordan's alternative theorem (see [Man94, HK82]), there exist $\bar{\lambda}_0 \geq 0$, $\bar{\lambda}_i \geq 0$, $i \in I(x_0)$, not all zero, such that

$$\bar{\lambda}_0 h(x) + \sum_{i \in I(x_0)} \bar{\lambda}_i h_i(x) \geq 0, \ \forall x \in cone(X - x_0). \tag{8}$$

Therefore, if $\bar{\lambda}_0 = 0$ then

$$\sum_{i \in I(x_0)} \bar{\lambda}_i h_i(x) \geq 0, \ \ \forall x \in cone(C - x_0). \tag{9}$$

By (CQ2) regularity condition, there is $\bar{x} \in cone(C - x_0)$ such that $(g_i')^\infty(x_0, \bar{x}) < 0$ for all $i \in I(x_0)$. This implies that (note that $\bar{\lambda}_i \geq 0$ for all $i \in I(x_0)$ and not all zero)

$$\sum_{i \in I(x_0)} \bar{\lambda}_i h_i(\bar{x}) \leq \sum_{i \in I(x_0)} \bar{\lambda}_i (g_i')^\infty(x_0, \bar{x}) < 0,$$

which contradicts (9). Hence, $\bar{\lambda}_0 \neq 0$ (and we can take $\bar{\lambda}_0 = 1$). With $x = \bar{r}$, (8) gives

$$h(\bar{r}) + \sum_{i \in I(x_0)} \bar{\lambda}_i h_i(\bar{r}) \geq 0.$$

Since $h(\bar{r}) \leq \Phi^{\bar{r}}(\bar{r}) = f'(x_0, \bar{r})$, $h_i(\bar{r}) \leq \Phi_i^{\bar{r}}(\bar{r}) = g_i'(x_0, \bar{r})$, and $\bar{\lambda}_i \geq 0$ for all $i \in I(x_0)$, we arrive at

$$f'(x_0, \bar{r}) + \sum_{i \in I(x_0)} \bar{\lambda}_i g_i'(x_0, \bar{r}) \geq 0.$$

Take $\bar{\lambda}_i(\bar{r}) = \bar{\lambda}_i$ for $i \in I(x_0)$ and $\bar{\lambda}_i(\bar{r}) = 0$ for all $i \notin I(x_0)$ and $\bar{\lambda}(\bar{r}) = (\bar{\lambda}_i(\bar{r}))_{i \in I}$. It is obvious that $\bar{\lambda}(\bar{r})$ satisfies the condition (DKT) at $r = \bar{r}$. The proof is complete in this case.

(ii) The proof for the case where (b) holds is the same as in the previous case, using Lemma 2.6 in [War91] instead of Lemma 1 (see the remark following Lemma 1).

(iii) The proof for the case where (a) holds is quite similar to that of (c). Take $\Phi^{\bar{r}}$, $\Phi_i^{\bar{r}}$ to be the upper approximates of f and g_i, $i \in I(x_0)$ (respectively) at x_0 in the direction $\bar{r}$ that exist by (CQ1). The inconsistency (7) implies the inconsistency of the following system:

$$\bar{x} \in cone(C - x_0), \ \ \Phi^{\bar{r}}(x) < 0, \ \ \Phi_i^{\bar{r}}(x) < 0, \ i \in I(x_0).$$

Then we get (8) with h is replaced by $\Phi^{\bar{r}}$ and h_i is replaced by $\Phi_i^{\bar{r}}$, $i \in I(x_0)$. If $\bar{\lambda}_0 = 0$ then

$$\sum_{i \in I(x_0)} \bar{\lambda}_i \Phi_i^{\bar{r}}(x) \geq 0, \forall x \in cone(C - x_0).$$

This is impossible since by (CQ1), $\Phi_i^{\bar{r}}(\bar{x}) < 0$ for all $i \in I(x_0)$ and $\bar{\lambda}_i \geq 0$, $(i \in I(x_0))$ not all zero. Hence $\bar{\lambda}_0 \neq 0$. The rest is the same as in (i). The proof is complete. $\square$

The relation between (CQ1), (CQ2) and the other regularity conditions, as well as the relation between (DKT) and some other Kuhn-Tucker condition will be discussed at the end of Section 3 in the context of quasidifferentiable programs.

2.2 Sufficient condition for optimality

We now prove that the directional Kuhn-Tucker condition (in Theorem 2) is also sufficient for optimality under an assumption on the invexity of (P). This notion of generalized convexity has been widely used in smooth as well as nonsmooth optimization problems (see [BRS83, Cra81, Cra86, Cra00, Han81, Sac00, SKL00, SLK03, YS93], ...). Our definition of invexity is slightly different from the others.

Definition 5. *Suppose that $\phi(\cdot)$, $\phi_i(\cdot)$, $i \in I(x_0)$ are positively homogenous functions defined on X such that*

$$f'(x_0, x) \leq \phi(x), \ \forall \ x \in X,$$
$$g_i'(x_0, x) \leq \phi_i(x), \ \forall \ x \in X, \ \forall i \in I(x_0).$$

The Problem (P) is called invex at x_0 on S with respect to $\phi(\cdot), \phi_i(\cdot), i \in I(x_0)$ if there exists a function $\eta : S \longrightarrow cone(C - x_0)$ such that the following holds:

$$f(x) - f(x_0) \geq \phi(\eta(x)), \ \forall x \in S,$$
$$g_i(x) - g_i(x_0) \geq \phi_i(\eta(x)), \ \forall x \in S, \ \forall i \in I(x_0).$$

If (P) is invex (at x_0 on S) with recpect to $f'(x_0, \cdot), g_i'(x_0, \cdot), i \in I(x_0)$ then it is called simply invex (the most important case).

Note that if f, g_i are differentiable at x_0 then the invexity of (P) (with respect to $f'(x_0, \cdot)$, $g_i'(x_0, \cdot)$, $i \in I(x_0)$) is exactly the one which appeared in [Han81, Cra81]. In Definition 5, if in addition, f, g_i are locally Lipschitz at x_0 and if we take $\phi(\cdot) = f^o(x_0, \cdot)$, $\phi_i(\cdot) = g_i^o(x_0, \cdot)$, $i \in I(x_0)$ then we come back to the definition of invexity appearing in [YS93, BRS83]. This also relates to the cone-invexity for locally Lipschitz functions, which was defined in [Cra86]. The following result was established in [Cra00] concerning feasible directions and for $X = \mathbb{R}^n$. Its proof is almost the same as in [Cra00, DT03] and so it will be omitted.

Theorem 3. *(Sufficient condition for optimality) Let $f, g_i, i \in I$ be directionally differentiable at x_0. If x_0 is a directional Kuhn-Tucker point of (P) and if (P) is invex at x_0 on S then x_0 is a global minimizer of (P).*

In view of Theorems 2 and 3, it is easy to obtain the following necessary and sufficient optimaity conditions with upper approximates:

Corollary 1. *For the problem (P), let x_0 is a feasible point and let ϕ, ϕ_i be upper approximates of f, g_i, $i \in I$ at x_0, respectively. Suppose that g_i is continuous at x_0 for all $i \notin I(x_0)$.*

(i) If x_0 is a local minimizer of (P) then there exist $\lambda_0 \geq 0$, $\lambda_i \geq 0$, $i \in I$, not all zero, such that

$$\lambda_0 \phi(x) + \sum_{i \in I} \lambda_i \phi_i(x) \geq 0, \quad \forall x \in \text{cone}(C - x_0); \quad \lambda_i g_i(x_0) = 0, \ \forall i \in I.$$

Moreover, if there exists $\bar{x} \in S$ such that $\phi_i(\bar{x}) < 0$ for all $i \in I(x_0)$ then $\lambda_0 \neq 0$ (and hence, one can take $\lambda_0 = 1$). That is, there exists $\lambda = (\lambda_i)_{i \in I} \in \mathbb{R}_+^m$ such that

$$\phi(x) + \sum_{i \in I} \lambda_i \phi_i(x) \geq 0, \quad \forall x \in \text{cone}(C - x_0); \quad \lambda_i g_i(x_0) = 0, \ \forall i \in I. \tag{10}$$

(ii) Conversely, if x_0 satisfies (10) (for some upper approximates ϕ, ϕ_i of f, g_i on cone $(C - x_0)$, respectively, and some $\lambda \in \mathbb{R}_+^m$) and if (P) is invex at x_0 on $S := C \cap \{x \in X | g_i(x) \leq 0, \ i = 1, 2, \cdots, m\}$ with respect to ϕ, ϕ_i, $i \in I(x_0)$ then x_0 is a global minimizer of (P).

Proof. (i) Since x_0 is a solution of (P) the following system of variable $\xi \in X$:

$$\xi \in \text{cone } (C - x_0), \quad f'(x_0, \xi) < 0, \quad g_i'(x_0, \xi) < 0, \quad i \in I(x_0)$$

is inconsistent. By the definition of upper approximates, the following system of convex functions is inconsistent.

$$\xi \in \text{cone } (C - x_0), \quad \phi(x) < 0, \quad \phi_i(x) < 0, \quad i \in I(x_0). \tag{11}$$

The rest of the proof is similar to those of Theorems 2 and 3.   □

It is worth noting that the conclusion of Corollary 1 still holds if in the definition of upper approximate (Definition 1) one replaces directional derivatives $g'(x_0, d)$ by the upper Dini derivative $g^+(x_0, d)$ of g at x_0 in the direction $d \in X$ which is defined by

$$g^+(x_0, d) := \limsup_{\lambda \to 0^+} \frac{g(x_0 + \lambda d) - g(x_0)}{\lambda}.$$

This can be seen by replacing (11) by the following:

$$\xi \in \text{cone } (C - x_0), \quad f^+(x_0, \xi) < 0, \quad g_i^+(x_0, \xi) < 0, \quad i \in I(x_0).$$

The Lagrange multipliers that exist in Corollary 1 are constants (independent from the directions). The price for this is that (10) is just based on the upper approximates of f and g_i at x_0 instead of $f'(x_0, \cdot)$, $g_i'(x_0, \cdot)$, $i \in I(x_0)$ as in the previous subsection (Theorem 2). However, for smooth problems (i.e., f, g_i are differentiable), or convex, or locally Lipschitz problems, condition (10) collapses to the standard optimality conditions. For instant, if f and g_i are convex then $f'(x_0, \cdot)$, $g_i'(x_0, \cdot)$, $i \in I(x_0)$ are convex and hence, by taking $\phi(\cdot) = f'(x_0, \cdot)$, $\phi_i(\cdot) = g_i'(x_0, \cdot)$, $i \in I(x_0)$, (10) is none other than

$$f'(x_0, x) + \sum_{i \in I(x_0)} \lambda_i g_i'(x_0, x) \geq 0, \ \forall x \in \text{ cone } (C - x_0)$$

(provided that there is $\bar{x} \in$ cone $(C - x_0)$ satisfying $g_i'(x_0, \bar{x}) < 0$ for all $i \in I(x_0)$). Note also that by separation theorem, this inequality is equivalent to

$$0 \in \partial f(x_0) + \sum_{i \in I(x_0)} \lambda_i \partial g_i(x_0) + N_C(x_0)$$

where $N_C(x_0)$ stands for the normal con of C at x_0 in the sense of convex analysis.

Example 1. Consider the following problem (P1)

$$\min \quad f(x)$$
$$\text{subject to} \quad g(x) \leq 0, \quad x = (x_1, x_2) \in C$$

where

$$C = \text{co } \{(0,0), (-1,-1), (-1,1)\}$$

and the functions $f, g : \mathbb{R}^2 \longrightarrow \mathbb{R}$ are defined by

$$f(x) = -x_2 + \sqrt{|x_1^2 - x_2^2|}, \quad g(x) = | \ |x_1| + x_2|.$$

Observe that

$$S = C \cap \{x \in \mathbb{R}^2 : g(x) \leq 0\} = \text{co } \{(0,0), (-1,-1)\} \subset \text{ cone } C.$$

Let $x_0 = (0,0)$. It is easy to see that $f(x_0) = g(x_0) = 0$, $f'(x_0, r) = f(r)$ and $g'(x_0, r) = g(r)$ for all $r = (r_1, r_2) \in \mathbb{R}^2$.

Set, for $r = (r_1, r_2) \in$ cone $(C - x_0)$,

$$\lambda(r) := \begin{cases} 0 & \text{if} \ -r_2 + \sqrt{|r_1^2 - r_2^2|} \geq 0, \\ -\frac{f(r)}{g(r)} & \text{if} \ -r_2 + \sqrt{|r_1^2 - r_2^2|} < 0 \end{cases}$$

(note that when $-r_2 + \sqrt{|r_1^2 - r_2^2|} < 0$, $g(r) \neq 0$). Then the following holds:

$$f'(x_0, r) + \lambda(r)g'(x_0, r) \geq 0, \quad \forall r \in \text{ cone } (C - x_0).$$

This means that $x_0 = (0,0)$ is a directional Kuhn-Tucker point of (P1).

On the other hand, it is clear that for each $x \in S$ (feasible set),

$$f(x) - f(x_0) = f'(x_0, x),$$
$$g(x) - g(x_0) = g'(x_0, x),$$

which proves that (P1) is invex at x_0 (with $\eta : S \longrightarrow$ cone $(C - x_0)$, $\eta(x) = x$). Thus, x_0 is a minimizer of (P1).

3 Special Cases and Applications

In this section we will show that for some special classes of problems such as composite nonsmooth problems with Gâteaux differentiablity or for problems where the directional derivatives are generalized subconvexlike, the generalized Lagrange multipliers can be chosen to be constants. The last part of this section is left for an application to a class of quasidifferentiable problems. Some examples are given to illustrate the significant of the results.

3.1 Problems with convexlike directional derivatives

Let D be a subset of X. Let $\Phi = (\phi_1, \phi_2, \cdots, \phi_m) : D \longrightarrow \mathbb{R}^m$. Recall that the map Φ is called *convexlike* (*subconvexlike*, resp.) if $\Phi(D) + \mathbb{R}_+^m$ is convex ($\Phi(D) + \mathrm{int}\mathbb{R}_+^m$ is convex, resp.). It is called *gerneralized subconvexlike* if $\mathrm{cone}\Phi(D) + \mathrm{int}\mathbb{R}_+^m$ is convex (see [HK82, Jey85, Sac02]).

It is well-known that the Gordan's alternative theorem still holds with convexlike, subconvexlike, generalized subconvexlike functions instead of convex ones (see [Jey85, Sac02] for more extensions). Namely, if $\Phi = (\phi_1, \phi_2, \cdots, \phi_m) : D \longrightarrow \mathbb{R}^m$ is generalized subconvexlike (convexlike, subconvexlike) on D then exactly one of the following assertions holds:

(i) $\exists x \in D$ such that $\phi_i(x) < 0$, $i = 1, 2, \cdots, m$,

(ii) $\exists \lambda = (\lambda_1, \lambda_2, \cdots, \lambda_m) \in \mathbb{R}_+^m$, $\lambda \neq 0$ such that $\sum_{i=1}^m \lambda_i \phi_i(x) \geq 0$, $\forall x \in D$.

Theorem 4. *Suppose that f and g_i, $i \in I(x_0)$ are directionally differentiable at x_0 and that g_i with $i \notin I(x_0)$ are continuous at x_0. Suppose further that the map $\Phi : \mathrm{cone}\,(C - x_0) \longrightarrow \mathbb{R}^{|I(x_0)|+1}$ defined by $\Phi(\xi) = (f'(x_0, \xi), g_i'(x_0, \xi))$, $i \in I(x_0)$ is generalized subconvexlike. If $x_0 \in S$ is a local minimizer of (P) then there exists $\lambda = (\lambda_0, \lambda_1, \ldots, \lambda_m) \in \mathbb{R}_+^{m+1}$, $\lambda \neq 0$ such that the following conditions hold.*

$$\lambda_0 f'(x_0, r) + \sum_{i=1}^m \lambda_i g_i'(x_0, r) \geq 0, \forall r \in \mathrm{cone}\,(C - x_0),$$

$$\lambda_i g_i(x_0) = 0, \quad \text{for all } i = 1, 2, \cdots, m.$$

Moreover, if there exists $\bar{x} \in \mathrm{cone}\,(C - x_0)$ such that $g_i'(x_0, \bar{x}) < 0$ for all $i \in I(x_0)$ then $\lambda_0 \neq 0$ (and hence, one can take $\lambda_0 = 1$).

Proof. It is easy to see that the optimality of x_0 implies the inconsistency of the following system of variable $\xi \in X$:

$$\xi \in \mathrm{cone}\,(C - x_0), \quad f'(x_0, \xi) < 0, \quad g_i'(x_0, \xi) < 0, \quad i \in I(x_0).$$

The existence of $\lambda_0 \geq 0$, $\lambda_i \geq 0$, $i \in I(x_0)$, not all zero, satisfying the conclusion of the theorem now follows from Gordan's theorem for generalized subconvexlike systems (setting $\lambda_i = 0$ for $i \notin I(x_0)$). The rest is obvious.   □

Let x_0 be a feasible point of (P) and let D be the set of all feasible directions of (P) from x_0. Set

$$M := \{(f'(x_0, d), (g_i'(x_0, d))_{i \in I(x_0)}) \mid d \in D\}.$$

We now apply Corollary 1 to derive an optimality condition (with constant Lagrange multipliers) for (P), which was established recently in [Cra00].

Corollary 2. *[Cra00] Let Σ be a closed convex cone contained in M. Denote $q := (f, (g_i)_{i \in I(x_0)})$. Assume that some d^* satisfies $q'(x_0, d^*) \in \Sigma$ and $g_i'(x_0, d^*) < 0$ for all $i \in I(x_0)$. Then there exists $\lambda = (\lambda_i)_{i \in I(x_0)} \in \mathbb{R}_+^{|I(x_0)|}$, dependent on Σ but not on $d \in \tilde{D}$, such that for each $d \in \tilde{D} := \{d \in D \mid \exists \eta \in \Sigma, q'(x_0, d) = \eta\}$,*

$$f'(x_0, d) + \sum_{i \in I(x_0)} \lambda_i g_i'(x_0, d) \geq 0.$$

Proof. By definition, for each $d \in \tilde{D}$, $q'(x_0, d) \in \Sigma$. Define

$$\Phi = (f'(x_0, \xi), (g_i'(x_0, \xi))_{i \in I(x_0)}) : \tilde{D} \longrightarrow \mathbb{R}^{|I(x_0)|+1}.$$

Then $\Phi(\tilde{D}) = \Sigma$ is a closed convex cone and hence, Φ is convexlike. The conclusion follows from Theorem 4 with $\tilde{D}$ playing the role of $\text{cone}(C - x_0)$.
□

3.2 Composite nonsmooth programming with Gâteaux differentiability

Let X, Y be Banach space and C be a closed convex subset of X. Consider the composite problem (CP):

(CP)   Minimize $f_0(F_0(x))$

subject to $x \in C$, $f_i(F_i(x)) \leq 0$, $i = 1, 2, \cdots, m$,

where $F_i : X \to Y$ is Gâteaux differentiable with Gâteaux derivative $F_i'(\cdot)$ and $f_i : Y \to \mathbb{R}$ is locally Lipschitz, $i = 0, 1, \cdots, m$.

Note that the Gâteaux differentiability of a map $F : X \to Y$ at a does not necessarily imply the continuity of F at a. The following simple example [IT79, p. 24] shows this. Let $f(x, y) = 1$ if $x = y^2$ and $y \neq 0$, $f(x, y) = 0$ otherwise. Then f is Gâteaux differentiable at $(0, 0)$ and $f'(0, 0) = 0$ while f is not continuous at 0.

Now let $x_0 \in C \cap \{x \in X \mid f_i(F_i(x)) \leq 0, \quad i = 1, 2, \cdots, m\}$, $I = \{1, \cdots, m\}$, $I_0 = \{0\} \cup I$ and let $I(x_0) = \{j \in I \mid f_j(F_j(x_0)) = 0\}$, $x_0 \in C$. We shall use the notation $(f \circ F)^+(a, d)$ to indicate the upper Dini derivative of $f \circ F$ at a in the direction d, which is defined by

$$(f \circ F)^+(a, d) := \limsup_{\lambda \to 0^+} \frac{f(F(a + \lambda d)) - f(F(a))}{\lambda}.$$

The following lemma is crucial for establishing optimality conditions for (CP).

Lemma 2. *Let $a \in X$. If $F : X \to Y$ is continuous and Gâteaux differentiable at a and $f : Y \to \mathbb{R}$ is locally Lipschitz at $F(a)$ then for any $d \in X$, there exists $v \in \partial f(F(a))$ such that*

$$(f \circ F)^+(a, d) = \langle v, \ F'(a)d \rangle .$$

Proof. By the definition of upper Dini derivative, there exists $(\lambda_n) \subset \mathbb{R}_+$, $\lambda_n \to 0$ such that

$$(f \circ F)^+(a, d) = \lim_{n \to \infty} \frac{f(F(a + \lambda_n d)) - f(F(a))}{\lambda_n}. \tag{12}$$

Assume that f is Lipschitz of rank K on a convex open neighborhood U of $F(a)$. Note that f is also locally Lipschitz at any point of U with the same rank K. Since F is continuous at a, without loss of generality, we can assume that for all n, $F(a + \lambda_n d) \in U$.

It follows from the mean-value theorem of Lebourg [Cla83, Theorem 2.3.7, p. 41], for each $n \in \mathbb{N}$, there exist $t_n \in (0, 1), v_n \in \partial f(z_n)$ such that

$$f(F(a + \lambda_n d)) - f(F(a)) = \langle v_n, \ F(a + \lambda_n d) - F(a) \rangle \tag{13}$$

where $z_n := F(a) + f(F(a + \lambda_n d) - F(a)) \in U$.

Note that $v_n \in Y^*$ and $\|v_n\| \leq K$. Hence we can assume $(v_n)_n$ weak* converges to v. Note also that when $n \to \infty$, we have $z_n \to F(a)$ and it follows from the weak* - closedness of the ∂f, we get $v \in \partial f(F(a))$. It now follows from (12) and (13) that

$$(f \circ F)^+(a, d) = \langle v, F'(a)d \rangle .$$

□

Following Lemma 2, if we set

$$\Psi(d) = \max_{v \in \partial f(F(a))} \langle v, F'(a)d \rangle$$

then $\Psi : X \to \mathbb{R}$ is a l.s.c. sublinear function (finite valued). Moreover,

$$(f \circ F)^+(a, d) \leq \Psi(d) \quad \text{for all} \quad d \in X.$$

This means that Ψ is an upper approximate of $f \circ F$ at a (see the remark that follows Corollary 1).

We are now in a position to give a necessary condition for optimality for (CP).

Theorem 5. *Assume that F_i is continuous and Gâteaux differentiable at a feasible point x_0 of (CP) and f_i is locally Lipschitz at $F_i(x_0)$, $i = 0, 1, \cdots, m$. If x_0 is a solution of (CP) then there exist $\lambda_0, \lambda_1, \cdots, \lambda_m \geq 0$, not all zero, $v_i \in \partial f_i(F_i(x_0))$, $i \in I_0$ such that*

$$[\lambda_0 F_0'(x_0)^* v_0 + \sum_{i=1}^m \lambda_i F_i'(x_0)^* v_i](x - x_0) \geq 0, \ \forall x \in C,$$

$$\lambda_i f_i(F_i(x_0)) = 0, \ \forall i \in I,$$

where $F_i'(x_0)^*$ is the adjoint operators of $F_i'(x_0)$.

Proof. We first notice that if x_0 is solution of (CP) then the following system has no solution $d \in X$:

$$d \in cone(C - x_0), \ \ (f_i \circ F_i)^+(x_0, d) < 0, \ i \in I(x_0) \cup \{0\}.$$

Let $\Psi_i(d) := \max\limits_{v_i \in \partial f_i(F_i(x_0))} \langle v_i, F_i'(x_0) d \rangle$. Then $(f_i \circ F_i)^+(x_0, d) \leq \Psi_i(d)$ for all $d \in cone(C - x_0)$. It follows from Corollary 1 that there exist λ_0, $\lambda_i \geq 0$, $i \in I(x_0)$, not all zero, such that

$$\lambda_0 \Psi_0(d) + \sum_{i \in I(a)} \lambda_i \Psi_i(d) \geq 0, \ \forall d \in cone \ (C - x_0). \tag{14}$$

Since $x_0 \in C$, $0 \in cone(C - x_0)$, the above inequality means that 0 is a minimizer of the convex problem

$$\text{Minimize} \qquad [\lambda_0 \Psi_0(d) + \sum_{i \in I(x_0)} \lambda_i \Psi_i(d)]$$

$$\text{subject to} \quad d \in cone(C - x_0).$$

This is equivalent to

$$0 \in \lambda_0 \partial \Psi_0(0) + \sum_{i \in I(x_0)} \lambda_i \partial \Psi_i(0) + N_C(x_0). \tag{15}$$

Note that for each $d \in X$, $i \in I(x_0) \cup \{0\}$,

$$\Psi_i(d) = \max\limits_{v \in \partial f_i(F_i(x_0))} \langle F_i'(x_0)^* v, d \rangle = \max\limits_{w \in B_i} \langle w, d \rangle = \sigma_{B_i}(d)$$

where $B_i := F_i'(x_0)^* [\partial f_i(F_i(x_0))]$ and σ_{B_i} is the support function of B_i. It follows from [Cla83, proposition 2.1.4, p. 29] that the set $F_i'(x_0)^* [\partial f_i(F_i(x_0))]$ is weak*-compact and we have

$$\partial \Psi_i(0) = F_i'(x_0)^* [\partial f_i(F_i(x_0))].$$

It follows from the last equality and (15) that there exist $v_i \in \partial f_i(F_i(x_0))$, $i \in I \cup \{0\}$ such that

$$\lambda_0 F_0'(x_0)^* v_0 + \sum_{i=1}^m \lambda_i F_i'(x_0)^* v_i](x - x_0) \geq 0, \ \forall x \in C.$$

The conclusion follows by setting $\lambda_i = 0$ if $i \notin I(x_0)$, $i \neq 0$.   □

We now give a necessary condition for (CP) in Kuhn-Tucker form.

Theorem 6. *Assume that all the conditions in Theorem 5 hold. Assume further that the regularity condition that there is $d_0 \in \text{cone}(C - x_0)$ satisfying $\Psi_i(d_0) < 0$, for all $i \in I(x_0)$ holds. If x_0 is a solution of (CP) then there exist $\lambda_i \geq 0$, $i \in I$, $v_i \in \partial f_i(F_i(x_0))$, $i \in I_0$ such that*

$$[F_0'(x_0)^* v_0 + \sum_{i=1}^m \lambda_i F'(x_0)^* v_i](x - x_0) \geq 0, \ \forall x \in C,$$

$$\lambda_i f_i(F_i(x_0)) = 0, \ \forall i \in I.$$

Proof. The proof is the same as that of Theorem 5. Note that if the regularity condition in the statement of the theorem holds then $\lambda_0 \neq 0$ in (14).   □

It is worth noting that the same conditions as in Theorems 5–6 were established in [Jey91] under the additional assumption that the maps F_i, $i \in I_0$ are locally Lipschitz.

The following example illustrates the significance of Theorems 5, 6.

Example 2. Consider the following problem (P2)

$$\text{Minimize} \quad f(F(x, y))$$
$$\text{subject to} \quad g(G(x, y)) \leq 0, (x, y) \in \mathbb{R}^2$$

where $f : \mathbb{R} \to \mathbb{R}$, $g : \mathbb{R} \to \mathbb{R}$, $F : \mathbb{R}^2 \to \mathbb{R}$, $G : \mathbb{R}^2 \to \mathbb{R}$ are the functions defined by

$$f(z) = z, \ g(z) = z, \ G(x, y) = x,$$

$$F(x, y) = \begin{cases} x^2 & \text{if} \quad y = 0, \\ 0 & \text{if} \quad y \neq 0. \end{cases}$$

Note that F is continuous at $x_0 = (0, 0)$, Gâteaux differentiable at this point and $F'(x_0) = (0, 0)$ but F is not locally Lipschitz $x_0 = (0, 0)$. It is easy to see that $x_0 = (0, 0)$ is a solution of (P2), and the necessary condition in Theorem 6 holds with $\lambda_0 = 1$, $\lambda_1 = 0$ (note also that for (P2) the regularity condition in Theorem 6 holds).

3.3 Quasidifferentiable problems

Quasidifferentiable functions are those of which the directional derivatives can be represented as a difference of two sublinear functions. The class of these functions covers all classes of differentiable functions, convex functions, DC-functions, $\cdots$. It was introduced by V.F. Demyanov and A.M. Rubinov ([DR80]) in 1980. Since then optimization problems with quasidifferentiable data have been widely investigated and developed by many authors (see [DJ97, DV81, DT03, EL87, Gao00, Gl92, LRW91, MW90, Sha84, Sha86, War91], ... See also [DPR86] for a discussion on the place and the role of quasidifferentiable functions in nonsmooth optimization). Many optimality conditions were introduced. Most of them are conditions that base on the subdifferentials and superdifferentials of the quasidifferentiable functions involved.

In this section we will apply the results obtained in Section 2 to quasi-differentiable programs. The relation between the directional Kuhn-Tucker condition and some other type of Kuhn-Tucker conditions that appeared in the literature is established. It is shown (by an example) that for quasidifferentiable problems (even in the finite dimensional case) the generalized Lagrange multipliers can not be constants.

Let X be a real Banach space and $x_0 \in X$. A function $f : X \longrightarrow \mathbb{R} \cup \{+\infty\}$ is called *quasidifferentiable* at x_0 if f is directionally differentiable at x_0 and if there are two weak* compact subsets $\underline{\partial} f(x_0)$, $\overline{\partial} f(x_0)$ of the topological dual X^* of X such that

$$f'(x_0, d) = \max_{\xi \in \underline{\partial} f(x_0)} \langle d, \xi \rangle + \min_{\xi \in \overline{\partial} f(x_0)} \langle d, \xi \rangle, \ \forall d \in X. \qquad (16)$$

The pair of sets $Df(x_0) := [\underline{\partial} f(x_0), \overline{\partial} f(x_0)]$ is called the quasidifferential of f at x_0 and $\underline{\partial} f(x_0)$, $\overline{\partial} f(x_0)$ are called the subdifferential and superdifferential of f at x_0, respectively. Note that the quasidifferential $Df(x_0)$ of f at x_0 is not uniquely defined (see [LRW91]). Note also that (16) can be written in the form

$$f'(x_0, d) = \max_{\xi \in \underline{\partial} f(x_0)} \langle d, \xi \rangle - \max_{\xi \in -\overline{\partial} f(x_0)} \langle d, \xi \rangle, \ \forall d \in X \qquad (17)$$

and hence, $f'(x_0, .)$ can be represented as a difference of two sublinear functions. In general, $f'(x_0, .)$ is not convex.

Throughout this subsection, the following lemma plays a key role.

Lemma 3. *If f is quasidifferentiable at x_0 then for any direction $\xi \in X$, there is an upper approximate of f at x_0 in the direction ξ.*

Proof. Let $\xi \in X$. Since $\overline{\partial} f(x_0)$ is weak* compact, there exists $\bar{v} \in \overline{\partial} f(x_0)$ such that

$$\langle \xi, \bar{v} \rangle = \min_{v \in \overline{\partial} f(x_0)} \langle \xi, v \rangle.$$

Let

$$\Phi^\xi(x) := \max_{v \in \underline{\partial} f(x_0)} \langle x, v \rangle + \langle x, \bar{v} \rangle. \tag{18}$$

It is easy to see that $\Phi^\xi(.)$ is sublinear, l.s.c., $f'(x_0, x) \leq \Phi^\xi(x)$ for all $x \in X$, and $f'(x_0, \xi) = \Phi^\xi(\xi)$, which proves $\Phi^\xi(.)$ to be an upper approximate of f in the direction ξ.   □

Consider the problem (P) defined in Section 1. Let S be the feasible set of (P) and $x_0 \in S$. We are now ready to get necessary and sufficient optimality conditions for (P).

Theorem 7. *(Necessary condition) For the problem (P), assume that f, g_i, $i \in I = \{1, 2, 3, \cdots, m\}$ are quasidifferentiable at x_0 and g_i is continuous at x_0 for all $i \notin I(x_0)$. If x_0 is a minimizer of (P) then*

$$\forall r \in cone(C - x_0), \ \exists \ (\lambda_0, \lambda_1, \cdots, \lambda_m) \in \mathbb{R}_+^{m+1} \setminus \{0\} \ satisfying$$
$$\lambda_0 f'(x_0, r) + \sum_{i \in I} \lambda_i g_i'(x_0, r) \geq 0, \ \lambda_i g_i(x_0) = 0, \ \forall i \in I.$$

Moreover, if one of the following conditions holds
(i) $dim X < +\infty$ and (P) is (CQ2) regular at x_0,
(ii) (P) is (CQ2) regular at x_0 and $g_i'(x_0, .)$ is l.s.c. for all $i \in I(x_0)$
then $\lambda_0 \neq 0$ and hence we can take $\lambda_0 = 1$ (i.e., x_0 is a directional Kuhn-Tucker point of (P)).

Proof. It follows from Lemma 3 that the functions f and g_i possess upper approximates at x_0 in any direction $\xi \in X$. The conclusion now follows from Theorem 2.   □

The following theorem is a direct consequence of Theorem 3 in Section 2.

Theorem 8. *(Sufficient condition) For the problem (P), assume that f, g_i, $i \in I = \{1, 2, 3, \cdots, m\}$ are quasidifferentiable at x_0 and g_i is continuous at x_0 for all $i \notin I(x_0)$. Assume further that x_0 is a directional Kuhn-Tucker point of (P). If (P) is invex at x_0 on the feasible set S then x_0 is a global solution of (P).*

It should be noted that both the necessary and sufficient optimality conditions for (P) established in Theorems 7, 8 do not depend on any specific choice of quasidifferentials of f and g_i, $i \in I(x_0)$.

The regularity conditions are of special interest in quasidifferentiable optimization. The above (CQ2) condition was introduced in [MW90]. It is prefered much since it does not depend on any specific choice of the quasidifferentials (see [LRW91]). In order to make some relation between our results and the others, we take a quick look at some other regularity conditions that appeared in the literature and for the sake of simplicity we consider the case where $C = X$.

(CQ3) $i \in I(x_0)$, $\forall v_i \in \overline{\partial} g_i(x_0)$, $0 \notin$ co $\bigcup_{i \in I(x_0)} (\partial g_i(x_0) + v_i)$.

(RC) There exists $\overline{x} \in X$ such that

$$\max_{v_i \in \underline{\partial} g_i(x_0)} \langle \overline{x}, v_i \rangle + \max_{w_i \in \overline{\partial} g_i(x_0)} \langle \overline{x}, w_i \rangle < 0, \quad \forall i \in I(x_0).$$

The (CQ3) condition was used in [Sac00] and [LRW91] while (RC) was introduced in [War91], both for the case where $X = \mathbb{R}^n$.

It was proved in [DT03] that in the finite dimensional case (CQ3) is equivalent to (RC). By Lemma 3, it is clear that (RC) implies (CQ1).

On the other hand, it was proved in [LRW91] that (RC) implies (CQ2) when $X = \mathbb{R}^n$. However, the proof (given in [LRW91]) goes through without any change for the case where X is a real Banach space. Briefly, the following scheme holds for quasidifferentiable problems:

$$(CQ3) \quad \Longleftrightarrow_{(X=\mathbb{R}^n)} \quad (RC) \quad \Longrightarrow \quad (CQ2)$$
$$\Downarrow$$
$$(CQ1)$$

The conclusion in Theorem 7 (also Theorem 8) was established in [DT03] for quasidifferentiable Problem (P) when $C = X = \mathbb{R}^n$ and under the (CQ3) (or the same, (RC)).

Due to the previous observation, Theorems 7 still holds if (RC) is assumed instead of (i) or (ii).

As mentioned above, the quasidifferentiable problems with inequality constraints of the form (P) have been studied by many authors. Various types of Kuhn-Tucker conditions were proposed to be necessary optimality conditions for (P) (under various assumptions and regularity conditions). A typical such condition is as follows:

$$-\overline{\partial} f(x_0) \subset \bigcup_{\substack{\overline{w}_i \in \overline{\partial} g_i(x_0) \\ i \in I(x_0)}} \left[\underline{\partial} f(x_0) + \sum_{i \in I(x_0)} \text{cone} \left(\partial g_i(x_0) + \overline{w}_i \right) \right]. \quad (19)$$

A point x_0 satisfies (19) is called a *Kuhn-Tucker point* of (P) (see [War91, LRW91]). The ondition (19) was established in [War91] as a necessary condition for a point $x_0 \in S$ to be a minimizer for (P) when $C = X = \mathbb{R}^n$ (under some reagularity condition). It was proved in [DT03] for $C = X = \mathbb{R}^n$ that if x_0 is a Kuhn-Tucker point of (P) then it is also a directional Kuhn-Tucker point of (P). This conclusion still holds (without any change in the proof) when X is a Banach space.

The following example shows that the two notions of the Kuhn-Tucker point and the directional Kuhn-Tucker point are not coincide, and that even for a simple nonconvex problem the generalized Lagrange multiplier can not be chosen to be a constant function. It also shows that one can use the directional Kuhn-Tucker condition to search for a minimizer.

Example 3. Consider the following problem (P3)

$$\text{min} \qquad f(x)$$
$$\text{subject to} \quad g(x) \leq 0, \quad x = (x_1, x_2) \in C \subset \mathbb{R}^2,$$

where $f, g : \mathbb{R}^2 \longrightarrow \mathbb{R}$ are functions defined by

$$f(x) := x_2,$$
$$g(x) := \begin{cases} x_1 + (x_1^2 + x_2^2)^{\frac{1}{2}} - x_2 & \text{if} \quad x_2 \geq 0, \\ x_1 + (x_1^2 + x_2^2)^{\frac{1}{2}} & \text{if} \quad x_2 < 0. \end{cases}$$

Let $x_0 = (0, 0) \in \mathbb{R}^2$.

(a) *Consider first the case where* $C := $ co $\{(0,0), (0,1), (1,-1)\}$.
   (i) It is clear that $S = C \cap \{x \in \mathbb{R}^2 \mid g(x) \leq 0\} = $ co $\{(0,0), (0,1)\} \subset$ cone $(C - x_0) = $ cone C, where S is the feasible set of (P3). It is also easy to check that x_0 *is a directional Kuhn-Tucker point* of (P3). The generalized Lagrange multiplier $\lambda : $ cone $C \longrightarrow \mathbb{R}_+$ can be chosen as follows ($r = (r_1, r_2) \in$ cone C):

$$\lambda(r) = \begin{cases} -\frac{r_2}{g(r)}, & \text{if} \quad r_2 < 0 \\ 0, & \text{if} \quad r_2 \geq 0. \end{cases} \tag{20}$$

Equivalently, the following inequality holds for all $r = (r_1, r_2) \in$ cone C:

$$f'(x_0, r) + \lambda(r)g'(x_0, r) \geq 0. \tag{21}$$

On the other hand, since $f'(x_0, r) = r_2$, $g'(x_0, r) = g(r)$, $f(x_0) = g(x_0) = 0$, it is easy to see that (P3) is invex at x_0 with $\eta : S \longrightarrow$ cone C, $\eta(x) = x$. Consequently, x_0 is a minimizer of (P3) due to Theorem 3.
   (ii) For (P3), the generalized Lagrange multiplier $\lambda : $ cone $C \longrightarrow \mathbb{R}_+$ *can not be chosen to be a constant function.* In fact, (21) is equivalent to

$$r_2 + \lambda(r)g(r) \geq 0. \tag{22}$$

This shows that for $r = (r_1, r_2) \in$ cone C with $r_2 < 0$ (then $g(r) > 0$), $\lambda(r)$ satisfies (22) if and only if $\lambda(r) \in [-\frac{r_2}{g(r)}, +\infty)$. So the multilipier $\lambda(r) = -\frac{r_2}{g(r)}$ which is chosen in (20) is the smallest possible number such that (22) holds.

We now take a sequence of directions $(r_n)_n \subset$ cone C with $r_n = (r_{1n}, r_{2n})$, $r_{2n} = -1$, for all $n \in \mathbb{N}$ and $r_{1n} \to -\infty$ as $n \to +\infty$. Then

$$-\frac{r_{2n}}{g(r_n)} = \frac{1}{r_{1n} + \sqrt{1 + r_{1n}^2}} = \sqrt{1 + r_{1n}^2} - r_{1n} \to +\infty \text{ as } n \to +\infty.$$

(b) The case where $C = \mathbb{R}^2$. The Problem (P3) with $C = \mathbb{R}^2$ was considered in [War91, Example 3.2], [LRW91, Example 3] and [DT03, Example 3.9]. It was proved in [LRW91] that x_0 *is not a Kuhn-Tucker point* of (P3). But it is shown in [DT03] that x_0 *is a directional Kuhn-Tucker point* of (P3). Moreover, similar observations as in the case (a) ((i) and (ii)) still hold. We now show another feature of the directional Kuhn-Tucker condition.

It is possible to search for the candidates for minimizers of (P3) by using the directional Kuhn-Tucker condition. Note that a point x is a directional Kuhn-Tucker point of (P3) if and only if for each $r = (r_1, r_2) \in \mathbb{R}^2$ the following system (linear in variable λ) has at least one solution λ:

$$
\begin{cases}
f'(x, r) + \lambda g'(x, r) \geq 0, \\
\lambda \geq 0, \\
\lambda g(x) = 0.
\end{cases}
\tag{23}
$$

Note also that $g(x) = 0$ iff $x_1 = 0$, $x_2 \geq 0$ or $x_1 \leq 0$, $x_2 = 0$. We consider various possibilities.

(α) If $x = (x_1, x_2) \in \mathbb{R}^2$ such that $g(x) \neq 0$ then λ must be zero and the first inequality in (23) becomes $r_2 \geq 0$ ($\lambda = 0$). This is impossible for all $r = (r_1, r_2) \in \mathbb{R}^2$.

(β) If $x = (x_1, x_2) \in \mathbb{R}^2$ with $x_1 = 0$, $x_2 > 0$ then $g(x) = 0$. Some elementary calculation gives $g'(x, r) = r_1$, $f'(x, r) = r_2$. The system (23) becomes

$$
\begin{cases}
r_2 + \lambda r_1 \geq 0, \\
\lambda \geq 0,
\end{cases}
$$

which has no solution λ when $r_1 < 0$ and $r_2 < 0$.

(γ) If $x = (x_1, x_2) \in \mathbb{R}^2$ with $x_1 < 0$, $x_2 = 0$ then $g(x) = 0$. Take $r = (r_1, r_2) \in \mathbb{R}^2$, $r_2 < 0$ then we get $f'(x, r) = r_2$ and $g'(x, r) = 0$. In this case, (23) is equivalent to

$$
\begin{cases}
r_2 + \lambda 0 \geq 0, \\
\lambda \geq 0,
\end{cases}
$$

which has no solution.

Therefore, every point $x \in \mathbb{R}^2 \setminus \{(0, 0)\}$ fails to be a directional Kuhn-Tucker point of (P3). As it is already known that $x_0 = (0, 0)$ is a directional Kuhn-Tucker point of (P3) and so it is a minimizer of (P3).

It is worth noting that in this case ($C = \mathbb{R}^2$), x_0 is not the unique solution of (P3). In fact, all points of the form $(x, 0)$ where $x \leq 0$ are solutions of (P3). However, these points, except $x_0 = (0, 0)$, are not directional Kuhn-Tucker points of (P3). This happens since (P3) does not satisfy regularity conditions stated in Theorem 2. This means that even for non-regular problems the directional Kuhn-Tucker condition can be used to find out solutions satisfying this condition (if any).

4 Directionally Differentiable Problems with DSL-approximates

In this section we will give some extension of the framework to some larger classes of problems. Namely, the class of problems for which the objective function and the functions appeared in the inequality constraints possess some upper DSL-approximates (in the sense introduced in [Sha86], [MW90]) at the minimum point. Let X be a Banach space.

Definition 6. *[Sha86] A function* $h : X \longrightarrow \mathbb{R}$ *is called a DSL-function if* h *is a difference of two sublinear functions. That is, there exist* $p, q : X \longrightarrow \mathbb{R}$ *which are sublinear and such that* $h(x) = p(x) - q(x)$ *for all* $x \in X$.

Note that a DSL-function can be represented in the form

$$h(x) = \max_{a \in A} \langle x, a \rangle + \min_{b \in B} \langle x, b \rangle, \ \forall x \in X. \tag{24}$$

where A, B are convex, compact subsets of X. Obviously, h is quasidifferentiable at 0 (the origin in X) and one can take $Dh(0) := [\underline{\partial}h(0), \overline{\partial}h(0)] = [A, B]$ (see the definition of $Dh(0)$ in section 3.3, and note that $Dh(0)$ is not uniquely defined).

Definition 7. *[Sha86] Let* $g : X \longrightarrow \mathbb{R}$ *be directionally differentiable at* x_0. *A function* $\phi : X \longrightarrow \mathbb{R}$ *is said to be an upper DSL-approximate of* g *at* x_0 *if* ϕ *is a DSL-function and if*

$$g'(x_0, x) \leq \phi(x), \ \forall x \in X. \tag{25}$$

Suppose now that X is a real Banach space and $g : X \longrightarrow \mathbb{R} \cup \{+\infty\}$ is directionally differentiable at x_0.

Consider the Problem (P) in Section 1 with $C = X$. As usual, let S be the feasible set of (P). Assume that $f, g_i, i \in I$ are directionally differentiable at $x_0 \in S$. Moreover, let f and g_i possess upper DSL-approximates $\phi, \phi_i, i \in I$, at x_0, respectively.

Note that each ϕ, ϕ_i has the form (24) and hence, for each $\xi \in X$, we can construct the functions $\Phi^\xi, \Phi_i^\xi, i \in I$, as in (18) (with A, B in (24) playing the role of $\underline{\partial}f(x_0), \overline{\partial}f(x_0)$). These functions are upper approximates of f and g_i, respectively. Moreover,

$$\phi(\xi) = \Phi^\xi(\xi); \quad \phi(x) \le \Phi^\xi(x), \; \forall x \in X$$
$$\phi_i(\xi) = \Phi_i^\xi(\xi); \quad \phi_i(x) \le \Phi_i^\xi(x), \; \forall x \in X, \; \forall i \in I. \tag{26}$$

Theorem 9. *Assume that (P) is (CQ2) regular. If x_0 is a minimizer of (P) then*

$$\forall r \in X \; \exists \, (\lambda_1, \cdots, \lambda_m) \in \mathbb{R}_+^m \; satisfying$$
$$\phi(r) + \textstyle\sum_{i \in I} \lambda_i \phi_i(r) \ge 0, \quad \lambda_i g_i(x_0) = 0, \quad \forall i \in I. \tag{27}$$

Theorem 10. *If $x_0 \in S$ satisfies (27) for some upper DSL-approximates ϕ, ϕ_i of f and g_i (at x_0) and if (P) is invex with respect to ϕ, ϕ_i, $i \in I(x_0)$ then x_0 is a global minimizer of (P).*

The proof of Theorem 10 is the same as that of Theorem 3 with ϕ, ϕ_i playing the role of $f'(x_0, .)$, $g_i'(x_0, .)$, $i \in I(x_0)$, respectively.

Proof. (for Theorem 9.) We follow almost the same argument as in the proof of Theorem 2 under the assumption (b).

Fix $r \in X$. Since x_0 is a minimizer of (P), the following system of variable $\xi \in X$ is inconsistent:

$$f'(x_0, \xi) < 0, \quad g_i'(x_0, \xi) < 0, \quad \forall i \in I(x_0). \tag{28}$$

Take Φ^r, Φ_i^r, $i \in I(x_0)$ to be the functions with the property (26) and with $\xi = r$. Lemma 1 then ensures the existence of h and h_i which are upper approximates of f and g_i, $i \in I(x_0)$, respectively, such that for all $x \in X$,

$$\begin{cases} h(x) \le \min\{\Phi^r(x), (f')^\infty(x_0, x)\}, \\ h_i(x) \le \min\{\Phi_i^r(x), (g_i')^\infty(x_0, x)\}, & \forall i \in I(x_0). \end{cases} \tag{29}$$

It follows from the inconsistency of (28) and the definition of upper appriximate functions that

$$h(x) < 0, \quad h_i(x) < 0, \; i \in I(x_0)$$

is inconsistent. In turn, Gordan's theorem leads to the existence of $\lambda_0 \ge 0$, $\lambda_i \ge 0$, $i \in I(x_0)$, not all zero, such that

$$\lambda_0 h(x) + \sum_{i \in I(x_0)} \lambda_i h_i(x) \ge 0, \; \forall x \in X. \tag{30}$$

If $\lambda_0 = 0$ then by (30), $\sum_{i \in I(x_0)} \lambda_i h_i(x) \ge 0$, for all $x \in X$. This is impossible because of (CQ2), (29) and the fact that λ_i, $i \in I(x_0)$ are nonnegative, not all zero. Therefore, $\lambda_0 \ne 0$ (take $\lambda_0 = 1$). We get from (30) for $x = r$,

$$h(r) + \sum_{i \in I(x_0)} \lambda_i h_i(r) \ge 0.$$

Combining this, (29), and (26), we get

$$\phi(r) + \sum_{i \in I(x_0)} \lambda_i \phi_i(r) \geq 0.$$

Then (27) follows by setting $\lambda_i = 0$ with $i \notin I(x_0)$.   □

We now show the relation between our results and the results in [Sha86]. In [Sha86] the author considered a problem with equality and inequality constraints but here we ignore the equality constraints. In [Sha86], the author considered the Problem (P) with $C = X = \mathbb{R}^n$, f and g_i, $i \in I$ are locally Lipschitz at point $x_0 \in S$ (S is the feasible set of (P)). The upper Dini directional derivative of a (locally Lipschitz) function g at x_0, denoted by $g^+(x_0, .)$. The upper DSL-approximate of a locally Lipschitz g was defined as in Definition 7 with $g'(x_0, x)$ was replaced by $g^+(x_0, x)$ in (25). Suppose that ϕ, ϕ_i are upper DSL-approximates of f, g_i, $i \in I$ (respectively) at x_0. It was established in [Sha86] that under the so-called *"nondegeneracy condition"* (regularity condition) with respect to ϕ_i, $i \in I(x_0)$:

$$\mathrm{cl}\ \{y \mid \phi_i(y) < 0, \forall i \in I(x_0)\} = \{y \mid \phi_i(y) \leq 0, \forall i \in I(x_0)\},$$

the following is necessary for x_0 to be a local minimizer of (P):

$$-\overline{\partial}\phi(0) \subset \bigcup_{\substack{\overline{w}_i \in \overline{\partial}\phi_i(0) \\ i \in I(x_0)}} \left[\underline{\partial}\phi(0) + \mathrm{cone} \bigcup_{i \in I(x_0)} \left(\underline{\partial}\phi_i(0) + \overline{w}_i \right) \right]. \tag{31}$$

Note that in (31) the inclusion holds for the quasidifferentials of upper DSL-approximates of f and g_i instead of those of f and g_i themselves as in (19). Note also that (31) can be found in [MW90] (as a special case) where it was proved under (CQ2) regular condition. The relation between the necessary optimaity conditions (31) and (27) is established below.

Theorem 11. *(31) implies (27).*

Proof. We first note that

$$\mathrm{cone} \bigcup_{i \in I(x_0)} \left(\underline{\partial}\phi_i(0) + \overline{w}_i \right) = \sum_{i \in I(x_0)} \mathrm{cone} \left(\underline{\partial}\phi_i(0) + \overline{w}_i \right).$$

Hence, (26) can be rewritten in the form

$$-\overline{\partial}\phi(0) \subset \bigcup_{\substack{\overline{w}_i \in \overline{\partial}\phi_i(0) \\ i \in I(x_0)}} \left[\underline{\partial}\phi(0) + \sum_{i \in I(x_0)} \mathrm{cone} \left(\underline{\partial}\phi_i(0) + \overline{w}_i \right) \right]. \tag{32}$$

Suppose (32) holds and r is an arbitrary point of X. Take

$$\overline{v} \in \operatorname{argmin}_{\xi \in \overline{\partial}\phi(0)} \langle r, \xi \rangle, \quad \overline{v}_i \in \operatorname{argmin}_{\xi_i \in \overline{\partial}\phi_i(0)} \langle r, \xi_i \rangle, \quad i \in I(x_0). \tag{33}$$

Then (32) implies that

$$0 \in \underline{\partial}\phi(0) + \overline{v} + \sum_{i \in I(x_0)} \operatorname{cone}\left(\underline{\partial}\phi_i(0) + \overline{v}_i\right).$$

This ensures the existence of $a \in \underline{\partial}\phi(0)$, $b_i \in \underline{\partial}\phi_i(0)$, and $\lambda_i \geq 0$, $i \in I(x_0)$ such that

$$0 = a + \overline{v} + \sum_{i \in I(x_0)} \lambda_i(b_i + \overline{v}_i).$$

Combining this and (33) we get

$$\begin{aligned}
\phi(r) + \sum_{i \in I(x_0)} \lambda_i \phi_i(r) &= \max_{v \in \underline{\partial}\phi(0)} \langle r, v \rangle + \min_{w \in \overline{\partial}\phi(0)} \langle r, w \rangle \\
&\quad + \sum_{i \in I(x_0)} \lambda_i \Big[\max_{\xi_i \in \underline{\partial}\phi_i(0)} \langle r, \xi_i \rangle + \min_{\eta_i \in \overline{\partial}\phi_i(0)} \langle r, \eta_i \rangle \Big] \\
&= \max_{v \in \underline{\partial}\phi(x_0)} \langle r, v \rangle + \langle r, \overline{v} \rangle \\
&\quad + \sum_{i \in I(x_0)} \lambda_i \Big[\max_{\xi_i \in \underline{\partial}\phi_i(0)} \langle r, \xi_i \rangle + \langle r, \overline{v}_i \rangle \Big] \\
&\geq \langle r, a \rangle + \langle r, \overline{v} \rangle + \sum_{i \in I(x_0)} \lambda_i \big[\langle r, b_i \rangle + \langle r, \overline{v}_i \rangle \big] \\
&\geq \langle r, a + \overline{v} + \sum_{i \in I(x_0)} \lambda_i (b_i + \overline{v}_i) \rangle \\
&\geq 0.
\end{aligned}$$

Set $\lambda_i = 0$ for $i \notin I(x_0)$. Then (28) holds since $r \in X$ is arbitrary. $\qquad \square$

Acknowledgement

The authors would like to thank the referees whose comments improved the paper. Work of the first author was supported partly by the project "Rought Analysis - Theory and Applications", Institute of Mathematics, Vietnam Academy of Science and Technology, Vietnam, and by the APEC postdoctoral Fellowship from the KOSEF, Korea. The second author was supported by the Brain Korea 21 Project in 2003.

References

[BRS83] Brandão, A.J.V., Rojas-Medar, M.A., Silva, G.N.: Invex nonsmooth alternative theorems and applications. Optimization, 48, 230–253 (2000)

[Cla83]   Clarke, F.H.: Optimization and Nonsmooth Analysis. Wiley, New York (1983)

[Cra81]   Craven, B.D.: Invex functions and constrained local minima. Bull. Austral. Math. Soc., **24**, 357 – 366 (1981)

[Cra86]   Craven, B.D.: Nondifferentiable optimization by smooth approximations. Optimization, **17**, 3–17 (1986)

[Cra00]   Craven, B.D.: Lagrange Multipliers for Nonconvex Optimization. Progress in Optimization. Kluwer Academic Publishers (2000)

[DJ97]   Demyanov, V.F., Jeyakumar, V.: Hunting for a smaller convex subdifferential. J. Global Optimization, **10**, 305–326 (1997)

[DPR86]   Demyanov, V.F., Polyakova, L.N., Rubinov, A.M.: Nonsmoothness and quasidifferentiability. Mathematical Programming Study **29**, 1–19 (1986)

[DR80]   Demyanov, V.F., Rubinov, A.M.: On quasidifferentiable functionals. Dokl. Acad. Sci. USSR, **250**, 21–25 (1980) (in Russian)

[DT03]   Dinh, N., Tuan, L.A.: Directional Kuhn-Tucker conditions and duality for quasidifferentiable programs. Acta Mathematica Vietnamica, **28**, 17 – 38 (2003)

[DV81]   Demyanov, V.F., Vasiliev, L.V.: Nondifferentiable optimization. Nauka, Moscow (1981) (in Russian).

[EL87]   Eppler, K., Luderer, B.: The Lagrange principle and quasidifferent calculus. Wiss. Z. Techn. Univ. Karl-Marx-Stadt., **29**, 187–192 (1987)

[Gao00]   Gao, Y.: Demyanov difference of two sets and optimality conditions of Lagrange multiplier type for constrained quasidifferentiable optimization. Journal of Optimization Theory and Applications, **104**, 177–194 (2000)

[Gl92]   Glover, B.M.: On quasidifferentiable functions and non-differentiable programming. Optimization, **24**, 253–268 (1992)

[Han81]   Hanson, M.A.: On sufficiency of the Kuhn-Tucker conditions. J. Math. Anal. Appl. **80**, 545–550 (1981)

[HK82]   Hayashi, M., Komiya, H.: Perfect duality for convexlike programs. Journal of Optimization Theory and Applications, **38**, 179–189 (1982)

[IT79]   Ioffe, A.D., Tikhomirov, V.M.: Theory of extremal problems. North-Holland, Amsterdam (1979)

[Jey85]   Jeyakumar, V.: Convexlike alternative theorems and mathematical programming. Optimization, **16**, 643–652 (1985)

[Jey91]   Jeyakumar, V.: Composite nonsmooth programming with Gâteaux differentiability. SIAM J. Optimization, **1** , 30–41 (1991)

[LRW91]   Luderer, B., Rosiger, R., Wurker, U.: On necessary minimum conditions in quasidifferential calculus: independence of the specific choice of quasidifferentials. Optimization, **22**, 643–660 (1991)

[Man94]   Mangasarian, O.L.: Nonlinear Programming. SIAM, Philadelphia (1994)

[MW90]   Merkovsky, R.R., Ward, D.E.: Upper DSL approximates and nonsmooth optimization. Optimization, **21**, 163–177 (1990)

[Sac00]   Sach, P.H.: Martin's results for quasidifferentiable programs (Draft) (2000)

[Sac02]   Sach, P.H.: Nonconvex alternative theorems and multiobjective optimization. Proceedings of the Korea-Vietnam Joint seminar: Mathematical Optimization Theory and Applications. November 30 - December 2, 2002. Pusan, Korea (2002)

[SKL00]   Sach, P.H., Kim, D.S., Lee, G.M.: Invexity as a necessary optimality condition in nonsmooth programs. Preprint 2000/30, Institute of Mathematics, Hanoi (2000)

[SLK03] Sach, P.H., Lee, G.M., Kim, D.S.: Infine functions, nonsmooth alternative theorems and vector optimization problems. J. Global Optimization, **27**, 51–81 (2003)

[Sha84] Shapiro, A.: On optimality conditions in quasidifferentiable optimization. SIAM J. Control and Optimization, **22**, 610–617 (1984)

[Sha86] Shapiro, A.: Quasidifferential calculus and first-order optimality conditions in nonsmooth optimization. Mathematical Programming Study, **29**, 56–68 (1986)

[War91] Ward, D.E.: A constraint qualification in quasidifferentiable programming. Optimization, **22**, 661–668 (1991)

[YS93]   Yen, N.D., Sach, P.H.: On locally Lipschitz vector-valued Invex functions. Bull. Austral. Math. Soc., **47**, 259–271 (1993)

Slice Convergence of Sums of Convex functions in Banach Spaces and Saddle Point Convergence

Robert Wenczel and Andrew Eberhard

Department of Mathematics
Royal Melbourne University of Technology
Melbourne, VIC 3001, Australia
robert.wenczel@rmit.edu.au, andy.eb@rmit.edu.au

Summary. In this note we provide various conditions under which the slice convergence of $f_v \to f$ and $g_v \to g$ implies that of $f_v + g_v$ to $f + g$, where $\{f_v\}_{v \in W}$ and $\{g_v\}_{v \in W}$ are parametrized families of closed, proper, convex function in a general Banach space X. This 'sum theorem' complements a result found in [EW00] for the epidistance convergence of sums. It also provides an alternative approach to the derivation of *some* of the results recently proved in [Zal03] for slice convergence in the case when the spaces are Banach spaces. We apply these results to the problem of convergence of saddle points associated with Fenchel duality of slice convergent families of functions.

2000 MR Subject Classification. Primary 49J52, 47N10; Secondary 46A20, 52A27

Key words: slice convergence, Young–Fenchel duality

1 Introduction

In this paper we provide alternative proofs of some recent results of Zalinescu [Zal03]. Some hold for the case when the underlying spaces are general Banach spaces and others only require the spaces to be normed linear. The paper [Zal03] was originally motivated by [WE99] and extended the results of this paper to the context of normed space and to the convergence of marginal or perturbation functions (rather than just sums of convex functions). In this paper we clarify to what degree we are able to deduce such results from the work of [EW00, WE99] by either modifications of the proofs of [WE99] or short deduction using the methods of [EW00, WE99].

The first results give conditions under which slice convergence of a sum $\{f_v + g_v\}_{v \in W}$ follows from the slice convergence of the two parametrized families $\{f_v\}_{v \in W}$ and $\{g_v\}_{v \in W}$. This result has a counterpart for epi–distance convergence which was proved by the authors in [EW00] and we refer to such results as sum theorems. We show that in the particular case of Banach spaces the corresponding result for slice convergence follows easily from the work in [WE99] and moreover so do the corresponding results for the so-called marginal or perturbation functions used to study duality of convex optimization problems which are studied in [Zal03]. Such results only hold under certain conditions which we will refer to as qualification assumptions due to their similarity (and connections) to constraint qualifications in convex optimization problems. The approach here is more aligned with that of [AR96] were the sum theorem is the primary point of departure.

The marginal or perturbation function is given by $h(y) := \inf_{x \in X} F(x, y)$ from which the primal (convex) problem corresponds to $h(0)$ and the dual problem corresponds to $-h^{**}(0) = \inf_{y^* \in Y^*} F^*(0, y^*) = \inf_{y^* \in Y^*} h^*(y^*)$. This leads to the consideration of the dual perturbation function $k(x^*) := \inf_{y^* \in Y^*} F^*(x^*, y^*)$ (see [Roc74, ET99]) and the consideration of the closedness and properness of $h(y)$ at $y = 0$. Letting $F, F_i \in \Gamma(X \times Y)$ $(i \in I)$ then as a framework for the study of stability of optimization problem one may study the variational convergence of $\{F_i(\cdot, 0)\}_{i \in I}$ to $F(\cdot, 0)$ and $\{F_i^*(0, \cdot)\}_{i \in I}$ to $F^*(0, \cdot)$ (see for example [AR96, Zal03]). Clearly this analysis is greatly facilitated when the variational convergence under consideration is generated by a topology for which the Fenchel conjugate is bi–continuous. Thus typically the so–called slice and epi–distance topologies are usually considered as we will also do in this paper. Once this is enforced the generality of this formulation allows one to obtain the sum theorem alluded to in the beginning of this introduction as well as many other stability results with respect to other operations on convex functions and sets (which preserve convexity). In this way the study of perturbation functions appears to be more general than the study of any one single operation (say, addition) of convex functions. Indeed this is only partly true in that when all spaces considered are Banach and the constraint qualification is imposed on the primal functions we will show that the slice stability of the perturbation function follows easily from sum theorems. When the qualification assumption is placed on the dual function we are able to deduce the main result in this direction of [Zal03] in a straightforward manner when all spaces are only normed (possibly not complete) linear spaces. It is also possible to treat the upper and lower slice (respectively, epi–distance) convergences separately as is done in [Zal03, Pen93, Pen02] and in part in [WE99]. There is an economy of statement gained by avoiding this and it will also avoid us reworking results in previously published papers. Consequently we will not do so in this paper.

Convex-concave bivariate functions are related to convex bivariate functions through partial conjugation (i.e. conjugation with respect to one of the variables). In this context we are led to the introduction of equivalence classes of saddle–functions which are uniquely associated with concave or convex parents (depending on the which variable is partially conjugated). Two bivariate functions are said to belong to the same equivalence class if they have the same convex and concave parents. Such members of the same equivalence class not only have the same saddle–point but so do all linear perturbations of these two functions. Thus when discussing the variational convergence of saddle–functions one is necessarily led to the study of the convergence of the equivalence class. We investigate saddle–point convergence of the associated saddle function. This allows one to investigate the convergence of approximate solutions of the perturbed Fenchel primal and dual optimization problems to solutions of the limiting problem. It may be shown that one can quite generally deduce the existence of an accumulation point of the approximating dual solutions.

2 Preliminaries

In this section we draw together a number of results and definitions. This is done to make the development self–contained. A reader conversant with set–convergence notions and the infimal convolution need only read the first part of this section, only returning to consult results and definitions as needed. A useful reference for much of the material of this section is [Bee93].

We will let $C(X)$ stand for the class of all nonempty closed convex subsets of a normed space X and $CB(X)$ the closed bounded convex sets. Place $d(a, B) = \inf\{ \|a - b\| \mid b \in B \}$, and $B_\rho = \{x \in X \mid \|x\| \le \rho\}$. Corresponding balls in the dual space X^* will be denoted B_ρ^*. The indicator function of a set A will be denoted δ_A, and $S(A, \cdot)$ shall denote the support function. We will use u.s.c. to denote upper–semicontinuity and l.s.c. to denote lower–semicontinuity. Recall that a function $f : X \to \overline{\mathbf{R}}$ is called closed, proper convex on X if and only if f is convex, l.s.c., is never $-\infty$, and not identically $+\infty$. The class of all closed proper convex functions on X is denoted by $\Gamma(X)$, and $\Gamma^*(X^*)$ denotes the class of all weak* closed proper convex functions on X^*. We shall use the notation $\overline{A}$ for the closure of a set A in a topological space (Z, τ) and, to emphasise the topology, we may write $\overline{A}^\tau$. For $x \in Z$, $\mathcal{N}_\tau(x)$ denotes the collection of all τ–neighborhoods of x. For a function $f : Z \to \overline{\mathbf{R}}$, the epigraph of f, denoted epi f, is the set $\{(x, \alpha) \in Z \times \mathbf{R} \mid f(x) \le \alpha\}$, and the strict epigraph epi$_s f$ is the set $\{(x, \alpha) \in Z \times \mathbf{R} \mid f(x) < \alpha\}$. The domain, denoted dom f is the set $\{x \in Z \mid f(x) < +\infty\}$. The (sub–)level set $\{x \in Z \mid f(x) \le \alpha\}$ (where $\alpha > \inf_Z f$) will be given the abbreviation $\{f \le \alpha\}$. Any product $X \times Y$ of normed spaces will always be understood to be endowed with the box norm $\|(x, y)\| = \max\{\|x\|, \|y\|\}$; any balls in such product spaces will always be with respect to the box norm. The natural projections from $X \times Y$ to X or Y will be denoted by P_X and P_Y respectively. We also will assume the following convention for products $Z \times \mathbf{R}$ where (Z, τ) is topological: We assume the product topology, where $\mathbf{R}$ has the usual topology, and for any subset

$C \subseteq Z \times \mathbf{R}$, its closure in this topology is written as $\overline{C}^{\tau}$. If $f : (Z, \tau) \to \overline{\mathbf{R}}$, its τ-l.s.c. hull, denoted $\overline{f}^{\tau}$, is defined by $\overline{f}^{\tau}(x) = \liminf_{x' \xrightarrow{\tau} x} f(x')$. The (extended) lower closure $\underline{\mathrm{cl}}_{\tau} f$ is defined to coincide with $\overline{f}^{\tau}$ if the latter does not take the value $-\infty$ anywhere, and to be identically $-\infty$ otherwise.

Definition 1. *Let $F{:}W \to 2^X$ be a multifunction from topological spaces W to X.*

1. $\limsup_{v \to w} F(v) = \bigcap_{V \in \mathcal{N}(w)} \overline{\bigcup_{v \in V} F(v)}$.
2. $\liminf_{v \to w} F(v) = \bigcap_{\{B \subseteq W | w \in \overline{B}\}} \overline{\bigcup_{v \in B} F(v)}$.
3. $F(\cdot)$ *is lower–semicontinuous at w iff $F(w) \subseteq \liminf_{v \to w} F(v)$.*

Remark 1. It is easily seen that this notion of lower–semicontinuity is equivalent to the classical formulation—namely: For any open set U intersecting $F(w)$ there is a neighborhood V of w for which $F(v) \cap U$ is nonempty for every v in V.

Remark 2. For metrizable X, the above definitions can be shown to have the equivalent forms:

1.
$$\limsup_{v \to w} F(v)$$
$$= \{x \in X \mid \exists \text{ a net } v_\beta \to w \text{ and } x_\beta \in F(v_\beta) \text{ with } x_\beta \to x\}$$
$$= \{x \in X \mid \liminf_{v \to w} d(x, F(v)) = 0\}$$

2.
$$\liminf_{v \to w} F(v)$$
$$= \{x \in X \mid \forall \text{ nets } v_\beta \to w, \exists x_\beta \to x \text{ with } x_\beta \in F(v_\beta) \text{ eventually}\}$$
$$= \{x \in X \mid \limsup_{v \to w} d(x, F(v)) = 0\}$$

with obvious analogs for nets of sets.

Definition 2. *Let A be a convex set in a topological vector space and $x \in A$. Then $\operatorname{cone} A := \cup_{\lambda > 0} \lambda A$ (the smallest convex cone containing A).*

The infimal convolution plays a central role in our development.

Definition 3. *Let f and g be closed convex functions on X into the extended reals. Then*
$$(f \Box g)(x) := \inf_{y \in X} (f(y) + g(x - y))$$
is called the inf-convolution.

It is well known that the strict epigraph of the inf–convolution is equal to the set–addition of the strict epigraphs of the individual functions:

$$\operatorname{epi}_s(f \square g) = \operatorname{epi}_s f + \operatorname{epi}_s g.$$

Also $\operatorname{dom}(f \square g) = \operatorname{dom} f + \operatorname{dom} g$; $\operatorname{epi} f \square g \supseteq \operatorname{epi} f + \operatorname{epi} g$, and

$$(f \square g)^* = f^* + g^*$$

where $f^*(x^*) = \sup_{x \in X}(\langle x, x^* \rangle - f(x))$ is the Young–Fenchel conjugate of f.

Lower semi–continuity of the epi-graphical multi-function $v \mapsto \operatorname{epi}_s(f_v \square g_v)$ may be deduced from that of its components using the following lemma, a proof of which may be found in [WE99].

Lemma 1. *If $F_1(\cdot)$ and $F_2(\cdot)$ are multi–functions l.s.c. at w then $F(v) := F_1(v) + F_2(v)$ is l.s.c. at w.*

We conclude this section with a summary of variational limit notions used in this paper. Let X and W be topological spaces, then for $x \in X$, $w \in W$, and $\{f_v\}_{v \in W}$ a collection of $\overline{\mathbf{R}}$-valued functions on X, define the lower and upper epi–limits by:

$$(\text{e-li}_{v \to w} f_v)(x) := \sup_{U \in \mathcal{N}(x)} \sup_{V \in \mathcal{N}(w)} \inf_{v \in V} \inf_{y \in U} f_v(y),$$

$$(\text{e-ls}_{v \to w} f_v)(x) := \sup_{U \in \mathcal{N}(x)} \inf_{V \in \mathcal{N}(w)} \sup_{v \in V} \inf_{y \in U} f_v(y).$$

It is well known [RW84] that these limits correspond to the Kuratowski(–Painlevé) limit of the epi-graph multifunction in the sense that

$$\operatorname{epi}(\text{e-ls}_{v \to w} f_v) = \liminf_{v \to w} \operatorname{epi} f_v,$$

$$\operatorname{epi}(\text{e-li}_{v \to w} f_v) = \limsup_{v \to w} \operatorname{epi} f_v. \tag{1}$$

These definitions and relations have natural counterparts for nets $\{f_\gamma\}_{\gamma \in I}$ of functions.

Definition 4. *Let $\{f_v\}_{v \in W}$ be a family of functions and τ a topology on X. We say that $\{f_v\}_{v \in W}$ is τ-epi–u.s.c. at $w \in W$ if for all x we have*

$$(\tau\text{-e-ls}_{v \to w} f_v)(x) \leq f_w(x)$$

and τ-epi–l.s.c. if for all x

$$f_w(x) \leq (\tau\text{-e-li}_{v \to w} f_v)(x)$$

where the epi–limits are taken with respect to the underlying topology τ.

We will say that $\{f_v\}_{v \in W}$ is strongly epi–u.s.c. when τ corresponds to the strong (norm) topology on X. In this case we will drop the reference to τ. Thus for an epi–u.s.c. family the epi–graphs of f_v are lower Kuratowski–convergent to epi f_w in the (strong) norm topology.

Definition 5. *A family of functions* $\{f_v\}_{v \in W}$ *in* $\overline{\mathbf{R}}^X$ *is epi–convergent to a function* f_w *(as* $v \to w$*) if it is both epi–u.s.c. and epi–l.s.c. at* w.

Since e-li$_{v \to w} f_v \leq$ e-ls$_{v \to w} f_v$ on X, the relation defining epi–convergence is in fact an equality.

Definition 6. *Let* $\{f_v\}_{v \in W}$ *be a family of functions on* X *and* $\{f_v^*\}_{v \in W}$ *the family of conjugate functions on* X^* *(for a normed space* X*). We denote the bounded–weak* upper epi-limit (as* $v \to w$*) of* $\{f_v^*\}_{v \in W}$ *by*

$$bw^*\text{-}\limsup_{v \to w} \text{epi } f_v^* := \{(x^*, \alpha) \in X^* \times \mathbf{R} \mid \exists \text{ nets } v_\beta \to w; \ (y_\beta^*, \alpha_\beta) \in \text{epi } f_{v_\beta}^*$$

$$\text{such that } \alpha_\beta \to \alpha; \ y_\beta^* \text{ norm bounded}; \ y_\beta^* \xrightarrow{w^*} x^*\}.$$

The above closely resembles the limit–superior of epigraphs, relative to the bounded–weak* topology on X^* (hence the terminology). The bounded–weak* topology is described in, for example, [Hol75]. For a family of sets $\{F(v)\}_{v \in W}$ we will also say that it is bw^*–upper–semicontinuous (at w) whenever $F(w) \supseteq$ bw^*-$\limsup_{v \to w} F(v)$

Definition 7. *[Bee92, Bee93] We say* $\{f_v\}_{v \in W}$ *in* $\Gamma(X)$ *is upper slice convergent to* $f \in \Gamma(X)$ *(as* $v \to w$*) if whenever* $v_\alpha \to w$ *is a convergent net and* $\{x_\alpha\}$ *a bounded net in* X *we have for each* $(y^*, \eta) \in \text{epi}_s f^*$ *that* $f_{v_\alpha}(x_\alpha) > \langle x_\alpha, y^* \rangle - \eta$ *eventually. If we also have that* $f_w \geq$ e-ls$_v f_v$*, then* f_v *is said to slice converge to* f_w.

A dual slice convergence on $\Gamma^*(X^*)$ may be defined, which ensures the bicontinuity of Fenchel conjugation. For our purposes, we work with an equivalent definition of dual slice convergence, as contained in the proposition to follow.

Again, analogous definitions follow for nets of functions. The following characterization of slice convergence is essentially contained in [WE99, Cor. 3.6].

Proposition 1. *For functions* $f_v \in \Gamma(X)$, f_v *slice-converges to* f_w *if and only if*

$$bw^*\text{-}\limsup_{v \to w} \text{epi } f_v^* \subseteq \text{epi } f_w^* \subseteq s^*\text{-}\liminf_{v \to w} \text{epi } f_v^*,$$

where s^* *denotes the norm topology on* X^*.

Note that this result gives a characterisation of dual slice convergence for the conjugate functions in $\Gamma^*(X^*)$.

From [Hol75] we have the following. Recall that a set A in a topological linear space X is *ideally convex* if for any bounded sequence $\{x_n\} \subseteq A$ and $\{\lambda_n\}$ of nonnegative numbers with $\sum_{n=1}^{\infty} \lambda_n = 1$, the series $\sum_{n=1}^{\infty} \lambda_n x_n$ either converges to an element of A, or else does not converge at all. Open or closed convex sets are ideally convex, as is any finite–dimensional convex set. In particular, if X is Banach, then such series always converge, and the definition of ideal convexity only requires that $\sum_{n=1}^{\infty} \lambda_n x_n$ be in A. From [Hol75, Section 17E] we have

Proposition 2. *For a Banach space X,*

1. *If $C \subseteq X$ is closed convex, it is ideally convex.*
2. *For ideally convex C, $\operatorname{int} \overline{C} = \operatorname{int} C$.*
3. *If A and B are ideally convex subsets of X, one of which is bounded, then $A - B$ is ideally convex.*

Proof. We prove the last assertion only; the rest can be found in the cited reference. Let $\{a_n - b_n\} \subseteq A - B$ be a bounded sequence, let $\lambda_n \geq 0$ be such that $\sum_{n=1}^{\infty} \lambda_n = 1$. Then $\{a_n\} \subseteq A$ and $\{b_n\} \subseteq B$ are both bounded, so $\sum_{n=1}^{\infty} \lambda_n a_n \in A$ and $\sum_{n=1}^{\infty} \lambda_n b_n \in B$ (both convergent). Thus $\sum_{n=1}^{\infty} \lambda_n (a_n - b_n) = \sum_{n=1}^{\infty} \lambda_n a_n - \sum_{n=1}^{\infty} \lambda_n b_n \in A - B$. $\qquad\square$

3 A Sum Theorem for Slice Convergence

We will now discuss the passage of slice convergence through addition. Such theorems will hereafter be referred to as sum theorems. In [WE99] was proved a sum theorem for slice convergence of $f_n + g_n$ (for convergent f_n, g_n) under the rather restrictive condition that the conjugates g_n^* have domains uniformly contained in a weak* locally compact cone. (This hypothesis arose from an attempt to derive a sufficient condition that acts on only *one* of the summands, whereas most such conditions are symmetric in both f_n and g_n.) In the normed-space context, [Zal03, Prop. 25 or Prop. 13] yields an extension of the results of [WE99], using a constraint-qualification more in the spirit of those usually appearing in sum theorems for variational convergences (for instance, in [AP90, Pen93, EW00]). In this Section, we show that in the Banach space context, the cited results of [Zal03] may also be derived using a slight modification of arguments appearing in [WE99].

Definition 8. *Following [Att86], define for $K \in \mathbf{R}$, and for functions f_v, g_v $(v \in W)$,*

$$H_K(X^*, v) := \{(x^*, y^*) \in X^* \times X^* \mid f_v^*(x^*) + g_v^*(y^*) \leq K, \ \|x^* + y^*\| \leq K\}.$$

We shall also need the related object in $X_v := \operatorname{span}(\operatorname{dom} f_v - \operatorname{dom} g_v)$ given by

Definition 9.

$$H_K(X_v^*, v) := \left\{ (x^*, y^*) \in X_v^* \times X_v^* \left| \begin{array}{c} (f_v|_{X_v})^*(x^*) + (g_v|_{X_v})^*(y^*) \le K, \\ \|x^* + y^*\|_{X_v^*} \le K \end{array} \right. \right\},$$

where the conjugate functions are computed relative to the subspace X_v.

The following lemma from [WE99] provides a criterion for the inf-convolution of conjugate functionals to be weak* lower semicontinuous.

Lemma 2. ([WE99, Lem. 4.1]) Let f_v and g_v be in $\Gamma(X)$ for a Banach space X, such that $H_K(X^*, v)$ is bounded for each $K \in \mathbf{R}$. Then $f_v^* \square g_v^* \in \Gamma^*(X^*)$.

The next lemma is elementary, and its proof will be omitted.

Lemma 3. Let f_v be in $\Gamma(X)$, with f_v slice converging to f_w, and $x_v \to x_w$ in norm, as $v \to w$. Then $f_v(x_v + \cdot)$ slice converges to $f_w(x_w + \cdot)$.

The following three lemmas provide bounds that will be of use in the next theorem.

Lemma 4. Let f_v and g_v be proper closed convex $\overline{\mathbf{R}}$-valued functions with $\operatorname{dom} f_v \subseteq X_v$ and $\operatorname{dom} g_v \subseteq X_v$ for all v in some set V. If, additionally, for some positive ρ, δ,

$$(\forall v \in V) \quad B_\delta \cap X_v \subseteq \{f_v \le \rho\} \cap B_\rho - \{g_v \le \rho\} \cap B_\rho \tag{2}$$

then for each $K > 0$,

$$\sup \left\{ \|(x^*, y^*)\|_{X_v^* \times X_v^*} \mid (x^*, y^*) \in H_K(X_v^*, v), \ v \in V \right\} < +\infty.$$

Proof. For $v \in V$, and $(x^*, y^*) \in H_K(X_v^*, v)$, the Fenchel Inequality gives (since $f_v|_{X_v}$, $f_g|_{X_v}$ are in $\Gamma(X)$)

$$K \ge (f_v|_{X_v})^*(x^*) + (g_v|_{X_v})^*(x^*) \ge \langle x^*, x \rangle + \langle y^*, y \rangle - f_v(x) - g_v(y)$$

for any $x \in \operatorname{dom} f_v$, $y \in \operatorname{dom} g_v$ ($\subseteq X_v$).

Let $\xi \in X_v \cap B_\delta$. From (2), $\xi = x - y$ where $x, y \in B_\rho$, $f_v(x) \le \rho$, $g_v(y) \le \rho$ whence (noting that x and y are in X_v also)

$$K \ge \langle x^*, \xi \rangle + \langle y^* + x^*, y \rangle - 2\rho$$
$$\ge \langle x^*, \xi \rangle - \rho(K + 2),$$

since $\|x^* + y^*\|_{X_v^*} \le K$ and $y \in \operatorname{dom} g_v \subseteq X_v$ with $\|y\| \le \rho$. This yields that $\|x^*\|_{X_v^*} \le \frac{1}{\delta}(K(1 + \rho) + 2\rho)$, from arbitrariness of $\xi \in B_\delta \cap X_v$. Also, $\|y^*\|_{X_v^*} \le \|y^* + x^*\|_{X_v^*} + \|x^*\|_{X_v^*} \le K + \|x^*\|_{X_v^*}$ thus giving a uniform bound on $H_K(X_v^*, v)$ for all v. $\square$

Lemma 5. *([WE99, Lem 4.2]) Let $\{f_v\}_{v \in W}$ be a family of proper closed convex extended–real–valued functions on a normed space X. Suppose that $f_w \geq$ e-ls $_{v \to w} f_v$ on X. Then for each $M > 0$,*

$$(\exists V' \in \mathcal{N}(w))(\exists \mu \in \mathbf{R})(\forall v \in V')(\forall \|x^*\| \leq M)(f_v^*(x^*) \geq \mu). \tag{3}$$

Lemma 6. *Let f_v, g_v be proper closed convex functions in $\Gamma(X)$ ($v \in W$). Suppose that $f_w \geq$ e-ls $_{v \to w} f_v$ on X. Then for any fixed $K > 0$ and $\gamma > 0$, there is a neighborhood V of w and a positive ρ for which*

$$(\forall v \in V)(\forall (x_1^*, x_2^*) \in H_K(X^*, v) \cap B_\gamma^*)(g_v^*(x_2^*) \leq \rho).$$

Proof. Supposing the contrary, there are nets $v_\beta \to w$, $(x_{1_\beta}^*, x_{2_\beta}^*) \in H_K(X^*, v_\beta)$ $\cap B_\gamma$ with $\lim_\beta g_{v_\beta}^*(x_{2_\beta}^*) = +\infty$. It then follows that $\lim_\beta f_{v_\beta}^*(x_{1_\beta}^*) = -\infty$, and since $\|x_{1_\beta}^*\| \leq \gamma$ for all β, we have contradicted the statement of Lemma 5. □

Before proving the first of our main theorems we make the following important observation for latter reference.

Lemma 7. *Let X be a Banach space and f_v and g_v ($v \in W$) be in $\Gamma(X)$. Assume that there exist $\delta > 0$, $\rho > 0$, V a neighborhood of w such that for all $v \in V$ ($v \neq w$)*

$$B_\delta \cap X_v \subseteq \{f_v \leq \rho\} \cap B_\rho - \{g_v \leq \rho\} \cap B_\rho \tag{4}$$

where $X_v := \text{span}(\text{dom } f_v - \text{dom } g_v)$. Then for $v \neq w$ in V we have

$$0 \in \text{int}_{\overline{X_v}}(\{f_v \leq \rho\} \cap B_\rho - \{g_v \leq \rho\} \cap B_\rho). \tag{5}$$

Proof. From the assumptions follows that $\text{dom } f_v \cap \text{dom } g_v$ is nonempty, and

$$(\text{int } B_\delta) \cap \overline{X_v} \subseteq \overline{B_\delta \cap X_v} \subseteq \overline{\{f_v \leq \rho\} \cap B_\rho - \{g_v \leq \rho\} \cap B_\rho}$$
$$= \overline{(\{f_v \leq \rho\} \cap B_\rho - \bar{x}_v) - (\{g_v \leq \rho\} \cap B_\rho - \bar{x}_v)}$$

where $\bar{x}_v$ is any member of $\text{dom } f_v \cap \text{dom } g_v$. Both $\{f_v \leq \rho\} \cap B_\rho - \bar{x}_v$ and $\{g_v \leq \rho\} \cap B_\rho - \bar{x}_v$ are bounded, ideally convex [Hol75] subsets of the Banach space $\overline{X_v}$. Hence, by Proposition 2, $\{f_v \leq \rho\} \cap B_\rho - \{g_v \leq \rho\} \cap B_\rho$ is also ideally convex in $\overline{X_v}$ and has the same interior (in $\overline{X_v}$) as does its $\overline{X_v}$-closure. Thus we obtain (5). □

Theorem 1. *Let X be a Banach space, let f_v and g_v ($v \in W$) be in $\Gamma(X)$, with the slice convergence $f_v \to f_w$ and $g_v \to g_w$. Assume that there exist $\delta > 0$, $\rho > 0$, V a neighborhood of w such that for all $v \in V$ ($v \neq w$) (4) holds. Also, assume that $\overline{f_w^* \square g_w^*}$ is proper and weak* lower-semicontinuous. Then $f_v + g_v$ slice converges to $f_w + g_w$.*

Proof. We use as template the proof of [WE99, Thm. 4.3]. We temporarily append the condition that X_v contain both dom f_v and dom g_v for v in V (and shall remove this later). Observe immediately from (5) that cone $(\text{dom } f_v - \text{dom } g_v)$ coincides with the closed subspace $\overline{X_v}$ so that $f_v^* \square g_v^* \in \Gamma^*(X^*)$ by [Att86, Thm. 1.1] (for $v \neq w$). Again, via the characterization given by Proposition 1 we seek to prove that $f_v^* \square g_v^*$ converges in the dual slice topology to $f_w^* \square g_w^*$. It is straightforward to deduce that $v \mapsto \text{epi } f_v^* \square g_v^*$ is strongly lower-semicontinuous at $v = w$ (see the opening paragraph of the proof of [WE99, Thm. 4.3]). To complete the proof, we require that

$$bw^*\text{-}\limsup_{v \to w} \text{epi } f_v^* \square g_v^* \subseteq \text{epi } \overline{f_w^* \square g_w^*}.$$

Let $(x^*, \alpha) \in bw^*\text{-}\limsup_{v \to w} \text{epi } f_v^* \square g_v^*$. Then there are nets $v_\beta \to w$, $(x_\beta^*, \alpha_\beta) \to^{w^*} (x^*, \alpha)$, and $K > 0$ with $(x_\beta^*, \alpha_\beta) \in B_K^* \cap \text{epi }_s f_{v_\beta}^* \square g_{v_\beta}^*$ for all β. For such β, there is $y_\beta^* \in X^*$ for which $K \geq \alpha_\beta > f_{v_\beta}^*(y_\beta^*) + g_{v_\beta}^*(x_\beta^* - y_\beta^*)$. Place $x_{1_\beta}^* := \widetilde{y_\beta^*|_{X_{v_\beta}}}$, a norm-preserving extension of $y_\beta^*|_{X_{v_\beta}} \in X_{v_\beta}^*$ (obtained, say, by the Hahn-Banach Theorem). Also, define $x_{2_\beta}^* := x_\beta^* - x_{1_\beta}^*$. Then $\|x_{1_\beta}^* + x_{2_\beta}^*\| = \|x_\beta^*\| \leq K$, and

$$
\begin{aligned}
K \geq \alpha_\beta &\geq f_{v_\beta}^*(y_\beta^*) + g_{v_\beta}^*(x_\beta^* - y_\beta^*) \\
&= f_{v_\beta}^*(x_{1_\beta}^*) + g_{v_\beta}^*(x_{2_\beta}^*) \qquad (\text{so } (x_{1_\beta}^*, x_{2_\beta}^*) \in H_K(X^*, v_\beta)) \\
&= (f_{v_\beta}|_{X_{v_\beta}})^*(x_{1_\beta}^*|_{X_{v_\beta}}) + (g_{v_\beta}|_{X_{v_\beta}})^*(x_{2_\beta}^*|_{X_{v_\beta}})
\end{aligned}
$$

(since X_v contains dom f_v and dom g_v, and on X_{v_β} we have $x_{1_\beta}^* \equiv y_\beta^*$ and $x_{2_\beta}^* \equiv x_\beta^* - y_\beta^*$). Thus $(x_{1_\beta}^*, x_{2_\beta}^*) \in H_K(X^*, v_\beta)$ and $(x_{1_\beta}^*|_{X_{v_\beta}}, x_{2_\beta}^*|_{X_{v_\beta}}) \in H_K(X_{v_\beta}^*, v_\beta)$, the latter since $\|x_{1_\beta}^*|_{X_{v_\beta}} + x_{2_\beta}^*|_{X_{v_\beta}}\|_{X_{v_\beta}^*} \leq \|x_{1_\beta}^* + x_{2_\beta}^*\| \leq K$. It now follows from Lemma 4 that $\|x_{1_\beta}^*|_{X_{v_\beta}}\| \leq \gamma'$ eventually in β for some $\gamma' > 0$. Since $x_{1_\beta}^* \in X^*$ is a norm-preserving extension of $x_{1_\beta}^*|_{X_{v_\beta}} = y_\beta^*|_{X_{v_\beta}} \in X_{v_\beta}^*$, we have $\|x_{1_\beta}^*\| = \|x_{1_\beta}^*|_{X_{v_\beta}}\|_{X_{v_\beta}^*} \leq \gamma'$ for all β, so $\|x_{2_\beta}^*\| \leq \|x_\beta^*\| + \|x_{1_\beta}^*\| \leq K + \gamma' := \gamma$. Thus,

$$
\begin{aligned}
(x_\beta^*, \alpha_\beta) &= (x_{1_\beta}^*, \alpha_\beta - g_{v_\beta}^*(x_{2_\beta}^*)) + (x_{2_\beta}^*, g_{v_\beta}^*(x_{2_\beta}^*)) \\
&\in \text{epi } f_{v_\beta}^* \cap (B_\gamma^{X^*} \times \mathbf{R}) + \text{epi } g_{v_\beta}^* \cap (B_\gamma^{X^*} \times \mathbf{R})
\end{aligned}
$$

for all β. We need some uniform bound on the $g_{v_\beta}^*(x_{2_\beta}^*)$. These follow from Lemma 5 (lower bounds) and Lemma 6 (upper bounds), the latter since $(x_{1_\beta}^*, x_{2_\beta}^*) \in H_K(X^*, v_\beta) \cap B_\gamma^*$ and $v_\beta \to w$. Thus, the $g_{v_\beta}^*(x_{2_\beta}^*)$ are eventually uniformly bounded in β, and

$$(x_\beta^*, \alpha_\beta) = (x_{1_\beta}^*, \alpha_{1_\beta}) + (x_{2_\beta}^*, \alpha_{1_\beta}) \in \text{epi } f_{v_\beta}^* + \text{epi } g_{v_\beta}^*$$

with the $x_{1_\beta}^*$, $x_{2_\beta}^*$, α_{1_β}, α_{2_β} all uniformly bounded in β.

We may now argue as in the final paragraph of the proof of [WE99, Thm. 4.3] to conclude that

$$(x^*, \alpha) = w^*\text{-}\lim_{\beta}(x_\beta^*, \alpha_\beta) \qquad \text{(on passing to subnets)}$$

$$= w^*\text{-}\lim_{\beta}(x_{1_\beta}^*, \alpha_{1_\beta}) + w^*\text{-}\lim_{\beta}(x_{2_\beta}^*, \alpha_{2_\beta})$$

$$\in bw^*\text{-}\limsup_{v \to w} \text{epi } f_v^* + bw^*\text{-}\limsup_{v \to w} \text{epi } g_v^*$$

$$\subseteq \text{epi } f_w^* + \text{epi } g_w^* \qquad \text{(from the slice convergence } f_v \to f_w, \, g_v \to g_w)$$

$$\subseteq \text{epi } f_w^* \square g_w^* \,.$$

This completes the proof for the case where $X_v \supseteq \text{dom } f_v \cup \text{dom } g_v$ for all v.

For the general case, let $\rho > \inf_X f_w$. Then, $v \mapsto \{f_w \le \rho\}$ is norm–l.s.c. at w since (see [Bee93]) $\{f_v \le \rho\}$ slice converges to $\{f_w \le \rho\}$ as $v \to w$. Thus on choosing some $x_w \in \{f_w \le \rho\}$, we have some $x_v \in \{f_v \le \rho\}$ with x_v strongly convergent to x_w as $v \to w$. Place $\hat{f}_v := f_v(x_v + \cdot)$ and $\hat{g}_v := g_v(x_v + \cdot)$. By Lemma 3, $\hat{f}_v$ and $\hat{g}_v$ slice converge to $\hat{f}_w$ and $\hat{g}_w$ respectively. Also, $0 \in \text{dom } \hat{f}_v$, whence X_v contains both $\text{dom } \hat{f}_v$, and $\text{dom } \hat{g}_v$, with $X_v = \text{span}\,(\text{dom } \hat{f}_v - \text{dom } \hat{g}_v)$. The form of the conditions in the theorem statement are not altered by passing from f_v, g_v to $\hat{f}_v$, $\hat{g}_v$, the only change being an increase in the value of ρ in the interiority condition. Thus we obtain the slice convergence $\hat{f}_v + \hat{g}_v \to \hat{f}_w + \hat{g}_w$. Translating the sum by $-x_v$, Lemma 3 yields the convergence

$$f_v + g_v = (\hat{f}_v + \hat{g}_v)(\cdot - x_v) \to (\hat{f}_w + \hat{g}_w)(\cdot - x_w) = f_w + g_w \,.$$

$$\square$$

It is well known (see, for instance, [AR96]) that results for sums, such as Theorem 1, imply convergence results for restrictions $F(\cdot, 0)$ of bivariate functions on product spaces $X \times Y$ (just apply a sum theorem to the combination $F + \delta_{X \times \{0\}}$) and that such results may be used to extend sum theorems to include an operator, that is, yield convergence of functions of the form $f + g \circ T$ where $T : X \to Y$ is a bounded linear operator. As discussed in [Zal03], convergence theorems for $F(\cdot, 0)$ may be used to derive theorems not only for sums, but also for other combinations of functions, such as $\max(f, g \circ T)$, and so, in a sense, results for sums are equivalent to results for sums with operator and equivalent to results on restrictions of bivariate functions. Thus, it is a matter of taste, or the intended application, that will dictate the choice of primary form to be considered.

We now use Theorem 1 to obtain a convergence theorem for restrictions of functions on product (Banach) spaces. (cf. [Zal03, Prop. 13] for the normed-space version)

Corollary 1. *Let X and Y be Banach spaces, let F_v ($v \in W$) be in $\Gamma(X \times Y)$ with F_v slice convergent to F_w. Assume that $0 \in P_Y(\text{dom } F_v)$ for all v, and, moreover, that there are $\delta > 0$, $\rho > 0$ and neighborhood V of w such that for all $v \in V$ (with $v \ne w$)*

$$B_\delta^Y \cap Y_v \subseteq P_Y(\{F_v \le \rho\} \cap B_\rho^{X \times Y}),$$

where $Y_v := \text{span}(P_Y(\text{dom } F_v)) \subseteq Y$, and that $h : X^* \to \overline{\mathbf{R}}$ given by $h(x^*) = \inf_{y^* \in Y^*} F_w^*(x^*, y^*)$ satisfies $\overline{h} = \overline{h}^{w^*}$. Then $F_v(\cdot, 0) \to F_w(\cdot, 0)$ in $\Gamma(X)$.

Proof. Note that since $h^* = F_w(\cdot, 0) \in \Gamma(X)$, it follows that h^{**}, and therefore $\overline{h}^{w^*}$, is in $\Gamma^*(X^*)$. Place $G_v = G := \delta_{X \times \{0\}} \in \Gamma(X \times Y)$, where $\delta_{X \times \{0\}}$ denotes indicator function of $X \times \{0\}$. We shall apply Theorem 1 to F_v and G_v, so we check its hypotheses.

Since $\{G_v \le \rho\} = X \times \{0\}$ for any $\rho \ge 0$, we have $\{F_v \le \rho\} \cap B_\rho^{X \times Y} - \{G_v \le \rho\} = X \times P_Y(\{F_v \le \rho\} \cap B_\rho^{X \times Y})$ and $\text{dom } F_v - \text{dom } G_v = X \times P_Y(\text{dom } F_v)$, whence $Z_v := \text{span}(\text{dom } F_v - \text{dom } G_v) = X \times Y_v$, implying

$$\begin{aligned}
B_\delta^{X \times Y} \cap Z_v = B_\delta^X \times (B_\delta^Y \cap Y_v) &\subseteq X \times (B_\delta^Y \cap Y_v) \\
&\subseteq X \times P_Y(\{F_v \le \rho\} \cap B_\rho^{X \times Y}) \\
&= \{F_v \le \rho\} \cap B_\rho^{X \times Y} - \{G_v \le \rho\}
\end{aligned}$$

for all $v \in V \setminus \{w\}$. Moreover, since $(F_w^* \square G_w^*)(x^*, y^*) = h(x^*)$ for $x^* \in X^*$, $y^* \in Y^*$, we see that properness of $\overline{h} = \overline{h}^{w^*}$ implies that $\overline{F_w^* \square G_w^*} = \overline{F_w^* \square G_w^*}^{w^*}$ and is proper. Thus, the conditions of Theorem 1 hold, from which follows the slice convergence of $F_v + \delta_{X \times \{0\}}$ to $F_w + \delta_{X \times \{0\}}$ in $\Gamma(X \times Y)$, which in turn implies that $F_v(\cdot, 0) \to F_w(\cdot, 0)$ in $\Gamma(X)$. $\square$

We can use Corollary 1 to obtain a version of Theorem 1 "with an operator". We start with an elementary lemma whose proof will be omitted. (This lemma is also a consequence of Lemmas 19,20 of [Zal03].)

Lemma 8. *Let $f_v \to f_w$ and $g_v \to g_w$ be slice convergent in $\Gamma(X)$ and $\Gamma(Y)$ respectively, and let $T_v \to T_w$ be a norm-convergent family of continuous linear operators mapping X into Y. Place $F_v(x, y) = f_v(x) + g_v(T_v x + y)$ for $(x, y) \in X \times Y$. Then F_v slice converges to F_w in $\Gamma(X \times Y)$.*

The next result now extends Theorem 1.

Corollary 2. *Let X and Y be Banach spaces. Let $f_v \to f_w$ and $g_v \to g_w$ under slice convergence in $\Gamma(X)$ and $\Gamma(Y)$ respectively, and let $T_v : X \to Y$ be continuous linear operators with $T_v \to T_w$ in operator norm. Assume that there exist a neighborhood V of w, and $\delta > 0$, $\rho > 0$ such that*

$$\forall v \in V \setminus \{w\} \qquad B_\delta \cap Y_v \subseteq T_v(\{f_v \le \rho\} \cap B_\rho) - \{g_v \le \rho\} \tag{6}$$

where $Y_v = \text{span}(\text{dom } g_v - T_v \text{ dom } f_v)$. Assume further that $h : X^* \to \overline{\mathbf{R}}$ defined by $h(x^*) := \inf_{y^* \in Y^*}(f_w^*(x^* - T_w^* y^*) + g_w^*(y^*))$ satisfies $\overline{h} = \overline{h}^{w^*}$.
Then $f_v + g_v \circ T_v$ slice converges to $f_w + g_w \circ T_w$.

Proof. Place $F_v(x, y) := f_v(x) + g_v(T_v x + y)$. Then F_v slice converges to F_w by Lemma 8. It is easily seen that $Y_v = \mathrm{span}\,(P_Y \mathrm{dom}\, F_v)$. If $y \in B_\delta \cap Y_v$, then $y = y_1 - T_v x$ with $g_v(y_1) \leq \rho$, $\|y_1\| \leq \rho'$ (for some $\rho' > \rho$), $f_v(x) \leq \rho$, $\|x\| \leq \rho$, so $\|(x, y)\| \leq \rho + \rho'$, $F_v(x, y) = f_v(x) + g_v(y_1) \leq \rho + \rho'$, thus yielding that $y \in P_Y(\{F_v \leq \rho + \rho'\} \cap B_{\rho + \rho'}^{X \times Y})$. Thus, there is a $\rho > 0$ such that $B_\delta \cap Y_v \subseteq P_Y(\{F_v \leq \rho\} \cap B_\rho^{X \times Y})$ for all $v \in V \backslash \{w\}$. We may then apply Corollary 1 to obtain $f_v + g_v \circ T_v = F_v(\cdot, 0) \to F_w(\cdot, 0) = f_w + g_w \circ T_w$. $\square$

Remark 3. The condition $0 \in \mathrm{sqri}\,(T_w \mathrm{dom}\, f_w - \mathrm{dom}\, g_w)$ can be shown to be equivalent to assuming the condition in (6) to hold at $v = w$. If this is assumed, then there follows, by a standard Fenchel duality result, that $h = (f_w + g_w \circ T_w)^*$ so h is weak*-closed and hence $\overline{h} = \overline{h}^{w^*}$. Indeed,

$$h(x^*) = -\sup_{y^*} \left[-(f - x^*)^*(-T^* y^*) - g^*(y^*) \right]$$

$$= -\inf_x (f - x^* + g \circ T)(x) = \sup_x \left[\langle x, x^* \rangle - (f + g \circ T)(x) \right]$$

$$= (f + g \circ T)^*(x^*).$$

Alternately, we may deduce the above by using [Zal03, Lemmas 15, 16] with $F(x, y) := f_w(x) + g_w(T_w x + y)$

In [Zal03] a number of qualification conditions are framed in the dual spaces. We consider some related results next.

Proposition 3. *Let X be normed and linear $\{f_v\}_{v \in W}$ and $\{g_v\}_{v \in W}$ be slice convergent families in $\Gamma(X)$ convergent to f_w and g_w, respectively, with $\overline{f_v \square g_v}$ proper for all v. Suppose in addition that for $F_v(x, y) := f_v(y) + g_v(x - y)$ we have*

$$\forall \rho > 0,\ \exists \bar\rho > 0,\ \exists V_\rho \in \mathcal{N}(w),\ \forall v \in V_\rho\ \forall s < \rho : \qquad \{\overline{f_v \square g_v} < s\} \cap B_\rho \subseteq P_X(\{F_v < s\} \cap (X \times B_{\bar\rho})). \tag{7}$$

Then $\{\overline{f_v \square g_v}\}_{v \in W}$ is slice convergent to $\overline{f_w \square g_w}$ as $v \to w$.

Proof. It is straightforward to deduce that $v \mapsto \mathrm{epi}\, \overline{f_v \square g_v}$ is strongly lower-semicontinuous at $v = w$ (see the opening paragraph of the proof of [WE99, Thm. 4.3]). For the upper slice convergence take

$$\eta > (f_w \square g_w)^* (x^*) = f_w^*(x^*) + g_w^*(x^*) \quad (\text{so } (x^*, 0, \eta) \in \mathrm{epi}_s F_w^*) \tag{8}$$

and $v_\alpha \to w$. Let $\{x_\alpha\}$ be bounded. Place $\rho = \sup_\alpha \{\|x^*\| \|x_\alpha\| - \eta, \|x_\alpha\|\}$. If $\overline{f_{v_\alpha} \square g_{v_\alpha}}(x_\alpha) \geq \rho$ we immediately have

$$\overline{f_{v_\alpha} \square g_{v_\alpha}}(x_\alpha) \geq \langle x^*, x_\alpha \rangle - \eta,$$

whence, without losing generality we may assume $\overline{f_{v_\alpha} \square g_{v_\alpha}}(x_\alpha) < \rho$. By redefining the index set for the above net if necessary, we may assert the existence of a net $\varepsilon_\alpha > 0$ tending to zero, such that $s_\alpha := \overline{f_{v_\alpha} \square g_{v_\alpha}}(x_\alpha) + \varepsilon_\alpha < \rho$.

Then we have $x_\alpha \in \{\overline{f_{v_\alpha} \Box g_{v_\alpha}} < s_\alpha\} \cap B_\rho$, implying (by (7)) the existence of $\|y_\alpha\| < \bar{\rho}$ with $F_v(x_\alpha, y_\alpha) < s_\alpha$. As noted earlier we always have $\{F_v\}_{v \in W}$ slice convergent to F_w. Also note that a simple calculation shows

$$F_w^*(x^*, y^*) = f_w^*(x^* + y^*) + g_w^*(x^*).$$

Thus (8) and the (upper) slice convergence of $\{F_v\}_{v \in W}$ (recalling that $F_w^*(x^*, 0) < \eta$) implies

$$\overline{f_{v_\alpha} \Box g_{v_\alpha}}(x_\alpha) + \varepsilon_\alpha = s_\alpha > F_{v_\alpha}(x_\alpha, y_\alpha) > \langle x^*, x_\alpha \rangle - \eta.$$

Since $\varepsilon_\alpha \to 0$ we arrive at the desired conclusion.   □

Now consider $f_v^* \Box g_v^*$ in the dual space X^*. We use the projection $P_{X^* \times \mathbf{R}}$: $(x^*, y^*, \beta) \mapsto (x^*, \beta)$.

Lemma 9. *Let X and Y be normed linear spaces and $\{F_v\}_{v \in W} \subseteq \Gamma(X \times Y)$, with $0 \in P_Y \operatorname{dom} F_v$ for all v . Place $h_v(x^*) = \inf_{y^* \in Y^*} F_v^*(x^*, y^*)$ and assume that $\{F_v\}_{v \in W}$ slice converges to F_w along with*

$$\forall \rho > 0, \ \exists \bar{\rho} > 0, \ \exists V_\rho \in \mathcal{N}(w), \ \forall v \in V_\rho :$$

$$\operatorname{epi}{}_s \overline{h_v}^{w^*} \cap B_\rho^* \subseteq P_{X^* \times \mathbf{R}} \left(\operatorname{epi}{}_s F_v^* \cap (X^* \times B_{\bar{\rho}}^* \times \mathbf{R}) \right) \qquad (9)$$

and also that the norm- and weak*-closures of h_w coincide. Then $\overline{h_v}^{w^*}$ (dual) slice converges to $\overline{h_w}^{w^*} = \overline{h_w}$ and $\{F_v(\cdot, 0)\}_{v \in W}$ slice converges to $F_w(\cdot, 0)$.

Proof. First we show that the multi-function $v \mapsto \operatorname{epi} \overline{h_v}^{w^*}$ is bounded–weak* upper–semicontinuous. Let $v_\alpha \to w$ be taken so that $(x_{v_\alpha}^*, \beta_{v_\alpha}) \in \operatorname{epi} h_{v_\alpha}^{**}$ weak* converge to (x^*, β) and $\|(x_{v_\alpha}^*, \beta_{v_\alpha})\|$ is bounded. Then we have $(x_{v_\alpha}^*, \beta_{v_\alpha} + \varepsilon_\alpha) \in \operatorname{epi}{}_s h_{v_\alpha}^{**}$ for any positive net $\varepsilon_\alpha \to 0$.

Then take $\rho = \max_\alpha \{\|(x_{v_\alpha}^*, \beta_{v_\alpha} + \varepsilon_\alpha)\|\}$ and apply (9) to deduce the existence of $y_{v_\alpha}^* \in Y^*$ such that $\|y_{v_\alpha}^*\| \leq \bar{\rho}$ and $(x_{v_\alpha}^*, y_{v_\alpha}^*, \beta_{v_\alpha} + \varepsilon_\alpha) \in \operatorname{epi}{}_s F_{v_\alpha}^*$. Take a weak* convergent subnet if necessary (and on reparametrizing) we may assume that $(x_{v_\alpha}^*, y_{v_\alpha}^*, \beta_{v_\alpha} + \varepsilon_\alpha) \to (x^*, y^*, \beta)$. Since $\{F_v\}_{v \in W}$ is slice convergent so is $\{F_v^*\}_{v \in W}$ and hence $\{\operatorname{epi} F_v^*\}_{v \in W}$ is bounded weak* upper semi–continuous. Hence we have $(x^*, y^*, \beta) \in \operatorname{epi} F_w^*$ implying $(x^*, \beta) \in P_{X^* \times \mathbf{R}}(\operatorname{epi} F_w^*) \subseteq \operatorname{epi} h_w^{**}$.

The slice convergence of $\{F_v^*\}_{v \in W}$ has been observed to follow from that of $\{F_v\}_{v \in W}$. Thus $v \mapsto \operatorname{epi} F_v^*$ is strongly lower semi–continuous. Next note that for any open set $O \subseteq X^* \times \mathbf{R}$ we have $P_{X^* \times \mathbf{R}}^{-1}(O) = O \times Y^*$ and so $P_{X^* \times \mathbf{R}}(\operatorname{epi} F_v^* \cap (O \times Y^*)) = P_{X^* \times \mathbf{R}}(\operatorname{epi} F_v^*) \cap O$. Hence $\{v \in W \mid P_{X^* \times \mathbf{R}}(\operatorname{epi} F_v^*) \cap O \neq \emptyset\} = \{v \in W \mid P_{X^* \times \mathbf{R}}(\operatorname{epi} F_v^* \cap (O \times Y^*)) \neq \emptyset\}$ which clearly coincides with the open set $\{v \in W \mid \operatorname{epi} F_v^* \cap (O \times Y^*) \neq \emptyset\}$ implying strong lower semi–continuity.

Finally note that $h_v^{**} = (F_v(\cdot, 0))^*$ and hence slice convergence of

$$\{F_v(\cdot, 0)\}_{v \in W}$$

follows from the bicontinuity of Fenchel conjugation.   □

This last result could be used to deduce the next result but instead we prefer to use a direct argument along the lines of argument in Theorem 1.

Theorem 2. *Let X be a normed linear space, let f_v and g_v ($v \in W$) be in $\Gamma(X)$, with the slice convergence $f_v \to f_w$ and $g_v \to g_w$ and $\operatorname{dom} f_v \cap \operatorname{dom} g_v \neq \emptyset$ for all v. Also, assume that $\overline{f_w^* \Box g_w^*}$ is proper and weak* lower-semicontinuous, $F_v(x, y) := f_v(x) + g_v(x + y)$ and*

$$\forall \rho > 0, \ \exists \bar{\rho} > 0, \ \exists V_\rho \in \mathcal{N}(w), \ \forall v \in V_\rho, \ \forall s < \rho:$$

$$\overline{\{\overline{f_v^* \Box g_v^*}^{w^*} < s\} \cap B_\rho^*} \subseteq \overline{P_X^*(\{F_v^* < s\} \cap (X^* \times B_{\bar{\rho}}))}^{w^*}. \tag{10}$$

Then $f_v + g_v$ slice converges to $f_w + g_w$.

Proof. As noted earlier, the strong lower-semicontinuity of $v \mapsto \operatorname{epi} \overline{f_v^* \Box g_v^*}^{w^*}$ at $v = w$ follows straightforwardly. For the other half of the convergence, let

$$(x^*, \alpha) \in bw^*\text{-}\limsup_{v \to w} \operatorname{epi} \overline{f_v^* \Box g_v^*}^{w^*}.$$

Then there are nets $v_\beta \to w$, $(x_\beta^*, \alpha_\beta) \to^{w^*} (x^*, \alpha)$, and $\rho > 0$ with $(x_\beta^*, \alpha_\beta) \in B_\rho^* \cap \operatorname{epi}_s \overline{f_{v_\beta}^* \Box g_{v_\beta}^*}^{w^*}$ for all β. By use of (10) we obtain a bounded net $\|y_\beta^*\| \leq \bar{\rho}$ such that $\alpha_\beta \geq F_{v_\beta}^*(x_{v_\beta}^*, y_{v_\beta}^*) = f_{v_\beta}^*(x_{v_\beta}^* - y_{v_\beta}^*) + g_{v_\beta}^*(y_{v_\beta}^*)$ and we may now argue as in the final part of the proof of [WE99, Theorem 4.3] to deduce that $(x^*, \alpha) \in \operatorname{epi} f_w^* \Box g_w^*$. □

We note that one could have framed a qualification assumption based on the assumption that $\overline{f_v^* \Box g_v^*}^{w^*} = \overline{f_v^* \Box g_v^*}$ for all $v \in W$ and the assumption of 10 without the weak star closure on the right hand side. A similar proof as above then obtains essentially [Zal03, Prop. 28].

We close this section with the observation that the argument of Corollary 1 also permits the deduction of epi-distance convergence results for perturbation functions from those for sums. (See, for example, [EW00] for detail on epi-distance convergence.)

Proposition 4. *Let X and Y be Banach, and F_n, F in $\Gamma(X \times Y)$ with $F_n \to F$ in epi-distance. Assume that $0 \in \operatorname{sqri}(P_Y \operatorname{dom} F)$, and that $Y_0 := \overline{\operatorname{span}}(P_Y \operatorname{dom} F)$ has closed algebraic complement Y_0' for which $Y_0' \cap Y_n = \{0\}$ eventually (where $Y_n := \overline{\operatorname{span}}(P_Y \operatorname{dom} F_n)$). Then $F_n(\cdot, 0)$ epi-distance converges to $F(\cdot, 0)$.*

Proof. As $0 \in \operatorname{sqri}(P_Y \operatorname{dom} F)$ we have $\operatorname{cone}(P_Y \operatorname{dom} F) = \overline{\operatorname{span}}(P_Y \operatorname{dom} F)$.

Place $G_n \equiv G := \delta_{X \times \{0\}}$. We apply [EW00, Thm 4.9] to F_n, F, G_n, G. Place $Z_n = \operatorname{span}(\operatorname{dom} F_n - \operatorname{dom} G_n)$ and $Z_0 = \operatorname{span}(\operatorname{dom} F - \operatorname{dom} G)$. Since $\operatorname{dom} F - \operatorname{dom} G = X \times P_Y \operatorname{dom} F$, we have $Z_0 = X \times Y_0$ and $Z_n = X \times Y_n$, with

$$\operatorname{cone}(\operatorname{dom} F - \operatorname{dom} G) = \operatorname{cone}(X \times P_Y \operatorname{dom} F) = X \times Y_0$$

therefore being closed in $X \times Y$. Also Z_0 has closed complement $Z_0' := \{0\} \times Y_0'$, and

$$\overline{Z_n} \times Z_0 = (X \times \overline{Y_n}) \cap (\{0\} \times Y_0') = \{0\} \times (\overline{Y_n} \cap Y_0') = \{0\}.$$

Hence the hypotheses of [EW00, Thm 4.9] are satisfied, yielding

$$F_n + \delta_{X \times \{0\}} \to F + \delta_{X \times \{0\}},$$

or equivalently, $F_n(\cdot, 0) \to F(\cdot, 0)$.   □

4 Saddle–point Convergence in Fenchel Duality

When discussing saddle point convergence we are necessarily lead to the study of equivalence classes of saddle–functions which are uniquely associated with concave or convex parents (depending on the which variable is partially conjugated). We direct the reader to the excellent texts of Rockafellar [Roc70, Roc74] for a detailed treatment of this phenomenon. The following is taken from [AAW88] from which we adapt results and proofs.

Definition 10. *Suppose that (X, τ) and (Y, σ) are two topological spaces and $\{K^n : X \times Y \to \overline{\mathbf{R}}, n \in \mathbf{N}\}$ is a sequence of bi–variate functions. Define:*

$$e_\tau / h_\sigma - ls\, K^n(x, y) = \sup_{\{y_n \to^\sigma y\}} \inf_{\{x_n \to^\tau x\}} \limsup_{n \to \infty} K^n(x_n, y_n)$$

$$h_\sigma / e_\tau - li\, K^n(x, y) = \inf_{\{x_n \to^\tau x\}} \sup_{\{y_n \to^\sigma y\}} \liminf_{n \to \infty} K^n(x_n, y_n).$$

Definition 11. *Suppose that (X, τ) and (Y, σ) are two topological spaces and $\{K^n : X \times Y \to \overline{\mathbf{R}}, n \in \mathbf{N}\}$ is a sequence of bivariate functions.*

1. *We say that they epi/hypo–converge in the extended sense to a function $K : X \times Y \to \overline{\mathbf{R}}$ if*

$$\underline{cl}_x(e_\tau / h_\sigma - ls\, K^n) \leq K \leq \overline{cl}^y(h_\sigma / e_\tau - li\, K^n)$$

   where $\underline{cl}_x$ denotes the extended lower closure with respect to x (and therefore w.r.t. τ) for fixed y and $\overline{cl}^y$ denotes the extended upper closure with respect to y (and therefore w.r.t. σ) for fixed x. Note that by definition, $\overline{cl}\, f := -\underline{cl}\, (-f)$.
2. *A point $(\bar{x}, \bar{y})$ is a saddle–point of a bivariate function $K : X \times Y \to \overline{\mathbf{R}}$ if for all $(x, y) \in X \times Y$ we have $K(\bar{x}, y) \leq K(\bar{x}, \bar{y}) \leq K(x, \bar{y})$.*

The interest in this kind of convergence stems from the following result (see [AAW88, Thm 2.4]).

Proposition 5. *Let us assume that* $\{K^n, K : (X, \tau) \times (Y, \sigma) \to \overline{\mathbf{R}}, n \in \mathbf{N}\}$ *are such that they epi/hypo–converge in the extended sense. Assume also that* $(\bar{x}_k, \bar{y}_k^*)$ *are saddle points of* K^{n_k} *for all* k *and* $\{n_k\}$ *is an increasing sequence of integers, such that* $\bar{x}_{n_k} \xrightarrow{\tau} \bar{x}$ *and* $\bar{y}_{n_k}^* \xrightarrow{\sigma} \bar{y}^*$. *Then* $(\bar{x}, \bar{y}^*)$ *is a saddle point of* K *and*

$$K(\bar{x}, \bar{y}^*) = \lim_{k \to \infty} K^{n_k}(\bar{x}_k, \bar{y}_k^*) .$$

The next result from [AAW88] uses sequential forms of the epi–limit functions, as per the following

Definition 12. *[AAW88, p 541] Let* (X, τ) *be topological,* $f_n : X \to \overline{\mathbf{R}}$. *Then*

$$(\tau\text{-seq-e-ls}_{n \to \infty} f_n)(x) := \inf_{\{x_n\} \xrightarrow{\tau} x} \limsup_{n \to \infty} f_n(x_n)$$

$$(\tau\text{-seq-e-li}_{n \to \infty} f_n)(x) := \inf_{\{x_n\} \xrightarrow{\tau} x} \liminf_{n \to \infty} f_n(x_n)$$

It can be shown that these reduce to the usual (topologically defined) forms if (X, τ) is first–countable, and that the above infima are achieved. We will need these alternate forms, for generally weak topologies on normed spaces are not first–countable.

Definition 13. *Let* (X, τ) *and* (X^*, τ^*) *be topological vector spaces. We shall say they are paired if there is a bilinear map* $\langle \cdot, \cdot \rangle : X \times X^* \to \mathbf{R}$ *such that the maps* $x^* \mapsto \langle \cdot, x^* \rangle$ *and* $x \mapsto \langle x, \cdot \rangle$ *are (algebraic) isomorphisms such that* $X^* \cong (X, \tau)^*$ *and* $X \cong (X^*, \tau^*)^*$ *respectively.*

It is readily checked that if (X, τ) and (X^*, τ^*) are paired, and so are (Y, σ) and (Y^*, σ^*), then $(X \times Y, \tau \times \sigma)$ is paired with $(X^* \times Y^*, \tau^* \times \sigma^*)$, with the pairing

$$\langle (x, y), (x^*, y^*) \rangle = \langle x, x^* \rangle + \langle y, y^* \rangle,$$

and similarly for other combinations of product spaces.

For any convex–concave saddle function $K : X \times Y^* \to \overline{\mathbf{R}}$, that is, where K is convex in the first argument and concave in the second, we may associate a convex and concave parent. These play a fundamental role in convex duality (see [Roc74]). These are defined respectively as:

$$F(x, y) = \sup_{y^* \in Y^*}[K(x, y^*) + \langle y, y^* \rangle]$$
$$G(x^*, y^*) = \inf_{x \in X}[K(x, y^*) - \langle x, x^* \rangle].$$

Subject to suitable closure properties on K, it follows that $G = -F^*$, and that K is a saddle function for the dual pair of optimization problems $\inf_X F(\cdot, 0)$ and $\sup_{X^*} G(\cdot, 0)$. One may also proceed in reverse, and show that for any closed convex function $F : X \times Y \to \overline{\mathbf{R}}$, if $G := -F^*$ relative to the natural pairing of $X \times Y$ with $X^* \times Y^*$, (these yielding the primal objective $F(\cdot, 0)$ and dual objective $G(0, \cdot)$), we have an interval of saddle functions, all equivalent in the sense that they possess the same saddle points, given by

$[\underline{K}, \overline{K}] := \{K : X \times Y^* \to \overline{\mathbf{R}} \mid K \text{ convex–concave}, \underline{K} \leq K \leq \overline{K} \text{ on } X \times Y^*\},$

where

$$\underline{K}(x, y^*) = \sup_{x^* \in X^*} [G(x^*, y^*) + \langle x, x^* \rangle]$$
$$\overline{K}(x, y^*) = \inf_{y \in Y} [F(x, y) - \langle y, y^* \rangle] .$$

Our focus will be on the Fenchel duality, where given the primal problem $\inf_X f + g$, we form $F(x, y) := f(x) + g(x + y)$, so that $G(x^*, y^*) = -f^*(x^* - y^*) - g^*(y^*)$ and the Fenchel dual takes the form $\sup_{y^* \in X^*} G(0, y^*) = \sup_{y^* \in X^*} -f^*(-y^*) - g^*(y^*)$ (cf. (12) below). Also, any $K \in [\underline{K}, \overline{K}]$ is a suitable saddle function for the Fenchel primal/dual pair and we shall use $K := \overline{K}$ in what follows.

The following result is taken from [AAW88] and requires no additional assumption.

Proposition 6. *Let* $(X, \tau), (X^*, \tau^*)$ *and* $(Y, \sigma), (Y^*, \sigma^*)$ *be paired topological vector spaces, with the pairings sequentially continuous; let* $\{F^n, F : X \times Y \to \overline{\mathbf{R}}, n \in \mathbf{N}\}$ *be a family of bivariate* $(\tau \times \sigma)$*–closed convex functions. Then, if* K^n*,* K *are members of the corresponding equivalence classes of bivariate convex–concave saddle functions,*

1.

$$(\tau \times \sigma)\text{-seq-e-ls}_{n \to \infty} F^n \leq F \quad \text{on } X \times Y$$
$$\text{implies} \quad \underline{\text{cl}}_x(e_\tau/h_{\sigma^*}\text{-ls } \overline{K}^n) \leq \underline{K} ;$$

2.

$$(\tau^* \times \sigma^*)\text{-seq-e-ls}_{n \to \infty} (F^n)^* \leq (F)^* \quad \text{on } X^* \times Y^*$$
$$\text{implies} \quad \overline{K} \leq \overline{\text{cl}}^{y^*} (h_{\sigma^*}/e_\tau\text{-li } \underline{K}^n) .$$

Proposition 7. *Suppose that* X *is a Banach space and*

$$\{f_n, f\}_{n=1}^\infty \quad \text{and} \quad \{g_n, g\}_{n=1}^\infty$$

be two families of proper closed, convex extended–real–valued functions slice–convergent to f *and* g*, respectively. Then*

$$K^n(x, y^*) = \inf_{y \in X} [f_n(x) + g_n(x + y) - \langle y, y^* \rangle]$$

epi/hypo–converges (in the extended sense) to

$$K(x, y^*) = \inf_{y \in X} [f(x) + g(x + y) - \langle y, y^* \rangle]$$

with respect to the strong topology on X *and the weak* topology on* X^**.*

Proof. From the slice convergence of f_n and g_n it is elementary exercise to show that $F_n(x, y) := f_n(x) + g_n(x + y)$ is slice-convergent to $F(x, y) = f(x) + g(x + y)$. From the bicontinuity of conjugation with respect to slice convergence, follows the dual slice convergence of $F_n^* \to F^*$. From the resulting strong epi-upper-semicontinuity for the F_n and F_n^* on $X \times X$ and $X^* \times X^*$ respectively,

$$F \geq (s \times s)\text{-e-ls}_{n \to \infty} F^n = (s \times s)\text{-seq-e-ls}_{n \to \infty} F^n \qquad \text{and}$$
$$F^* \geq (s^* \times s^*)\text{-e-ls}_{n \to \infty} (F^n)^* = (s^* \times s^*)\text{-seq-e-ls}_{n \to \infty} (F^n)^*$$
$$\geq (w^* \times w^*)\text{-seq-e-ls}_{n \to \infty} (F^n)^*,$$

where s and s^* stand for the respective norm topologies on X and X^*. Now apply Proposition 6.   □

We note the following for later reference. For $u \in X$, write

$$\varphi(u) := \inf_{x \in X} \{f(x) + g(x + u)\} = \widehat{(f \Box \widehat{g})}(u), \qquad (11)$$

and similarly for φ_n, where for any function ψ, $\widehat{\psi}(x) := \psi(-x)$. Note that $\text{dom}\,\varphi = \text{dom}\,g - \text{dom}\,f$ and similarly for φ_n. The operation $\psi \mapsto \widehat{\psi}$ commutes with conjugation and with slice limits, the verification of this being an elementary exercise. From [Roc74] we have the following: Calling $\inf_X (f + g)$ the primal problem, and $\inf_X (f_n + g_n)$ the approximate problems, then $-\varphi^*$ and $-\varphi_n^*$ are the associated dual objective functionals, and:

$$(\bar{x}, \bar{y}^*) \quad \text{is a saddle-point of } K \text{ iff}$$
$$\varphi(0) = (f + g)(\bar{x}) = \inf_X (f + g) = \sup_{X^*} -\varphi^* = -\varphi^*(\bar{y}^*),$$

and similarly for φ_n and the saddle-points $(\bar{x}_n, \bar{y}_n^*)$ of K^n. On taking conjugates of φ_n we obtain

$$\varphi_n^* = (\widehat{f_n \Box \widehat{g_n}})^* = (\widehat{f_n^* + \widehat{g_n}^*}) = \widehat{f_n^*} + g_n^*$$

and so the dual problem becomes

$$\sup_{X^*} -\varphi_n^* = -(f_n^* \Box g_n^*)(0) = -\inf_{y^* \in X^*} (f_n^*(-y^*) + g_n^*(y^*)). \qquad (12)$$

The next result tackles the problem of finding convergent sequences of dual variables. (Note that Proposition 5 makes no claim about such existence).

Corollary 3. *Suppose that X is a separable Banach space and $\{f_n, f\}_{n=1}^\infty$ and $\{g_n, g\}_{n=1}^\infty$ be two families of proper closed, convex extended-real-valued functions slice-convergent to f and g respectively. Let K^n, K be the associated saddle-functions as in Proposition 7. Assume also the following:*

1. $\exists \delta > 0$, $\rho > 0$ such that for all large $n \in \mathbb{N}$,

$$B_\delta \cap M_n \subseteq \{f_n \leq \rho\} \cap B_\rho - \{g_n \leq \rho\} \cap B_\rho$$

where $M_n := \overline{\mathrm{span}\,(\mathrm{dom}\,f_n - \mathrm{dom}\,g_n)}$

2. $\overline{f^* \Box g^*}$ is proper w^*-lsc.

Then if $(\bar{x}_n, \bar{y}_n^*)$ are saddle–points of K^n for each n and the $\bar{x}_n$ has a strong limit $\bar{x}$, and the saddle–values are bounded below, then K has a saddle–point $(\bar{x}, \bar{y}^*)$ that is a $(s \times w^*)$–limit of saddlepoints $(\bar{x}_n, \widetilde{(\bar{y}_n^*)}|_{M_n})$ of a subsequence of the K^n, with $K(\bar{x}, \bar{y}^*)$ the limit of the corresponding saddle–function values.

(Here 's' stands for the norm topology on X and $\widetilde{(\bar{y}_n^*)}|_{M_n}$ denotes any norm-preserving extension (via Hahn-Banach Theorem, for example) to X^* of the restriction of $\bar{y}_n^*$ to M_n).

Proof. The proof follows from Propositions 7 and 5, on showing that the $\widetilde{(\bar{y}_n^*)}|_{M_n}$ are norm–bounded in X^*, so that weak*–convergent subsequences are available and and are the required dual variables.

Since the sublevel–sets of f_n are themselves slice convergent [Bee93], there are $x_n \in \mathrm{dom}\,f_n$ converging to some $x \in \mathrm{dom}\,f$. Place $\check{f}_n(\cdot) := f_n(x_n + \cdot)$, $\check{g}_n(\cdot) := g_n(x_n + \cdot)$, with analogous definitions for $\check{f}$ and $\check{g}$ as translates by x. Then $0 \in \mathrm{dom}\,\check{f}_n$, implying that $\mathrm{dom}\,\check{f}_n \cap \mathrm{dom}\,\check{g}_n \subseteq M_n$.

Let $\check{\varphi}_n$ be the value function corresponding to $\check{f}_n$ and $\check{g}_n$ via (11). Similarly, denote the corresponding saddle function by $\check{K}^n$. Then we immediately observe that $\check{\varphi}_n^* = \varphi_n^*$, from which follows that

$(\bar{x}_n, \bar{y}_n^*)$ is a saddlepoint of K^n iff $(\bar{x}_n - x_n, \bar{y}_n^*)$ is a saddlepoint of $\check{K}^n$,

since $(\bar{x}_n, \bar{y}_n^*)$ are an optimal pair for the primal and dual problems if and only if $(\bar{x}_n - x_n, \bar{y}_n^*)$ are optimal for the problems based on the translated functions $\check{f}_n, \check{g}_n$. Evidently the optimal values are not affected by this translation, so we also obtain that $\check{K}^n(\bar{x}_n - x_n, \bar{y}_n^*) = K^n(\bar{x}_n, \bar{y}_n^*)$. Hence the saddle–values of $\check{K}^n$ are also bounded below. As M_n contains both $\mathrm{dom}\,\check{f}_n$ and $\mathrm{dom}\,\check{g}_n$ (recall this follows from $0 \in \mathrm{dom}\,\check{f}_n$), we obtain

$$\check{K}^n(\bar{x}_n - x_n, \bar{y}_n^*) = \check{K}^n(\bar{x}_n - x_n, \widetilde{\bar{y}_n^*|_{M_n}}) ,$$

which follows from $\check{\varphi}_n^*(\bar{y}_n^*) = \check{f}_n^*(-\bar{y}_n^*) + \check{g}_n^*(\bar{y}_n^*) = \check{f}_n^*(-\widetilde{\bar{y}_n^*|_{M_n}}) + \check{g}_n^*(\widetilde{\bar{y}_n^*|_{M_n}}) = \check{\varphi}_n^*(\widetilde{\bar{y}_n^*|_{M_n}})$, since M_n contains the domains of $\check{f}_n$ and $\check{g}_n$. Letting $-\alpha \in \mathbf{R}$ be a lower bound for the saddle–values of K^n (and hence of $\check{K}^n$), we have for all n large, that $(-\bar{y}_n^*|_{M_n}, \bar{y}_n^*|_{M_n}) \in H_\alpha(M_n^*, n)$ (where the latter set is defined relative to the translated functions $\check{f}_n, \check{g}_n$), since

$$(\check{f}_n|_{M_n})^*(-\bar{y}_n^*|_{M_n}) + (\check{g}_n|_{M_n})^*(\bar{y}_n^*|_{M_n}) = \check{f}_n^*(-\bar{y}_n^*) + \check{g}_n^*(\bar{y}_n^*)$$
$$= -\check{K}^n(\bar{x}_n - x_n, \bar{y}_n^*) < \alpha .$$

By Lemma 4, the $\|\bar{y}_n^*|_{M_n}\|$ are norm-bounded in M_n^* for all large n. Then the sequence of norm-preserving extensions $\bar{z}_n^* := \widetilde{\bar{y}_n^*|_{M_n}} \in X^*$ is also norm-bounded and hence has a weakly* convergent subsequence $\bar{z}_n^* \to \bar{z}^*$. For each n, $(\bar{x}_n - x_n, \bar{z}_n^*)$ a saddlepoint for $\check{K}^n$, so $(\bar{x}_n, \bar{z}_n^*)$ is one for K^n. By Propositions 7 and 5, $(\bar{x}, \bar{z}^*)$ is a saddlepoint for K, with value the limit of the saddle–values along the sequence.   □

References

[Att84] Attouch, H.: Variational Convergence for Functions and Operators. Applicable Mathematics Series, Pitman, London (1984)

[Att86] Attouch, H. Brézis, H.: Duality for the sum of convex functions in general Banach spaces. In: Barroso, J. (ed) Aspects of Mathematics and its Applications, 125–133. Elsevier Sc. Publ. (1986)

[AR96] Aze, D., Rahmouni, A.: On Primal-Dual stability in convex optimization. Journal of Convex Analysis, **3**, 309–327 (1996)

[AAW88] Azé, D., Attouch, H., Wets, R.J.-B.: Convergence of convex–concave saddle functions: applications to convex programming and mechanics. Ann. Inst. Henri Poincaré, **5**, 537–572 (1988)

[AP90] Azé, D., Penot, J.-P.: Operations on convergent families of sets and functions. Optimization, **21**, 521–534 (1990)

[Bee92] Beer, G.: The slice topology: a viable alternative to mosco convergence in non–reflexive spaces. Nonlinear Analysis: Theory, Methods and Applications, **19**, 271–290 (1992)

[Bee93] Beer, G.: Topologies on closed and closed convex sets. Mathematics and its Applications, **268**, Kluwer Acad. Publ. (1993)

[BL92] Borwein, J.M., Lewis, A.S.: Partially–finite convex programming. Mathematical Programming, **57**, 15–83 (1992)

[EW00] Eberhard, A., Wenczel, R.: Epi–distance convergence of parametrised sums of convex functions in non-reflexive spaces. J. Conv. Anal., **7**, 47–71 (2000)

[ET99] Ekeland, I., Témam, R.: Convex Analysis and Variational Problems. SIAM Classics in Applied Mathematics, **28** (1999)

[Hol75] Holmes, R.B.: Geometric Functional Analysis and its Applications. Springer–Verlag Graduate Texts in Mathematics **24** (1975)

[Pen93] Penot, J. -P.: Preservation of persistence and stability under intersection and operations. J. Optim. Theory & Appl., **79**, 525–561 (1993)

[Pen02] Penot, J-P, Zălinescu, C.: Continuity of usual operations and variational convergence. personal communication, 30/04/02 (2002)

[RW84] Rockafellar, R.T., Wets, J.-B.: Variational systems, an introduction. In: Salinetti, G. (ed) Multifunctions and Integrands. Springer–Verlag Lecture Notes in Mathematics, **1091**, 1–54 (1984)

[Roc70] Rockafellar, R.T.: Convex Analysis. Princeton University Press (1970)

[Roc74] Rockafellar, R.T.: Conjugate Duality and Optimization. SIAM publ. (1974)

[WE99] Wenczel, R.B., Eberhard, A.C.: Slice convergence of parametrised sums of convex functions in nonreflexive spaces. Bull. Aust. Math. Soc., **60**, 429–458 (1999)

[Zal03] Zălinescu, C.: Slice convergence for some classes of convex functions. J. Nonlinear and Convex Analysis, **4**, (2003)

Topical Functions and their Properties in a Class of Ordered Banach Spaces

Hossein Mohebi

Department of Mathematics
Shahid Bahonar University of Kerman
Kerman, Iran
hmohebi@mail.uk.ac.ir;
CIAO, School of Information Technology and Mathematical Sciences
University of Ballarat
Ballarat, VIC 3353, Australia
h.mohebi@ballarat.edu.au

Summary. We study topical functions in a class of ordered Banach spaces and show that these functions are abstract convex with respect to a certain set of elementary functions and obtain an explicit formula for their subdifferential. We give characterizations of the Fenchel-Moreau conjugate and the conjugate of type Lau of topical functions. We also present necessary and sufficient conditions for plus-weak Pareto points of a closed downward set in terms of separation from outside points.

2000 MR Subject Classification. Primary: 26B25, 52A41; Secondary: 46B42

Key words: Topical function, Downward set, Fenchel-Moreau conjugation, Conjugation of type Lau, Plus-weak Pareto point, Subdifferential, Ordered Banach space

1 Introduction

A function $f : \mathbb{R}^n \longrightarrow \mathbb{R}^m$ is called topical if this function is increasing ($x \geq y \implies f(x) \geq f(y)$) and plus-homogeneous ($f(x + \lambda \mathbf{1}) = f(x) + \lambda \mathbf{1}$ for all $x \in \mathbb{R}^n$ and all $\lambda \in \mathbb{R}$), where $\mathbf{1}$ is the vector of the corresponding dimension with all coordinates equal to one. These functions are studied in [GG98, Gun98, Gun99, GK95, RS01, Sin02] and they have many applications in various parts of applied mathematics (see [Gun98, Gun99]).

In this paper we study topical functions $f : X \longrightarrow \bar{\mathbb{R}}$ defined on an ordered Banach space X. We show that the topical functions $f : X \longrightarrow \bar{\mathbb{R}}$

are characterized by the fact that the Fenchel-Moreau conjugate function and the conjugate function of type Lau admits a very simple explicit description. Most of these results have been obtained by A. Rubinov and I. Singer in finite dimensional case (see [RS01, Sin02]). In this paper, we obtain these results in ordered Banach spaces without using the concepts of lattice theory.

The structure of the paper is as follows. In Section 2, we recall main definitions and prove some results related to downward sets and topical functions. We also show that a topical function is abstract convex. Characterizations of plus-weak Pareto points for a closed downward set are investegated in Section 3. In Section 4, we study the subdifferential of a topical function and we present the characterizations of plus-weak Pareto points of a closed downward set in terms of separation from outside points. In Section 5, we give characterizations of a topical function in terms of its Fenchel-Moreau conjugate and biconjugate with respect to a certain set of elementary functions. In section 6, we first give characterizations of topical functions in terms of the conjugate of type Lau. Next, we show that for topical functions, the conjugate of type Lau and the Fenchel-Moreau conjugate coincide.

2 Preliminaries

Let X be a Banach space with the norm $\|.\|$ and let C be a closed convex cone in X such that $C \cap (-C) = \{0\}$ and int $C \neq \emptyset$. We assume that X is equipped with the order relation $\geq$ generated by $C : x \geq y$ if and only if $x - y \in C$ ($x, y \in X$). Moreover, we assume that C is a normal cone. Recall that a cone C is called *normal* if there exists a constant $m > 0$ such that $\|x\| \leq m\|y\|$, whenever $0 \leq x \leq y$, and $x, y \in X$. Let $\mathbf{1} \in$ int C and let

$$B = \{x \in X : -\mathbf{1} \leq x \leq \mathbf{1}\}. \tag{1}$$

It is well known and easy to check that B can be considered as the unit ball of a certain norm $\|.\|_1$, which is equivalent to the initial norm $\|.\|$. Assume without loss of generality that $\|.\| = \|.\|_1$.

We study in this paper topical functions and downward sets. Recall (see [Sin87]) that a subset W of X is said to be *downward*, if $w \in W$ and $x \in X$ with $x \leq w$, then $x \in W$. A function $f : X \longrightarrow \overline{\mathbb{R}} := [-\infty, +\infty]$ is called *topical* if this function is increasing ($x \geq y \implies f(x) \geq f(y)$) and plus-homogeneous ($f(x + \lambda \mathbf{1}) = f(x) + \lambda$ for all $x \in X$ and all $\lambda \in \mathbb{R}$). The definition of a topical function in finite dimensional case can be found in [RS01].

For any subset W of X, we shall denote by int W, cl W, and bd W the interior, the closure and the boundary of W, respectively.

For a non-empty subset W of X and $x \in X$, define

$$d(x, W) = \inf_{w \in W} \|x - w\|.$$

Recall (see [Sin74]) that a point $w_0 \in W$ is called a *best approximation* for $x \in X$ if

$$\|x - w_0\| = d(x, W).$$

Let $W \subset X$. For $x \in X$, denote by $P_W(x)$ the set of all best approximations of x in W :

$$P_W(x) = \{w \in W : \|x - w\| = d(x, W)\}.$$

It is well-known that $P_W(x)$ is a closed and bounded subset of X. If $x \notin W$ then $P_W(x)$ is located in the boundary of W.

For $x \in X$ and $r > 0$, by (1), we have

$$B(x, r) := \{y \in X : \|x - y\| \le r\} = \{y \in X : x - r\mathbf{1} \le y \le x + r\mathbf{1}\}. \quad (2)$$

Let $\varphi : X \times X \longrightarrow \mathbb{R}$ be a function defined by

$$\varphi(x, y) := \sup\{\lambda \in \mathbb{R} : \lambda\mathbf{1} \le x + y\} \quad \forall\, x, y \in X. \quad (3)$$

It follows from (1) that the set $\{\lambda \in \mathbb{R} : \lambda\mathbf{1} \le x + y\}$ is non-empty and bounded from above (by $\|x + y\|$). Clearly this set is closed. It follows from the definition of φ that the function φ enjoys the following properties:

$$-\infty < \varphi(x, y) \le \|x + y\| \text{ for each } x, y \in X \quad (4)$$

$$\varphi(x, y)\mathbf{1} \le x + y \text{ for all } x, y \in X \quad (5)$$

$$\varphi(x, y) = \varphi(y, x) \text{ for all } x, y \in X; \quad (6)$$

$$\varphi(x, -x) = \sup\{\lambda \in \mathbb{R} : \lambda\mathbf{1} \le x - x = 0\} = 0 \text{ for all } x \in X. \quad (7)$$

For each $y \in X$, define the function $\varphi_y : X \longrightarrow \mathbb{R}$ by

$$\varphi_y(x) := \varphi(x, y) \quad \forall\, x \in X. \quad (8)$$

The function φ_y defined by (8) is topical (see [MR05]).

Let S be a set and $L = \{h : S \longrightarrow \bar{\mathbb{R}} : h \text{ is a function}\}$ be a set of functions. We recall (see [Rub00, Sin87]) that a function $f : S \longrightarrow \bar{\mathbb{R}}$ is called *abstract convex* with respect to L, or, briefly, *L-convex*, if there exists a subset L_0 of L such that

$$f(s) = \sup_{h \in L_0,\ h \le f} h(s) \quad (s \in S).$$

Proposition 1. *Let $f : X \longrightarrow \mathbb{R}$ be a topical function. Then f is Lipschitz continuous.*

Proof. Let $x, y \in X$ be arbitrary. Since by (2) we have

$$-\|x - y\|\mathbf{1} \le x - y \le \|x - y\|\mathbf{1},$$

it follows that

$$y - \|x - y\|\mathbf{1} \le x \le y + \|x - y\|\mathbf{1}.$$

Since by hypothesis f is topical, we get

$$f(y) - \|x - y\| \le f(x) \le f(y) + \|x - y\|,$$

and hence

$$|f(x) - f(y)| \le \|x - y\|. \tag{9}$$

Thus, f is Lipschitz continuous. □

Corollary 1. *The function φ_y defined by (8) is Lipschitz continuous.*

Proof. It follows from Proposition 1. □

Corollary 2. *The function φ defined by (3) is continuous.*

Proof. It follows from (9). □

Proposition 2. *Let $f : X \longrightarrow \bar{\mathbb{R}}$ be a topical function. Then the following assertions are true:*
1) If there exists $x \in X$ such that $f(x) = +\infty$, then $f \equiv +\infty$.
2) If there exists $x \in X$ such that $f(x) = -\infty$, then $f \equiv -\infty$.

Proof. 1) Suppose that there exists $x \in X$ such that $f(x) = +\infty$, and let $y \in X$ be arbitrary. Let $\lambda = \varphi(-x, y)$, where φ is the function defined by (3). Then by (4) we have $\lambda \in \mathbb{R}$. In view of (5), it follows that $\lambda \mathbf{1} \le y - x$, and so $x + \lambda \mathbf{1} \le y$. Since f is a topical function, we conclude that $f(x) + \lambda \le f(y)$. This implies that $f(y) = +\infty$.
2) Assume that there exists $x \in X$ such that $f(x) = -\infty$, and let $y \in X$ be arbitrary. Let $\lambda = \varphi(x, -y)$, where φ is the function defined by (3). Then by (4) we have $\lambda \in \mathbb{R}$. In view of (5), it follows that $\lambda \mathbf{1} \le x - y$, and so $y + \lambda \mathbf{1} \le x$. Since f is a topical function, we conclude that $f(y) \le f(x) - \lambda$. This implies that $f(y) = -\infty$, which completes the proof. □

It follows from Proposition 2, for any topical function $f : X \longrightarrow \bar{\mathbb{R}}$, either we have $\operatorname{dom} f = X$ or $f \equiv +\infty$, where $\operatorname{dom} f := \{x \in X : f(x) < +\infty\}$.

In the following we denote by X_φ the set of all functions φ_l ($l \in X$) defined by (8). That is:

$$X_\varphi = \{\varphi_l := \varphi(., l) : l \in X\}. \tag{10}$$

Theorem 1. *Let φ be the function defined by (3). Then for a function $f : X \longrightarrow \bar{\mathbb{R}}$ the following assertions are equivalent:*
1) f is a topical function.
2) For each $y \in X$ there exists $l_y \in X$ such that

$$\varphi_{l_y}(x) \le f(x) \quad \forall\, x \in X, \quad \text{and} \quad \varphi_{l_y}(y) = f(y).$$

3) f is X_φ-convex, where X_φ is defined by (10).

Proof. 1) $\implies$ 2). Suppose that f is a topical function and let $y \in X$ be arbitrary. Define

$$l_y := f(y)\mathbf{1} - y \in X. \tag{11}$$

Now, let $x \in X$ be arbitrary and $\lambda := \varphi(x, -y)$. Then by (5) we have $\lambda\mathbf{1} \leq x - y$, and so $y + \lambda\mathbf{1} \leq x$. Using (11) and that $\varphi(x, .)$ and f are topical functions, we obtain

$$f(x) \geq f(y + \lambda\mathbf{1}) = f(y) + \lambda = f(y) + \varphi(x, -y)$$

$$= \varphi(x, f(y)\mathbf{1} - y) = \varphi(x, l_y) = \varphi_{l_y}(x).$$

Also, by using (7) we have

$$\varphi_{l_y}(y) = \varphi(y, f(y)\mathbf{1} - y) = \sup\{\lambda \in \mathbb{R} : \lambda\mathbf{1} \leq y + f(y)\mathbf{1} - y\}$$

$$= \sup\{\lambda \in \mathbb{R} : \lambda\mathbf{1} \leq f(y)\mathbf{1}\} = \sup\{\alpha + f(y) \in \mathbb{R} : \alpha\mathbf{1} \leq 0\}$$

$$= \sup\{\alpha \in \mathbb{R} : \alpha\mathbf{1} \leq 0\} + f(y) = 0 + f(y) = f(y).$$

Hence, we have 2).

2) $\implies$ 3). Assume that 2) holds. Then we have

$$f(x) = \sup_{y \in X} \varphi_{l_y}(x) \quad (x \in X),$$

and hence f is X_φ-convex.

3) $\implies$ 1). Assume that 3) holds. First, note that it is easy to check that every supremum of topical functions defined on X is a topical function. Since every function φ_l ($l \in X$) defined by (8) is topical, it follows from the hypothesis that f is a topical function, which completes the proof. $\square$

Corollary 3. *Every topical function* $f : X \longrightarrow \bar{\mathbb{R}}$ *is lower semi-continuous.*

3 Plus-Minkowski gauge and plus-weak Pareto point for a downward set

We start with the following definition, which is given in [MRS02], [RS01] for the finite dimensional case.

Definition 1. *Let W be a downward subset of X. The function $\rho_W : X \longrightarrow \bar{\mathbb{R}}$ defined by*

$$\rho_W(x) = \inf\{\lambda \in \mathbb{R} : x \in \lambda\mathbf{1} + W\} \quad (x \in X)$$

is called the plus-Minkowski gauge of the set W.

The following proposition has been proved in finite dimensional case (see [RS01]). However, the same proof is valid in the case under consideration.

Proposition 3. *Let W be a downward subset of X. Then ρ_W is a topical function.*

In the sequel, we give a definition of plus-weak Pareto points.

Definition 2. *Let W be a closed downward subset of X. A point $w \in W$ is called a plus-weak Pareto point of W if $(\lambda 1 + w) \notin W$ for all $0 < \lambda \in \mathbb{R}$.*

Lemma 1. *Let W be a closed downward subset of X and $w \in W$ be arbitrary. Then w is a plus-weak Pareto point of W if and only if $\rho_W(w) = 0$.*

Proof. Let
$$D_w = \{\lambda \in \mathbb{R} : w \in \lambda 1 + W\} \quad (w \in W).$$

Then we have

$$w \text{ is a plus} - weak \text{ } Pareto \text{ } point \text{ } of \text{ } W \Longleftrightarrow (\lambda 1 + w) \notin W \quad \forall \, \lambda > 0$$
$$\Longleftrightarrow w \notin -\lambda 1 + W \quad \forall \, \lambda > 0$$
$$\Longleftrightarrow -\lambda \notin D_w \quad \forall \, \lambda > 0$$
$$\Longleftrightarrow \lambda \notin D_w \quad \forall \, \lambda < 0$$
$$\Longleftrightarrow \lambda \in D_w \quad \forall \, \lambda \geq 0$$
$$\Longleftrightarrow \rho_W(w) = \inf D_w = 0.$$

$\square$

Lemma 2. *Let W be a closed downward subset of X and $w \in W$ be arbitrary. Then the following assertions are equivalent:*
1) w is a plus-weak Pareto point of W.
2) $w \in \operatorname{bd} W$.

Proof. 1) $\Longrightarrow$ 2). Assume 1) holds and if possible that $w \notin \operatorname{bd} W$. Then, $w \in \operatorname{int} W$. It follows that there exists $\varepsilon > 0$ such that

$$V := \{x \in X : \|x - w\| \leq \varepsilon\} \subset W.$$

This implies, by (2), that $w + \varepsilon 1 \in W$. Hence w is not a plus-weak Pareto point of W. This is a contradiction.
2) $\Longrightarrow$ 1). Suppose that 2) holds. We claim that $\lambda 1 + w \notin W$ for all $\lambda > 0$. Assume if possible that there exists $\lambda_0 > 0$ such that $\lambda_0 1 + w \in W$. Let

$$V = \{x \in X : \|x - w\| < \lambda_0\}$$

be a neighbourhood of w. It follows from (2) that

$$V = \{x \in X : w - \lambda_0 1 < x < w + \lambda_0 1\}.$$

Since W is a downward set and $\lambda_0 1 + w \in W$, we conclude that $V \subset W$. Hence, $w \in \operatorname{int} W$. This is a contradiction. Thus, the claim is true, and so w is a plus-weak Pareto ponit of W, which completes the proof. $\square$

Proposition 4. *Let* $0 \neq x \in X$ *and* $R = \{\alpha \mathbf{1} + x : \alpha \geq 0\}$. *Let* W *be a closed downward subset of* X. *Then* $|R \cap \mathrm{bd}\, W| \leq 1$, *where* $|A|$ *denotes the cardinality of the set* A.

Proof. If $R \cap \mathrm{bd}\, W = \emptyset$, then $|R \cap \mathrm{bd}\, W| = 0 < 1$. Now, suppose that $R \cap \mathrm{bd}\, W \neq \emptyset$. We may assume that $x \in R \cap \mathrm{bd}\, W$. Thus, $x \in \mathrm{bd}\, W \subset W$. It follows from Lemma 2 that x is a plus-weak Pareto point of W, and so $\lambda \mathbf{1} + x \notin W$ for all $\lambda > 0$.

On the other hand, assume if possible that there exists $\lambda_0 < 0$ such that $\lambda_0 \mathbf{1} + x \in \mathrm{bd}\, W$. Then, by Lemma 2, $\lambda_0 \mathbf{1} + x$ is a plus-weak Pareto point of W. Hence for $-\lambda_0 > 0$, we have $[-\lambda_0 \mathbf{1} + (\lambda_0 \mathbf{1} + x)] \notin W$. That is, $x \notin W$. This is a contradiction. It follows that $\lambda \mathbf{1} + x \in W$ only for $\lambda = 0$. Consequently, $R \cap \mathrm{bd}\, W = \{x\}$, and hence $|R \cap \mathrm{bd}\, W| = 1$. $\qquad\square$

4 X_φ-subdifferential of a topical function

Definition 3. *Let* $f : X \longrightarrow \bar{R}$ *be a topical function and* φ *be the function defined by (3). Define the* X_φ-*subdifferential* $\partial_{X_\varphi} f(x)$ *of* f *at a point* $y \in X$ *by*

$$\partial_{X_\varphi} f(y) = \{l \in X : \varphi_l(x) \leq f(x) \ \forall \, x \in X, \text{ and } \varphi_l(y) = f(y)\}, \qquad (12)$$

where X_φ *is defined by (10).*

Lemma 3. *Let* $f : X \longrightarrow R$ *be a topical function and let* $y \in X$. *Then*

$$\partial_{X_\varphi} f(y) = \{l \in X : \varphi_l(y) \geq f(y), \text{ and } f(-l) = 0\}.$$

Hence, in particular, $(f(y)\mathbf{1} - y) \in \partial_{X_\varphi} f(y)$.

Proof. Let

$$D = \{l \in X : \varphi_l(y) \geq f(y), \text{ and } f(-l) = 0\}$$

and let $l \in \partial_{X_\varphi} f(y)$ be arbitrary. Then, by (12), we have $\varphi_l(y) \geq f(y)$. This implies, by (5), that $y + l \geq \varphi_l(y)\mathbf{1} \geq f(y)\mathbf{1}$, and so $y \geq f(y)\mathbf{1} - l$. Since f is a topical function, it follows that $f(y) \geq f(y) + f(-l)$. Thus, $f(-l) \leq 0$.

On the other hand, by (7), we have $f(-l) \geq \varphi_l(-l) := \varphi(-l, l) = 0$. Hence, $f(-l) = 0$. Therefore, $l \in D$. Conversely, assume $l \in D$ and if possible that there exists $x \in X$ such that $\varphi_l(x) > f(x)$. This implies that there exists $\lambda > 0$ such that $\varphi_l(x) > f(x) + \lambda$, and so by (5) we get $x > (f(x) + \lambda)\mathbf{1} - l$. Since f is a topical function and that $f(-l) = 0$, it follows that

$$f(x) \geq f(x) + \lambda + f(-l) = f(x) + \lambda.$$

This is a contradiction. Thus we conclude that $\varphi_l(x) \leq f(x)$ for all $x \in X$, and hence, in particular, we have $\varphi_l(y) \leq f(y)$. Consequently, since $l \in D$, we obtain $\varphi_l(y) = f(y)$, and so $l \in \partial_{X_\varphi} f(y)$.

Finally, let $l_0 = f(y)\mathbf{1} - y$. Since $\varphi(y, .)$ is a topical function and (7) holds, it follows that

$$\varphi_{l_0}(y) = \varphi(y, l_0) = \varphi(y, f(y)\mathbf{1} - y) = f(y) + \varphi(y, -y) = f(y).$$

Also, we have $f(-l_0) = f(y) - f(y) = 0$. We conclude that $l_0 \in D = \partial_{X_\varphi} f(y)$, which completes the proof.   □

Remark 1. If W is a downward subset of X and ρ_W is its plus-Minkowski gauge function, then

$$\{x \in X : \rho_W(x) < 0\} \subset W \subset \{x \in X : \rho_W(x) \leq 0\}.$$

Indeed, if $x \in \{x \in X : \rho_W(x) < 0\}$, then there exists $\lambda < 0$ such that $x \in \lambda\mathbf{1} + W$. Since $x < x - \lambda\mathbf{1}$, $x - \lambda\mathbf{1} \in W$ and W is a downward set, it follows that $x \in W$. Also, note that if W is a closed downward subset of X, then

$$W = \{x \in X : \rho_W(x) \leq 0\}.$$

Lemma 4. *Let W be a proper closed downward subset of X, $w \in W$ be a plus-weak Pareto ponit of W and $l \in X$. Assume that φ is the function defined by (3). Then the following assertions are equivalent:*
1) $l \in \partial_{X_\varphi} \rho_W(w)$.
2) $\sup_{y \in W} \varphi(y, l) \leq 0 = \varphi(w, l)$.

Proof. Since w is a plus-weak Pareto point of W, it follows from Lemma 3.1 that $\rho_W(w) = 0$.
1) $\implies$ 2). Suppose that 1) holds. Then, by Definition 3 and Remark 1, we have

$$\varphi(y, l) \leq \rho_w(y) \leq 0 \quad \forall y \in W$$

and $\varphi(w, l) = \varphi_l(w) = \rho_W(w) = 0$. Hence, $\sup_{y \in W} \varphi(y, l) \leq 0 = \varphi(w, l)$.
2) $\implies$ 1). Assume that 2) holds. Let $y \in X$ and $x = y - \rho_W(y)\mathbf{1}$. Since, by Proposition 3, ρ_W is a topical function, it follows that $\rho_W(x) = 0$. In view of Remark 1, we have $x \in W$. Thus, by hypothesis, $\varphi(x, l) \leq 0$. This implies that $\varphi_l(y) \leq \rho_W(y)$ for all $y \in X$. Also, we have $\varphi_l(w) := \varphi(w, l) = 0 = \rho_W(w)$. Hence, by Definition 3, $l \in \partial_{X_\varphi} \rho_W(w)$, which completes the proof.   □

Theorem 2. *Let W be a closed downward subset of X, $x_0 \in X \setminus W$, $w_0 \in W$ and $r_0 = \|x_0 - w_0\|$. If there exists $l \in X$ such that*

$$\varphi(w, l) \leq 0 \leq \varphi(y, l) \quad \forall w \in W, y \in B(x_0, r_0).$$

Then w_0 is a plus-weak Pareto point of W.

Proof. Since $r_0 = \|x_0 - w_0\|$, then $w_0 \in B(x_0, r_0)$. Also, we have $w_0 \in W$. It follows by hypothesis that $\varphi(w_0, l) = 0$. Now, assume if possible that w_0 is not a plus-weak Pareto point of W. Then there exists $\lambda_0 > 0$ such that

$\lambda_0 1 + w_0 \in W$, and hence by hypothesis, $\varphi(\lambda_0 1 + w_0, l) \leq 0$. This implies, since $\varphi(., l)$ is a topical function, that

$$0 \geq \varphi(\lambda_0 1 + w_0, l) = \lambda_0 + \varphi(w_0, l) = \lambda_0 + 0 = \lambda_0.$$

This is a contradiction. □

Remark 2. If W is a closed downward subset of X and $x_0 \in X$, then the least element $g_0 = x_0 - r1$ of the set $P_W(x_0)$ exists (see [MR05], Proposition 3.2), where $r = d(x_0, W)$.

Theorem 3. *Let W be a closed downward subset of X, $x_0 \in X \setminus W$ and $g_0 = x_0 - r_0 1$ be the least element of the set $P_W(x_0)$, where $r_0 = d(x_0, W)$. Then the following assertions are equivalent:*
1) g_0 is a plus-weak Pareto point of W.
2) There exists $l \in X$ such that

$$\varphi(w, l) \leq 0 \leq \varphi(y, l) \quad \forall w \in W, \ y \in B(x_0, r_0).$$

Proof. 1) $\implies$ 2). Suppose that 1) holds. Let $l = -g_0$ and $y \in B(x_0, r_0)$ be arbitrary. Since $g_0 = x_0 - r_0 1$, it follows that g_0 is also the least element of $B(x_0, r_0)$. Hence, $g_0 \leq y$, and so by (7) and that $\varphi(., l)$ is a topical function, we have

$$0 = \varphi(g_0, l) \leq \varphi(y, l) \quad \forall y \in B(x_0, r_0). \tag{13}$$

On the other had, by hypothesis g_0 is a plus-weak Pareto point of W. In view of Lemma 1, we have $\rho_W(g_0) = 0$. It follows from Lemma 3 that $l = -g_0 = \rho_W(g_0)1 - g_0 \in \partial_{X_\varphi} \rho_W(g_0)$. Thus, by Lemma 4, we have

$$\varphi(w, l) \leq 0 \quad \forall w \in W. \tag{14}$$

Therefore, (13) and (14) imply 2).
2) $\implies$ 1). Assume that 2) holds. Since $g_0 \in P_W(x_0)$ and $r_0 = d(x_0, W)$, it follows that $r_0 = \|x_0 - g_0\|$. Therefore, In view of Theorem 2, we have g_0 is a plus-weak Pareto point of W, which completes the proof. □

Corollary 4. *Let W be a closed downward subset of X, $x_0 \in X \setminus W$ and $g_0 = x_0 - r_0 1$ be the least element of the set $P_W(x_0)$, where $r_0 = d(x_0, W)$. Then there exists $l \in X$ such that*

$$\varphi(w, l) \leq 0 \leq \varphi(y, l) \quad \forall w \in W, \ y \in B(x_0, r_0).$$

Proof. Since $g_0 \in P_W(x_0)$ and $P_W(x_0) \subset \text{bd}\, W$, then $g_0 \in \text{bd}\, W$, and so by Lemma 2, g_0 is a plus-weak Pareto point of W. Hence, by Theorem 3, there exists $l \in X$ such that

$$\varphi(w, l) \leq 0 \leq \varphi(y, l) \quad \forall w \in W, \ y \in B(x_0, r_0),$$

and the proof is complete. □

The following example shows that every plus-weak Pareto point of a closed downward set W need not separate W and ball $B(x_0, r_0)$.

Example 1. Let $X = \mathbb{R}^2$ with the maximum norm $\|x\| = \max_{1 \leq i \leq 2} |x_i|$ and

$$C = \{(x_1, x_2) \in \mathbb{R}^2 : x_1 \geq 0, \ x_2 \geq 0\}.$$

Let

$$W = \{(w_1, w_2) \in \mathbb{R}^2 : \min\{w_1, w_2\} \leq 1\},$$

$x_0 = (2, 2) \in X \setminus W$ and $w_0 = (1, 3)$. It is clear that C is a closed convex normal cone in X, W is a closed downward subset of X and $w_0 \in \text{bd}\,W$. Also, we have $\mathbf{1} = (1, 1) \in \text{int}\,C$. We have $d(x_0, W) = 1 = \|x_0 - g_0\|$, where $g_0 = (1, 1)$ is the least element of the set $P_W(x_0)$. Since $w_0 \in \text{bd}\,W$, it follows from Lemma 2 that w_0 is a plus-weak Pareto point of W, and we have also $r_0 := \|x_0 - w_0\| = 1 = d(x_0, W)$.

Now, let $l = -w_0$ and $w = (w_1, w_2) \in W$ be arbitrary. Then we have

$$\varphi(w, l) = \varphi(w, -w_0) = \sup\{\lambda \in \mathbb{R} : \lambda\mathbf{1} \leq w - w_0\}$$

$$= \sup\{\lambda \in \mathbb{R} : \lambda \leq \min\{w_1 - 1, w_2 - 3\}\} \leq 0, \qquad (15)$$

and

$$\varphi(x_0, l) = \varphi(x_0, -w_0) = \sup\{\lambda \in \mathbb{R} : \lambda\mathbf{1} \leq x_0 - w_0\}$$

$$= \sup\{\lambda \in \mathbb{R} : \lambda \leq -1\} = -1 < 0. \qquad (16)$$

Therefore, (15) and (16) show that $-w_0$ does not separate W and $B(x_0, r_0)$.

Theorem 4. *Let W be a closed downward subset of X, $x_0 \in X \setminus W$, $w_0 \in W$ and $r_0 = \|x_0 - w_0\|$. If there exists $l \in X$ such that*

$$\varphi(w, l) \leq 0 \leq \varphi(y, l) \quad \forall\, w \in W, \ y \in B(x_0, r_0). \qquad (17)$$

Then, $w_0 \in P_W(x_0)$. Moreover, if (17) holds with $l = -w_0$, then, $w_0 = \min P_W(x_0) := x_0 - r\mathbf{1}$, where $r = d(x_0, W)$.

Proof. Let $g_0 = x_0 - r\mathbf{1}$ be the least element of the set $P_W(x_0)$. It is clear that $g_0 \leq x_0$. Now, assume if possible that $w_0 \notin P_W(x_0)$. Then, $r < r_0$. Choose $\lambda \in \mathbb{R}$ such that $1 - r_0 r^{-1} < \lambda < 0$, and let $w = \lambda x_0 + (1 - \lambda)g_0$. Since $g_0 \leq x_0$, it follows that $w - g_0 = \lambda(x_0 - g_0) \leq 0$, and so $w \leq g_0$. Since W is a downward set and $g_0 \in W$, we conclude that $w \in W$. Also, we have

$$\|x_0 - w\| = \|x_0 - \lambda x_0 - (1 - \lambda)g_0\| = (1 - \lambda)\|x_0 - g_0\| = (1 - \lambda)r < r_0,$$

and hence $w \in B(x_0, r_0)$. This implies by hypothesis that $\varphi(w, l) = 0$. But, on the other hand, since $\varphi(., l)$ is a topical function, we have

$$\varphi(w, l) = \varphi(g_0 + \lambda(x_0 - g_0), l) = \varphi(g_0 + r\lambda\mathbf{1}, l)$$

$$= \varphi(g_0, l) + r\lambda \leq 0 + r\lambda = r\lambda < 0.$$

This is a contradiction. Hence, $w_0 \in P_W(x_0)$.

Finally, Suppose that (17) holds with $l = -w_0$. Then by the above $w_0 \in P_W(x_0)$, and so $r = r_0$. Thus, we have $g_0 \in B(x_0, r_0)$, and hence $0 \leq \varphi(g_0, l) = \varphi(g_0, -w_0)$. In view of (5), we get $0 \leq \varphi(g_0, -w_0)\mathbf{1} \leq g_0 - w_0$. This implies that $w_0 \leq g_0$. Since g_0 is the least element of the set $P_W(x_0)$, it follows that $w_0 = g_0$, which completes the proof. $\qquad\square$

Theorem 5. *Let W be a closed downward subset of X, $x_0 \in X \setminus W$, $w_0 \in W$ and $r_0 = \|x_0 - w_0\|$. Then the following assertions are equivalent:*
1) $w_0 \in P_W(x_0)$.
2) There exists $l \in X$ such that

$$\varphi(w, l) \leq 0 \leq \varphi(y, l) \quad \forall\, w \in W,\, y \in B(x_0, r_0). \tag{18}$$

Proof. 1) $\implies$ 2). Suppose that 1) holds and $r := d(x_0, W)$. Then $r = r_0$. Since, by Lemma 2, $g_0 = x_0 - r_0\mathbf{1}$ the least element of the set $P_W(x_0)$ is a plus-weak Pareto point of W, it follows from Theorem 3 that there exists $l \in X$ such that

$$\varphi(w, l) \leq 0 \leq \varphi(y, l) \quad \forall\, w \in W,\, y \in B(x_0, r_0).$$

The implication 2) $\implies$ 1) follows from Theorem 4. $\qquad\square$

5 Fenchel-Moreau conjugates with respect to φ

Recall (see [Rub00, Sin87]) that if V and W are sets and $\theta : V \times W \longrightarrow \mathbb{R}$ is a coupling function, then for a function $f : V \longrightarrow \bar{\mathbb{R}}$ the *Fenchel-Moreau conjugate function of f with respect to θ* is the function $f^{c(\theta)} : W \longrightarrow \bar{\mathbb{R}}$ defined by

$$f^{c(\theta)}(w) := \sup_{v \in V}\{\theta(v, w) - f(v)\} \quad (w \in W). \tag{19}$$

We point out that $(-\infty)^{c(\theta)} = +\infty$ and $(+\infty)^{c(\theta)} = -\infty$.

Also, we recall that the dual of any mapping $u : \bar{\mathbb{R}}^V \longrightarrow \bar{\mathbb{R}}^W$ is the mapping $u' : \bar{\mathbb{R}}^W \longrightarrow \bar{\mathbb{R}}^V$ defined by

$$h^{u'}(v) = \inf_{f \in \bar{\mathbb{R}}^V,\, f^u \leq h} f \quad (h \in \bar{\mathbb{R}}^W), \tag{20}$$

where for any mapping $u : \bar{\mathbb{R}}^V \longrightarrow \bar{\mathbb{R}}^W$ and any $f \in \bar{\mathbb{R}}^V$ we write f^u instead of $u(f)$, and for a set A, $\bar{\mathbb{R}}^A$ denotes the set of all functions $g : A \longrightarrow \bar{\mathbb{R}}$.

In the sequel, we define the coupling function $\psi : X \times X \longrightarrow \mathbb{R}$ by

$$\psi(x, y) := \inf\{\lambda \in \mathbb{R} : x + y \leq \lambda\mathbf{1}\} \quad \forall\, x, y \in X. \tag{21}$$

It follows from (1) that the set $\{\lambda \in \mathbb{R} : x + y \leq \lambda \mathbf{1}\}$ is non-empty and bounded from below (by $-\|x + y\|$). Clearly this set is closed. It follows from the definition of ψ that it enjoys the following properties:

$$-\|x + y\| \leq \psi(x, y) < +\infty \text{ for each } x, y \in X \tag{22}$$

$$x + y \leq \psi(x, y)\mathbf{1} \text{ for all } x, \ y \in X \tag{23}$$

$$\psi(x, y) = \psi(y, x) \text{ for all } x, \ y \in X; \tag{24}$$

$$\psi(x, -x) = \inf\{\lambda \in \mathbb{R} : 0 = x - x \leq \lambda \mathbf{1}\} = 0 \text{ for all } x \in X. \tag{25}$$

For each $y \in X$, define the function $\psi_y : X \longrightarrow \mathbb{R}$ by

$$\psi_y(x) := \psi(x, y) \quad \forall\, x \in X. \tag{26}$$

It is not difficult to show that the function ψ_y is topical and Lipschitz continuous and consequently, ψ is continuous (see Proposition 1 and its corollaries).

Definition 4. *Let W be a non-empty subset of X and $\theta : X \times X \longrightarrow \mathbb{R}$ be a coupling function. We define the plus-polar set of W by*

$$W^\circ = \{x \in X : \theta(x, w) \leq 0, \ \forall w \in W\},$$

and the plus-bipolar set of W by

$$W^{\circ\circ} = (W^\circ)^\circ = \{x \in X : \theta(x, w) \leq 0, \ \forall w \in W^\circ\}.$$

Clearly, $X^\circ = \emptyset$, and by definition, $\emptyset^\circ = X$.

Theorem 6. *Let φ be the function defined by (3). Then for a function $f : X \longrightarrow \bar{\mathbb{R}}$, the following assertions are equivalent:*
1) f is topical.
2) We have

$$f^{c(\varphi)}(x) = -f(-x) \quad (x \in X).$$

Proof. 1) $\implies$ 2). Assume that f is a topical function. Let $x, \ y \in X$ be arbitrary. It follows from (5) that $\varphi(x, y)\mathbf{1} \leq x + y$, and hence $x \geq \varphi(x, y)\mathbf{1} - y$. Since f is a topical function, we conclude that

$$\varphi(x, y) - f(x) \leq -f(-y) \quad (x, \ y \in X),$$

and so

$$f^{c(\varphi)}(y) = \sup_{x \in X}\{\varphi(x, y) - f(x)\} \leq -f(-y) \quad (y \in X). \tag{27}$$

Also, by definition of the Fenchel-Moreau conjugate function of f and (7), we have

$$f^{c(\varphi)}(y) = \sup_{x \in X}\{\varphi(x,y) - f(x)\} \geq \varphi(-y,y) - f(-y) = -f(-y) \quad (y \in X).$$
$$(28)$$

Hence (27) and (28) imply 2).

2) $\implies$ 1). Suppose that 2) holds. Then we have

$$f(x) = -f^{c(\varphi)}(-x) \quad (x \in X).$$

It is not difficult to show that for any function $f : X \longrightarrow \bar{\mathbb{R}}$, $f^{c(\varphi)}$ is a topical function, and hence we conclude that f is a topical function, which completes the proof. $\qquad\square$

The proof of the following theorem is similar to that in finite dimensional case (see [RS01]).

Theorem 7. *Let $f : X \longrightarrow \bar{\mathbb{R}}$ be a plus-homogeneous function and $\theta : X \times X \longrightarrow \mathbb{R}$ be a coupling function such that $\theta(.,y)$ $(y \in X)$ is a topical function. Then*

$$f^{c(\theta)}(y) = \sup_{x \in X,\ f(x)=0} \theta(x,y) = \sup_{x \in S_0(f)} \theta(x,y) \quad (y \in X).$$

Corollary 5. *Let $f : X \longrightarrow \bar{\mathbb{R}}$ be a plus-homogeneous function and $\theta : X \times X \longrightarrow \mathbb{R}$ be a coupling function such that $\theta(.,y)$ $(y \in X)$ is a topical function. Then*

$$S_0(f^{c(\theta)}) = S_0(f)^\circ.$$

Proof. The proof follows from Definition 4 and Theorem 7. $\qquad\square$

Remark 3. We recall (see [Rub00]) that if X is a set and $\theta : X \times X \longrightarrow \mathbb{R}$ is a coupling function such that

$$\theta(x,y) = \theta(y,x) \quad (x,\ y \in X),$$

that is, θ is symmetric. Then the Fenchel-Moreau conjugate mapping $c(\theta) : \bar{\mathbb{R}}^X \longrightarrow \bar{\mathbb{R}}^X$ of (19), is self-dual. That is, $c(\theta) = c(\theta)'$.

We recall (see [Rub00]) that if V and W are sets and $\theta : V \times W \longrightarrow \mathbb{R}$ is a coupling function, then for a function $f : V \longrightarrow \bar{\mathbb{R}}$ the *Fenchel-Moreau biconjugate function of f with respect to θ*, is the function $f^{c(\theta)c(\theta)'} : V \longrightarrow \bar{\mathbb{R}}$ defined by

$$f^{c(\theta)c(\theta)'}(v) := (f^{c(\theta)})^{c(\theta)'}(v) \quad (v \in V).$$

For the proof of the following theorem see [RS01] in finite dimensional case. The same proof is valid in the case under consideration.

Theorem 8. *Let φ be the function defined by (3). Then for a function $f : X \longrightarrow \bar{\mathbb{R}}$, the following assertions are equivalent:*

1) f is topical.

2) We have

$$f^{c(\varphi)c(\varphi)'}(x) = f(x) \quad (x \in X).$$

Proposition 5. *Let φ and ψ be the functions defined by (3) and (21), respectively. Let $f : X \longrightarrow \bar{I\!R}$ be a plus-homogeneous function. Then the following assertions are true:*

1) We have

$$f^{c(\psi)c(\psi)'}(x) = \sup_{y \in S_0(f)^o} \psi(x,y) = f^{c(\varphi)c(\psi)'}(x) \quad (x \in X).$$

2) We have

$$f^{c(\varphi)c(\varphi)'}(x) = \sup_{y \in S_0(f)^o} \varphi(x,y) = f^{c(\psi)c(\varphi)'}(x) \quad (x \in X).$$

3) We have

$$S_0(f^{c(\psi)c(\psi)'}) = S_0(f^{c(\psi)c(\varphi)'}) = S_0(f^{c(\varphi)c(\psi)'})$$

$$= S_0(f^{c(\varphi)c(\varphi)'}) = S_0(f)^{oo}.$$

Proof. 1). It is easy to check that $f^{c(\psi)}$ and $f^{c(\varphi)}$ are topical functions. Since ψ and φ are symmetric coupling functions, It follows from Remark 3 that $c(\psi) = c(\psi)'$ and $c(\varphi) = c(\varphi)'$. Therefore, by Theorem 7 and Corollary 5, we conclude that

$$f^{c(\psi)c(\psi)'}(x) = (f^{c(\psi)})^{c(\psi)}(x)$$

$$= \sup_{y \in S_0(f^{c(\psi)})} \psi(x,y) = \sup_{y \in S_0(f)^o} \psi(x,y) \quad (x \in X),$$

and

$$f^{c(\varphi)c(\psi)'}(x) = (f^{c(\varphi)})^{c(\psi)}(x)$$

$$= \sup_{y \in S_0(f^{c(\varphi)})} \psi(x,y) = \sup_{x \in S_0(f)^o} \psi(x,y) \quad (x \in X),$$

which proves 1). The proof of statement 2) is similar to the proof of statement 1).

3). We apply Corollary 5 to the functions $f^{c(\psi)}$, $f^{c(\varphi)}$ and f, it follows that

$$S_0(f^{c(\psi)c(\psi)'}) = S_0((f^{c(\psi)})^{c(\psi)}) = (S_0(f^{c(\psi)}))^o = (S_0(f)^o)^o = S_0(f)^{oo},$$

and

$$S_0(f^{c(\varphi)c(\psi)'}) = S_0((f^{c(\varphi)})^{c(\psi)}) = (S_0(f^{c(\varphi)}))^o = (S_0(f)^o)^o = S_0(f)^{oo}.$$

By a similar proof, we have

$$S_0(f^{c(\psi)c(\varphi)'}) = S_0(f)^{oo} = S_0(f^{c(\varphi)c(\varphi)'}),$$

which completes the proof. $\qquad\square$

6 Conjugate of type Lau with respect to φ

Recall (see [Sin87]) that if V and W are sets and $\triangle : 2^V \longrightarrow 2^W$ is any duality, then for a function $f : V \longrightarrow \bar{\mathbb{R}}$ the *conjugate of type Lau of f with respect to* $\triangle$, is the function $f^{L(\triangle)} : W \longrightarrow \bar{\mathbb{R}}$ defined by

$$f^{L(\triangle)}(w) := - \inf_{v \in V, \, w \in W \setminus \triangle(\{v\})} f(v) \quad (w \in W). \tag{29}$$

If $\theta : V \times W \longrightarrow \bar{\mathbb{R}}$ is a coupling function, then for the conjugate of type Lau $f^{L(\triangle_\theta)}$ with respect to the θ-duality $\triangle_\theta : 2^V \longrightarrow 2^W$ defined by

$$\triangle_\theta(G) := \{w \in W : \theta(g, w) \leq 0, \ \forall g \in G\} \quad (G \subset V),$$

which will be also called the conjugate of type Lau with respect to θ, and denoted by $f^{L(\theta)}$, we have

$$f^{L(\theta)}(w) = f^{L(\triangle_\theta)}(w) = - \inf_{v \in V, \, \theta(v, w) > 0} f(v) \quad (f : V \longrightarrow \bar{\mathbb{R}}, \ w \in W). \tag{30}$$

Remark 4. We recall (see [Sin87]) that if V and W are sets, then for any duality $\triangle : 2^V \longrightarrow 2^W$ and any function $f : V \longrightarrow \bar{\mathbb{R}}$, the lower level set $S_\lambda(f^{L(\triangle)})$ $(\lambda \in \mathbb{R})$ has the following form:

$$S_\lambda(f^{L(\triangle)}) = \cap_{v \in V, \, f(v) \leq -\lambda} \triangle(\{v\}).$$

Remark 5. Note that since C is a closed convex normal cone in X and $1 \in \text{int} \, C$, it is not difficult to show that

$$\varphi(x, y) > 0 \Longleftrightarrow x + y \in \text{int} \, C \quad (x, \, y \in X),$$

where φ is the function defined by (3).

Therefore, for the coupling function $\varphi : X \times X \longrightarrow \mathbb{R}$ defined by (3) and a function $f : X \longrightarrow \bar{\mathbb{R}}$, it follows from (30) and Remark 5 that

$$f^{L(\varphi)}(y) = - \inf_{x \in X, \, \varphi(x, y) > 0} f(x) = - \inf_{x \in X, \, x + y \in \text{int} \, C} f(x) \quad (y \in X). \tag{31}$$

Proposition 6. *Let $\theta : X \times X \longrightarrow \mathbb{R}$ be a coupling function such that $\theta(x, .)$ $(x \in X)$ is an increasing and lower semi-continuous function. Then for any function $f : X \longrightarrow \bar{\mathbb{R}}$, the conjugate of type Lau $f^{L(\theta)} : X \longrightarrow \bar{\mathbb{R}}$ is an increasing and lower semi-continuous function.*

Proof. Let $y, \, z \in X$ and $y \leq z$. Since $\theta(x, .)$ $(x \in X)$ is an increasing function, it follows that

$$A := \{x \in X : \theta(x, y) > 0\} \subset B := \{x \in X : \theta(x, z) > 0\}.$$

This implies, by (31), that

$$f^{L(\theta)}(y) = -\inf_{x \in A} f(x) \le -\inf_{x \in B} f(x) = f^{L(\theta)}(z).$$

Hence, $f^{L(\theta)}$ is an increasing function.

Finally, it follows from Remark 4 that

$$S_\lambda(f^{L(\theta)}) = \cap_{x \in X,\, f(x) \le -\lambda} \{y \in X : \theta(x,y) \le 0\} = \cap_{x \in X,\, f(x) \le -\lambda} E_x \quad (\lambda \in \mathbb{R}),$$

where $E_x := \{y \in X : \theta(x,y) \le 0\}$ $(x \in X)$. Since $\theta(x,.)$ $(x \in X)$ is lower semi-continuous, we have E_x is a closed set in X, and hence $S_\lambda(f^{L(\theta)})$ is closed for each $\lambda \in \mathbb{R}$. Thus, $f^{L(\theta)}$ is lower semi-continuous, which completes the proof. $\qquad\square$

Lemma 5. *Let $\theta : X \times X \longrightarrow \mathbb{R}$ be a coupling function such that $\theta(x,.)$ $(x \in X)$ is an increasing and lower semi-continuous function. Let $f : X \longrightarrow \bar{\mathbb{R}}$ be any function such that*

$$f^{L(\theta)}(x) = -f(-x) \quad (x \in X).$$

Then f is increasing and upper semi-continuous.

Proof. This is an immediate consequence of Proposition 6 and that the function $h(x) := -f(-x)$ $(x \in X)$ is topical, whenever f is a topical function. $\qquad\square$

Corollary 6. *Let φ and ψ be the functions defined by (3) and (21), respectively. Let $f : X \longrightarrow \bar{\mathbb{R}}$ be any function such that*

$$f^{L(\varphi)}(x) = -f(-x) \quad (x \in X),$$

or

$$f^{L(\psi)}(x) = -f(-x) \quad (x \in X).$$

Then f is increasing and upper semi-continuous.

Theorem 9. *Let φ be the function defined by (3). Then for a function $f : X \longrightarrow \bar{\mathbb{R}}$, the following assertions are equivalent:*
1) We have

$$f^{L(\varphi)}(x) = -f(-x) \quad (x \in X).$$

2) f is increasing and upper semi-continuous.

Proof. The implication 1) $\implies$ 2) follows from Corollary 6.
2) $\implies$ 1). Assume that 2) holds. By (31) and that $\mathrm{cl}\,(\mathrm{int}\,C) = C$, we have

$$f^{L(\varphi)}(x) = -\inf_{y \in X,\, x+y \in \mathrm{int}\,C} f(y) = -\inf_{y \in X,\, x+y \in C} f(y) \quad (x \in X). \tag{32}$$

Now, let $x \in X$ be fixed and $y \in X$ be such that $x + y \in C$. Then $y \ge -x$. Since f is increasing, we have $f(y) \ge f(-x)$, and so $-f(y) \le -f(-x)$. In view of (32), we get

$$f^{L(\varphi)}(x) = - \inf_{y \in X,\ x+y \in C} f(y) = \sup_{y \in X,\ x+y \in C} (-f(y)) \leq -f(-x). \qquad (33)$$

We also have f is upper semi-continuous. It follows that $-f$ is lower semi-continuous, and hence by (32), we obtain

$$f^{L(\varphi)}(x) = - \inf_{y \in X,\ x+y \in C} f(y)$$

$$= \sup_{y \in X,\ x+y \in C} (-f(y)) = \sup_{y \in X,\ y \geq -x} (-f(y)) \geq -f(-x). \qquad (34)$$

Therefore, (33) and (34) imply 1), which completes the proof. $\qquad \square$

We recall (see [Sin02, Lemma 3.3]) that if V is a set and $\theta : V \times V \longrightarrow \bar{\mathbb{R}}$ is a symmetric coupling function, then the conjugate of type Lau $L(\theta) : \bar{\mathbb{R}}^V \longrightarrow \bar{\mathbb{R}}^V$ of (29) is self-dual, that is, $L(\theta) = L(\theta)'$. Also, if V and W are sets and $\triangle : 2^V \longrightarrow 2^W$ is any duality, then the biconjugate of type Lau of a function $f : V \longrightarrow \bar{\mathbb{R}}$ with respect to $\triangle$, is the function $f^{L(\triangle)L(\triangle)'} : V \longrightarrow \bar{\mathbb{R}}$ defined by $f^{L(\triangle)L(\triangle)'} := (f^{L(\triangle)})^{L(\triangle)'}$ (see [Sin87]). In particular, for the function φ defined by (3) and the φ-duality $\triangle_\varphi : 2^X \longrightarrow 2^X$ of (30), we have $f^{L(\varphi)L(\varphi)'} = (f^{L(\varphi)})^{L(\varphi)'}$, where $f : X \longrightarrow \bar{\mathbb{R}}$ is a function.

Theorem 10. *Let φ be the function defined by (3). Then for a function $f : X \longrightarrow \bar{\mathbb{R}}$ the following assertions are equivalent:*
1) We have

$$f^{L(\varphi)L(\varphi)'} = f.$$

2) f is increasing and lower semi-continuous.

Proof. 1) $\Longrightarrow$ 2). Suppose that 1) holds. Since φ is a symmetric coupling function, we have $L(\varphi) = L(\varphi)'$, and so $f = f^{L(\varphi)L(\varphi)'} = (f^{L(\varphi)})^{L(\varphi)}$. It follows from Proposition 6 with $\theta = \varphi$ that f is increasing and lower semi-continuous.
2) $\Longrightarrow$ 1). Assume that 2) holds. Since $L(\varphi) = L(\varphi)'$, by (31) and that $\mathrm{cl}\,(\mathrm{int}\,C) = C$, we have

$$f^{L(\varphi)L(\varphi)'}(y) = (f^{L(\varphi)})^{L(\varphi)}(y) = - \inf_{x \in X,\ x+y \in \mathrm{int}\,C} f^{L(\varphi)}(x)$$

$$= \sup_{x \in X,\ x+y \in \mathrm{int}\,C} (-f^{L(\varphi)}(x)) = \sup_{x \in X,\ x+y \in C} (-f^{L(\varphi)}(x))$$

$$= \sup_{x \in X,\ x \geq -y} (-f^{L(\varphi)}(x)) \quad (y \in X). \qquad (35)$$

Now, let $y \in X$ be fixed and $x \in X$ be such that $x \geq -y$. Since by Proposition 6, $f^{L(\varphi)}$ is an increasing function, it follows that $-f^{L(\varphi)}(x) \leq -f^{L(\varphi)}(-y)$, and so in view of (35) and (31) and that $\mathrm{cl}\,(\mathrm{int}\,C) = C$, we get

$$f^{L(\varphi)L(\varphi)'}(y) \leq -f^{L(\varphi)}(-y) = \inf_{x \in X,\ x-y \in \text{int}\, C} f(x)$$

$$= \inf_{x \in X,\ x-y \in C} f(x) = \inf_{x \in X,\ x \geq y} f(x) \leq f(y) \quad (y \in X). \tag{36}$$

On the other hand, by (35) and (31) and that f is increasing and lower semi-continuous, we obtain

$$f^{L(\varphi)L(\varphi)'}(y) = \sup_{x \in X,\ x \geq -y} (-f^{L(\varphi)}(x))$$

$$= \sup_{x \in X,\ x \geq -y} \inf_{z \in X,\ z+x \in \text{int}\, C} f(z) = \sup_{x \in X,\ x \geq -y} \inf_{z \in X,\ z+x \in C} f(z)$$

$$= \sup_{x \in X,\ x \geq -y} \inf_{z \in X,\ z \geq -x} f(z) \geq \sup_{x \in X,\ x \geq -y} f(-x) \geq f(y) \quad (y \in X). \tag{37}$$

Hence the result follows from (36) and (37), which completes the proof.   □

The proof of the following theorem is similar to that in finite dimensional case (see [Sin02]).

Theorem 11. *Let φ be the function defined by (3). Then for any topical function $f : X \longrightarrow \bar{I\!R}$, we have*

$$f^{L(\varphi)}(x) = f^{c(\varphi)}(x) \quad (x \in X).$$

References

[GG98]   Gaubert, S., Gunawardena, J.: A non-linear hierarchy for discrete event dynamical systems. Proc. 4th Workshop on discrete event systems, Calgiari, Technical Report HPL-BRIMS-98-20, Hewlett-Packard Labs. (1998)

[Gun98] Gunawardena, J.: An introduction to idempotency. Cambridge University Press, Cambridge (1998)

[Gun99] Gunawardena, J.: From max-plus algebra to non-expansive mappings: a non-linear theory for discrete event systems. Theoretical Computer Science, Technical Report HPL-BRIMS-99-07, Hewlett-Packard Labs. (1999)

[GK95]   Gunawardena, J., Keane, M.: On the existence of cycle times for some non-expansive maps. Technical Report HPL-BRIMS-95-003, Hewlett-Packard Labs. (1995)

[MRS02] Martinez-Legaz, J.-E., Rubinov, A.M., Singer, I.: Downward sets and their separation and approximation properties. Journal of Global Optimization, **23**, 111–137 (2002)

[MR05]   Mohebi, H., Rubinov, A.M.: Best approximation by downward sets with applications. Journal of Analysis in Theory and Applications, (to appear) (2005)

[Rub00] Rubinov, A.M.: Abstarct Convexity and Global Optimization. Kluwer Academic Publishers, Boston/Dordrecht/London (2000)

[RS01]   Rubinov, A.M., Singer, I.: Topical and sub-topical functions, downward
          sets and abstract convexity. Optimization, **50**, 307-351 (2001)
[Sin74]   Singer, I.: The theory of best approximation and functional analysis. Re-
          gional Conference Series in Applied Mathematics, **13** (1974)
[Sin87]   Singer, I.: Abstract Convex Analysis. Wiley-Interscience, New York (1987)
[Sin02]   Singer, I.: Further application of the additive min-type coupling function.
          Optimization, **51**, 471–485 (2002)

Part III

Applications

Dynamical Systems Described by Relational Elasticities with Applications

Musa Mammadov, Alexander Rubinov, and John Yearwood

CIAO, School of Information Technology and Mathematical Sciences
University of Ballarat
Ballarat, VIC 3353, Australia
m.mammadov@ballarat.edu.au, a.rubinov@ballarat.edu.au,
j.yearwood@ballarat.edu.au

Summary. In this paper we describe a new method for modelling dynamical systems assuming that the information about the system is presented in the form of a data set. The main idea is to describe the relationships between two variables as influences of the changes of one variable on another. The approach introduced was examined in data classification and global optimization problems.

Key words: Dynamical systems, elasticity, data classification, global optimization.

1 Introduction

In [Mam94] a new approach for mathematical modeling of dynamical systems was introduced. This approach was further developed in [Mam01a]-[MYA04] and has been applied to solving many problems, including data classification and global optimization. This paper gives a systematic survey to this approach.

The approach is based on non-functional relationship between two variables which describes the *influences* of the *change* (increase or decrease) of one variable on the *change* of the other variable. It can be considered as a certain analog of elasticity used in the literature (see, for example, [Int71]). We shall refer to this relationship between variables as *relational elasticity* (*fuzzy derivative*, in [Mam94, Mam01b, MY01]).

In [MM02] the notion of influence (of one state on another state) as a measure of the non-local contribution of a state to the value function at other states was defined. Conditional probability functions were used in this definition, but the idea behind this notion is close to the notion of influence used in [Mam94]. The calculations undertaken have shown that ([Mam01a, MY01]) this definition of the influence provides better results than if we use conditional probability.

As mentioned in [MM02] the notion of influence is also closely related to *dual variables* (or *shadow prices* in economics) for some problems (see, for example, [Gor99]).

We now describe some situations, where the notion of relational elasticity can be applied. Classical mathematical analysis, which is based on the notion of functional dependance, is suitable for examination of many situations, where influence of one variable on another can be explicitly described. The theory of probabilities is used in the situation, where such a dependance is not clear. However, this theory does not include many real-world situations. Indeed, probability can be used for examination of situations, which repeat (or can be repeated) many times. The attempts to use probability theory in uncertain situations, which can not be repeated many times, may lead to great errors.

We consider here only real-valued variables (some generalizations to vector-valued variables are also possible, however we do not consider them in the current paper). One of the main properties of a real-valued variable is monotonicity. We define the notion of influence by the increase or decrease of one variable on the increase or decrease of the other. We can consider the change of a variable as a result of activity of some unknown forces. In many instances our approach can be used for finding resulting state without explicit description of forces. Although the forces are unknown, this approach allows us to predict their action and as a result, to predict the behavior of the system and/or give a correct forecast. In this paper we undertake an attempt to give some description of forces acting on the system through the influences between variables and to describe dynamical systems generated by these forces.

The suggested approach of description of relationships between variables has been successfully applied to data classification problems (see [Mam01a]-[MY01], and references therein). In this paper we will only concentrate on some applications of dynamical systems, generated by this approach, and trajectories to these systems.

In Section 5, we examine the dynamical systems approach to data classification by introducing a simple classification algorithm. Using dynamical system ideas (trajectories) makes results, obtained by such a simple algorithm, comparable with the results obtained by other algorithms, designed for the purpose of data classification. The main idea behind this algorithm is close to some methods used in Nonlinear Support Vector Machines (see, for example, [Bur98]) where the domain is mapped to another space using some nonlinear (mainly, quadratic) mappings. In our case the transformation of the domain is made using the forces acting at each point of the domain.

The main application of this dynamical systems approach is to global optimization problems. In Section 6, we describe a global optimization algorithm based on this approach. The algorithm uses a new global search mechanism based on dynamical systems generated by the given objective function. The results, obtained for many test examples and some difficult practical problems ([Mam04, MYA04]), have shown the efficiency of this global search mechanism.

2 Relationship between two variables: relational elasticity

Let us consider two objects and assume that the states of these objects can be described by the scalar variables x and y. Increases and decreases of these variables indicates changes in the objects. The relationship between x and y will be defined by changes in both directions: increase and decrease.

We define the influence of y on x as follows: consider for instance the following event: y increases. As a result of this event x may either increase or decrease. To determine the influence we have to define the degree of these events. So we need to have the following expressions:

1) *the degree of the increase of x when y increases;*
2) *the degree of the decrease of x when y increases.*

Obviously, the expressions *increase* and *decrease* should be precisely defined in applications. For example if we say that y *increases* then we should determine: a) by how much? and b) during what time? These factors mainly depend on the problem under consideration and the nature of variables. For example, if we consider an economic system and y stands for the National Product, then we can take one year (or a month, etc) as the *time* interval, and for the *increase* we can take the relative increase of y. In some applications we do not need to determine the *time*. We denote the events y increases and y decreases by $y \uparrow$ and $y \downarrow$, respectively.

The key point in expressions 1) and 2) is the *degree*. Of course the degree of these events depends on the initial state (point) (x, y). For example, we can describe it by fuzzy sets on the plane (x, y); that is, at every initial point (x, y) the degree can be defined as a number in the interval $[0,1]$. In general, we will assume that the degree is a function of (x, y) with non-negative values.

We denote the degrees corresponding to 1) and 2) by $d(y \uparrow x \uparrow)$ and $d(y \uparrow x \downarrow)$, or by $\xi_1(x, y)$ and $\xi_2(x, y)$, respectively. We assume that the case $\xi_1(x, y) = 0$ corresponds to the lowest influence.

Similarly we can define the degree of decrease and increase of x when y decreases. They will be described by functions $\xi_3(x, y)$ and $\xi_4(x, y)$: $\xi_3 = d(y \downarrow x \downarrow)$, $\xi_4 = d(y \downarrow x \uparrow)$.

Note that in applications the functions $\xi_i(x, y)$ can be computed in quite different ways. For example, assume that there is a functional relation $y = f(x)$ and the directional derivative $f'_+(x)$ exists at the point x. In this case, we can define $\xi_1(x, y) = f'_+(x)$ and $\xi_2(x, y) = 0$ if $f'_+(x) > 0$, and $\xi_1(x, y) = 0$ and $\xi_2(x, y) = -f'_+(x)$ if $f'_+(x) < 0$. However, if the relation between variables is presented in the form of some finite set of observations (for example, in terms of applications to global optimization, it might be a set of some local minimum points found so far) we need to develop special techniques for computing the functions $\xi_i(x, y)$ (see Section 3)

Therefore, the functions ξ_i, $i = 1, 2, 3, 4$ completely describe the influence of the variable y on x in terms of changes. We will call it the relational elasticity between the two variables and denote it by $\partial x / \partial y$.

Let $\xi(x, y) = (\xi_1(x, y), \xi_2(x, y), \xi_3(x, y), \xi_4(x, y))$. So we have $\partial x / \partial y = \xi(x, y)$, where $\xi_1(x, y)$, $\xi_2(x, y)$, $\xi_3(x, y)$ and $\xi_4(x, y)$ are non-negative valued functions.

By analogy we define $\partial y / \partial x$ as an influence of x on y. Let $\partial y / \partial x = \eta(x, y)$, where $\eta = (\eta_1, \eta_2, \eta_3, \eta_4)$, and $\eta_1 = d(x \uparrow y \uparrow)$, $\eta_2 = d(x \uparrow y \downarrow)$, $\eta_3 = d(x \downarrow y \downarrow)$, $\eta_4 = d(x \downarrow y \uparrow)$.

Thus, the relationship between variables x and y will be described in the following form:

$$\partial x / \partial y = \xi(x, y), \quad \partial y / \partial x = \eta(x, y). \tag{1}$$

The examples of relationships presented below show that the system (1) covers quite a large range of relations including those that can not be described by some functions (or even set-valued mappings).

1. A homotone relationship. Assume that $\xi_1(x, y) \gg \xi_2(x, y)$, $\xi_3(x, y) \gg \xi_4(x, y)$ and $\eta_1(x, y) \gg \eta_2(x, y)$, $\eta_3(x, y) \gg \eta_4(x, y)$.

This case can be considered as a homotone relationship, because the influence of the increase (or decrease) of one variable on another is, mainly, directed in the same direction: increase (or decrease).

2. An antitone relationship. Assume that $\xi_1(x, y) \ll \xi_2(x, y)$, $\xi_3(x, y) \ll \xi_4(x, y)$ and $\eta_1(x, y) \ll \eta_2(x, y)$, $\eta_3(x, y) \ll \eta_4(x, y)$.

This case can be considered as an antitone relationship, because the influence of the increase (or decrease) of one variable on another is, mainly, directed in the inverse direction: decrease (or increase).

3. Assume that the influence of y on x such that $\partial x / \partial y = (a, a, a, a)$, where $a \geq 0$. In this case the variable x may increase or decrease with the same degree and these changes do not depend on y. We can say that the influence of y on x is quite indefinite.

4. Let $\partial x / \partial y = (a, 0, 0, a)$, $(a \geq 0)$. In contrast to case 3, in this case the influence of y on x is quite definite; every change in y increases x.

5. Let $\partial x / \partial y = (a, 0, b, 0)$, where $a, b > 0$ and $a \gg b$. This is a special case (known as hysteresis) of a homotone relationship considered above, where as y increases x increases strongly and when y decreases then x decreases not as strongly. If such a relationship is valid at all points (x, y) then the dependence between these variables can not be described by some mappings, like $y = y(x)$ or $x = x(y)$.

More complicated relationships arise when all the components in $\partial x / \partial y$ are not zero. This is the case that we have when dealing with real problems where the information about the systems is given in the form of some datasets.

3 Some examples for calculating relational elasticities

In this section we give some examples to demonstrate the calculation of relational elasticities. Note that we can suggest quite different methods according to the problem under consideration. In this paper, we examine the introduced notions in the context of global optimization and data classification problems. Accordingly, we consider the case when the relationship between variables x and y is given in the form of a dataset and we present some formulae to calculate relational elasticities which will be used in the applications below.

M.1. Consider data $\{(x^m, y^m), \ m = 1, ..., M\}$. To calculate relational elasticities first we have to define the events "increase" and "decrease". Here we suggest two techniques. For the sake of definiteness, we consider only the variable x.

Let x^0 be the initial point.

Global approach. If $x^m > x^0$ ($x^m < x^0$, respectively) we say that for the observation m the variable x increases (decreases, respectively).

Remark 3.1. In some cases it might be useful to define the increase (decrease, respectively) of x for the observation m by $x^m > x^0 + \delta$ ($x^m < x^0 - \delta$, respectively), where $\delta > 0$.

Local approach. Take any number $\varepsilon > 0$. If $x^m \in (x^0, x^0 + \varepsilon)$ ($x^m \in (x^0 - \varepsilon, x^0)$, respectively) we say that for the observation m the variable x increases (decreases, respectively).

Note that in the second case we follow the notion of the derivative in classical mathematics as a local notion.

Now we give two methods, related to the global and local approaches, for calculating a relational elasticity $\partial y / \partial x = (\eta_1, \eta_2, \eta_3, \eta_4)$ at the initial point (x^0, y^0). We set

$$\eta_1 = M_{11}/(M_1 + 1), \quad \eta_2 = M_{12}/(M_1 + 1),$$
$$\eta_3 = M_{13}/(M_2 + 1), \quad \eta_4 = M_{14}/(M_2 + 1). \tag{2}$$

For the global approach the numbers $M_1, M_{11}, M_{12}, M_2, M_{13}, M_{14}$ stand for the number of points (x^m, y^m), satisfying $x^m > x^0$, $x^m > x^0$ and $y^m > y^0$, $x^m > x^0$ and $y^m < y^0$, $x^m < x^0$, $x^m < x^0$ and $y^m < y^0$, $x^m < x^0$ and $y^m > y^0$, respectively. In the local approach we use $x^m \in (x^0, x^0 + \varepsilon)$ and $x^m \in (x^0 - \varepsilon, x^0)$ instead of $x^m > x^0$ and $x^m < x^0$.

Note that according to Remark 3.1 we could define the changes of the variable y by taking any small number $\delta > 0$. For instance, we could take $y^m > y^0 + \delta$ instead of $y^m > y^0$.

M.2. Now we present a method for calculating relational elasticities which will be used in the applications to global optimization.

Consider an objective function $f(x) : \ R^n \rightarrow R$ and assume that the values of the function have been calculated at some points; that is,

$f^m = f(x_1^m, x_2^m, ..., x_n^m)$, $m = 1, ..., M$. Therefore, we have data $A = \{(x_1^m, x_2^m, ..., x_n^m, f^m) : m = 1, ..., M\}$. We can refer to these points as "local" minimum points found so far. Let $x^0 = (x_1^0, x_2^0, ..., x_n^0)$ be the "best" point among these; that is, $f^0 = f(x^0) \geq f^m$ for all m.

We will consider the relations between f and each particular variable, say x_i, at the initial point x^0. Clearly, in data A the event $f \downarrow$ (that is, f decreases) will not occur. Therefore, we set $d(x_i \uparrow, f \downarrow) = 0$, $d(x_i \downarrow, f \downarrow) = 0$, $d(f \downarrow, x_i \uparrow) = 0$, $d(f \downarrow, x_i \downarrow) = 0$. We need to calculate the values $d(x_i \uparrow, f \uparrow)$, $d(x_i \downarrow, f \uparrow)$, $d(f \uparrow, x_i \uparrow)$, and $d(f \uparrow, x_i \downarrow)$.

We denote by $\| \cdot \|$ the Euclidian distance and let $\Delta x_i^m = x_i^m - x_i^0$, $\Delta f^m = f(x^m) - f(x^0)$, $m = 1, ..., M$. Then we set:

$$d(x_i \uparrow, f \uparrow) = \frac{1}{|X_i^+|} \sum_{m \in X_i^{++}} \frac{\Delta f^m}{\Delta x_i^m} \cdot \alpha_i^m; \quad d(x_i \downarrow, f \uparrow) = \frac{1}{|X_i^+|} \sum_{m \in X_i^{-+}} \frac{\Delta f^m}{|\Delta x_i^m|} \cdot \alpha_i^m;$$

$$d(f \uparrow, x_i \uparrow) = \frac{1}{|F_i^+|} \sum_{m \in F_i^{++}} \frac{\Delta x_i^m}{\Delta f^m} \cdot \alpha_i^m; \quad d(f \uparrow, x_i \downarrow) = \frac{1}{|F_i^+|} \sum_{m \in F_i^{+-}} \frac{|\Delta x_i^m|}{\Delta f^m} \cdot \alpha_i^m;$$

where $X_i^+ = \{m; \ \Delta x_i^m > 0\}$; $X_i^{++} = \{m; \ \Delta x_i^m > 0, \ \Delta f^m > 0\}$;
$X_i^- = \{m; \ \Delta x_i^m < 0\}$;
$X_i^{-+} = \{m; \ \Delta x_i^m < 0, \ \Delta f^m > 0\}$; $F_i^+ = \{m; \ \Delta f^m > 0 > 0\}$;
$F_i^{++} = \{m; \ \Delta f^m > 0, \ \Delta x_i^m > 0\}$; $F_i^{+-} = \{m; \ \Delta f^m > 0, \ \Delta x_i^m < 0\}$.
The coefficients $\alpha_i^m = (|\Delta x_i^m|/\|x^m - x^0\|)^2$ are used to indicate the contribution of the coordinate i in the change $\|x^m - x^0\|$. Clearly, $\alpha_1^m + ... + \alpha_n^m = 1$ for all m.

4 Dynamical systems

In this section we present some notions introduced in [Mam94] which have been used for studying the changes in the system.

Consider a system which consists of two variables x and y, and assume that at every point (x, y) the relationship between them is presented by relational elasticities (1); that is:

$$\partial x / \partial y = \xi(x, y), \quad \partial y / \partial x = \eta(x, y).$$

In this case we say that a Dynamical System is given. Here we study the changes of these variables using only the information obtained from relational elasticities. In this way the notion of *forces* introduced below will play an important role.

Definition 1. *At given point* (x, y) : *the quantities* $F(x \uparrow) = \eta_1 \xi_1 + \eta_2 \xi_4$ *and* $F(x \downarrow) = \eta_3 \xi_3 + \eta_4 \xi_2$ *are called the forces acting from y on the increase and decrease of x, respectively; the quantity* $F(x) = F(x \uparrow) + F(x \downarrow)$ *is called the force acting from y on x. By analogy, the forces* $F(y), F(y \uparrow), F(y \downarrow)$ *acting from x on y are defined:* $F(y) = F(y \uparrow) + F(y \downarrow)$, $F(y \uparrow) = \xi_1 \eta_1 + \xi_2 \eta_4$, $F(y \downarrow) = \xi_3 \eta_3 + \xi_4 \eta_2$.

The main sense of this definition, for example for $F(x \uparrow)$, becomes clear from the expression

$$F(x \uparrow) = d(x \uparrow y \uparrow)d(y \uparrow x \uparrow) + d(x \uparrow y \downarrow)d(y \downarrow x \uparrow).$$

From Definition 1 we obtain

Proposition 1. *At every point* (x, y) *the forces* $F(x)$ *and* $F(y)$ *are equal:*

$$F(x) = F(y).$$

This proposition states that, the size of the force on x equals the size of the force on y. It can be considered as a generalization of Newton's Third Law of Motion. To explain this statement, and, also, the reasonableness of Definition 1, we consider one example from Mechanics.

Assume that there are two particles, placed on a line, and x, y are their coordinates. Let $x < y$. Then, in terms of gravitational influences, we would have

$$\partial x/\partial y = (\xi_1, 0, 0, \xi_4), \quad \partial y/\partial x = (0, \eta_2, \eta_3, 0);$$

where $\xi_1, \xi_4, \eta_2, \eta_3 > 0$. Then, from Definition 1 it follows that

$$F(x \downarrow) = 0, \ F(y \uparrow) = 0, \text{ and } F(x \uparrow) = F(y \downarrow) = \eta_2 \xi_4 = d(x \uparrow y \downarrow) \, d(y \downarrow x \uparrow).$$

This is the Newton's Third Law of Motion.

Now, we assume that the influences $d(x \uparrow y \downarrow)$ and $d(y \downarrow x \uparrow)$ are proportional to the masses m_1 and m_2 of the particles, and are inverse-proportional to the distance $r = |x - y|$ between them; that is, $d(x \uparrow y \downarrow) = C_1 m_1/r$ and $d(y \downarrow x \uparrow) = C_2 m_2/r$. Then, from Definition 1 we have

$$F(x \uparrow) = F(y \downarrow) = C_1 C_2 \cdot \frac{m_1 m_2}{r^2}.$$

This is consistent with the Newton's Law of Gravity.

The main characteristic of non-mechanical systems is that, all values $F(x \downarrow)$, $F(x \uparrow)$, $F(y \downarrow)$ and $F(y \uparrow)$ may be non-zero. This might be, in particular, as a result of outside influences (say, some other variable z has an influence on x and y). This is the main factor that complicates the description (modelling) relationships between variables and makes it difficult to study the changes in the system.

Let the inequality $F(x \uparrow) > F(x \downarrow)$ hold at the point (x, y). In this case we can say that there are superfluous forces acting for the increase of variable x. If $F(x \uparrow) = F(x \downarrow)$ then these forces are balanced. So we can introduce

Definition 2. *The point* (x, y) *is called a stationary point if*

$$F(x \uparrow) = F(x \downarrow), \ F(y \uparrow) = F(y \downarrow);$$

and an absolutely stationary point if

$$F(x \uparrow) = F(x \downarrow) = F(y \uparrow) = F(y \downarrow) = 0.$$

Proposition 2. *Assume that relational elasticities at the point (x, y) are calculated such that*

$$\xi_1 + \xi_2 = \xi_3 + \xi_4 = 1, \tag{3}$$

$$\eta_1 + \eta_2 = \eta_3 + \eta_4 = 1. \tag{4}$$

Then the point (x, y) is an absolutely stationary point if and only if one of the following conditions holds:

i. $$\partial x / \partial y = (1, 0, 1, 0), \quad \partial y / \partial x = (0, 1, 0, 1); \tag{5}$$

ii. $$\partial x / \partial y = (0, 1, 0, 1), \quad \partial y / \partial x = (1, 0, 1, 0). \tag{6}$$

Proof. From conditions $F(x \uparrow) = F(x \downarrow) = F(y \uparrow) = F(y \downarrow) = 0$ we have

$$\eta_1 \xi_1 = 0; \tag{7}$$

$$\eta_3 \xi_3 = 0; \tag{8}$$

$$\eta_2 \xi_4 = 0; \tag{9}$$

$$\eta_4 \xi_2 = 0; \tag{10}$$

Consider two cases.

1). Let $\xi_1 = 0$. In this case we have

$$\xi_2 = 1 \overset{(10)}{\Rightarrow} \eta_4 = 0 \overset{(4)}{\Rightarrow} \eta_3 = 1 \overset{(8)}{\Rightarrow}$$

$$\xi_3 = 0 \overset{(3)}{\Rightarrow} \xi_4 = 1 \overset{(9)}{\Rightarrow} \eta_2 = 0 \overset{(4)}{\Rightarrow} \eta_1 = 1 \Rightarrow (6).$$

2). Let $\eta_1 = 0$. In this case we have

$$\eta_2 = 1 \overset{(9)}{\Rightarrow} \xi_4 = 0 \overset{(3)}{\Rightarrow} \xi_3 = 1 \overset{(8)}{\Rightarrow}$$

$$\eta_3 = 0 \overset{(4)}{\Rightarrow} \eta_4 = 1 \overset{(10)}{\Rightarrow} \xi_2 = 0 \overset{(3)}{\Rightarrow} \xi_1 = 1 \Rightarrow (5).$$

□

This proposition shows that if (x, y) is an absolutely stationary point then the influences x on y and y on x are inverse. In this case, the state (x, y) can not be changed without outside forces (say the change can be generated as an influence of some other variable z on x and y).

It is not difficult to prove the following propositions.

Proposition 3. *Assume that at the point (x, y) there is a homotone relationship between x and y and*

$$\partial x / \partial y = (\xi_1(x, y), 0, \xi_3(x, y), 0), \quad \partial y / \partial x = (\eta_1(x, y), 0, \eta_3(x, y), 0).$$

If $\xi_1(x, y)\, \eta_1(x, y) = \xi_3(x, y)\, \eta_3(x, y)$ then (x, y) is a stationary point.

Proposition 4. *Assume that at the point (x, y) there is an antitone relationship between x and y and*

$$\partial x / \partial y = (0, \xi_2(x, y), 0, \xi_4(x, y)), \quad \partial y / \partial x = (0, \eta_2(x, y), 0, \eta_4(x, y)).$$

If $\xi_2(x, y)\, \eta_4(x, y) = \xi_4(x, y)\, \eta_2(x, y)$ then (x, y) is a stationary point.

In this case we can say that there are no internal forces creating the changes in the system. Changes in the system may arise only as a result of outside forces.

4.1 Trajectories of the system (1)

In this section we study trajectories of the system (1). We define a trajectory (x_t, y_t), $(t = 0, 1, 2, ...)$, of the system (1) using the notion of forces acting between x and y. At every point (x, y) the forces $F(x \uparrow)$, $F(x \downarrow)$, $F(y \uparrow)$, $F(y \downarrow)$ are defined as in Definition 1.

Different methods can be used for calculation of trajectories. We present here two methods which will be used in the applications below.

Consider a variable ξ and let $\Delta\xi(t) = F(\xi(t) \uparrow) - F(\xi(t) \downarrow)$. In the first method we define a trajectory as follows:

$$\xi(t + 1) = \xi(t) + \alpha \cdot \text{Sign}(\Delta\xi(t)); \tag{11}$$

where

$$\text{Sign}(a) = \begin{cases} 1 & \text{if } a > 0; \\ 0 & \text{if } a = 0; \\ -1 & \text{if } a < 0. \end{cases}$$

In the second method we set

$$\xi(t + 1) = \xi(t) + \alpha \cdot \Delta\xi(t). \tag{12}$$

The difference between these formulae is that, in (12) the variables are changed with different steps along the direction $\Delta\xi(t)$, whilst, in (11) all the variables are changed with the same step $\alpha > 0$.

Consider an example.

Example 1. Consider a domain $D = \{(a, b) : a \in [0, 10], b \in [1, 10]\}$. Assume that the field of forces in the domain D is defined by the data $\{(x, y)\}$ presented in Table 1. Using this data, we can calculate forces acting at each

Table 1. Data used in Example 4.1

x	1	1	2	2	2	3	3	4	4	5	5	6	6	7
y	2	4	3	4	5	3	4	2	4	2	5	3	4	4

point $(x, y) \in D$ and, then, we can calculate trajectories to system (1). First we calculate the values of relational elasticities $\partial y / \partial x$ and $\partial x / \partial y$ by the local approach taking $\varepsilon = 1.1$ (see Section 3). Then, we generate trajectories taking $\alpha = (0.5)^k$ and different initial points. We consider two cases $k = 0$ and $k \geq 1$.

1. Let $k = 0$. Consider a trajectory $(x(t), y(t))$ starting from the initial point $(x(0), y(0)) = (2, 2)$. We have $(x(1), y(1)) = (3, 3)$ and $(x(2m), y(2m)) = (4, 4)$, $(x(2m + 1), y(2m + 1)) = (5, 3)$ for $m \geq 1$. Therefore, the set $P_1 = \{(4, 4), (5, 3)\}$ is a limit cycle of the trajectory $(x(t), y(t))$. Now consider other trajectories starting from different initial points (a, b), $a, b \in \{0, 1, ..., 10\}$. Each trajectory has one of the following three limit cycles: P_1, $P_2 = \{(2, 3), (3, 4)\}$, $P_3 = \{(5, 4), (4, 3)\}$. Thus, the domain D is divided into 3 parts so that all trajectories, starting from one of these parts, have the same limit cycle.

2. Let $k \geq 1$. In this case we observe that each trajectory has one of the following limit cycles:

$$P_1^k = \{(4, 4), (4 - (0.5)^k, 4 - (0.5)^k)\}, \ P_2^k = \{(3, 4), (3 - (0.5)^k, 4 - (0.5)^k)\}.$$

Therefore, in this case, the domain D can divided into two parts, as well as data presented in Table 1. Clearly, if $k \to \infty$ then $P_1^k \to \{(4, 4)\}$, $P_2^k \to \{(3, 4)\}$ in the Hausdorff metric.

We observe that there are three sets P_1, P_2, P_3 for $k = 0$ and two sets P_1^k, P_2^k for $k \geq 1$, which are the limit cycles for all trajectories. This means that the turnpike property is true for this example (see [MR73]). Thus, the idea of describing dynamical systems in the form of (1) and the study of trajectories to this system can be used in different problems. In the next section, we check this approach in data classification problems. As a domain D we take the heart disease and liver disorder databases.

5 Classification Algorithm based on a dynamical systems approach

Consider a database $A \subset R^n$, which consists of two classes: A^1 and A^2. We denote by $J = \{1, 2, ..., n\}$ the set of features.

The first stage of the data classification is the scaling phase. In this phase the data is considered to be measured on an m level scale. We did not use any of the known methods (for example, [DKS95]) for discretizing continuous attributes. Here we treat all the attributes uniformly i.e. we simply considered m levels for each attribute. Intervals related to these levels are defined only by using the training set and, therefore, the scaled values of the features of the observation depends on the training set.

Scaling. Take any number $m \in \{1, 2, ...\}$. First for every feature $j \in J$ we calculate the maximum and minimum quantities among all points of the set $A = A^1 \cup A^2$. Let a_j^1 and a_j^2 be the maximum and minimum quantities, respectively. Then any observation $x = (x_1, ..., x_n)$ is transformed into $y = (y_1, ..., y_n)$ by the formula

$$y_j = \begin{cases} 1 & \text{if } x_j \leq a_j^2; \\ p & \text{if } x_j \in \left[a_j^2 + \alpha_j(p-1), \ a_j^2 + \alpha_j p\right), \ p = 1, \ldots, m; \\ m & \text{if } x_j \geq a_j^1, \end{cases}$$

where $\alpha_j = (a_j^1 - a_j^2)/m$.

As a result, all the observation $x = (x_1, ..., x_n)$ are transformed into vectors $y = (y_1, ..., y_n)$, with coordinates $y_j \in \{1, 2, 3, ..., m\}$. Every new observation (test example) will also be scaled by this formula. After this scaling the database A is transformed into a set, which will be denoted by $\mathcal{A}$. The set $\mathcal{A}$ consists of two scaled classes $\mathcal{A}^1$ and $\mathcal{A}^2$ which are the transformation of the classes A^1 and A^2, respectively.

This scaling is not linear and so the structure of the sets can essentially be changed after this scaling. This is why we use different numbers m in the classification. Note that for small numbers m the classes $\mathcal{A}^1$ and $\mathcal{A}^2$ may not be disjoint even if the sets B and D are disjoint. The minimal number m for which these classes are disjoint is $m=19$ for the liver-disorder database and is $m=4$ for the heart disease database.

Therefore, we have a scaled (with m subdivisions) database $\mathcal{A} \subset R^K$ which consists of two classes $\mathcal{A}^1$ and $\mathcal{A}^2$. By $a^i = (a_1^i, a_2^i, ..., a_n^i)$ $(i = 1, 2)$ we denote the centroid of the class $\mathcal{A}^i$. Let x^{ts} be a test point.

For classification we use a very simple method which consists of two ordered rules. The point x^{ts} is predicted to belong to the class $\mathcal{A}^i$ if:

First Rule: $\mathcal{R}_1(x^{ts}) = \mathcal{A}^i$; that is, $x^{ts} \approx a$ for some $a \in \mathcal{A}^i$ and $x^{ts} \neq b$ for all $b \in \mathcal{A}^j, j \neq i$; otherwise go to the second rule.

Second Rule: $\mathcal{R}_2(x^{ts}) = \mathcal{A}^i$; that is, $\|x^{ts} - a^i\| < \|x^{ts} - a^j\|$, $j \neq i$.

Here we use the Euclidean norm $\| \cdot \|$ in R^n. The notations $x^{ts} \approx a$ and $x^{ts} \neq b$ are used in the following sense:

$$\max_{j=1,n} |x_j^{ts} - a_j| < \eta \quad \text{and} \quad \max_{j=1,n} |x_j^{ts} - b_j| > \eta;$$

where $\eta > 0$ is a given tolerance. Since the set $\mathcal{A}$ consists of the vectors with integer coordinates we take $\eta = 1/3$ in the calculations below.

Clearly we can not expect a good performance from such a simple algorithm (see the results presented in Table 2 for $T = 0$), but considering

trajectories starting from test points, we can increase the accuracy of classification. The results obtained in this way are even comparable with the results
obtained by other classification algorithms (see Table 3).

We define the field of forces in R^n using the set $\mathcal{A}$ which contains all
training examples from both classes. At a given point $x = (x_1, ..., x_n)$ the
relational elasticities are calculated for each pair of features (i, j) by the global
approach (see Section 3). Let $F(x_j \to x_i \downarrow)$ and $F(x_j \to x_i \uparrow)$ be the forces
acting from the feature j to decrease and increase, respectively, the feature i
at the point x. Then the resulting forces on the feature i is defined as a sum
of all these forces; that is,

$$F(x_i \downarrow) = \sum_{j \in J, j \neq i} F(x_j \to x_i \downarrow), \quad F(x_i \uparrow) = \sum_{j \in J, j \neq i} F(x_j \to x_i \uparrow). \quad (13)$$

Then, given new (test) point x^{ts}, we calculate (as in Example 1) a trajectory $\tilde{x}(t)$ $(t = 0, 1, 2, ..., T)$ started from this point. We use a step $\alpha = 0.25$.
To decrease the influence of circulating effects on the transform the trajectory
$\tilde{x}(t)$ to $x(t)$ by taking middle points of each of the last 5 steps; that is,

$$x(t) = \begin{cases} \frac{\tilde{x}(0) + \tilde{x}(1) + ... \tilde{x}(t)}{t+1} & \text{if } t < 4; \\ \frac{\tilde{x}(t-4) + \tilde{x}(t-3) + ... \tilde{x}(t)}{5} & \text{if } t \geq 4. \end{cases}$$

Table 2. Accuracy for test set for the heart disease and liver-disorder databases
with 10-fold cross validation obtained by Algorithm F

T	0	2	4	6	8	10	12	14	16	18	20
Heart	80.0	80.0	80.3	80.7	81.0	81.4	81.4	81.7	81.7	81.7	82.1
Liver	60.6	60.3	63.8	63.8	67.1	67.6	68.5	69.7	69.7	69.4	70.9

T	22	24	26	28	30	32	34	36	38	40	42
Heart	82.1	82.4	82.8	82.4	82.4	82.4	82.1	82.8	82.4	83.1	83.1
Liver	70.3	70.9	70.9	70.6	70.3	70.0	70.0	71.8	71.2	70.6	70.6

Classification Algorithm (F).

Step 1. Set $t = 0$.

Step 2. If $\mathcal{R}_1(\tilde{x}(t)) = \mathcal{A}^i$ then the example x^{ts} is predicted to belong to
the class $\mathcal{A}^i$. Otherwise we set $t = t + 1$. If $t \leq T$ go to Step 2, otherwise go
to Step 3.

Step 3. If $\mathcal{R}_2(x(T)) = \mathcal{A}^i$ then the example x^{ts} is predicted to belong
to the class $\mathcal{A}^i$. Otherwise the program terminates and the test point x^{ts} is
unclassified.

We apply this algorithm to the heart disease and liver disorder databases
taking the consecutive scaling numbers $m = 20, 21, ...40$. We use 10-fold cross
validation.

Table 3. Results for the heart disease and liver-disorder databases with 10-fold cross validation obtained by other methods

	Heart		Liver	
Algorithm	p_{tr}	p_{ts}	p_{tr}	p_{ts}
HMM	87.5	82.8	72.2	66.6
PMM	91.4	82.2	74.9	68.4
RLP	84.5	83.5	69.0	66.9
SVM $\| \cdot \|_1$	85.3	84.6	67.8	64.0
SVM $\| \cdot \|_\infty$	85.8	82.5	68.7	64.6
SVM $\| \cdot \|_2^2$	84.7	75.9	60.2	61.0

Note that in this application the choice of a combination of features is very important. The combination of features should form, in some sense, a minimal closed system in which the influences of the features on each other contain complete information about the process under consideration (disease in our case). For example, using two "similar" features can contribute noise because of summing (13). In this paper we did not try to find an optimal combination of features. Our aim is to find some combination of features for which the summing (13) does not create so much noise. For the heart disease database we use all 13 features, for the liver disorder database good results are obtained when we take just three features - the third, fourth and fifth. The result obtained for the test set for different time periods T are presented in Table 2. The accuracy for the training set is stable: 100.0 for the heart disease database and 99.7 for the liver disorder database and, so we did not present them in Table 2. The results show that when T increases, more test points become closer to the centroid of their own class. As a result, the accuracy of classification becomes sufficiently high. To have some idea about the level of accuracy that could be achieved in these domains, in Table 3, we present results obtained by other methods: HMM - Hybrid misclassification minimization ([CM95]), PMM - Parametric misclassification minimization ([Man94]), RLP - Robust linear programming ([BM92]), SVM $\| \cdot \|_1$, SVM $\| \cdot \|_\infty$, SVM $\| \cdot \|_2^2$ - Support vector machines algorithms with 1-norm, ∞-norm and 2-norm ([BM98]).

6 Algorithm for global optimization

In this section we apply the approach described above to global optimization problems. More detailed information about this application can be found in [Mam04].

We consider the following unconstrained continuous optimization problem

$$\text{minimize}\quad f(x) \tag{14}$$

$$\text{s.t.} \quad x \in R^n, \ a_i \le x_i \le b_i, \ i = 1, ..., n. \tag{15}$$

For the convenience, we will use the symbols $LocDD$, $LineSearch$ and $LocOpt$ defined below.

$LocDD$. Given point x, we denote by $l = (l_1, ..., l_n) = LocDD(x)$ a local descent direction from this point. It can be calculated in different ways. In the calculations below, it is calculated as follows. let $\varepsilon > 0$ be a given small number. Take any coordinate $i \in \{1, ..., n\}$, and calculate the values of the objective function f; let $a_0 = f(x_1, ..., x_n)$, $a_1 = f(x_1, ..., x_i - \varepsilon, ..., x_n)$, $a_2 = f(x_1, ..., x_i + \varepsilon, ..., x_n)$. Then we set $l_i = 0$ if $a_1 \ge a_0$ and $a_2 \ge a_0$; $l_i = a_0 - a_2$ if $a_1 \ge a_0$ and $a_2 < a_0$; $l_i = a_0 - a_1$ if $a_1 < a_0$ and $a_2 \ge a_0$. If $a_1 < a_0$ and $a_2 < a_0$ then we set $l_i = a_0 - a_2$ if $a_0 - a_2 \ge a_0 - a_1$; and $l_i = a_0 - a_1$ if $a_0 - a_2 < a_0 - a_1$.

$LineSearch$. Given point x and direction l, we denote by $LineSearch\,(l)$ the best point on the line $x + tl$, $t \ge 0$. In the calculations below, we apply inexact line search, taking $t = m\gamma$, $(m = 0, 1, ...)$, where $\gamma > 0$ is a some small step.

$LocOpt$. For the local minimization we could use different methods. In this paper we apply a *direct search* method called *local variations*. This is an efficient local optimization technique that does not explicitly use derivatives and can be applied to non-smooth functions. A good survey of direct search methods can be found in [KLT03].

The algorithm contains the following steps.

Step 1. Let L be a given integer, and $k \in \{0, 1, 2, ..., L - 1\}$. For each k we define the box

$$B_k = \{x \in R^n, \ a_i^k \le x_i \le b_i^k, \ i = 1, ..., n\};$$

where $\delta_i = (b_i - a_i)/2L$ and $a_i^k = a_i + k\delta_i$, $b_i^k = b_i - k\delta_i$.

Step 2. For each box B_k, we find a minima x^k, $k = 1, 2, ..., L - 1$.

Step 3. Let $x^* = \arg\min\{f(x^k), \ k = 1, 2, ..., L - 1\}$. We refine the point x^* by local optimization and get the global solution $x_{min} = LocOpt(x^*)$.

Now, given box B_k, we describe the procedure of finding a good solution x^k.

1. To apply the methods of dynamical systems, described above, we need to have a corresponding dataset. In other words, we need to generate some initial points and calculate values of the objective function at these points. Different methods can be used for the choice of initial points. In the algorithm described here, we generate initial points from the vertices of boxes B_k.

Let $A = \{x^1, ..., x^m\}$ be the set of initial points.

2. Given point x we find $x^* = LocOpt(LineSearch(LocDD(x)))$ which means that

- we calculate the local descent direction $l = LocDD(x)$ at the point x;
- then we find the best point $y = LineSearch(l)$ on the line l;
- and, finally, we refine the point y by local optimization and get the point $x^* = LocOpt(y)$.

We apply this procedure for each initial point from the set A and obtain the set $A(0) = \{x^{*,1}, ..., x^{*,m}\}$, where

$$x^{*,i} = LocOpt(LineSearch(LocDirection(x^i))), \quad i = 1, ..., m.$$

Let

$$x^*(0) = \arg\min\{f(x) : x \in A(0)\}.$$

3. The set $A(0)$ together with the values of the objective function allows us to generate a dynamical system. Our aim in this step is to find some "good" point $x^*(1)$ and add it to the set $A(0)$.

Let $t = 0$ and the point $x^*(t)$ be the "best" point in the set $A(t)$.

The main part of the algorithm is to determine a direction, say $F(t)$, at the point $x^*(t)$, which can provide a better solution $x^*(t+1)$. We can consider $F(t)$ as a global descent direction. For this aim, using the set $A(t)$, we calculate the forces acting on $f \uparrow$ at the point $x^*(t)$ from each variable $i \in \{1, ..., n\}$. We set $F(t) = (F_1(t), ..., F_n(t))$ where the components $F_i(t) = F(i \to f \uparrow)$ are calculated at the point $x^*(t)$ (see Definition 1). Then we define a point $x(t+1)$ by formula (12); that is, we consider the vector $-F(t)$ as a descent direction and set

$$x(t + 1) = x^*(t) - \alpha^*(t)F(t) \tag{16}$$

where the step $\alpha^*(t)$ is calculated as

$$\alpha^*(t) = \arg\min\{f(x^*(t) - \alpha F(t)) : \tag{17}$$

$$\alpha \in \{(\alpha_1, ..., \alpha_n) : \alpha_i = \frac{l}{M}(b_i - a_i), \ l = 1, ..., M\}\}. \tag{18}$$

Clearly $x(t+1) \neq x^*(t)$. Then we calculate $x^0(t+1) = LocOpt(x(t+1))$, and set $A(t + 1) = A(t) \cup \{x^0(t + 1)\}$. The next "good" point $x^*(t + 1)$ is defined as the best point in the set $A(t+1)$; that is, $x^*(t+1) = x^0(t+1)$, if our search was successful ($f(x^0(t + 1)) < f(x^*(t))$), and $x^*(t + 1) = x^*(t)$, if it was not.

We continue this procedure and obtain a trajectory $x^*(t)$, $t = 1, 2, ...$, starting from initial point $x^*(0)$. The process is terminated at the point $x^*(T)$, if either $F(T) = 0$ or $T > T^*$, where T^* is a priori given number. We note that $F(t) = 0$ means that $x^*(t)$ is a stationary point.

Therefore, $x^k = x^*(T)$ is a minimum point for the box B_k.

In the calculations below we take $L = 10$, $M = 100$ and $T^* = 20$. It is clear that, the results obtained can be refined by choosing larger L, M, T^*.

We call this algorithm AGOP ([Mam04]). For the calculation of direction $F(t)$, we need to determine the influences $d(x_i \uparrow, f \uparrow)$, $d(x_i \downarrow, f \uparrow)$, $d(f \uparrow, x_i \uparrow)$ and $d(f \uparrow, x_i \downarrow)$. For this aim, we use the methods introduced in Section 3. Therefore, we will consider two versions of the algorithm AGOP; the version AGOP(F) which uses the method **M.1** described in Section 3 and the version AGOP(D) which uses the method **M.2**.

There are many different methods and algorithms developed for global optimization problems (see, for example, [MPV01, PR02, Pin95] and references therein). Here, we mention some of them and note some aspects.

The algorithm AGOP takes into account some relatively "poor" points for further consideration. This is what many other methods do, such as Simulated Annealing ([Glo97, Loc02], Genetic Algorithms ([Smi02]) and Taboo Search ([CK02, Glo97]). The choice of a descent (good) direction is the main part of each algorithm. Instead of using a stochastic search (as in the algorithms mentioned), AGOP uses the formula (16), where the direction $F(t)$ is defined by relational elasticities.

Note that the algorithm AGOP has quite different settings and motivations compared with the methods that use so called "dynamical search" (see [PWZ02] and references therein). Our method of a search has some ideas in common with the heuristic method which attempts to estimate the "overall" convexity characteristics of the objective function ([DPR97]). This method does not work well when the postulated quadratic model is unsuitable. The advantage of our approach is that we do not use any approximate underestimations (including convex underestimations).

The methods that we use in this paper, are quite different from the homotopy and trajectory methods ([Die95, For95]), which attempt to visit (enumerate) all stationary points (local optimas) of the objective function, and, therefore, cannot be fast for high dimensional problems. The algorithm AGOP attempts to jump over local minimum points trying to find "deeper" points that do not need to be a local minima.

7 Results of numerical experiments

Numerical experiments have been carried out on a Pentium III PC with 800 MHz main processor. We use the following notations:

n - is the number of variables;

f_{min} - is the minimum value obtained;

f_{best} - is the global minimum or the best known result;

t (sec) - is the CPU time in seconds;

N_f - is the number of function evaluation.

We used 24 well known test problems (the list of test problems can be found at [Mam04]). The results obtained by algorithms AGOP(F) and AGOP(D) are presented in Table 4. We observe that the version AGOP(F) is more stable in finding global minima in all cases, meanwhile the version AGOP(D) has failed in two cases (for the Rastrigin function). In Table 5, we present the elapsed times and the number of function evaluations for functions with large number of variables obtained by AGOP(F).

The results obtained have shown the efficiency of the algorithm. For instance, for some of the test examples (where the number of variables could be chosen arbitrarily), the number of variables is increased up to 3000, and the time of processing was between 2 (for Rastrigin and Ackley's functions) and 12 (for Michalewicz function) minutes. We could not find comparable results in the literature. For instance, in [LL05] (Genetic Algorithms), the problems for Rastrigin, Griewank and Ackley's functions are solved for up to 1000 variables only, with the number of function evaluations [337570, 574561], 563350 and [548306, 686614], respectively (3 digit accuracy was the goal to be achieved). In our case, we have the number of function evaluations 174176, 174124 and 185904, respectively (see Table 4), with the complete global search.

8 Conclusions and future work

In this paper we developed a method to describe a relationship between two variables based on the notion of relational elasticities. Some methods for calculation of the relational elasticities are presented. We defined dynamical systems by using the relational elasticities and made some brief analysis of trajectories of such dynamical systems with applications to data classification and global optimization problems. The results obtained show that the relational elasticities can be considered a sound mathematical method to describe a relationship between two variables.

One of the main problems of our future investigation is to study a relationship between more than two variables. In this paper we simply used either formula (13), where the forces acting on some variable are summed, or the method described in **M.2**, Section 3. It will be very useful to define the influence of a combination of variables on some other variable.

We introduced a global optimization algorithm that can be used to handle functions with a large number of variables for solving continuous unconstrained optimization problems. The algorithm can be developed for solving continuous constrained optimization problems where special penalty functions and non-linear Lagrange-type functions (see [RY03]) are involved. In fact, the methodology that we use can be adapted for discrete optimization, because the determination of forces does not need a continuous state space. Therefore, the development of algorithms for solving discrete (unconstrained and constrained) optimization problems will be our future work.

Table 4. The results obtained by AGOP for non-convex continuously differentiable functions

Function	n	$f_{Gl.Min}$	AGOP(F)	AGOP(D)
Ackleys	2	0	0.000048	0.000048
Ackleys	1000	0	0.000459	0.000241
Ackleys	3000	0	0.000516	0.000495
Bohachevsky 1	2	0	0	0
Bohachevsky 2	2	0	0	0
Bohachevsky 3	2	0	$5.5753 \cdot 10^{-7}$	$3.1066 \cdot 10^{-5}$
Branin	2	0	$1.5445 \cdot 10^{-7}$	$1.5445 \cdot 10^{-7}$
Camel	2	-1.03163	-1.03162842	-1.03162844
Easom	2	-1	-0.9999998	-0.9999999
Golds. and Price	2	3	3.00000037	3.00000037
Griewank	2	0	$7.38 \cdot 10^{-8}$	$7.83 \cdot 10^{-8}$
Griewank	1000	0	$4.248 \cdot 10^{-5}$	$3.784 \cdot 10^{-5}$
Griewank	3000	0	$4.431 \cdot 10^{-5}$	$9.917 \cdot 10^{-3}$
Hansen	2	-176.5417	-176.54179	-176.54179
Hartman	3	-3.86278	-3.86278	-3.86278
Hartman	6	-3.32237	-3.322368	-3.3223678
Levy Nr.1	2	0	$1.309 \cdot 10^{-9}$	$1.309 \cdot 10^{-9}$
Levy Nr.1	1000	0	$1.433 \cdot 10^{-6}$	$1.433 \cdot 10^{-6}$
Levy Nr.1	3000	0	$3.875 \cdot 10^{-6}$	$3.875 \cdot 10^{-6}$
Levy Nr.2	2	0	$6.618 \cdot 10^{-9}$	$6.618 \cdot 10^{-9}$
Levy Nr.2	1000	0	$1.434 \cdot 10^{-8}$	$1.434 \cdot 10^{-8}$
Levy Nr.2	3000	0	$1.292 \cdot 10^{-8}$	$1.292 \cdot 10^{-8}$
Levy Nr.3	4	-11.5044	-11.5044	-11.5044
Levy Nr.3	1000	-11.5044	-11.395	-11.395
Levy Nr.3	3000	-11.5044	-11.395	-11.5044
Michalewicz	2	-1.8013	-1.8013	-1.8013
Michalewicz	5	-4.687	-4.6876581	-4.6876577
Michalewicz	10	-9.660	-9.6601482	-9.6601481
Michalewicz	1000	N/A	-957.0770	-964.1458
Michalewicz	3000	N/A	-2859.124	-2859.124
Rastrigin	2	0	$1.016 \cdot 10^{-9}$	$2.525 \cdot 10^{-6}$
Rastrigin	1000	0	$1.440 \cdot 10^{-8}$	323.362
Rastrigin	3000	0	$2.159 \cdot 10^{-4}$	1546.17
Schaffer Nr.1	2	0	0	0
Schaffer Nr.2	2	0	$4.845 \cdot 10^{-8}$	$4.845 \cdot 10^{-8}$
Shekel-5	4	-10.15320	-10.15319	-10.15319
Shekel-7	4	-10.40294	-10.40294	-10.40294
Shekel-10	4	-10.53641	-10.5364045	-10.5364045
Shubert Nr.1	2	-186.7309	-186.7309	-186.7309
Shubert Nr.2	2	-186.7309	-186.7309	-186.3406
Shubert Nr.3	2	-24.06250	-24.062498	-24.062498

Table 5. Elapsed times and the number of function evaluations for AGOP(F)

Function	n	f_{Best}	f_{min}	t (sec)	N_f
Ackleys	1000	0	0.000459	21.23	185904
Ackleys	3000	0	0.000516	145.67	530154
Griewank	1000	0	$4.248 \cdot 10^{-5}$	42.74	174124
Griewank	3000	0	$4.431 \cdot 10^{-5}$	367.09	555123
Levy Nr.1	1000	0	$1.433 \cdot 10^{-6}$	22.07	163724
Levy Nr.1	3000	0	$3.875 \cdot 10^{-6}$	201.06	463924
Levy Nr.2	1000	0	$1.434 \cdot 10^{-8}$	46.75	165724
Levy Nr.2	3000	0	$1.292 \cdot 10^{-8}$	380.01	463724
Levy Nr.3	1000	-11.5044	-11.395	24.68	182522
Levy Nr.3	3000	-11.5044	-11.395	174.62	573514
Michalewicz	1000	N/A	-957.0770	68.08	257265
Michalewicz	3000	N/A	-2859.124	715.60	955907
Rastrigin	1000	0	$1.440 \cdot 10^{-8}$	20.69	174176
Rastrigin	3000	0	$2.159 \cdot 10^{-4}$	162.07	509125

References

[BM92]   Bennett, K.P., Mangasarian, O.L.: Robust linear programming discrimination of two linearly inseparable sets. Optimization Methods and Software, **1**, 23–34 (1992)

[BM98]   Bradley, P.S., Mangasarian, O.L.: Feature selection via concave minimization and support vector machines.In: Shavlik, J. (ed) Machine Learning Proceedings of the Fifteenth International Conference (ICLML'98), 82–90. Morgan Kaufmann, San Francisco, California (1998)

[Bur98]   Burges, J.C.: A tutorial on support vector machines for pattern recognition. Data Mining and Knowledge Discovery, **2** 121–167 (1998) (http://svm.research.bell-labs.com/SVMdoc.html)

[CM95]   Chen, C., Mangasarian, O.L.: Hybrid misclassification minimization. Mathematical Programming Technical Report, **95-05**, University of Wisconsin (1995)

[CK02]   Cvijovic, D., Klinovski, J.: Taboo search: an approach to the multiple-minima problem for continuous functions. In: Pardalos, P., Romeijn, H. (eds) Handbook of Global Optimization, **2**, Kluwer Academic Publishers (2002)

[Die95]   Diener, I.: Trajectory methods in global optimization. In: Horst, R., Pardalos, P. (eds) Handbook of Global Optimization, Kluwer Academic Publishers (1995)

[DPR97]   Dill, K.A., Phillips, A.T., Rosen, J.M.: Molecular structure prediction by global optimization. In: Bomze, I.M. et al (eds) Developments in Global Optimization, Kluwer Academic Publishers (1997)

[DKS95]   Dougherty, J., Kohavi, R., Sahami, M.: Supervised and unsupervised discretization of continuous features. ICML-95 (1995)

[For95]   Forster, W.: Homotopy methods. In: Horst, R., Pardalos, P. (eds) Handbook of Global Optimization, Kluwer Academic Publishers (1995)

384   M.A. Mammadov et al.

[Glo97]   Glover, F., Laguna, M.: Taboo search. Kluwer Academic Publishers (1997)
[Gor99]   Gordon, G.J.: Approximate solutuions to Markov decision processes. Ph.D. Thesis, CS department, Carnegie Mellon University, Pittsburgh, PA (1999)
[Int71]   Intriligator, M.D.: Mathematical Optimization and Economic Theory, Prentice-Hall, Englewood Cliffs (1971)
[LL05]   Lazauskas, L: http://solon.cma.univie.ac.at/ neum/glopt/results/ga.html – Some Genetic Algorithms Results (collected by Leo Lazauskas) (2005)
[KLT03]   Kolda, T.G., Lewis, R.M., Torczon, V.: Optimization by direct search: new perspectives on some classical and modern methods, *SIAM Review*, **45**, 385–482 (2003)
[Loc02]   Locatelli, M.: Simulated annealing algorithms for continuous global optimization. In: Pardalos, P., Romeijn, H. (eds) Handbook of Global Optimization, **2**, Kluwer Academic Publishers (2002)
[MR73]   Makarov, V.L., Rubinov, A.M.: Mathematical theory of economic dynamics and equilibria, Nauka, Moscow (1973) (English trans.: Springer-Verlag, New York, 1977)
[Mam94]   Mamedov, M.A.: Fuzzy derivative and dynamic systems. In: Proc. of the Intern. Conf. On Appl. of Fuzzy systems, ICAFS-94, Tabriz (Iran), Oct. 17-19, 122–126 (1994)
[Mam01a] Mammadov, M.A.: Sequential separation of sets with a given accuracy and its applications to data classification. In: Proc. of the 16-th National Conference the Australian Society for Operations Research in conjuction with Optimization Day, 23-27 Sep., Mclarens on the Lake Resort, South Australia (2001)
[Mam01b] Mammadov, M.A.: Fuzzy derivative and its applications to data classification. The 10-th IEEE International Conference on Fuzzy Systems, Melbourne, 2-5 Dec. (2001)
[Mam04]   Mammadov, M.A.: (2004). A new global optimization algorithm based on dynamical systems approach. In: Rubinov, A., Sniedovich, M. (eds) Proceedings of The Sixth International Conference on Optimization: Techniques and Applications (ICOTA6), University of Ballarat, Australia, Dec. 2004, Article index number 198 (94th article); Also in: Research Report 04/04, University of Ballarat, Australia (2004) (http://www.ballarat.edu.au/ard/itms/publications/researchPapers.shtml)
[MRY01]   Mammadov, M.A., Rubinov, A.M., Yearwood, J.: Sequential separation of sets and its applications to data classification. In: Proc. of the Post-gr. ADFA Conf. On Computer Science, 14 Jul. 2001, Canberra, Australia, 75–80 (2001)
[MSY04]   Mammadov, M.A., Saunders, G., Yearwood, J.: A fuzzy derivative approach to classification of outcomes from the ADRAC database. International Transactions in Operational Research, **11**, 169–179 (2004)
[MY01]   Mammadov, M.A., Yearwood, J.: An induction algorithm with selection significance based on a fuzzy derivative. In: Abraham, A., Koeppen, M. (eds) Hybrid Information Systems, 223–235. Physica-Verlag, Springer (2001)
[MYA04]   Mammadov, M.A., Yearwood, J., Aliyeva, L.: (2004). Multi label classification and drug-reaction associations using global optimization techniques. In: Rubinov, A., Sniedovich, M. (eds) Proceedings of The Sixth

International Conference on Optimization: Techniques and Applications (ICOTA6),University of Ballarat, Australia, Dec. 2004, Article index number 168 (76th article) (2004)

[Man94] Mangasarian, O.L.: Misclassification minimization. Journal of Global Optimization, **5**, 309–323 (1994)

[MPV01] Migdalas, A., Pardalos, P., Varbrand P.: From Local to Global Optimization. Nonconvex Optimization and Its Applications, **53**, Kluwer Academic Publishers (2001)

[MM02] Munos, R., Moore, A.: Variable resolution discretization in optimal control. Machine Learning, **49**, 291–323 (2002)

[PR02] Pardalos, P., Romeijn, H. (eds): Handbook of Global Optimization, **2**, Kluwer Academic Publishers (2002)

[Pin95] Pinter, J. (ed): Global Optimization in Action. Kluwer Academic Publishers (1995)

[PWZ02] Pronzato, L., Wynn, H., Zhigljausky, A.A.: An introduction to dynamical search. In: Pardalos, P., Romeijn, H. (eds) Handbook of Global Optimization, **2**, Kluwer Academic Publishers (2002)

[RY03] Rubinov, A.M., Yang, X.Q.: Lagrange-type Functions in Constrained Non-convex Optimization. Applied Optimization, Volume **85**. Kluwer Academic Publishers (2003)

[Smi02] Smith, J.: Genetic algorithms. In: Pardalos, P., Romeijn, H. (eds) Handbook of Global Optimization, **2**, Kluwer Academic Publishers (2002)

Impulsive Control of a Sequence of Rumour Processes

Charles Pearce[1], Yalcin Kaya[2], and Selma Belen[1]

[1] School of Mathematics
   The University of Adelaide
   Adelaide, SA 5005, Australia
   cpearce@maths.adelaide.edu.au, sbelen@ankara.baskent.edu.tr
[2] School of Mathematics and Statistics
   University of South Australia
   Mawson Lakes, SA 5095, Australia; Departamento de Sistemas e Computação
   Universidade Federal do Rio de Janeiro
   Rio de Janeiro, Brazil
   Yalcin.Kaya@unisa.edu.au

Summary. In this paper we introduce an impulsive control model for a sequence of rumour processes evolving in a given population. Each rumour process begins with a broadcast, the recipients of which begin to spread that rumour. The recipients of the first broadcast are termed the subscribers. The second and subsequent broadcasts are either to the subscribers (Scenario 1) or to all individuals who have at any time to date been spreaders (Scenario 2). The objective is to time the second and subsequent broadcasts so as to minimise the final proportion of ignorants. It is shown that with either scenario the optimal time for each broadcast after the first is when the proportion of spreaders in the rumour process begun by the previous broadcast reaches zero. Results are presented concerning dependence on initial conditions, as well as graphical illustration of the controlled rumour processes under each scenario.

Key words: rumours, information spread, impulsive optimal control

2000 MR Subject Classification. 60J75, 60J27, 91D99

1 Introduction

Stochastic rumour models were introduced by Daley and Kendall [DK65], who considered a single initial spreader introducing a rumour into a closed population. Initially the remainder of the population do not know the rumour and as such are termed ignorants. The members of the population meet one another with uniform mixing. A spreader–ignorant interaction converts the ignorant into a spreader. When two spreaders interact, they stop spreading the rumour and become stiflers. A spreader–stifler interaction results in the spreader becoming a stifler. Otherwise interactions leave the roles of individuals unchanged. With a change of time scale, this model may be converted into one in which those interactions effecting a change do so only with probability p $(0 < p < 1)$.

Daley and Kendall reported a striking result later refined as follows: with one initial spreader, the proportion of the population never to hear the rumour converges almost surely to ≈ 0.2031878 of the population size as the latter tends to infinity. The constant is the solution of a certain transcendental equation. The same constant arises with a variant stochastic model of Maki and Thompson [MT73]. With both models the number of spreaders eventually becomes zero and the process stops.

A rigorous treatment of this result requires surprisingly delicate analysis, and much of the ensueing literature has been involved with technical questions arising from this. See, for example, [Bar72, Gan00, Pit90, Sud85, Wat87].

Rumour models can be used to describe a number of phenomena, such as the dissemination of information, disinformation or memes, and changes in political persuasion and the stock market. They are therefore of some practical significance. It is difficult to dispute that rumours have an important impact on stock prices. A stochastic model of the so-called pervasive rumour phenomenon in the stock markets can be found in [Bom03]. Reference [DMC01] studies the customer behaviour and marketing implications of *urban legends* and rumours. There have also been studies of rumour models over ordered networks [FPRU90, Zan01]. References [AH98, DP03, OT77] deal with processes with more than one rumour at any given time.

Many broad questions pertaining to stochastic rumours have still to be addressed, partly because the technical questions tend to be rather more difficult than those for the related stochastic epidemic. It is remarkable that while the time–dependent behaviour of the general stochastic epidemic was determined in the 1960s, that of the Daley–Kendall and Maki–Thompson rumour models was not elucidated until 2000 [Pea00]. The effect of varying from unity the initial number of spreaders has also been investigated only recently (Belen and Pearce [BP04]). The perhaps surprising result was discovered that even when the proportion of the initial population who are spreaders tends to unity, the proportion of the initial ignorants who never hear the rumour does not tend to zero.

In an age of mass communication, it is natural to consider the initiation of a rumour by means of television, radio or the internet (Frost [Fro00]). We may use the term broadcast to refer to such an initiation. In a companion paper [BKP05] a model with two broadcasts is envisaged. A control ingredient is incorporated, the timing of the second broadcast.

This paper presents a generalisation of this model to a general number $n > 1$ of broadcasts, with the intention of reducing the final proportion of the population never hearing the rumour. The rumour process is started by a broadcast to a subpopulation, the *subscribers*, who commence spreading the

rumour. We wish to determine when to effect subsequent broadcasts $2, 3, \ldots, n$
so as to minimise the final proportion of ignorants in the population.

Two basic scenarios are considered. In the first, the recipients of each
broadcast are the fixed group of subscribers: a subscriber who had become a
stifler becomes activated again as a subscriber spreader. In the second, the
recipients of any subsequent broadcast are those individuals who have been
spreaders at any time during the rumour initiated by the immediately previous
broadcast.

To obtain some results without becoming too enmeshed in probabilistic
technicalities, we follow Daley and Kendall and, after an initial discrete description of the population, describe the process in the continuum limit corresponding to a total population tending to infinity. Exactly the same formulation occurs in the continuum limit if one starts with the Maki–Thompson
formulation. The resultant differential equations with each scenario can be
expressed in state–space form, with the upward jump in spreaders at each
broadcast epoch constituting an impulsive control input. Since we are dealing with an optimal control problem, a natural approach would be to employ
a Pontryagin–like maximum principle furnishing necessary conditions for an
extremum of an impulsive control system (see, for example, Blaquière [Bla85]
and Rempala and Zapcyk [RZ88]). However, because of the tractability of the
dynamical system equations, we are able to solve the given impulsive control
problem without resorting to this theory.

In Section 2 we review the Daley–Kendall model and related results and
introduce two useful preliminary results. In Section 3 we solve the control
problem with Scenario 1 and in Sections 4 and 5 treat first– and second–order
monotonicity properties associated with the solution. In Section 6 we solve
the control problem for the somewhat more complicated Scenario 2. Also we
perform a corresponding analysis of the first–order monotonicity properties
for Scenario 2. Finally, in Section 7, we compare the two scenarios.

2 Single–Rumour Process and Preliminaries

The Daley–Kendall model considers a population of n individuals with three
subpopulations, ignorants, spreaders and stiflers. Denote the respective sizes
of these subpopulations by i, s and r. There are three kinds of interactions
which result in a change in the sizes of the subpopulations. The transitions
arising from these interactions along with their associated probabilities are as
tabulated. The other interactions do not result in any changes to the subpopulations.

We now adopt a continuum formulation appropriate for $n \to \infty$. Let $i(\tau)$,
$s(\tau)$, $r(\tau)$ denote respectively the proportions of ignorants, spreaders and
stiflers in the population at time $\tau \geq 0$. The evolution of the limiting form of
the model is prescribed by the deterministic dynamic equations

Interaction	Transition	Probability
$i \rightleftharpoons s$	$(i, s, r) \longmapsto (i-1, s+1, r)$	$is \, d\tau + o(d\tau)$
$s \rightleftharpoons s$	$(i, s, r) \longmapsto (i, s-2, r+2)$	$s(s-1)/2 \, d\tau + o(d\tau)$
$s \rightleftharpoons r$	$(i, s, r) \longmapsto (i, s-1, r+1)$	$sr \, d\tau + o(d\tau)$

$$\frac{di}{d\tau} = -i\,s, \tag{1}$$

$$\frac{ds}{d\tau} = -s\,(1 - 2i), \tag{2}$$

$$\frac{dr}{d\tau} = s(1 - i) \tag{3}$$

with initial conditions

$$i(0) = \alpha > 0, \; s(0) = \beta > 0 \text{ and } r(0) = \gamma \geq 0 \text{ satisfying } \alpha + \beta + \gamma = 1. \tag{4}$$

The dynamics and asymptotics of the continuum rumour process are treated by Belen and Pearce [BP04]. Under (4) i is a strictly decreasing function of time during the course of a rumour and we may reparametrise and regard i as the independent variable. Define the limiting value $\zeta := \lim_{\tau \to \infty} i(\tau)$. For our present purpose, the pertinent discussion of [BP04] may be summarised as follows.

Theorem 1. *In the rumour process prescribed by (1)–(4),*
(a) i is strictly decreasing with time with limiting value ζ satisfying

$$0 < \zeta < 1/2;$$

(b) ζ is the smallest positive solution to the transcendental equation

$$\frac{\zeta}{\alpha} e^{2(\alpha - \zeta)} = e^{-\beta}; \tag{5}$$

(c) s is ultimately strictly decreasing to limit 0.

The limiting case $\alpha \longrightarrow 1$, $\beta \longrightarrow 0$, $\gamma = 0$ is the classical situation treated by Daley and Kendall. In this case (5) becomes

$$\zeta \, e^{2(1 - \zeta)} = 1.$$

This is the equation used by Daley and Kendall to determine that in their classical case $\zeta \approx 0.2031878$.

It is interesting to look at the case when $\alpha \longrightarrow 0$, in other words when there are almost no initial ignorants in the population. For this purpose we introduce a new variable

$$\theta = \theta(\tau) := \frac{i(\tau)}{\alpha},$$

the ratio of the proportion of ignorants at time τ to the initial proportion. Note that $\theta(0) = 1$. We define also $\eta := \zeta/\alpha$, the limiting value of θ for $\tau \to \infty$. Then (5) reads as

$$\eta\, e^{2\alpha(1-\eta)} = e^{-\beta} \ .$$

For $\alpha \longrightarrow 0$, this becomes

$$\eta = e^{-\beta} \ .$$

If $\beta \longrightarrow 0$ too, that is, when there are almost no initial spreaders in the population, we get $\eta = 1$, that is, the proportion of the initial ignorant population remains unchanged. However if $\beta \longrightarrow 1$, then

$$\eta = 1/e \approx 0.368 \ .$$

Thus even when there is a small initial proportion of ignorants and a large initial proportion of spreaders, about 36.8% of the ignorant population never hear the rumour. This result is given in [BP04].

We shall make repeated use of the following theorem, which plays the role of a basis result for subsequent inductive arguments. Here we are examining the variation of ζ with respect to one of α, β, γ subject to (4), with another of α, β, γ being fixed.

Theorem 2. *Suppose (4) holds in a single–rumour process. Then we have the following.*
(a) For β fixed, ζ is strictly increasing in α for $\alpha \leq 1/2$.
(b) For β fixed, ζ is strictly decreasing in α for $\alpha \geq 1/2$.
(c) For γ fixed, ζ is strictly increasing in α.
(d) For α fixed, ζ is strictly increasing in β.

This is [BP04, Theorem 3], except that the statements there corresponding to (a) and (b) are for $\alpha < 1/2$ and $\alpha > 1/2$ respectively. The extensions to include $\alpha = 1/2$ follow trivially from the continuity of ζ as a function of α.

It is also convenient to articulate the following lemma, the proof of which is immediate.

Lemma 1. *For $x \in [0, 1/2]$, the map $x \mapsto xe^{-2x}$ is strictly increasing.*

3 Scenario 1

We now address a compound rumour process in which $n > 1$ broadcasts are made under Scenario 1. We shall show that the final proportion of the population never hearing a rumour is minimised when and only when the second and subsequent broadcasts are made at the successive epochs at which $s = 0$ occurs. We refer to this procedure as control policy $\mathcal{S}$. It is convenient to consider separately the cases $0 < \alpha \leq 1/2$ and $\alpha > 1/2$. Throughout this and the following two sections, ξ denotes the final proportion of the population hearing none of the sequence of rumours.

Theorem 3. *Suppose (4) holds with $0 < \alpha \leq 1/2$, that Scenario 1 applies and $n > 1$ broadcasts are made. Then*
(a) ξ is minimised if and only if the control process S is adopted;
(b) for β fixed, ξ is a strictly increasing function of α under control policy S.

Proof. Let $\mathcal{T}$ be an optimal control policy, with successive broadcasts occurring at times $\tau_1 \leq \tau_2 \leq \ldots \leq \tau_n$. We denote the proportion of ignorants in the population at τ_k by i_k $(k = 1, \ldots, n)$, so that $i_1 = \alpha$. Since i is strictly decreasing during the course of each rumour and is continuous at a broadcast epoch, we have from applying Theorem 1 to each broadcast in turn that

$$i_1 \geq i_2 \geq \cdots \geq i_n > \xi > 0, \tag{6}$$

all the inequalities being strict unless two consecutive broadcasts are simultaneous.

Suppose if possible that $s > 0$ at time $\tau_n - 0$. Imagine the broadcast about to be made at this epoch were postponed and s allowed to decrease to zero before that broadcast is made. Denote by ξ' the corresponding final proportion of ignorants in the population. Since i decreases strictly with time, the final broadcast would then occur when the proportion of ignorants had a value

$$i'_n < i_n. \tag{7}$$

In both the original and modified systems we have that $s = \beta$ at $\tau_n + 0$. By Theorem 2(a), (7) implies $\xi' < \xi$, contradicting the optimality of policy $\mathcal{T}$. Hence we must have $s = 0$ at $\tau_n - 0$ and so by Theorem 1 that

$$\frac{1}{2} > i_n > \xi.$$

Applying Theorem 2(a) again, to the last two broadcasts, gives that i_n is a strictly increasing function of i_{n-1} and that ξ is strictly increasing in i_n. Hence ξ is strictly increasing in i_{n-1}.

If $n = 2$, we have nothing left to prove, so suppose $n > 2$. We shall derive the desired results by backward induction on the broadcast labels. We suppose that for some k with $2 < k \leq n$ we have

(i) $s = 0$ at time $\tau_j - 0$ for $j = k, k+1, \ldots, n$;
(ii) ξ is a strictly increasing function of i_{k-1}.

To establish the inductive step, we need to show that $s = 0$ at $\tau_{k-1} - 0$ and that ξ is a strictly increasing function of i_{k-2}. The previous paragraph provides a basis $k = n$ for the backward induction.

If $s > 0$ at $\tau_{k-1} - 0$, then we may envisage again modifying the system, allowing s to reduce to zero before making broadcast $k - 1$. This entails that, if there is a proportion i'_{k-1} of ignorants in the population at the epoch of that broadcast, then

$$0 < i'_{k-1} < i_{k-1} .$$

By (ii) this gives $\xi' < \xi$ and hence contradicts the optimality of $\mathcal{T}$, so we must have $s = 0$ at $\tau_{k-1} - 0$. Theorem 2(a) now yields that i_{k-1} is a strictly increasing function of i_{k-2}, so that by (ii) ξ is a strictly increasing function of i_{k-2}. Thus we have the inductive step and the theorem is proved.   □

For the counterpart result for $\alpha > 1/2$, it will be convenient to extend the notation of Theorem 2 and use $\zeta(i)$ to denote the final proportion of ignorants when a single rumour beginning with state $(i, \beta, 1 - i - \beta)$ has run its course.

Theorem 4. *Suppose (4) holds with $\alpha > 1/2$, that Scenario 1 applies and $n > 1$ broadcasts are made. Then*
(a) ξ is minimised if and only if the control process S is adopted;
(b) for fixed β, ξ is a strictly decreasing function of α under control policy S.

Proof. First suppose that $i_n \geq 1/2$. By Theorem 1 and (6), this necessitates that $s > 0$ at time $\tau_2 - 0$. If we withheld broadcast 2 until $s = 0$ occurred, the proportion i'_2 of ignorants at that epoch would then satisfy

$$i'_2 = \zeta(i_1) \leq \zeta(i_n) = \xi < 1/2.$$

The relations between consecutive pairs of terms in this continued inequality are given by the definition of ζ, Theorem 2(b), the definition of ζ again, and Theorem 1 applied to broadcast n.

Hence policy S would give rise to ξ' satisfying

$$\xi' < i'_n \leq i'_2 \leq \xi ,$$

contradicting the optimality of $\mathcal{T}$. Thus we must have $i_n < 1/2$ and so

$$i_1 \geq i_2 \geq \cdots \geq i_k \geq 1/2 > i_{k+1} \geq \cdots \geq i_n > \xi$$

for some k with $1 \leq k < n$.

Suppose if possible $k > 1$. Then arguing as above gives

$$i'_2 = \zeta(i_1) \leq \zeta(i_k) \leq i_{k+1} < 1/2 .$$

The second inequality will be strict unless $s = 0$ at time $\tau_{k+1} - 0$. This leads to

$$i'_3 = \zeta(i'_2) \leq \zeta(i_{k+1}) \leq i_{k+2} < 1/2,$$

and proceeding recursively we obtain

$$i'_{n-k+1} \leq i_n < 1/2$$

and so

$$i'_{n-k+2} \le \xi .$$

Thus we have $\xi' < \xi$, again contradicting the optimality of $\mathcal{T}$. Hence we must have $k = 1$, and so

$$i_1 > 1/2 \ge i_2 \ge i_3 \ge \cdots \ge i_n > \xi .$$

Consider an optimally controlled rumour starting from state $(i_2, \beta, 1 - i_2 - \beta)$. By Theorem 3(b), ξ is a strictly increasing function of i_2. For $\mathcal{T}$ to be optimal, we thus require that i_2 be determined by letting the initial rumour run its full course, that is, that $s = 0$ at $\tau_2 - 0$. This yields Part (a). Since $\alpha > 1/2$, Theorem 2(b) gives that, with control policy $\mathcal{S}$, i_2 is a strictly decreasing function of α. Part (b) now follows from the fact that ξ is a strictly increasing function of i_2. □

Remark 1. For an optimal sequence of n broadcasts under Scenario 1, Theorems 1, 3 and 4 provide

$$\frac{i_k}{i_{k-1}} e^{2(i_{k-1}-i_k)} = e^{-\beta} \quad for \quad 1 < k \le n \tag{8}$$

and

$$\frac{\xi}{i_n} e^{2(i_n-\xi)} = e^{-\beta}. \tag{9}$$

Multiplying these relations together yields

$$\frac{\xi}{\alpha} e^{2(\alpha-\xi)} = e^{-n\beta},$$

which may be rewritten as

$$\xi e^{-2\xi} = \alpha e^{-(2\alpha+n\beta)}. \tag{10}$$

By Lemma 1, the left–hand side is a strictly increasing function of ξ for $\xi \in [0, 1/2]$. Hence (10) determines ξ uniquely.

Remark 2. Equations (8), (9) may be recast as

$$i_k e^{-2i_k} = i_{k-1} e^{-(\beta+2i_{k-1})} \quad for \quad 2 \le k \le n \tag{11}$$

and

$$\xi e^{-2\xi} = i_k e^{-(\beta+2i_k)} . \tag{12}$$

Consider the limiting case $\beta \to 0$ and $\gamma \to 0$, which gives the classical Daley–Kendall limit of a rumour started by a single individual. Since $i_k \le 1/2$ for $2 \le k \le n$ and $\xi \le 1/2$, we have by Lemma 1 that in fact

$$i_k = \xi \quad for \quad 2 \le k \le n.$$

If $\alpha \le 1/2$, then the above equality actually holds for $1 \le k \le n$. This is also clear intuitively: in the limit $\beta \to 0$ the reactivation taking place at the second and subsequent broadcast epochs does not change the system physically. This cannot occur for $\beta > 0$, which shows that when the initial broadcast is to a perceptible proportion of the population, as with the mass media, the effects are qualitatively different from those in the situation of a single initial spreader.

The behaviour of i_k with $n = 5$ broadcasts is depicted in Figure 1(a) with the traditional choice $\gamma = 0$. In generating the graphs, Equation 11 has been solved with initial conditions $\beta = 0, 0.2, 0.4, 0.6, 0.8, 1$. The figure illustrates Remark 2.

4 Monotonicity of ξ

In this section we examine the dependence of ξ on the initial conditions for Scenario 1. Equation (10) can be expressed as

$$n\beta + 2\,(\alpha - \xi) + \ln\xi - \ln\alpha = 0. \tag{13}$$

A single broadcast may be regarded as an instantiation of Scenario 1 with $n = 1$. The outcome is independent of the control policy. This enables us to derive the following extension of Theorem 2 to $n \geq 1$ broadcasts, ξ taking the role of ζ. We examine the variation of ξ with respect to one of α, β, γ subject to and one of α, β, γ being fixed. For example, if β is fixed then we can consider the variation of ξ with respect to α subject to the constraint $\alpha + \gamma = 1 - \beta$ supplied by (4). For clarity we adopt the notation $(\partial\xi/\partial\alpha)_\beta$ for the derivative of ξ with respect to α for fixed β subject to $\alpha + \gamma = 1 - \beta$. We use corresponding notation for the other possibilities arising with permutation of α, β, γ.

Theorem 5. *Suppose (4) holds with $n \geq 1$. Then under Scenario 1 we have the following.*

(a) For β fixed, ξ is strictly increasing in α for $\alpha \leq 1/2$ and strictly decreasing in α for $\alpha \geq 1/2$.
(b) For α fixed, ξ is strictly decreasing in β.
(c) For γ fixed, ξ is strictly increasing in α.

Proof. The case $n = 1$ is covered by Theorem 2, so we may assume that $n \geq 2$. Also Part (a) is simply a restatement of Theorem 3(b) and Theorem 4(b).

For parts (b) and (c), we use the fact that $\xi < 1/2$. Implicit differentiation of (13) yields

$$\left(\frac{\partial\xi}{\partial\beta}\right)_\alpha = -\frac{n\xi}{1 - 2\xi} < 0$$

and

$$\left(\frac{\partial\xi}{\partial\alpha}\right)_\gamma = \frac{\xi}{\alpha}\frac{1 + (n - 2)\alpha}{1 - 2\xi} > 0$$

for any $n > 1$, which yield (b) and (c) respectively. $\square$

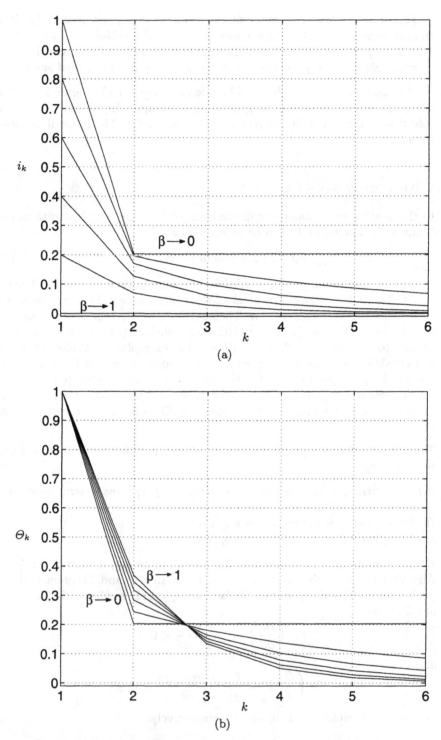

Fig. 1. An illustration of Scenario 1 with $\alpha + \beta = 1$ and five broadcasts. In successive simulations β is incremented by 0.2. For visual convenience, linear interpolation has been made between values of i_k (resp. Θ_k) for integral values of k.

The following result provides an extension to Corollary 4 of [BP04] to $n \geq 1$.

Corollary 1. *For any $n \geq 1$, we have $\bar{\xi} := \sup \xi = 1/2$. This occurs in the limiting case $\alpha = \gamma \to 1/2$ with $\beta \to 0$.*

Proof. From Theorem 5(c) we have for fixed $\gamma \geq 0$ that $\bar{\xi}$ is approached in the limit $\alpha = 1 - \gamma$ with $\beta = 0$. By Theorem 5(a), we have in the limit $\beta = 0$ that $\bar{\xi}$ arises from $\alpha = 1/2$. This gives the second part of the corollary.

From (13), $\bar{\xi}$ satisfies

$$1 - 2x + \ln(2x) = 0.$$

It is shown in Corollary 4 of [BP04] that this equation has the unique positive solution $x = 1/2$. The first part follows. $\square$

Figure 1(a) provides a graphical illustration of Theorem 5(c) for $\gamma \to 0$. For $\gamma = 0$, the initial state is given by a single parameter $\alpha = i_1 = 1 - \beta$.

Define $\Theta_k = i_k/\alpha$ for $1 \leq k \leq n$ and $\eta = \Theta_{n+1} = \xi/\alpha$. Then $\Theta_1 = 1$ and

$$\Theta_{k+1} e^{-2\alpha \Theta_{k+1}} = e^{-(2\alpha + k\beta)}, \quad 1 \leq k \leq n-1, \tag{14}$$

$$\eta e^{-2\alpha \eta} = e^{-(2\alpha + n\beta)}. \tag{15}$$

Remark 3. Put $w = -2\xi$. Then (15) gives

$$w e^w = -2\alpha e^{-(2\alpha + n\beta)},$$

the solution of which is given by the so-called Lambert w function ([CHJK96, BP04]). A direct application of the Lagrange series expression given in [BP04] provides

$$w = \sum_{j=1}^{\infty} \frac{\left(2\alpha e^{-(2\alpha + n\beta)}\right)^j}{j!} j^{j-1}.$$

Remark 4. In the case $\alpha \longrightarrow 0$ of a vanishing small proportion of initial ignorants, we have by (15) that

$$\eta = e^{-n\beta}. \tag{16}$$

Thus the ratio of the final proportion of ignorants to those at the beginning decays exponentially at a rate equal to the product of the number n of broadcasts and the proportion β of initial spreaders. Two subcases are of interest.

(i) The case $\beta \longrightarrow 0$ represents a finite number of spreaders in an infinite population. Almost all of the initial population consists of stiflers, that is, $\gamma \longrightarrow 1$, and we have $\eta = 1$. No matter how many broadcasts are made, the proportion of ignorants remains unchanged.

(ii) In the case of $\beta \longrightarrow 1$ almost all of the initial population consists of spreaders, and we obtain $\eta = e^{-n}$.

Consider Equation (16) again. For $0 < \beta < 1$, as well as for $\beta \longrightarrow 1$, we have that $\eta \longrightarrow 0$ as $n \longrightarrow \infty$.

The behaviour of Θ_k for the standard case $\gamma = 0$ is illustrated in Figure 1(b), for which we solve (14) with various initial conditions for 5 broadcasts. This brings out the variation with β more dramatically. The graph illustrates in particular Remark 4(ii). The curves pass through (1,1), since $i_1 = \alpha$ implies $\Theta_1 = 1$.

Remark 5. Given initial proportions α of ignorants and β of subscribers, with $0 < \beta < 1$ or with $\beta \longrightarrow 1$, the required number n of broadcasts to achieve a target proportion η or less of ignorants can be obtained through (15) as

$$k = \left\lceil -\frac{1}{\beta} \left[\ln(\eta) + 2\alpha \left(1 - \eta\right) \right] \right\rceil .$$

For example, consider the conventional case of $\gamma = 0$. Given 20% initial spreaders ($\beta = 0.2$) in the infinite population, in order to reduce the initial number of ignorants by 90% (that is, to reduce to a level where $\eta < 0.1$) at least five broadcasts are needed (see also Figure 1(b)). The same target is achieved in three broadcasts if the initial spreaders comprise 60% of the population ($\beta = 0.6$).

For $n \geq 1$, Equation (15) can be rewritten as

$$n\beta + 2\alpha \left(1 - \eta\right) + \ln \eta = 0 . \tag{17}$$

Theorem 6. *Suppose $n \geq 1$ and (4) applies. Then under Scenario 1:*

(a) for β fixed, η is strictly decreasing in α;
(b) for α fixed, η is strictly decreasing in β;
(c) for γ fixed, η is strictly decreasing in α for $n = 1$ and strictly increasing in α for $n \geq 2$.

Proof. We use the facts that $\eta < 1/2$ and $\xi = \alpha\eta < 1/2$. Implicit differentiation of (17) gives

$$\left(\frac{\partial \eta}{\partial \alpha}\right)_\beta = -\frac{2\eta(1 - \eta)}{1 - 2\alpha\eta} < 0 ,$$

$$\left(\frac{\partial \eta}{\partial \beta}\right)_\alpha = -\frac{n\eta}{1 - 2\alpha\eta} < 0 ,$$

which furnish (a) and (b) respectively.
    Similarly

$$\left(\frac{\partial \eta}{\partial \alpha}\right)_\gamma = -\frac{\eta(2 - n - 2\eta)}{1 - 2\alpha\eta} .$$

For $n = 1$ the numerator on the right is positive and so $(\partial \eta / \partial \alpha)_\gamma < 0$. For $n \geq 2$ the numerator is negative and $(\partial \eta / \partial \alpha)_\gamma > 0$. This completes the proof. $\square$

A graphical illustration of Theorem 6(c) for $\gamma = 0$ is given in Figure 1(b). Theorem 6(c) can be re–expressed as saying that, for fixed γ, $\eta = \Theta_{n+1}$ is increasing in β for $n = 1$ and decreasing for $n > 1$. This is reflected in the graphs almost having a point of concurrence between $k = 2$ and $k = 3$.

We may interpolate between integer values of k by extending (13) to define ξ for nonintegral $n \geq 1$, rather than by employing linear interpolation. Doing this yields exact concurrence of the interpolated curves. To see this, suppose we write (13) as

$$n(1 - \alpha - \gamma) + 2\alpha(1 - \Theta_{n+1}) + \ln \Theta_{n+1} = 0. \tag{18}$$

For $\gamma \geq 0$ given, if this curve passes through a point $(n+1, \Theta_{n+1})$ independent of α we must have

$$2\alpha(1 - \Theta_{n+1}) - n\alpha = \text{ constant .}$$

This necessitates

$$n = 2(1 - \Theta_{n+1}) \tag{19}$$

and so from (18) that

$$n(1 - \gamma) + \ln \Theta_{n+1} = 0. \tag{20}$$

Clearly (19) and (20) are together also sufficient for there to be a point of concurrence.

Elimination of n between (19) and (20) provides

$$2(1 - \Theta_{n+1})(1 - \gamma) + \ln \Theta_{n+1} = 0. \tag{21}$$

Denote by η_0 the value of ζ for a (single) rumour in the limit $\beta \to 0$ and the same fixed value of γ as in the repeated rumour. We have

$$2(1 - \gamma)(1 - \eta_0) + \ln \eta_0 = 0. \tag{22}$$

From (21) and (22) we can identify $\Theta_{n+1} = \eta_0$ and (19) then yields $n = 2(1 - \eta_0)$. We thus have a common point of intersection $(3 - 2\eta_0, \eta_0)$. In particular, for the traditional choice $\gamma = 0$, we have $\eta_0 \approx 0.203$ and the common point is approximately $(2.594, 0.203)$, a point very close to the cluster of points in Figure 1(b).

5 Convexity of ξ

We now address second–order monotonicity properties of ξ as a function of α, β, γ in Scenario 1. The properties derived are new for $n = 1$ as well as for $n \geq 2$. First we establish two results, of some interest in their own right, which will be useful in the sequel.

Theorem 7. *Suppose (4) holds with $n \geq 1$ and Scenario 1 applies. For $0 < x < 1$ and $\omega > 0$ define*

$$h(x, \omega) := \omega + 2(2x - 1) + \ln(1 - x) - \ln x.$$

Then

(a) $h(x, \omega) = 0$ *defines a unique*

$$x = \phi(\omega) \in (1/2, 1);$$

(b) h *is strictly increasing in* ω;
(c) $\xi > 1 - \alpha \iff \alpha > \phi(n\beta)$ *and* $\xi < 1 - \alpha \iff \alpha < \phi(n\beta)$.

Proof. We have

$$\frac{\partial h}{\partial x} = -\frac{(1 - 2x)^2}{x(1 - x)} \leq 0,$$

with equality if and only if $x = 1/2$, so $h(\cdot, \omega)$ is strictly decreasing on $(0, 1)$. Also $h(1/2, \omega) = \omega > 0$ and $h(x, \omega) \to -\infty$ as $x \to 1-$. Part (a) follows.

The relation $h(x, \omega) = 0$ may be written as

$$-\omega = 2(2x - 1) + \ln(1 - x) - \ln x.$$

Part (b) is an immediate consequence, since the right–hand side is a strictly decreasing function of x on $(0, 1)$.

Since h is strictly decreasing in x, we deduce from (a) that

$$h(x, \omega) > 0 \quad \text{for} \quad x < \phi(\omega) \quad \text{and} \quad h(x, \omega) < 0 \quad \text{for} \quad x > \phi(\omega). \quad (23)$$

For $y \in (0, 1)$ put

$$g(\alpha, \omega, y) := \omega + 2(\alpha - y) + \ln y - \ln \alpha.$$

We have readily that $\partial g / \partial y$ is positive for $y < 1/2$ and negative for $y > 1/2$, so g is strictly increasing in y for $y < 1/2$ and strictly decreasing in y for $y > 1/2$. Also $g \to \omega > 0$ as $y \to \alpha$ and $g \to -\infty$ as $y \to 0$, whence $g(\alpha, n\beta, \xi) = 0$ defines a unique $\xi \in (0, \alpha \wedge 1/2)$. We have

$$\xi \lessgtr 1 - \alpha \quad \text{according as} \quad g(\alpha, n\beta, 1 - \alpha) \gtrless 0.$$

But $g(\alpha, n\beta, 1 - \alpha) = h(\alpha, n\beta)$. Part (c) now follows immediately from (23).   □

Corollary 2. *Under the conditions of the preceding theorem with $n = 1$,*

$$\xi \gtrless \alpha/2 \quad \text{according as} \quad \gamma \gtrless 1 - \ln 2.$$

Proof. The argument of the theorem gives that

$$\xi \lessgtr \alpha/2 \quad \text{according as} \quad g(\alpha, \beta, \alpha/2) \gtrless 0,$$

that is,

$$\xi \lessgtr \alpha/2 \quad \text{according as} \quad \beta + \alpha - \ln 2 \gtrless 0.$$

The stated result follows from $\alpha + \beta + \gamma = 1$.   □

Theorem 8. *Suppose (4) holds with $n \geq 1$ and Scenario 1 applies. Then*

(a) for α fixed, ξ is strictly convex in β;
(b) for β fixed, ξ is strictly concave in α for $\alpha \in (0, \phi(n\beta))$ and strictly convex for $\alpha \in [\phi(n\beta), 1)$;
(c) for γ fixed, ξ is strictly convex in α if $n \geq 2$ or $n = 1$ and $\gamma > 1 - \ln 2$;
(d) for γ fixed, ξ is strictly concave in α if $n = 1$ and $\gamma < 1 - \ln 2$.

Proof. Implicit differentiation of 13 twice with respect to β yields

$$\left(\frac{1}{\xi} - 2\right) \left(\frac{\partial^2 \xi}{\partial \beta^2}\right)_\alpha = \frac{1}{\xi^2} \left(\frac{\partial \xi}{\partial \beta}\right)_\alpha^2 > 0,$$

which yields (a). Similarly

$$\left(\frac{1}{\xi} - 2\right) \left(\frac{\partial^2 \xi}{\partial \alpha^2}\right)_\beta = \frac{1}{\xi^2} \left(\frac{\partial \xi}{\partial \alpha}\right)_\beta^2 - \frac{1}{\alpha^2}$$

$$= \frac{1}{\alpha^2} \left[\left(\frac{1 - 2\alpha}{1 - 2\xi}\right)^2 - 1\right].$$

The expression in brackets has the same sign as

$$-(\alpha - \xi)[1 - (\alpha + \xi)],$$

that is, the opposite sign to $1 - (\alpha + \xi)$. By Theorem 7(c), the expression in brackets is thus negative if $\alpha < \phi(n\beta)$ and positive if $\alpha > \phi(n\beta)$, whence part (b).

Also by implicit differentiation of (13) twice with respect to α,

$$\left(\frac{1}{\xi} - 2\right) \left(\frac{\partial^2 \xi}{\partial \alpha^2}\right)_\gamma = \frac{1}{\xi^2} \left(\frac{\partial \xi}{\partial \alpha}\right)_\gamma^2 - \frac{1}{\alpha^2}, \tag{24}$$

and a single differentiation gives

$$\frac{1}{\xi} \left(\frac{\partial \xi}{\partial \alpha}\right)_\gamma - \frac{1}{\alpha} = n - 2 + 2 \left(\frac{\partial \xi}{\partial \alpha}\right)_\gamma. \tag{25}$$

By Theorem 5(c), the right–hand side of (25) is positive for $n \geq 2$, so the right–hand side of (24 must be positive and therefore so also the left–hand side, whence we have the first part of (c).

To complete the proof, we wish to show that for $n = 1$ the right–hand side of (25) is positive for $\gamma > 1 - \ln 2$ and negative for $\gamma < 1 - \ln 2$. Since

$$2 \left(\frac{\partial \xi}{\partial \alpha}\right)_\gamma - 1 = \frac{2\xi - \alpha}{\alpha(1 - 2\xi)},$$

the desired result is established by Corollary 2, completing the proof.   □

6 Scenario 2

Theorem 9. *Suppose (4) holds and $n > 1$ broadcasts are made under Scenario 2. Then*
(a) ξ is minimised if and only if control policy S is adopted;
(b) for fixed γ, ξ is a strictly increasing function of α under control policy S.

Proof. The argument closely parallels that of Theorem 3. The proof follows verbatim down to (7). We continue by noting that in either the original or modified system $r = \gamma$ at time $\tau_n + 0$. By Theorem 2(c), (7) implies $\xi' < \xi$, contradicting the optimality of control policy T. Hence we must have $s = 0$ at time $\tau_n - 0$. The rest of the proof follows the corresponding argument in Theorem 3 but with Theorem 2(c) invoked in place of Theorem 2(a).   □

Remark 6. The determination of ξ under Scenario 2 with control policy S is more involved than that under Scenario 1. For $1 \leq k \leq n$, set $\beta_k = s(\tau_k + 0)$. Then $i_k + \beta_k = 1 - \gamma = \alpha + \beta$, so that Theorem 1 yields

$$\frac{i_k}{i_{k-1}} \, e^{2(i_{k-1} - i_k)} = e^{-(\alpha + \beta - i_{k-1})} \qquad for \quad 1 < k \leq n+1,$$

where we set $i_{n+1} := \xi$. We may recast this relation as

$$i_k \, e^{-2i_k} = i_{k-1} \, e^{-(\alpha + \beta + i_{k-1})} \qquad for \quad 1 < k \leq n+1. \tag{26}$$

Since $i_k, \xi \in (0, 1/2)$ for $1 < k \leq n$, Lemma 1 yields that (26) determines $i_2, i_3, \ldots, i_n, \xi$ uniquely and sequentially from $i_1 = \alpha$.

Figure 2(a), obtained by solving (26), depicts the behaviour of i_k with $n = 5$ for the standard case of $\gamma = 0$. The initial values $\beta = 0, 0.2, 0.4, 0.6, 0.8, 1$ have been used to generate the graphs.

As with Scenario 1, we examine the dependence of ξ on the initial conditions. Equation (26) can be rewritten as

$$\beta + \alpha + i_{k-1} - 2i_k + \ln i_k - \ln i_{k-1} = 0, \qquad 1 < k \leq n+1 . \tag{27}$$

We now give the following result as a companion to Theorem 5. As before, a single broadcast may be regarded as an instantiation of Scenario 2 with $n = 1$.

Theorem 10. *Suppose (4) holds and Scenario 2 applies with $n \geq 1$. Then we have the following.*

(a) For α fixed, ξ is strictly decreasing in β.
(b) For γ fixed, ξ is strictly increasing in α.

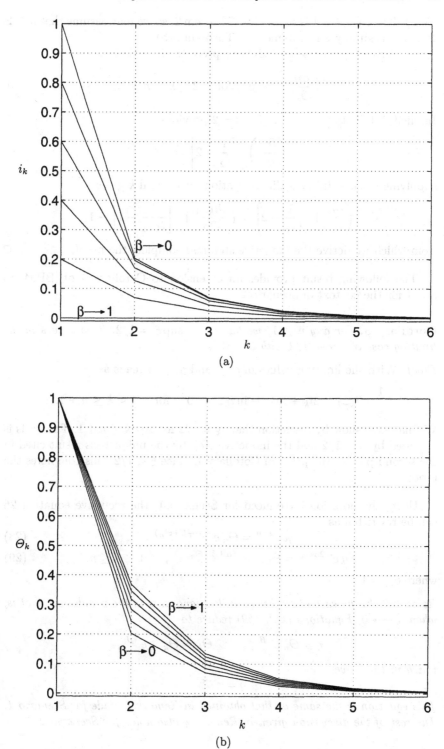

Fig. 2. An illustration of Scenario 2 with $\alpha+\beta = 1$ and five broadcasts. In successive simulations β is incremented by 0.2. For visual convenience, linear interpolation has been made between values of i_k (resp. Θ_k) for integral values of k.

Proof. The case $n = 1$ is covered by Theorem 2, so we may assume that $n \geq 2$. Part (b) is simply a restatement of Theorem 9(b).

To derive (a), we use an inductive proof to show that

$$\left(\frac{\partial i_k}{\partial \beta}\right)_\alpha < 0 \quad \text{for} \quad 2 \leq k \leq n+1.$$

Implicit differentiation of (27) for $k = 2$ provides

$$\left(\frac{\partial i_2}{\partial \beta}\right)_\alpha \left[\frac{1}{i_2} - 2\right] = -1,$$

supplying a basis. Implicit differentiation for general k gives

$$\left(\frac{\partial i_k}{\partial \beta}\right)_\alpha \left[\frac{1}{i_k} - 2\right] = \left(\frac{\partial i_{k-1}}{\partial \beta}\right)_\alpha \left[\frac{1}{i_{k-1}} - 1\right] - 1,$$

from which we derive the inductive step and complete the proof.   □

The following result provides an extension to Corollary 4 of [BP04] to $n \geq 1$ for the context of Scenario 2.

Corollary 3. *For any $n \geq 1$, we have $\bar{\xi} := \sup \xi = 1/2$. This occurs in the limiting case $\alpha = \gamma \to 1/2$ with $\beta \to 0$.*

Proof. With the limiting values of α, β and γ, (27) reads as

$$\frac{1}{2} + i_{k-1} - 2i_k + \ln i_k - \ln i_{k-1} = 0 \quad \text{for} \quad 1 < k \leq n+1.$$

We may now show by induction that $i_k = 1/2$ for $1 \leq k \leq n+1$. The basis is provided by $\alpha = 1/2$ and the inductive step by the uniqueness result cited in the second part of the proof of Corollary 1. since $\xi < 1/2$, this completes the proof.   □

Using the notation introduced for Scenario 1, the recursive equation 26 can be rewritten as

$$\eta e^{-2\alpha\eta} = \Theta_n e^{-(\alpha+\beta+\Theta_n)}, \tag{28}$$

$$\Theta_k e^{-2\alpha\Theta_k} = \Theta_{k-1} e^{-(\alpha+\beta+\Theta_{k-1})}, \quad 1 < k \leq n, \tag{29}$$

where $\Theta_1 = 1$.

Remark 7. In the case of almost no initial ignorants in the population, that is, when $\alpha \longrightarrow 0$, Equations (28), (29) reduce to

$$\eta = \Theta_n e^{-\beta}, \qquad \Theta_k = \Theta_{k-1} e^{-\beta},$$

which in turn give

$$\eta = e^{-n\beta}$$

This equation is the same as that obtained in Remark 4 made for Scenario 1. The rest of the discussion given in Remark 4 also holds for Scenario 2.

Figure 2(b) illustrates the above remark for $\alpha + \beta \to 1$. As with Figure 1(b), Figure 2(b) shows more dramatically the dependence on β: for a given initial value α, we have for each $k > 1$ that Θ_k increases with β, the relative and absolute effects being both less marked with increasing k.

Remark 8. The required number n of broadcasts necessary to achieve a target proportion ϵ or less of ignorants may be evaluated by solving (28)–(29) recursively to obtain the smallest positive integer n for which $\eta \leq \epsilon$.

7 Comparison of Scenarios

We now compare the eventual proportions ξ and ξ^* respectively of the population never hearing a rumour when n broadcasts are made under control policy S with Scenarios 1 and 2. For clarity we use the superscript * to distinguish quantities pertaining to Scenario 2 from the corresponding quantities for Scenario 1.

Theorem 11. *Suppose (4) holds and that a sequence of n broadcasts is made under control policy S. Then*
(a) if $n > 2$, we have

$$i_k^* < i_k \quad for \quad 2 < k \leq n;$$

(b) if $n \geq 2$, we have

$$\xi^* < \xi.$$

Proof. From (11), (12) (under Scenario 1) and (26) (under Scenario 2), ξ may be regarded as i_{n+1} and ξ^* as i_{n+1}^*, so it suffices to establish Part (a). This we do by forward induction on k.

Suppose that for some $k > 2$ we have

$$i_{k-1}^* \leq i_{k-1}. \tag{30}$$

A basis is provided by the trivial relation $i_2^* = i_2$. We have the defining relations

$$i_k^* e^{-2i_k^*} = i_{k-1}^* e^{-(\alpha + \beta + i_{k-1}^*)} \tag{31}$$

and

$$i_k e^{-2i_k} = i_{k-1} e^{-(\beta + 2i_{k-1})}. \tag{32}$$

The inequality

$$i_{k-1}^* < \alpha$$

may be rewritten as

$$\beta + 2i_{k-1}^* < \alpha + \beta + i_{k-1}^*,$$

so that

$$e^{-(\alpha+\beta+i^*_{k-1})} < e^{-(\beta+2i^*_{k-1})} .$$

Hence we have using (31) that

$$i^*_k e^{-2i^*_k} < i^*_{k-1} e^{-(\beta+2i^*_{k-1})} .$$

Lemma 1 and (30) thus provide

$$i^*_k e^{-2i^*_k} < i_{k-1} e^{-(\beta+2i_{k-1})} .$$

By (32) and a second application of Lemma 1 we deduce that $i^*_k < i_k$, the desired inductive step. This completes the proof.   □

Theorem 11 can be verified for the case of $\gamma = 0$ by comparing the graphs in Figures 1(a) and 2(a).

Acknowledgement

Yalcin Kaya acknowledges support by a fellowship from CAPES, Ministry of Education, Brazil (Grant No. 0138-11/04), for his visit to Department of Systems and Computing at the Federal University of Rio de Janeiro, during which part of this research was carried out.

References

[AH98]   Aspnes, J., Hurwood, W.: Spreading rumours rapidly despite an adversary. Journal of Algorithms, **26**, 386–411 (1998)

[Bar72]   Barbour, A.D.: The principle of the diffusion of arbitrary constants. J. Appl. Probab., **9**, 519–541 (1972)

[BKP05]   Belen, S., Kaya, C.Y., Pearce, C.E.M.: Impulsive control of rumours with two broadcasts. ANZIAM J. (to appear) (2005)

[BP04]   Belen, S., Pearce, C.E.M.: Rumours with general initial conditions. ANZIAM J., **45**, 393–400 (2004)

[Bla85]   Blaquière, A.: Impulsive optimal control with finite or infinite time horizon, J. Optimiz. Theory Applic., **46**, 431–439 (1985)

[Bom03]   Bommel, J.V.: Rumors. Journal of Finance, **58**, 1499–1521 (2003)

[CHJK96] Corless, R.M., Hare, D.E.G., Jeffrey, D.J., Knuth, D.E.: On the Lambert W function. Advances in Computational Mathematics, **5**, 329–359 (1996)

[DK65]   Daley, D.J., Kendall, D.G.: Stochastic rumours. J. Inst. Math. Applic., **1**, 42–55 (1965)

[DP03]   Dickinson, R.E., Pearce, C.E.M.: Rumours, epidemics and processes of mass action: synthesis and analysis. Mathematical and Computer Modelling, **38**, 1157–1167 (2003)

[DMC01]   Donavan, D.T., Mowen, J.C., Chakraborty, G.: Urban legends: diffusion processes and the exchange of resources. Journal of Consumer Marketing, **18**, 521–533 (2001)

[FPRU90] Feige, U., Peleg, D., Rhagavan, P., Upfal, E.: Randomized broadcast in networks. Random Structures and Algorithms, **1**, 447–460 (1990)

[Fro00]   Frost, C.: Tales on the internet: making it up as you go along. ASLIB Proc., **52**, 5–10 (2000)

[Gan00]   Gani, J.: The Maki-Thompson rumour model: a detailed analysis. Environmental Modelling and Software, **15**, 721–725 (2000)

[MT73]   Maki, D.P., Thompson, M.: Mathematical Models and Applications. Prentice–Hall, Englewood Cliffs (1973)

[OT77]   Osei, G.K., Thompson, J.W.: The supersession of one rumour by another. J. App. Prob., **14**, 127–134 (1977)

[Pea00]   Pearce, C.E.M.: The exact solution of the general stochastic rumour. Math. and Comp. Modelling, **31**, 289–298 (2000)

[Pit90]   Pittel, B.: On a Daley–Kendall model of random rumours. J. Appl. Probab., **27**, 14–27 (1990)

[RZ88]   Rempala, R., Zabczyk, J.: On the maximum principle for deterministic impulse control problems. J. Optim. Theory Appl., **59**, 281–288 (1988)

[Sud85]   Sudbury, A.: The proportion of the population never hearing a rumour. J. Appl. Probab., **22**, 443–446 (1985)

[Wat87]   Watson, R.: On the size of a rumour. Stoch. Proc. Applic., **27**, 141–149 (1987)

[Zan01]   Zanette, D.H.: Critical behaviour of propagation on small–world networks. Physical Review E, **64**, 050901(R), 4 pages (2001)

Minimization of the Sum of Minima of Convex Functions and Its Application to Clustering

Alexander Rubinov, Nadejda Soukhoroukova, and Julien Ugon

CIAO, School of Information Technology and Mathematical Sciences
University of Ballarat
Ballarat, VIC 3353, Australia
a.rubinov@ballarat.edu.au, n.soukhoroukova@ballarat.edu.au,
jugon@students.ballarat.edu.au

Summary. We study functions that can be represented as the sum of minima of convex functions. Minimization of such functions can be used for approximation of finite sets and their clustering. We suggest to use the local discrete gradient (DG) method [Bag99] and the hybrid method between the cutting angle method and the discrete gradient method (DG+CAM) [BRZ05b] for the minimization of these functions. We report and analyze the results of numerical experiments.

Key words: sum-min function, cluster function, skeleton, discrete gradient method, cutting angle method

1 Introduction

In this paper we introduce and study a class of sum-min functions. This class $\mathcal{F}$ consists of functions of the form

$$F(x_1, \ldots, x_k) = \sum_{a \in A} \min(\varphi_1(x_1, a), \varphi_2(x_2, a), \ldots \varphi_k(x_k, a)),$$

where A is a finite subset of a finite dimensional space and the function $x \mapsto \varphi_i(x, a)$ is convex for each i and $a \in A$. In particular, the cluster function (see, for example, [BRY02]) and Bradley-Mangasarian function [BM00] belong to $\mathcal{F}$. We also introduce the notion of a skeleton of the set A, which is a version of Bradley-Mangasarian approximation of a finite set. The search for skeletons can be carried out by a constrained minimization of a certain function belonging to $\mathcal{F}$.

We point out some properties of functions $F \in \mathcal{F}$. In particular we show that these functions are DC (difference of convex) functions.

Functions $F \in \mathcal{F}$ are nonsmooth and nonconvex. If the set A is large enough then these functions have a large number of shallow local minima.

Some functions $F \in \mathcal{F}$ (in particular, cluster functions) have a saw-tooth form. The minimization of these functions is a challenging problem. We consider both local and global minimization of functions $F \in \mathcal{F}$. We suggest to use the derivative-free discrete gradient (DG) method [Bag99] for local minimization of these functions. For global minimization we use the hybrid method between DG and the cutting angle method (DG+CAM)[BRZ05a, BRZ05b] and the commercial software GAMS (LGO solver), see [GAM05, LGO05] for more information.

These methods were applied to the minimization of two types of functions from $\mathcal{F}$: cluster functions C_k (generalized cluster functions $\tilde{C}_k$) and skeleton functions L_k (generalized skeleton functions $\tilde{L}_k$). These functions are used for finding clusters in datasets (unsupervised classification).

The notion of clustering is relatively flexible (see [JMF99, BRSY03] for more information). The goal of clustering is to group points in a dataset in a way that representatives of the same group (the same cluster) are similar to each other. There are different notions of similarity. Very often it is assumed that similar points have similar coordinates because each coordinate represents measurements of the same characteristic. The functions $C_k, \tilde{C}_k, L_k, \tilde{L}_k$ can be used to represent the dissimilarity of obtained systems of clusters. Therefore, a clustering system which gives a minimum of a chosen dissimilarity function is considered as a desired clustering system. Different dissimilarity functions lead to different approaches to clustering, therefore different clustering results can be obtained by the minimization of functions $F \in \mathcal{F}$.

We report results of numerical experiments and analyze these results.

2 A class of sum-min functions

2.1 Functions represented as the sum of minima of convex functions

Consider finite dimensional vector space $\mathbb{R}^n$ and $\mathbb{R}^m$. Let $A \subset \mathbb{R}^n$ be a finite set and let k be a positive integer. Consider a function F defined on $(\mathbb{R}^m)^k$ by

$$F(x_1, \ldots, x_k) = \sum_{a \in A} \min(\varphi_1(x_1, a), \varphi_2(x_2, a), \ldots \varphi_k(x_k, a)), \qquad (1)$$

where $x \mapsto \varphi_i(x, a)$ is a convex function defined on $\mathbb{R}^m$ ($i = 1, \ldots, k$, $a \in A$). We do not assume that this function is smooth. We denote the class of functions of the form (1) by $\mathcal{F}$.

The search for some geometric characteristics of a finite set can be accomplished by minimization (either unconstrained or constrained) of functions from $\mathcal{F}$, (see, for example [BRY02, BM00]). Location problems (see, for example, [BLM02]) also can be reduced to the minimization of functions from $\mathcal{F}$.

The minimization of function $F \in \mathcal{F}$ is a *min-sum-min* problem. We also can consider *min-max-min* problems with the objective function

$$\tilde{F}(x_1, \ldots, x_k) = \max_{a \in A} \min(\varphi_1(x_1, a), \varphi_2(x_2, a), \ldots \varphi_k(x_k, a)).$$

Using sum-min function F we take into account the contribution of each point $a \in A$ to a characteristic of the set A, which is described by means of functions $\varphi_i(x, a)$. This is not true if we consider $\tilde{F}$. From this point of view, the minimization of sum-min functions is preferable for examination of many characteristics of finite sets.

2.2 Some properties of functions belonging to $\mathcal{F}$.

Let $F \in \mathcal{F}$, that is

$$F(x_1, \ldots, x_k) = \sum_{a \in A} \min_{i=1,\ldots,k} \varphi_i(x_i, a),$$

where $x \mapsto \varphi_i(x_i, a)$ is a convex function. Then F enjoys the following properties:

1. F is quasidifferentiable ([DR95]). Moreover, F is DC (the difference of convex functions). Indeed, we have (see for example [DR95], p.108):

$$F(x) = f_1(x) - f_2(x), \qquad x = (x_1, \ldots, x_k),$$

where

$$f_1(x) = \sum_{a \in A} \sum_{i=1}^{k} \varphi_i(x_i, a)$$

$$f_2(x) = \sum_{a \in A} \max_{i=1,\ldots,k} \sum_{j \neq i} \varphi_j(x_j, a).$$

Both f_1 and f_2 are convex functions. The pair $DF(x) = (\partial f_1(x), -\partial f_2(x))$ is a quasidifferential [DR95] of F at a point x. Here ∂f stands for the convex subdifferential of a convex function f.

2. Since F is DC, it follows that this function is locally Lipschitz.

3. Since F is DC it follows that this function is semi-smooth.

We can use quasidifferentials of a function $F \in \mathcal{F}$ for a local approximation of this function near a point x. Clarke subdifferential also can be used for local approximation of F, since F is locally Lipschitz.

3 Examples

We now give some examples of functions belonging to class $\mathcal{F}$. In all the examples, datasets are denoted as finite sets $A \subset \mathbb{R}^n$, that is as sets of n-dimensional points (also denoted observations).

3.1 Cluster functions and generalized cluster functions

Assume that a finite set $A \subset \mathbb{R}^n$ consists of k clusters. Let $X = \{x_1, \ldots, x_k\} \subset (\mathbb{R}^n)^k$. Consider the distance $d(X, a) = \min\{\|x_1 - a\|, \ldots \|x_k - a\|\}$ between the set X and a point (*observation*) $a \in A$. (It is assumed that $\mathbb{R}^n$ is equipped with a norm $\| \cdot \|$.) The deviation of X from A is the quantity $d(X, A) = \sum_{a \in A} d(X, a)$. Let $\bar{X} = \{\bar{x}_1, \ldots \bar{x}_k\}$ be a solution to the problem:

$$\min_{x_1, \ldots, x_k \in \mathbb{R}^n} \sum_{a \in A} \min\{\|x_1 - a\|, \ldots \|x_k - a\|\}.$$

Then $\bar{x}_1, \ldots, \bar{x}_k$ can be considered as the centres of required clusters. (It is implicitly assumed that these are point-centred clusters.) *If the cluster centres are known each point is assigned to the cluster with the nearest centre.* Assume that N is the cardinality of set A. The function

$$C_k(x_1, \ldots, x_k) \equiv \frac{1}{N} d(X, A) = \frac{1}{N} \sum_{a \in A} \min(\|x_1 - a\|, \ldots, \|x_k - a\|) \qquad (2)$$

is called a cluster function. This function has the form (1) with $\varphi_i(x, a) = \|x - a\|$ for each $a \in A$ and $i = 1, \ldots, k$. The cluster function was examined in [BRY02]. Some numerical methods for its minimization were suggested in [BRY02].

The cluster function has a saw-tooth form and the number of teeth drastically increases as the number of addends in (2) increases. This leads to the increase of the number of shallow local minima and saddle points. If the norm $\| \cdot \|$ is a polyhedral one, say $\| \cdot \| = \| \cdot \|_1$, then the cluster function is piece-wise linear with a very large number of different linear pieces. The restriction of the cluster function to a one-dimensional line has the form of a saw with a huge amount of teeth of different size but of the same slope.

Let $(m_a)_{a \in A}$ be a family of positive numbers. Function

$$\tilde{C}_k(x_1, \ldots, x_k) = \frac{1}{N} \sum_{a \in A} m_a \min(\|x_1 - a\|, \ldots, \|x_k - a\|) \qquad (3)$$

is called a *generalized cluster function*. Clearly $\tilde{C}_k$ has the form (1). The structure of this function is similar to the structure of cluster function, however different teeth of generalized cluster function can have different slopes.

Clusters constructed according to centres, obtained as a result of the cluster function minimization are called *centre-based clusters*.

3.2 Bradley-Mangasarian approximation of a finite set

If a finite set A consists of flat parts it can be approximated by a collection of hyperplanes. Such kind of approximation was suggested by P.S. Bradley and O.L. Mangasarian [BM00]. Assume that we are looking for a collection of k hyperplanes $H_i = \{x : [l_i, x] = c_i\}$ approximating the set A. (Here $[l, x]$ stands for the inner product of vectors l and x.) The following optimization problem was considered in [BM00]:

$$\text{minimize} \sum_{a \in A} \min_{i=1,\ldots,k} ([l_i, a] - c_i)^2 \text{ subject to } \|l_i\|_2 = 1, \quad i = 1,\ldots,k. \quad (4)$$

Here $\min_{i=1,\ldots,k} ([l_i, a] - c_i)^2$ is the square of 2-norm distance between a point a and the nearest hyperplane from the given collection. Function

$$G((l_1, c_1), \ldots, (l_k, c_k)) = \sum_{a \in A} \min_{i=1,\ldots,k} ([l_i, a] - c_i)^2$$

can be represented in the form (1):

$$G((l_1, c_1), \ldots, (l_k, c_k)) = \sum_{a \in A} \min_{i=1\ldots,k} \varphi((l_i, c_i), a),$$

where $\varphi((l, c), a) = ([l, a] - c)^2$.

3.3 Skeleton of a finite set of points

We now consider a version of Bradley-Mangasarian definition, where the distances to hyperplanes are used instead of the squares of these distances. Assume that $\mathbb{R}^n$ is equipped with a norm $\|\cdot\|$. Let A be a finite set of points. Consider vectors $l_1, \ldots, l_k$ with $\|l_i\|_* \equiv \max_{\|x\|=1}[l, x] = 1$ and numbers c_i $(i = 1, \ldots, k)$. Let $H_i = \{x : [l_i, x] = c_i\}$ and $H = \cup_i H_i$. Then the distance between the set H_i and a point a is $d(a, H_i) = |[l_i, a] - c_i|$ and the distance between the set H and a is

$$d(a, H) = \min_i |[l_i, a] - c_i|. \quad (5)$$

The deviation of X from A is

$$\sum_{a \in A} d(a, H) \equiv \sum_{a \in A} \min_i |[l_i, a] - c_i|.$$

The function

$$L_k((l_1, c_1), \ldots, (l_k, c_k)) = \sum_{a \in A} \min_i |[l_i, a] - c_i| \quad (6)$$

is of the form (1). Consider the following constrained *min-sum-min* problem

$$\min \sum_{a \in A} \min_i |[l_i, a] - c_i| \text{ subject to } \|l_j\| = 1, \ c_j \in \mathbb{R} \ (j = 1, \ldots, k) \quad (7)$$

A solution of this problem will be called a *k-skeleton* of the set A. The function in (7) is called the *skeleton function*.

More precisely, k-skeleton is the union of k hyperplanes $\{x : [l_i, x] = c_i\}$, where $((l_1, c_1), \ldots, (l_k, c_k))$ is a solution of (7). *If the skeletons are known, each point is assigned to the cluster with the nearest skeleton.* It is difficult to find a global minimizer of (7), so sometimes we can consider the union of hyperplanes that is formed by a local solution of (7) as a skeleton.

Clusters constructed according to skeletons, obtained as a result of the skeleton function minimization are called *skeleton-based clusters*.

The concept of *shape* of a finite set of points was introduced and studied in [SU05]. By definition, the shape is a minimal (in a certain sense) ellipsoid, which contains the given set. A technique to find an ellipsoidal shape is then proposed in the same paper. In many instances the geometric characterization of a set A can be viewed as the intersection between its shape, describing its external boundary, and its skeleton, describing its internal aspect.

A comparative study of Bradley-Mangasarian approximation and skeletons was undertaken in [GRZ05]. It was shown there that skeletons are quite different from Bradley-Mangasarian approximation, even for simple sets.

3.4 Illustrative examples

We now give two illustrative examples.

Example 1. Consider the set depicted in Fig. 1

Fig. 1. Clusters based on centres

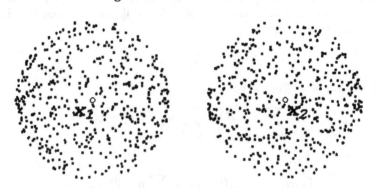

Clearly this set consists of two clusters, the centers of these clusters (points x_1 and x_2) can be found by the minimization of the cluster function. The skeleton of this set hardly depends on the number k of hyperplanes (straight lines). For each k this skeleton cannot give a clear presentation on the structure of the set.

Fig. 2. Clusters based on skeletons

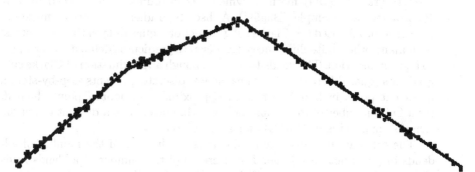

Example 2. Consider now the set depicted in Fig. 2.

It is difficult to say how many point-centred clusters has this set. Its description by means of such clusters cannot clarify its structure. At the same time this structure can be described by the intersection of its skeleton consisting on three straight lines and its shape. It does not make sense to consider k-skeletons of the given set with $k > 3$.

4 Minimization of sum-min functions belonging to class $\mathcal{F}$

Consider function F defined by (1):

$$F(x_1, \ldots, x_k) = \frac{1}{N} \sum_{a \in A} \min(\varphi_1(x_1, a), \varphi_2(x_2, a), \ldots \varphi_k(x_k, a)),$$

$$x_i \in \mathbb{R}^n, \ i = 1, \ldots, k.$$

where $A \subset \mathbb{R}^n$ is a finite set. This function depends on $n \times k$ variables. In real-world applications $n \times k$ is a large enough number and the set A contains some hundreds or thousands points. In such a case function F has a huge amount of shallow local minimizers that are very close to each other. The minimization of such functions is a challenging problem.

In this paper we consider both local and global minimization of sum-min functions from $\mathcal{F}$. First we discuss possible local techniques for the minimization.

The calculation of even one of the Clarke subgradients and/or a quasidifferential of function (1) is a difficult task, so methods of nonsmooth optimization based on subgradient information (quasidifferential information) at each iteration are not effective for the minimization of F. It seems that derivative-free methods are more effective for this purpose.

For the local minimization of functions (1) we propose to use the so-called discrete gradient (DG) method, which was introduced and studied by Adil Bagirov (see for example, [Bag99]). A discrete gradient is a certain finite difference approximated the Clarke subgradient or a quasidifferential. In contrast with many other finite differences, the discrete gradient is defined with respect to a given direction. This leads to a good enough approximation of Clarke subgradients (quasidifferentials). DG calculates discrete gradients step-by-step; if a current point in hands is not an approximate stationary point then after a finite number of iterations the algorithm calculates a descent direction. Armijo's method is used in DG for a line search.

The calculation of discrete gradients is much easier if the number of addends in (1) is not very large. The decrease of the number of addends leads also to a drastic diminishing of the number of shallow local minima. Since the number of addends is equal to the number of points in the dataset, we conclude that the results of the application of DG for minimization of (1) significantly depend on the size of the set A.

The discrete gradient method is a local method, which may terminate in a local minimum. In order to ascertain the quality of the solution reached, it is necessary to apply global methods. Here we call global method a method that does not get trapped on stationary points, and can leave local minima to a better solution.

Various combinations between local and global techniques have recently been studied (see, for example [HF02, YLT04]).

We use a combination of the DG and the cutting angle method (DG+CAM) in our experiments. We call this method the *hybrid global method*.

These two techniques (DG and DG+CAM) have been included in a new optimization software (CIAO-GO) created recently at the Centre for Informatics and Applied Optimization (CIAO) at the University of Ballarat, see [CIA05] for more information. This version of the CIAO-GO software (Centre for Informatics and Applied Optimization-Global Optimization) allows one to use four different solvers

1. DG,
2. DG multi start,
3. DG+CAM,
4. DG+CAM multi start.

Working with this software users have to input

- an objective function (for minimization),
- an initial point for optimization,
- upper and lower bounds for variables,
- constraints and a penalty constant (in the case of constrained optimization), constraints can be represented as equalities and inequalities,
- maximal running time,
- maximal number of iterations.

"Multi start" option in CIAO-GO means that the program starts from the initial point chosen by a user and also generates 4 additional random initial points. The final result is the best obtained result. The additional initial points are generated by CIAO-GO from the corresponding feasible region (or close to the feasible region).

As a global optimization technique we use the General Algebraic Modeling System (GAMS), see [GAM05] for more information. We use the Lipschitz global optimizer (LGO) solver [LGO05] from Pinter Consulting Services [Pin05].

5 Minimization of generalized cluster function

In this section we discuss applications DG, DG+CAM and the LGO solver for minimization of generalized cluster functions. We propose several approaches for selecting initial points.

5.1 Construction of generalized cluster functions

Consider a set $A \subset \mathbb{R}^n$ that contains N points. Choose $\varepsilon > 0$. Then choose a random vector $b^1 \in A$ and consider subset $A_{b^1} = \{a \in A : \|a - b^1\| < \varepsilon\}$ of the set A. Take randomly a point $b^2 \in A_1 = A \setminus A_{b^1}$. Let $A_{b^2} = \{a \in A_1 : \|a - b^2\| < \varepsilon\}$ and $A_2 = A_1 \setminus A_{b^2}$. If the set A_{j-1} is known, take randomly $b^j \in A_{j-1}$, define set A_{b^j} as $\{a \in A_{j-1} : \|a - b^j\| < \varepsilon\}$ and define set A_j as $A_{j-1} \setminus A_{b^j}$. The result of the described procedure is the set $B = \{b^j\}_{j=1}^{N_B}$, which is a subset of the original dataset A. The vector b^j is a representative for the whole group of vectors, removed on the step j.

If m_j is the cardinality of A_{b^j} then the generalized cluster function corresponding to B

$$\tilde{C}_k(x^1, \ldots, x^k) = \frac{1}{N} \sum_j m_j \min(\|x^1 - b^j\|, \ldots, \|x^k - b^j\|)$$

can be used for finding centers of clusters of the set A.

The size of the dataset B obtained as the result of the described procedure is the most important parameter, so we shall use this parameter for characterization of B.

It can be proved (see [BRSY03]) that this function does not differ by more than ε from the original cluster function.

Remark 1. We can use the same idea to construct the *generalized skeleton function.*

Remark 2. Unfortunately, it is very difficult to know a priori the value for ε which allows one to remove a certain proportion of observations. In our experiments we had to try several values for ε before we found suitable ones.

5.2 Initial points

Most methods of local optimization are very sensitive to the choice of an initial point. In this section we suggest a choice of initial points which can be used for the minimization of cluster functions and generalized cluster functions.

Consider a set $A \subset \mathbb{R}^n$ that contains N points. Assume that we want to find k clusters in A. In this case an initial point is a vector $x \in \mathbb{R}^{n \times k}$. The structure of the problem under consideration leads to different approaches to the choice of initial points. We suggest the following four approaches.

k-meansL_1 initial point The k-meansL_1 method is a version of the well-known k-means method (see, for example, [MST94]), where $\| \cdot \|_1$ is used instead of $\| \cdot \|_2$. (We use $\| \cdot \|_1$ in numerical experiments, this is the reason for consideration of k-meansL_1 instead of k-means.) We use the following procedure in order to sort N observations into k clusters:

1. Take any k observations as the centres of the first k clusters.
2. Assign the remaining $N - k$ observations to one of the k clusters on the basis of the shortest distance (in the sense of $\| \cdot \|_1$ norm) between an observation and the mean of the cluster.
3. After each observation has been assigned to one of the k clusters, the means are recomputed (updated).

Stopping criterion: there is no observation, which moves from one cluster to another.

Note that results of this procedure depend on the choice of an initial observation.

We apply this algorithm for original dataset A and then the result point $x \in \mathbb{R}^{n \times k}$ is considered as an initial point for minimization of generalized cluster function generated by the dataset B.

Uniform initial point The application of optimization methods to clustering requires a certain data processing. In particular, a scaling procedure should be applied. In our experiments we convert a given dataset to a dataset with the mean-value 1 for each feature (coordinate). In such a case we can choose the point $x = (1, 1, \ldots, 1) \in \mathbb{R}^{n \times k}$ as initial one. We shall call it the *uniform initial point*.

Ordered initial point Recall that m_j indicates the cardinality of the set of points $A_{b^j} \in A$, which are represented by a point $b^j \in B$. It is natural to consider the collection of the heaviest k points as an initial vector for the minimization of generalized cluster function $\tilde{C}$. To formalize this, we rearrange the points so that the numbers $m_j, j = 1, \ldots, N_B$ decrease and take the first k points from this rearranged dataset. Thus, in order to construct an initial point we choose the k observations with the largest values for weights m_j from the dataset B.

Uniform-ordered initial point This initial point is a hybrid between the Uniform and the Ordered initial points. It contains the heaviest $k - 1$ observations and the barycentre (each coordinate is 1).

6 Numerical experiments with generalized cluster function

For numerical experiments we use two types of datasets, namely the original dataset A and a small dataset B obtained by the procedure described in Subsection 5.1. We compare results obtained for B with the results obtained for the entire original dataset A.

6.1 Datasets

We carried out numerical experiments with two well-known test datasets (see [MST94]):

- Letters dataset (20000 observations, 26 classes, 16 features). This dataset consists of samples of 26 capital letters, printed in different fonts; 20 different fonts were considered and the location of the samples was distributed randomly within the dataset.
- Pendigits dataset (10992 observations, 10 classes, 16 features). This dataset was created by collecting 250 samples from 44 writers. These writers are asked to write 250 digits in random order inside boxes of 500 by 500 tablet pixel resolution.

Both Letters and Pendigit datasets have been used for testing different methods of supervised classification (see [MST94] for details). Since we use these datasets only for construction of generalized cluster function, we consider them as datasets with unknown classes.

6.2 Numerical experiments: description

We are looking for three and four clusters in both Letters and Pendigits datasets. Dimension of optimization problems is equal to 48 in the case of 3 clusters and 64 in the case of 4 clusters. We consider two small sub-databases of the Letters dataset (Let1, 353 points, approximately 2% of the original dataset; and Let2, 810 points, approximately 4% of the original dataset) and two small sub-sets of the Pendigits dataset (Pen1, 216 points, approximately 2% of the original dataset; and Pen2, 426 points, approximately 4% of the original dataset).

We apply local techniques (discrete gradient method) and global techniques (a combination between discrete gradient and cutting angle method and LGO solver) to minimize the generalized cluster function. Then we need to estimate the results obtained. We can use different approaches for this estimation. One of them is based on comparison of values of cluster function C_k constructed with respect to the centers obtained in the original dataset A and with respect to the centers obtained in its small sub-dataset B. We compare the cluster function values, started from different initial points in original datasets and their approximations.

We use the following procedure.

Let A be an original dataset and B be its small sub-dataset. First, the centres of clusters in B should be found by an optimization technique. Then we evaluate the cluster function values in A using the obtained points as the centers of clusters in A. Using this approach we can find out how the results of the minimization depend on initial points and how far we can go in the process of dataset reduction.

In our research we use 4 types of initial points, described in section 5.2. These initial points have been carefully chosen and the results obtained starting from these initial points are better than the results obtained starting from random initial points. Therefore, we present the results obtained for these 4 types of initial points rather than the results obtained starting from random initial points generated, for example, by "multi start" option.

6.3 Results of numerical experiments

Local optimization

First of all we have to point out that we have two groups of initial points

- Group 1: Uniform initial point and k-meansL_1 initial point,
- Group 2: Ordered initial point and Uniform-ordered initial point.

Initial points from Group 1 are the same for an original dataset and for all its reduced versions. Initial points from Group 2 are constructed according to their weights. Points in original datasets have the same weights which are equal to 1.

Remark 3. Because the weights can vary for different reductions of the dataset, the Ordered initial points for Let1 and Let2 do not necessarily coincide. The same is true for the Uniform-ordered initial points. The same observation applies to the Pendigits dataset and its reduced versions Pen1 and Pen2.

Our next step is to compare results obtained starting from different initial points in the original datasets and in their approximations. In our experiments we use two different kinds of function: the cluster function and the generalized cluster function. Values for the cluster function and the generalized cluster function are the same for original datasets because each point has the same weight which is equal to 1. In the case of reduced datasets we produce our numerical experiments in corresponding approximations of original datasets and calculate two different value: the cluster function value and the generalized function value. The **cluster function value** is the value of the cluster function calculated in the corresponding original dataset according to the centres found in the reduced dataset. The **generalized cluster function value** is the value of the generalized cluster function calculated in the reduced dataset according to the centres found in the same reduced dataset. Normally a cluster function value (calculated according to the centres found reduced datasets) is larger than a generalized cluster function value calculated according to the same centres and the corresponding weights, because optimization techniques have been actually applied to minimize the generalized cluster in the corresponding reduced dataset. In Tables 1-2 we present the results of our numerical experiments obtained for DG and DG+CA starting from the Uniform initial point.

It is also very important to remember that *a better result in a reduced dataset is not necessarily better for the original one.* For example, in the case of the Pen1 dataset, 3 clusters, the Uniform initial point the generalized function value is lower for DG+CAM than for DG, however the cluster function value is lower for DG than for DG+CAM. We observe the same situation in some other examples.

Table 1. Cluster function and generalized cluster function: DG, Uniform initial point

Dataset	Size	Cluster function value	Generalized cluster function value	Cluster function value	Generalized cluster function value
		3 clusters		4 clusters	
Pen1	216	6.4225	5.5547	5.7962	4.8362
Pen2	426	6.3844	5.8132	5.7725	5.0931
Pendigits	10992	6.3426	6.3426	5.7218	5.7218
Let1	353	4.3059	3.3859	4.1200	3.1611
Let2	810	4.2826	3.7065	4.0906	3.5040
Letters	20000	4.2494	4.2494	4.0695	4.0695

Our actual goal is to find clusters in the original datasets, therefore it is important to compare *cluster function values calculated in original datasets* according to obtained centres. Centres can be obtained from our numerical experiments with both types of datasets: original datasets and reduced datasets. It is one of the possible ways to test the efficiency of the proposed approach: substitution of original datasets by their smaller approximations.

Tables 3-8 represent cluster function values obtained in our numerical experiments starting from the k-meansL_1, Ordered and Uniform-ordered initial point. We do not present the obtained generalized function values because this function can not be used as a measure of the quality of clustering.

Table 2. Cluster function and generalized cluster function: DG+CAM, Uniform initial point

Dataset	Size	Cluster function value	Generalized cluster function value	Cluster function value	Generalized cluster function value
		3 clusters		4 clusters	
Pen1	216	6.4254	5.5546	5.7943	4.8353
Pen2	426	6.3843	5.8131	5.7718	5.0931
Pendigits	10992	6.3426	6.3426	5.7218	5.7218
Let1	353	4.3059	3.3859	4.1208	3.1600
Let2	810	4.2828	3.7061	4.0909	3.5020
Letters	20000	4.2494	4.2494	4.0695	4.0695

Recall that reduced datasets are approximations of corresponding original datasets. Decreasing the number of observations we reduce the complexity of our optimization problems but obtain less precise approximations. Therefore, our goal is to find some balance between the reduction of the complexity of optimization problems and the quality of obtained results. In some cases (mostly initial point from Group 2, see Remark 3 for more information) the results obtained on larger approximations of original datasets (more precise approximations) are worse than the results obtained on smaller approximations of original datasets (less precise approximations). For example, Pen1 and Pen2 for initial point from Group 2 (3 and 4 clusters).

Table 3. Cluster function: DG, k-meansL_1 initial point

Dataset	Size	Cluster function value 3 clusters	Cluster function value 4 clusters
Pen1	216	6.4272	5.8063
Pen2	426	6.3840	5.7704
Pendigits	10992	6.3409	5.7217
Let1	353	4.3087	4.1241
Let2	810	4.2816	4.1013
Letters	20000	4.2495	4.0726

Remark 4. In the original datasets, it is not relevant to consider the Ordered and Uniform-ordered initial points, because all the points have the same weight.

Summarizing the results of the numerical experiments (cluster function, local and hybrid global techniques, 4 special kinds of initial points) we can draw out the following conclusions:

Table 4. Cluster function: DG+CAM, k-meansL_1 initial point

Dataset	Size	Cluster function value 3 clusters	Cluster function value 4 clusters
Pen1	216	6.4278	5.8063
Pen2	426	6.3841	5.7723
Pendigits	10992	6.3409	5.7217
Let1	353	4.3087	4.1262
Let2	810	4.2824	4.1014
Letters	20000	4.2495	4.0726

Table 5. Cluster function: DG, Ordered initial point

Dataset	Size	Cluster function value 3 clusters	Cluster function value 4 clusters
Pen1	216	6.4188	5.8226
Pen2	426	6.6534	5.9047
Let1	353	4.3228	4.2049
Let2	810	4.3843	4.1112

Table 6. Cluster function: DG+CAM, Ordered initial point

Dataset	Size	Cluster function value 3 clusters	Cluster function value 4 clusters
Pen1	216	6.4171	5.8201
Pen2	426	6.6536	5.9047
Let1	353	4.3228	4.2045
Let2	810	4.3843	4.1107

Table 7. Cluster function: DG, Uniform-ordered initial point

Dataset	Size	Cluster function value 3 clusters	Cluster function value 4 clusters
Pen1	216	6.4188	5.7921
Pen2	426	6.6514	5.8718
Let1	353	4.2910	4.1225
Let2	810	4.2828	4.1129

1. DG and DG+CAM applied to the same datasets produce almost identical results if initial points are the same,
2. DG and DG+CAM applied to the same datasets starting from different initial points (4 proposed initial points) produce very similar results in most of the examples,

Table 8. Cluster function: DG+CAM, Uniform-ordered initial point

Dataset	Size	Cluster function value 3 clusters	Cluster function value 4 clusters
Pen1	216	6.4171	5.7945
Pen2	426	6.6492	5.8715
Let1	353	4.2905	4.1233
Let2	810	4.2828	4.1130

3. in some cases the results obtained on smaller approximations of original datasets are better than the results obtained on larger approximations of original datasets.

Global optimization: LGO solver

First we present the results obtained by the LGO solver (global optimization). We use the Uniform initial point. The results are in Table 9.

In almost all the cases (except Pendigits 3 clusters) the results for reduced datasets are better than for original datasets. It means that the cluster function is too complicate for the solver as an objective function and it is more efficient to use generalized cluster functions generated on reduced datasets. It is beneficial to use reduced datasets in the case of the LGO solver from two points of view

1. computations with reduced datasets allow one to reach a better minimizer;
2. computational time is significantly less for reduced datasets than for original datasets.

It is also obvious that the software failed to reach a global minimum. We suggest that the LGO solver has been developed for a broad class of optimization problems. However, the solvers included in CIAO-GO are more efficient for minimization of the sum of minima of convex functions, especially if the number of components in sums is large.

Remark 5. The LGO solver was not used in the experiments on skeletons.

7 Skeletons

7.1 Introduction

The problem of grouping (clustering) points by means of skeletons is not so widely studied as it is in the case of cluster function based models. Therefore, we would like to start with some examples produced in not very large datasets (no more than 1000 observations). In this subsection we formulate

Table 9. Cluster function: LGO solver

Dataset	Size	Cluster function value 3 clusters	Cluster function value 4 clusters
Pen1	216	6.4370	5.8029
Pen2	426	6.4122	5.7800
Pendigits	10992	6.3426	7.1859
Let1	353	4.3076	4.1426
Let2	810	4.2829	4.1191
Letters	20000	5.8638	4.2064

the problems of finding skeletons mathematically, discuss applications of DG and DG+SA to finding skeletons with respect to $\|\cdot\|_1$ and and give graphical implementation to obtained results (for examples with no more than 3 features).

The search for skeletons can be done by solving constrained minimization problem (7).

Both algorithms are designed for unconstrained problems so we use a penalty function in order to convert problem (7) to the unconstrained minimization. The corresponding unconstrained problem has the form:

$$\min_{(l_1,b_1)\in\mathbb{R}^{n+1},...,(l_k,b_k)\in\mathbb{R}^{n+1}} \sum_{q\in Q} \min_i |[l_i, a^q] - b_i| + R_p \sum_{i=1}^k |\|l_i\|_1 - 1|, \qquad (8)$$

where R_p is a penalty parameter.

Finally, the algorithms were applied starting from 3 different initial points, and the best solution found was selected. The 3 different points used in the example are:

- $P_1 = \begin{cases} l_i = \frac{1}{\sqrt{N}}(1,...,1) \\ b_i = 1 \end{cases}$

- $P_2 = \begin{cases} l_i = (1, 0, .., 0) \\ b_i = 1 \end{cases}$

- $P_3 = \begin{cases} l_i = \frac{1}{\sqrt{N-1}}(0, 1..., 1) \\ b_i = 1 \end{cases}$

The problem has been solved for different sets of points, selected from 3 different well known datasets: the Heart disease database (13 features, 2 classes: 160 observations are from the first class and 197 observations are from the second class), the Diabetes database (8 features, 2 classes: 500 observations are from the first class and 268 observations are from the second class) and the Australian credit cards database (14 features, 2 classes: 383 observations are from the first class and 307 observations are from the second class), see also [MST94] and references therein. Each of these datasets was submitted first to the feature selection method described in [BRY02].

The value of the objective function was considerably decreased by both methods. However, the discrete gradient method often gives a local solution which is very close to the initial point, while the hybrid gives a solution which is further and better. In the tables the distance considered is the Euclidean distance between the solution obtained and the initial solution, and the value considered is the value of the objective function at this solution.

Table 10. Australian credit card database with 2 hyperplanes skeletons

		DG method		hybrid method	
	Initial point	value	distance	value	distance
Class 1	1	22.9804	10.668	6.11298	7.98738
	2	25.5102	2.81543	13.2263	5.91397
	3	6.10334	4.40741	6.10334	4.40741
Class 2	1	0.473317	5.00549	0.473317	5.00549
	2	3.029	2.14784	0.222154	2.13944
	3	6.87897	6.06736	4.73828	6.74424
computation time		54 sec		664 sec	

Table 11. Diabetes database with 3 hyperplanes skeletons

		DG method		hybrid method	
	Initial point	value	distance	value	distance
Class 1	1	28.5856	6.78624	28.1024	6.79326
	2	39.3925	11.4668	28.2417	11.7711
	3	33.2006	3.09434	31.4624	2.31922
Class 2	1	22.2806	2.3755	22.2806	2.3755
	2	30.346	56.7222	19.5574	8.76914
	3	23.0529	1.61649	22.9495	1.76052
computation time		212 sec		1521 sec	

The different examples show that although sometimes the hybrid method does not improve the result obtained with the discrete gradient method, in some other cases the result obtained is much better than when the discrete gradient method is used. However the computations times it induces are much greater than the simple use of the discrete gradient method. The diabetes dataset has 3 features, after feature selection (see [BRY02]). This allows us to plot graphically some of the results obtained during the computations.

We can observe that the hybrid method does not necessarily give an optimal solution. Even with the hybrid method the initial point is very important. Figure 3 however, confirms that the solutions obtained are usually very good, and represent correctly the set of points. The set of points studied here is

Fig. 3. 2nd class for the diabetes database, with 2 hyperplanes

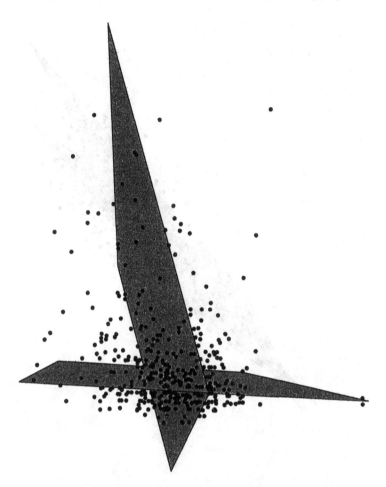

constituted by a big mass of points, and some other points spread around. It is interesting to remark that the hyperplanes intersect around the same place - where the big mass is situated - and take different directions, to be the closer possible to the spread points.

Figure 4 shows the complexity of the diabetes dataset.

7.2 Numerical experiments: description

We are looking for three and four clusters in both Letters and Pendigits datasets. Dimension of optimization problems is equal to 51 in the case of

Fig. 4. Diabetes database, with 1 hyperplane per class

3 skeletons and 68 in the case of 4 skeletons. We use the same sub-datasets as in section 6 (Pen1, Pen2, Let1, Let2).

We apply local techniques (DG and DG+CAM) for minimization of the generalized skeleton function. Then we use a procedure which is similar to the one we use for the cluster function to estimate the obtained results. First, we find skeletons in original datasets (or in reduced datasets). Then we evaluate the skeleton function values in original datasets using the obtained skeletons.

For the skeleton function the problem of constructing a good initial point has not been studied yet. Therefore, in our numerical experiments as an initial point we choose a feasible point. We also use "multi start" option to compare results obtained starting from different initial points.

7.3 Numerical experiments: results

In this subsection we present the results obtained for the skeleton function. Our goal is to find the centres in original datasets, therefore *we do not present the generalized skeleton function values.* Table 12 and Table 13 present the values of the skeleton function evaluated in the corresponding *original datasets* (Pendigits and Letters respectively) according to the skeletons obtained as optimization results reached in datasets from the first column of the tables. We use two different optimization methods: DG and DG+CAM and two different types of initial points: "single start" (DG or DG+CAM) and "multi start" (DGMULT or DG+CAMMULT).

Table 12. Skeleton function: Pendigits

Number of seletons	Dataset	Size	Skeleton function values			
			DG	DGMULT	DG+CAM	DG+CAMMULT
3	Pen1	216	2137.00	1287.58	1832.97	1320.00
	Pen2	426	735.00	735.47	735.47	735.47
	Pendigits	10992	567.20	567.20	567.20	566.55
4	Pen1	216	1223.16	1315.68	1194.65	1180.79
	Pen2	426	1360.16	946.74	1322.46	946.74
	Pendigits	10992	905.56	905.56	905.56	661.84

Table 13. Skeleton function: Letters

Number of seletons	Dataset	Size	Skeleton function values			
			DG	DGMULT	DG+CAM	DG+CAMMULT
3	Let1	353	1548.30	1548.30	1545.58	1545.58
	Let2	810	2201.75	1475.77	2171.01	1608.14
	Letters	20000	1904.71	1904.71	1904.71	964.37
4	Let1	353	1566.69	1566.69	1531.99	1531.99
	Let2	810	2030.20	2030.20	1892.31	1892.31
	Letters	20000	964.37	850.14	850.14	850.14

The most important conclusion to the results is that in the case of the skeleton function the best optimization results (the lowest value of the skeleton function) *have been reached in the experiments with the original datasets.* It means that the *proposed cleaning procedure is not as efficient in the case of skeleton function as it is in the case of the clustering function.* However, in the case of the clustering function the initial points for the optimization methods have been chosen after some preliminary study. It can happen that an efficient choice of initial points leads to better optimization results for both kinds of datasets: original and reduced.

Recall that (7) is a constrained optimization problem with equality constraints. This problem is equivalent to the following constrained optimization problem with inequality constraints

$$\min \sum_{a \in A} \min_i |[l_i, a] - c_i| \text{ subject to } \|l_j\| \geq 1, \quad c_j \in \mathbb{R} \ (j = 1, \ldots, k). \quad (9)$$

In our numerical experiments we use both formulations (7) and (9). In most of the experiments the results obtained for (7) are better than for (9) but computational time is much higher for (7) than for (9). It is recommended, however, to use the formulation (9) if, for example, experiments with (7) produce empty skeletons.

7.4 Other experiments

Another set of numerical experiments has been carried out on the both objective functions. Although of little interest from the point of view of the optimization itself, to the authors' opinion it may bring some more light on the clustering part.

The objective functions (2) and (7) has been minimized using two different methods: the discrete gradient method described above, and a hybrid method between the DG method and the well known simulated annealing method. This command is described with details in [BZ03].

The basic idea of the hybrid method is to alternate the descent method to obtain a local minima and the simulated annealing method to escape this minimum. This reduces drastically the dependency of the local method on an initial point, and ensures that the method reaches a "good" minimum.

Numerical experiments were carried out on the Pendigit and Letters datasets for the generalized cluster function using different size dataset approximations. The results have shown that the hybrid method reached a sensibly comparable value as the other methods, although the algorithm had to leave up to 50 local minima. This can be explained by the large number of local minima in the objective function, each close to one another.

The skeleton function was minimized for the Heart Disease and the Diabetes datasets. The same behaviour can be observed. As the results of these experiments were not drawing any major conclusion, they are not shown here.

Numerical experiments have shown that while considerably faster than the simulated annealing method, the hybrid method is still fairly slow to converge.

8 Conclusions

8.1 Optimization

In this paper, a particular type of optimization problems has been presented. The objective function of these problems is the sum of mins of convex functions. This type of problems appears quite often in the area of data analysis, and two examples have been solved.

The generalized cluster function has been minimized for two datasets, using three different methods: the LGO global optimization software included in GAMS, the discrete gradient method and a combination between this method and the cutting angle method.

The last two methods have been started from carefully selected initial points and from a random initial point.

The LGO software failed most of the time to reach even a good solution. This is due to the fact that the objective function has a very complex structure. This method was limited in time, and may have reached the global solution, had it been given a limitless amount of time.

Similarly, the local methods failed to reach the solution when started from a random point. The reason is the large amount of local minima in the objective function which prevent local methods to reach a good solution.

However the discrete gradient method, for all the examples, reached a good solution for at least one of the initial point. The combination reached a good solution for all of the initial points.

This shows that for such types of functions, presenting a complex structure and many local minima, most global methods will fail. However, well chosen initial points will lead to a deep local minimum. Because the local methods are much faster than global ones, it is more advantageous to start the local method from a set of carefully chosen initial points to reach a global minimum.

The application of the combination between the discrete gradient and the cutting angle methods appears to be a good alternative, as it is not very dependant on the initial point, while reaching a good solution in a limited time.

The second set of experiments was carried out over the hyperplanes function. This function having been less studied in the literature, it is harder to draw definite conclusions. However, the experiments show very clearly that the local methods once again strongly depend on the initial point. Unfortunately it is harder to devise a good initial point for this objective function.

8.2 Clustering

From the clustering point of view, two different similarity functions have been minimized. The first one is a variation of the widely studied cluster function, where the points are weighted. The second one is a variation of the Bradley-Mangasarian function, where distances from the hyperplanes are taken instead of their square.

A method for reducing the size of the dataset, ε-cleaning, has been devised and applied. Different values for epsilon lead to different sizes of datasets. Numerical experiments have been carried out for different values of epsilon, leading to very small (2% and 4%) datasets.

For the generalized cluster function, this method proves to be very successful: even for very small datasets, the function value obtained is very satisfactory. When the method was solved using the global method LGO, the results obtained for the reduced dataset were almost always better than those obtained for the original dataset. The reason is that the larger the dataset, the larger number of local minima for the objective function. When the dataset is reduced, what is lost in measurement quality is gained by the strong simplification of the function. Because each point in the reduced dataset acts already as a centre for its neighbourhood, minimizing the generalized cluster function is equivalent to group these "mini" clusters into larger clusters.

It has to be noted that there is not a monotone correspondence between the value of the generalized cluster function for the reduced and the original dataset. It may happen that a given solution is better than another one for the reduced dataset, and worse for the original. Thus we cannot conclude that the solution can be reached for the reduced dataset. However, the experiments show that the solution found for the reduced dataset is always good.

For the skeletons function, however, this method is not so successful. Although this has to be taken with precautions, as the initial points for this function could not be devised so carefully as for the cluster function, one can expect such behavior: the reduced dataset is actually a set of cluster centres. The skeleton approach is based on the assumption that the clusters in the dataset can be represented by hyperplanes, while the cluster approach assumes that the clusters are represented by centres.

The experiments show the significance of the choice of the initial point to reach good clusters. While random points did not allow any method to reach a good solution, all initial points selected upon the structure of the dataset lead the combination DG-CAM to the solution.

Since for the cluster function we are able to provide some good initial points, but not for the skeleton function, unless the structure of the dataset is known to correspond to some skeletons, we would recommend to use the centre approach.

Finally the comparison between the results obtained by the two different methods has to be relativized: experiments having shown the importance of initial points, it is difficult to draw definitive conclusions from the results obtained for the skeleton approach.

However, there seems to be a relationship between the classes and the clusters obtained by both approaches, some classes being almost absent from certain clusters. Further investigations should be carried out in this direction, and classification processes based on these approaches could be proposed.

Acknowledgements

The authors are very thankful to Dr. Adil Bagirov for his valuable comments.

References

[Bag99]   Bagirov, A.M.: Derivative-free methods for unconstrained nonsmooth optimization and its numerical analysis. Investigacao Operacional, **19**, 75–93 (1999)

[BRSY03] Bagirov, A.M., Rubinov, A.M., Soukhoroukova, N., Yearwood, J.: Unsupervised and Supervised Data Classification Via Nonsmooth and Global Optimization. Sociedad de Estadistica e Investigacion Operativa, Top, **11**, 1–93 (2003)

[BRY02]   Bagirov, A.M., Rubinov, A.M., Yearwood, J.: A global optimization approach to classification. Optimization and Engineering, **3**, 129–155 (2002)

[BRZ05a] Bagirov, A., Rubinov, A., Zhang, J.: Local optimization method with global multidimensional search for descent. Journal of Global Optimization (accepted) (http://www.optimization-online.org/DB_FILE/2004/01/808.pdf)

[BRZ05b] Bagirov, A., Rubinov, A., Zhang, J.: A new multidimensional descent method for global optimization. Computational Optimization and Applications (Submitted) (2005)

[BZ03]   Bagirov, A.M., Zhang, J.: Hybrid simulating annealing method and discrete gradient method for global optimization. In: Proceedings of Industrial Mathematics Symposium, Perth (2003)

[BBM03]   Beliakov, G., Bagirov, A., Monsalve, J.E.: Parallelization of the discrete gradient method of non-smooth optimization and its applications. In: Proceedings of the 3rd International Conference on Computational Science. Springer-Verlag, Heidelberg, **3**, 592–601 (2003)

[BM00]   Bradley, P.S., Mangasarian, O.L.: k-Plane clustering. Journal of Global Optimization, **16**, 23–32 (2000)

[BLM02]   Brimberg, J., Love, R.F., Mehrez, A.: Location/Allocation of queuing facilities in continuous space using minsum and minmax criteria. In: Pardalos, P., Migdalas, A., Burkard, R. (eds) Combinatorial and Global Optimization. World Scientific (2002)

[DR95]   Demyanov, V., Rubinov, A.: Constructive Nonsmooth Analysis. Peter Lang (1995)

[GRZ05]   Ghosh, R., Rubinov, A.M., Zhang, J.: Optimisation approach for clustering datasets with weights. Optimization Methods and Software, **20** (2005)

[HF02]   Hedar, A.-R., Fukushima, M.: Hybrid simulated annealing and direct search method for nonlinear unconstrained global optimization. Optimization Methods and Software, 17, 891–912 (2002)

[JMF99]   Jain, A.K., Murty, M.N., Flynn, P.J.: Data clustering: a review. ACM Computing Surveys, **31**, 264–323 (1999)

[Kel99]   Kelly, C.T.: Detection and remediatio of stagnation in the Nelder-Mead algorithm using a sufficient decreasing condition. SIAM J. Optimization, **10**, 43–55 (1999)

[MST94]   Michie, D., Spiegelhalter, D.J., Taylor, C.C. (eds): Machine Learning, Neural and Statistical Classification. Ellis Horwood Series in Artificial Intelligence, London (1994)

[SU05]   Soukhoroukova, N., Ugon, J.: A new algorithm to find a shape of a finite set of points. Proceedings of Conference on Industrial Optimization, Perth, Australia (Submitted) (2005)

[YLT04] Yiu, K.F.C., Liu, Y., Teo, K.L.: A hybrid descent method for global opti-
         mization. Journal of Global Optimization, **28**, 229–238 (2004)
[GAM05] http://www.gams.com/
[LGO05] http://www.gams.com/solvers/lgo.pdf
[Pin05] http://www.dal.ca/ jdpinter/
[CIA05] http://www.ciao-go.com.au/index.php

Analysis of a Practical Control Policy for Water Storage in Two Connected Dams

Phil Howlett[1], Julia Piantadosi[1], and Charles Pearce[2]

[1] Centre for Industrial and Applied Mathematics
University of South Australia
Mawson Lakes, SA 5095, Australia
phil.howlett@unisa.edu.au, julia.piantadosi@unisa.edu.au
[2] School of Mathematics
University of Adelaide
Adelaide, SA 5005, Australia
cpearce@maths.adelaide.edu.au

Summary. We consider the management of water storage in two connected dams. The first dam is designed to capture stormwater generated by rainfall. Water is pumped from the first dam to the second dam and is subsequently supplied to users. There is no direct intake of stormwater to the second dam. We assume random generation of rainfall according to a known probability distribution and wish to find practical pumping policies from the capture dam to the supply dam in order to minimise overflow. Within certain practical policy classes each specific policy defines a large sparse transition matrix. We use matrix reduction methods to calculate the invariant state probability vector and the expected overflow for each policy. We explain why the problem is more difficult when the inflow probabilities are time dependent and suggest an alternative procedure.

1 Introduction

The mathematical literature on storage dams, now half a century old, developed largely from the seminal work of Moran [Mor54, Mor59] and his school (see, for example, [Gan69, Yeo74, Yeo75]). Moran was motivated by specific practical problems faced by the Snowy Mountain Authority in Australia in the 1950s. Our present study is likewise motivated by a specific practical problem at Mawson Lakes in South Australia relating to a pair of dams in tandem.

The mathematical analysis of dams has proved technically more difficult than that of their discrete counterpart, queues. In order to deal with the complexity of a tandem system, we treat a discretised version of the problem and adopt the matrix–analytic methodology of Neuts and his school (see [LR99, Neu89] for a modern exposition). The Neuts' methodology is well suited for handling processes with a bivariate state space, here the contents of the two dams.

A further new feature in this study is the incorporation of control. For recent work on control in the context of a dam, see [Abd03] and the references therein. The present article is preliminary and raises issues of both practical and theoretical interest.

In Section 2 we formulate the problem in matrix–analytic terms and in Section 3 provide an heuristic for the determination of an invariant probability measure for the process. This depends on the existence of certain matrix inverses. Section 4 sketches a purely algebraic procedure for establishing the existence of these inverses. In Section 5 we show how this can be simplified and systematised using a probabilistic analysis based on modern machinery of the matrix–analytic approach. In Section 6 we describe briefly how these results enable us to determine expected long–term overflow, which is needed for the analysis of control procedures. We conclude in Section 7 with a discussion of extensions of the ideas presented in the earlier sections.

2 Problem formulation

We assume a discrete state model and let the first and second components of

$$z = z(t) \in [0, h] \times [0, k] \subseteq \mathbb{Z}^2$$

denote respectively the number of units of water in the first and second dams at time t. We assume a stochastic intake to the capture dam where p_r denotes the probability that r units of water will enter the dam on any given day and a regular demand from the supply dam of 1 unit per day. To begin we assume that $p_r > 0$ for all $r = 0, 1, 2, \ldots$ and we will also assume that these probabilities do not depend on time. The first assumption is a reasonable assumption in practice but the latter assumption is certainly not reasonable over an extended period of time. We revise these assumptions later in the paper.

We consider a class of practical pumping policies where the pumping decision depends only on the contents of the first dam. Choose an integer $m \in [1, h]$ and pump m units from the capture dam to the supply dam each day when the capture dam contains at least m units. For an intake r there are two basic transition patterns

- $(z_1, 0) \rightarrow (\zeta_1, 0)$
- $(z_1, z_2) \rightarrow (\zeta_1, z_2 - 1)$

where $\zeta_1 = \min\{[z_1 + r], h\}$ for $z_1 < m$, and two basic transition patterns

- $(z_1, 0) \rightarrow (\zeta_1^*, m)$
- $(z_1, z_2) \rightarrow (\zeta_1^*, \zeta_2^*)$

where $\zeta_1^* = \min\{[z_1 - m + r], h\}$ and where $\zeta_2^* = \min\{[z_2 - 1 + m], k\}$, for $z_1 \geq m$. These transitions have probability p_r. The variable m is the control variable for a class of practical control policies but in this paper we assume m is fixed and suppress any notational dependence on m.

We now set up a suitable Markov chain to describe the process. In terms of matrix–analytic machinery, it turns out to be more convenient to use the ordered pair (z_2, z_1) for the state of the process rather than the seemingly more natural (z_1, z_2). This we do for the remainder of the article. We now order the states as

$$(0,0), \ldots, (0,h), (1,0), \ldots, (1,h), \ldots, (k,0), \ldots, (k,h).$$

The first component (that is, the content of dam 2) we refer to as the *level* of the process and the second component (the content of dam 1) as the *phase*.

The one–step transition matrix

$$P \in \mathbb{R}^{(h+1)(k+1) \times (h+1)(k+1)}$$

then has a simple block structure

$$
P =
\begin{array}{c}
\begin{array}{cccccccccccc}
0 & 1 & \cdot & \cdot & m & \cdot & \cdot & \cdot & \cdot & \cdot & \cdot & k
\end{array} \\
\left[
\begin{array}{cccccccccccc}
A & 0 & \cdot & \cdot & B & 0 & \cdot & \cdot & \cdot & 0 & 0 & 0 \\
A & 0 & \cdot & \cdot & B & 0 & \cdot & \cdot & \cdot & 0 & 0 & 0 \\
0 & A & \cdot & \cdot & 0 & B & \cdot & \cdot & \cdot & 0 & 0 & 0 \\
\cdot & \cdot & \cdot & \cdot & \cdot & \cdot & \cdot & \cdot & \cdot & \cdot & \cdot & \cdot \\
\cdot & \cdot & \cdot & \cdot & \cdot & \cdot & \cdot & \cdot & \cdot & \cdot & \cdot & \cdot \\
0 & 0 & \cdot & \cdot & A & 0 & \cdot & \cdot & \cdot & 0 & 0 & 0 \\
0 & 0 & \cdot & \cdot & 0 & A & \cdot & \cdot & \cdot & 0 & 0 & 0 \\
\cdot & \cdot & \cdot & \cdot & \cdot & \cdot & \cdot & \cdot & \cdot & \cdot & \cdot & \cdot \\
\cdot & \cdot & \cdot & \cdot & \cdot & \cdot & \cdot & \cdot & \cdot & \cdot & \cdot & \cdot \\
0 & 0 & \cdot & \cdot & 0 & 0 & \cdot & \cdot & \cdot & B & 0 & 0 \\
0 & 0 & \cdot & \cdot & 0 & 0 & \cdot & \cdot & \cdot & 0 & B & 0 \\
0 & 0 & \cdot & \cdot & 0 & 0 & \cdot & \cdot & \cdot & 0 & 0 & B \\
\cdot & \cdot & \cdot & \cdot & \cdot & \cdot & \cdot & \cdot & \cdot & \cdot & \cdot & \cdot \\
0 & 0 & \cdot & \cdot & 0 & 0 & \cdot & \cdot & \cdot & A & 0 & B \\
0 & 0 & \cdot & \cdot & 0 & 0 & \cdot & \cdot & \cdot & 0 & A & B
\end{array}
\right]
\end{array}
$$

where

$$A \text{ and } B \in \mathbb{R}^{(h+1) \times (h+1)}.$$

On the one hand we have

$$A = \begin{bmatrix} A_{11} & A_{12} \\ 0 & 0 \end{bmatrix}$$

where

$$A_{11} = \begin{bmatrix} p_0 & p_1 & \cdot & p_{m-2} & p_{m-1} \\ 0 & p_0 & \cdot & p_{m-3} & p_{m-2} \\ \cdot & \cdot & \cdot & \cdot & \cdot \\ 0 & 0 & \cdot & p_0 & p_1 \\ 0 & 0 & \cdot & 0 & p_0 \end{bmatrix}$$

and

$$A_{12} = \begin{bmatrix} p_m & p_{m+1} & \cdots & p_{h-1} & p_h^+ \\ p_{m-1} & p_m & \cdots & p_{h-2} & p_{h-1}^+ \\ \cdot & \cdot & \cdots & \cdot & \cdot \\ p_1 & p_2 & \cdots & p_{h-m} & p_{h-m+1}^+ \end{bmatrix}$$

where we have defined $p_r^+ = p_r + p_{r+1} + \cdots$ and on the other hand

$$B = \begin{bmatrix} 0 & 0 \\ B_{21} & B_{22} \end{bmatrix}$$

where

$$B_{21} = \begin{bmatrix} p_0 & p_1 & \cdot & p_{m-2} & p_{m-1} \\ 0 & p_0 & \cdot & p_{m-3} & p_{m-2} \\ \cdot & \cdot & \cdot & & \cdot \\ 0 & 0 & \cdot & p_0 & p_1 \\ 0 & 0 & \cdot & 0 & p_0 \\ 0 & 0 & \cdot & 0 & 0 \\ \cdot & \cdot & \cdot & \cdot & \cdot \\ 0 & 0 & \cdot & 0 & 0 \end{bmatrix}$$

and

$$B_{22} = \begin{bmatrix} p_m & p_{m+1} & \cdots & p_{h-1} & p_h^+ \\ p_{m-1} & p_m & \cdots & p_{h-2} & p_{h-1}^+ \\ \cdot & \cdot & \cdots & \cdot & \cdot \\ p_0 & p_1 & \cdots & p_{h-m-1} & p_{h-m}^+ \\ 0 & p_0 & \cdots & p_{h-m-2} & p_{h-m-1}^+ \\ \vdots & \vdots & \ddots & \vdots & \vdots \\ 0 & 0 & \cdots & p_{m-1} & p_m^+ \end{bmatrix}$$

3 Intuitive calculation of the invariant probability

We consider an intuitive calculation of the invariant probability measure π. If we write

$$\pi = (\pi_0, \pi_1, \ldots, \pi_k) \in \mathbb{R}^{(h+1)(k+1)}$$

then the equation $\pi = \pi P$ can be rewritten as a linear system

$$\pi_0 = \pi_0 A + \pi_1 A \tag{1}$$
$$\pi_i = \pi_{i+1} A \quad (1 \leq i < m) \tag{2}$$
$$\pi_m = \pi_0 B + \pi_1 B + \pi_{m+1} A \tag{3}$$
$$\pi_i = \pi_{i-m+1} B + \pi_{i+1} A \quad (m < i < k) \tag{4}$$
$$\pi_k = \pi_{k-m+1} B + \cdots + \pi_k B. \tag{5}$$

We wish to know if this system has a unique solution. In a formal sense we observe that the sequence of non-negative vectors

$$\{\pi_i\}_{i=0}^k \subseteq \mathbb{R}^{h+1}$$

satisfy the recurrence relations

$$\pi_i = \pi_{i+1} V_i \quad (0 \leq i < k) \tag{6}$$

where the sequence of matrices

$$\{V_i\}_{i=0}^k \subseteq \mathbb{R}^{(h+1)\times(h+1)}$$

is defined as follows. Let

$$V_0 = A(I - A)^{-1} \tag{7}$$
$$V_i = A, \quad (0 < i < m) \tag{8}$$
$$V_m = A\left[I - A^{m-1}(I - A)^{-1}B\right]^{-1} \tag{9}$$
$$V_i = A\left[I - W_{i-1,\, i-m+1}B\right]^{-1} \quad (m < i < k) \tag{10}$$

where

$$W_{i,\, \ell} := V_i V_{i+1} \ldots V_\ell \quad (i \geq \ell) \tag{11}$$

provided the required inverse matrices exist and let

$$V_k := \left[I + \sum_{\ell=1}^{m-1} W_{k-1,\, k-\ell}\right] B. \tag{12}$$

The vector π_k is a scalar multiple of the invariant probability measure for the transition matrix V_k. We conclude that the invariant probability measure π for the transition matrix P is unique if and only if the associated invariant probability measure

$$\pi_k^* := \pi_k/(\pi_k \cdot 1)$$

for the transition matrix V_k is uniquely defined. We have established the following rudimentary result.

Theorem 1. *If the sequence of matrices*

$$\{V_i\}_{i=0}^k \subseteq \mathbb{R}^{(h+1)\times(h+1)}$$

is well defined by the formulae (7)–(12) then there exists an invariant measure π for the transition matrix P. The measure is not necessarily unique.

4 Existence of the inverse matrices

Provided $p_r > 0$ for all $r \leq h$ the matrix A_{11} is strictly sub-stochastic with

$$A_{11} \cdot 1 = \begin{bmatrix} 1 - p_m^+ \\ \vdots \\ 1 - p_1^+ \end{bmatrix} < 1.$$

It follows that $(I - A_{11})^{-1}$ is well defined and hence

$$(I - A)^{-1} = \begin{bmatrix} (I - A_{11})^{-1} & A_{12}(I - A_{11})^{-1} \\ 0 & I \end{bmatrix}$$

is also well defined. It is necessary to begin with an elementary but important result. This result, and other later results in this section, have already been established by Piantadosi [Pia04] but for convenience we present details of the more elementary proofs to indicate our general method of argument.

Lemma 1. *If $p_r > 0$ for all $r = 0, 1, \ldots$ then*

$$(I - A)^{-1}B \cdot 1 = 1 \quad and \quad A^{m-1}(I - A)^{-1}B \cdot 1 \leq A^{m-1} \cdot 1$$

and the matrix $V_m = A[I - A^{m-1}(I - A)^{-1}B]^{-1}$ is well defined.

Proof. Note that $A \cdot 1 + B \cdot 1 = 1$ implies $B \cdot 1 = (I - A) \cdot 1$ and hence

$$(I - A)^{-1}B \cdot 1 = 1.$$

Now

$$\begin{aligned}
A^{m-1}(I - A)^{-1}B \cdot 1 &= A^{m-1} \cdot 1 \\
&= \begin{bmatrix} A_{11}^{m-1} & A_{11}^{m-2}A_{12} \\ 0 & 0 \end{bmatrix} \begin{bmatrix} 1 \\ 1 \end{bmatrix} \\
&= \begin{bmatrix} A_{11}^{m-2}[A_{11} \cdot 1 + A_{12} \cdot 1] \\ 0 \end{bmatrix} \\
&= \begin{bmatrix} A_{11}^{m-2} \\ 0 \end{bmatrix} \\
&< 1.
\end{aligned}$$

Hence $V_m = A[I - A^{m-1}(I - A)^{-1}B]^{-1}$ is well defined. □

To establish the existence of the remaining inverse matrices it is necessary to establish some important identities.

Lemma 2. *The (JP) identities*

$$\left[I + \sum_{\ell=1}^{m-1} W_{i-1, \; i-\ell} \right] B \cdot 1 = 1$$

are valid for $i = m + 1, \ldots, k - 1$ and hence the matrix $V_i = A[I - W_{i-1, \; i-m+1}B]^{-1}$ is well defined.

Proof. For details of the rather long and difficult proof we refer the reader to Piantadosi [Pia04] where the notation dictates that the identities are described and established in two parts as the (JP) identities of the first and second kind. The complexity of these identities is masked in the current paper by notational sophistication.   □

5 Probabilistic analysis

In practice the matrix P can be expected to be irreducible. First we establish the following simple sufficient condition for this to be the case.

Theorem 2. *Suppose A, B have the forms displayed above and that $k > m$. If*
(i) $m > 1$ *and*
(ii) $p_0, p_1, \ldots, p_{h-1}, p_h^+ > 0$,
then the matrix P is irreducible.

Proof. We use the notation $P_{(i,j),(r,s)}$ to refer to the element in the matrix P describing the transition from state (i,j) to state (r,s) and we write $A = [a_{j,s}]$ and $B = [b_{j,s}]$ to denote the individual elements of A and B. To prove irreducibility, it suffices to show that, for any state (i,j), there is a path of positive probability from state (k,h) to state (k,h) and a path of positive probability from state (i,j) to state (k,h).

The former may be seen as follows. For $i = k$ with $h - m \leq j \leq h$,

$$P_{(k,h),(k,j)} = b_{h,j} > 0$$

by (ii), so there is a path consisting of a single step. For $i = k$ with $0 \leq j < h - m$, there exists a positive integer ℓ such that $h - m \leq j + \ell(m+1) \leq h$. One path of positive probability from (k,h) to (k,j) consists of the consecutive steps

$$(k,h) \to (k, j + \ell(m+1)) \to (k, j + (\ell-1)(m+1)) \to \ldots \to (k,j).$$

Finally, for $i < k$, one such path is obtained by passing from (k,h) to $(k,0)$ as above and then proceeding

$$(k,0) \to (k-1,0) \to \ldots \to (i+1,0) \to (i,j).$$

We now consider passage from (i,j) to (k,h). For $j = 0$, (i,j) has one–step access to $(0,h)$ (if $i = 0$) or to $(i-1,h)$ (if $i > 0$), while for $j > 0$, (i,j) has one–step access to $(i+m,h)$ (if $i = 0$), to $(i+m-1,h)$ (if $0 < i < k-m+1$) or to (k,h) (if $k - m + 1 < i \leq k$). Putting these results together shows that each state (i,j) has a path of positive probability to (k,h).

By the results of the two previous paragraphs, the chain is irreducible.   □

Next we derive invertibility results for some key $(h+1) \times (h+1)$ matrices. While this can be effected purely in terms of matrix arguments, a shorter derivation is available employing probabilistic arguments, based on successive censorings of a Markov chain.

Theorem 3. *Suppose conditions* (i) *and* (ii) *of Theorem 2 apply. Then there exists a sequence* $\{V_i\}_{0 \leq i \leq k}$ *of* $(h+1) \times (h+1)$ *matrices defined by equations* (7), (8), (9), (10), (11) *and* (12). *The matrices* $V_0, \ldots, V_{k-1}$ *are invertible.*

Proof.   It suffices to show that the formulae (7), (8), (9) hold and that for $k > m+1$ the formula (10) is valid. Let C_0 be a Markov chain of the same form as P but with k replaced by $K \geq 2k$. By Theorem 2, C_0 is irreducible and finite and so positive recurrent. Denote by C_i $(1 \leq i \leq k)$ the Markov chain formed by censoring out levels $0, 1, \ldots, i-1$, that is, observing C_0 only when it is in the levels $i, i+1, \ldots, K$. The chain C_i must also be irreducible and positive recurrent. For $0 \leq i \leq k$, denote by P_i the one–step transition matrix of C_i and by Q_i its leading block. Then Q_i is the sub-stochastic one–step transition matrix of a Markov chain D_i whose states form level i of C_0. Since C_i is recurrent, the states of D_i must all be transient and so $\sum_{n=0}^{\infty} Q_i^n < \infty$. Hence $I - Q_i$ is invertible for $0 \leq i < k$.

We shall show that the matrices

$$V_i := A(I - Q_i)^{-1} \quad (0 \leq i < k)$$

satisfy the conditions of the enunciation. Nonnegativity of V_i is inherited from that of Q_i. We have (7) and (8) immediately, since $Q_0 = A$ and we have easily that $Q_i = 0$ for $0 < i < m$. We now address (9) and (10).

One–step transitions in D_m arise from paths of two types. In the first, the process passes in sequence through levels $m, m-1, \ldots, 0$. These give rise to a one–step transition matrix $A^{m-1}B$. Paths of the second type are the same except that they spend one or more time points in level 0 between occupying levels 1 and m. These give rise to a one–step transition matrix

$$A^m \sum_{n=0}^{\infty} A^n B = A^m (I - A)^{-1} B.$$

Thus enumerating all paths yields

$$Q_m = A^{m-1}B + A^m(I - A)^{-1}B = A^{m-1}(I - A)^{-1}B,$$

which provides (9). From similar enumerations of paths, the leading row of P_n may be derived to be

$$Q_m \ A^{m-2}B \ A^{m-3}B \ldots AB \ B \ 0 \ldots 0.$$

The other rows of P_m are given by rows $m+1, m+2, \ldots, K$ of P_0 restricted to columns $m, m+1, \ldots, K$. The first two rows of P_m are then

$$Q_m \ A^{m-2}B \ A^{m-3}B \ldots AB \ B \ 0 \ldots 0$$
$$A \quad 0 \quad 0 \quad \ldots \quad 0 \quad 0 \ B \ldots 0,$$

from which we derive

$$Q_{m+1} = A \left[I - Q_m\right]^{-1} A^{m-2}B = V_m V_{m-1} \ldots V_2$$

and that the leading row of P_{m+1} is

$$Q_{m+1} \ V_m A^{m-3}B \ V_m A^{m-4}B \ \ldots \ V_m AB \ V_m B \ B \ 0 \ldots 0.$$

Using the notation in equation (11) we can write

$$Q_{m+1} = W_{m,2}B$$

and the leading row of P_{m+1} may be expressed as

$$Q_{m+1} \ W_{m,3}B \ W_{m,4}B \ \ldots \ W_{m,m}B \ B \ 0 \ldots 0.$$

We may use this as a basis ($i = m + 1$) for an inductive proof that for $m < i \le k$

$$Q_i = W_{i-1,i-m+1}B$$

and the leading row of P_i is

$$Q_i \ W_{i-1,i-m+1}B \ W_{i-1,i-m+2}B \ \ldots \ W_{i-1,i-1}B \ B \ 0 \ldots 0.$$

Suppose these hold for some i satisfying $m < i < k$. Since the two leading rows of P_i are

$$Q_i \ W_{i-1,i-m+2}B \ W_{i-1,i-m+3}B \ \ldots \ W_{i-1,i-1}B \ B \ 0 \ \ldots \ 0$$
$$A \quad 0 \quad 0 \quad \ldots \quad 0 \quad 0 \ B \ldots 0,$$

we have

$$Q_{i+1} = A \left[I - Q_i\right]^{-1} W_{i-1,i-m+2}B = V_i W_{i-1,i-m+2}B = W_{i,i-m+2}B$$

and, since $V_i W_{i-1,\ell} = W_{i,\ell}$, that the leading row of P_{i+1} is

$$Q_{i+1} \ W_{i,i-m+1}B \ W_{i,i-m+2}B \ \ldots \ W_{i,i-1}B \ W_{i,i}B \ B \ 0 \ldots 0,$$

providing the inductive step. $\square$

Under assumptions (i) and (ii) of Theorem 2, we may now proceed to the determination of the invariant measure $\pi = (\pi_0, \pi_1, \ldots, \pi_k)$ of the block–entry discrete–time Markov chain P. The relation $\pi = \pi P$ yields the block component equations (1), (2, (3), (4) and (5). The evaluation of π may be effected by the following.

Theorem 4. *Suppose that conditions* (i) *and* (ii) *of Theorem 2 apply. Then the probability vectors* π_i *satisfy the recurrence relations* (6) *and* π_k *is the invariant measure of the matrix* V_k *defined by* (12). *The measure* π *is unique.*

Proof.   For $i = 0$, (6) follows from (1) and (7). For $0 < i < m$, (6) is immediate from (2) and (8). These two parts combine to provide

$$\pi_0 = \pi_m A^m (I - A)^{-1} \quad \text{and} \quad \pi_1 = \pi_m A^{m-1},$$

so that (3) may be cast as

$$\pi_m \left[I - A^{m-1} (I - A)^{-1} B \right] = \pi_{m+1} A.$$

Equation (6) for $i = m$ follows from (9).

We have now shown that (6) holds for $0 \leq i \leq m$, from which

$$\pi_2 = \pi_{m+1} V_m V_{m-1} \ldots V_2.$$

Hence (4) with $i = m + 1$ yields

$$\pi_{m+1} \left[I - V_{m+1} \ldots V_3 B \right] = \pi_{m+2} A.$$

By (11), this is (6) for $i = m + 1$, which supplies a basis for a derivation of the remainder of the theorem by induction. For the inductive step, suppose that (6) holds for $i = m + 1, \ldots, q$ for some q with $m < q < k$. Then from (4),

$$\pi_{q+1} = \pi_{q+1} V_q V_{q-1} \ldots V_{q-m+2} B + \pi_{q+2} A.$$

By (11), this is simply (6) with $i = q + 1$, and so we have established the inductive step.

As a direct consequence we have

$$\pi_i = \pi_k V_{k-1} \ldots V_i = \pi_k W_{k-1, \, i}$$

for $0 \leq i < k$ so that (5) implies

$$\pi_k = \pi_k \left[I + \sum_{i=k-m+1}^{k-1} W_{k-1, \, i} \right] B = \pi_k V_k,$$

by definition. Hence π_k is an invariant measure of V_k. Any invariant measure π_k of V_k induces via (6) a distinct invariant measure π for P. Since the irreducibility of P guarantees it has a unique invariant measure (to a scale factor), π_k is unique invariant up to a scale factor. This completes the proof.   □

6 The expected long-term overflow

Using the invariant probability measure π we can calculate the expected overflow of water from the system. Let $(i, j) \in [0, k] \times [0, h]$ denote the collection of all possible states. The expected overflow is calculated by

$$J = \sum_{i=0}^{k} \sum_{j=0}^{h} \left[\sum_{r=0}^{\infty} f[(i,j)|r]p_r \right] \pi_{ij}$$

where π_{ij} is the invariant probability of state (i, j) at level i and phase j and $f[(i, j)|r]$ is the overflow from state (i, j) when r units of stormwater enter the system. Note that we have ignored pumping cost and other costs which are likely to be factors in a real system.

7 Extension of the fundamental ideas

The assumption that $p_r > 0$ for all $r = 0, 1, \ldots$ is convenient and is usually true in practice but many of the general results remain true with weaker assumptions. Let us suppose that the system is balanced. That is we assume that the expected daily supply is equal to the daily demand. Thus we assume that

$$0 \cdot p_0 + 1 \cdot p_1 + 2 \cdot p_2 + \cdots = 1.$$

Since

$$p_0 + p_1 + p_2 + \cdots = 1$$

it follows that the condition $p_0 = 0$ would imply that $p_1 = 1$ and $p_r = 0$ for all $r \geq 2$. This condition is not particularly interesting and suggests that the assumption $p_0 > 0$ is a reasonable assumption. If we assume also that $p_0 < 1$ then it is clear that there is some $r > 1$ such that $p_r > 0$.

By using a purely algebraic approach Piantadosi [Pia04] effectively established the following result.

Theorem 5. *If $p_0 > 0$ and $p_m^+ > 0$ then there is at least one finite cycle with non-zero invariant probability that includes all levels $0, 1, \ldots, k$ of the second dam. All states have access to this cycle in finite time with finite probability and hence are either transient with invariant probability zero or else are part of a single maximal cycle.*

Proof. (Outline)   In this paper we have tried to look beyond a simply algebraic view. For this reason we suggest an alternative proof. Let $p_0 = \delta > 0$. If $p_m^+ > 0$ then there is some $r \geq m$ with $p_r = \epsilon > 0$. Our argument here assumes $r > m$. Choose p so that $0 \leq h - pm < m$ and choose s so that $(s + 1)r - (p + s)m \geq 0$ and $s(m - 1) + 1 \geq k$ and t so that $t \geq p + k$ and consider the elementary cycle

$$(0, h - pm) \rightarrow (0, h - pm + r) \rightarrow (m, h - (p+1)m + 2r) \rightarrow$$
$$(2m - 1, h - (p+2)m + 3r) \rightarrow \cdots \rightarrow (k, h) \rightarrow \cdots \rightarrow (k, h) \rightarrow (k, h - m) \rightarrow$$
$$\cdots \rightarrow (k, h - pm) \rightarrow (k - 1, h - pm) \rightarrow \cdots \rightarrow (0, h - pm) \rightarrow$$
$$\cdots \rightarrow (0, h - pm)$$

for the state (i, j) of the system. We have $s + 1$ consecutive inputs of r units followed by t consecutive inputs of 0 units. The cycle has probability $p_r{}^{s+1}p_0{}^t = \epsilon^{s+1}\delta^t$. It is obvious that the state (k, h) is accessible in finite time with finite probability from any initial state (i, j). It follows that all states are either transient or are part of a unique irreducible cycle. Of course the irreducible cycle must include the elementary cycle. Hence there is a unique invariant probability

$$\pi = (\pi_0, \pi_1, \ldots, \pi_k)$$

where the invariant probability π_i for level i is non-zero for all $i = 0, \ldots, k$. All transient states have zero probability and all states in the cycle have non-zero probability. □

Observe that by adding together the separate equations (1), (2), (3), (4) and (5) for the vectors $\pi_0, \ldots, \pi_k$ we obtain the equation

$$(\pi_0 + \cdots + \pi_k)(A + B) = (\pi_0 + \cdots + \pi_k).$$

Therefore

$$\rho = \pi_0 + \cdots + \pi_k$$

is an invariant probability for the stochastic matrix

$$S = A + B.$$

Indeed a little thought shows us that S is the transition matrix for the phase j of the state vector. By analysing these transitions we can shed some light on the structure of the full irreducible cycle for the original system.

We have another interesting result.

Theorem 6. *If $p_0 = \delta > 0$ and $p_r = \epsilon > 0$ for some $r > m$ and if $\gcd(m, r) = 1$ then for every phase $j = 0, 1, 2, \ldots, h$ we can find non-negative integers $p = p(j)$ and $q = q(j)$ such that*

$$pr - qm = j$$

and the chain with transition matrix $S = A + B$ is irreducible.

Proof. (Outline)   We suppose only that $p_0 > 0$ and $p_r > 0$ for some $r > m$. In the following phase transition diagram we suppose that

$$r - m < m, \quad 2r - 3m < m, \quad \cdots$$

and note that the following phase transitions are possible for j with non-zero probability.

$$0 \to [0 \cup r]$$

$$r \to [(r - m) \cup (2r - m)]$$

$$(r - m) \to [(r - m) \cup (2r - m)]$$
$$(2r - m) \to [(2r - 2m) \cup (3r - 2m)]$$

$$(2r - 2m) \to [(2r - 3m) \cup (3r - 3m)]$$
$$(3r - 2m) \to [(3r - 3m) \cup (4r - 3m)]$$

$$(2r - 3m) \to [(2r - 3m) \cup (3r - 3m)]$$

$$\vdots$$

If $\gcd(m, r) = 1$ then it is clear by extending the above transition table that every phase $j \in [0, h]$ is accessible in finite time with finite probability.   □

This result means that the unique irreducible cycle for the (i, j) chain generated by P which already includes all possible levels $i \in [0, k]$ also includes all possible phases $j \in [0, h]$ although not necessarily all states (i, j).

In practice the input probabilities are likely to depend on time. Because there is a natural yearly cycle for rainfall we have used the notation $[t] = (t - 1) \bmod 365 + 1$ and $p_r = p_r([t])$ for all $r = 0, 1, 2, \ldots$ and all $t \in \mathbb{N}$. The transition from day t to day $t + 1$ is described by a matrix $P = P([t])$ with the same block structure as before but with elements that vary from day to day throughout the year. The transition from day t to day $t + 365$ is described by

$$x(t + 365) = x(t)R([t])$$

where the matrix $R([t])$ is defined by

$$R([t]) = P([t]) \cdots P(1)P(365) \cdots P([t + 1]).$$

In principle we can calculate an invariant probability $\pi([t])$ for each matrix $R([t])$ and it is easy to show that successive invariant probabilities are related by the equation

$$\pi([t + 1]) = \pi([t])P([t]).$$

However, although all $P([t])$ have the same block structure this structure is not preserved in the product matrix $R([t])$ and it is not clear that matrix reduction methods can be used in the calculation of $\pi([t])$. It is obvious that the invariant probabilities for the phase j on day $[t]$ can be calculated from

$$\rho([t]) = \rho([t])S([t]) \cdots S(1)S(365) \cdots S([t]+1)$$

where $S([t]) = A([t]) + B([t])$. Unfortunately knowledge of $\rho([t])$ does not help us directly to calculate $\pi([t])$. In general terms the existence of a unique invariant probability is associated with the idea of a contraction mapping. Define

$$T = \{x \in \mathbb{R}^n \mid x = (x_0, \ldots, x_k) \geq 0 \text{ where } x_j \in \mathbb{R}^{h+1} \text{ and } x_0 \cdot 1 + \cdots + x_k \cdot 1 = 1\}.$$

For each $t = 1, 2, \ldots$ we suppose that the mapping $\varphi_{[t]} : T \mapsto T$ is defined by

$$\varphi_{[t]}(x) = xP([t])$$

for each $x \in T$. We have the following conjecture.

Conjecture 1. For each $t = 1, 2, \ldots$ let $p_0([t]) > 0$ and suppose that for some $r = r([t]) > m$ with $\gcd(r, m) = 1$ we have $p_r([t]) > 0$. Then

$$[\varphi_{[t]}]^{k-1}(T) \subseteq int(T)$$

and there is a unique invariant measure $\pi([t])$ with

$$\varphi_{[t]}(\pi([t])) = \pi([t]).$$

If this conjecture is true then the iteration given by

$$x^{(1)} = \frac{1}{k+1}(\rho(1), \ldots, \rho(1)) \in \mathbb{R}^{(h+1)(k+1)} = \mathbb{R}^n$$

with

$$x^{(t+1)} = x^{(t)}P([t])$$

for each $t = 1, 2, \ldots$ should satisfy

$$x^{(t)} \to x([t])$$

as $t \to \infty$. Because the contraction operates in the same structural way for every value of $[t]$ we expect that convergence will occur quite seamlessly. This is demonstrated in the following simple example. There is no reason to expect the convergence to be slower in the case where we have a product of a larger number of matrices.

Example 1. Let $[t] = (t-1) \bmod 2 + 1$ with $R(1) = P(1)P(2)$ and $R(2) = P(2)P([3]) = P(2)P(1)$ where

$$P([t]) = \begin{bmatrix} A([t]) & 0 & B([t]) & 0 \\ A([t]) & 0 & B([t]) & 0 \\ 0 & A([t]) & 0 & B([t]) \\ 0 & 0 & A([t]) & B([t]) \end{bmatrix}$$

for each $[t] = 1, 2$ and

$$A(1) = \begin{bmatrix} 0.5 & 0.25 & 0.125 & 0.125 \\ 0 & 0.5 & 0.25 & 0.25 \\ 0 & 0 & 0 & 0 \\ 0 & 0 & 0 & 0 \end{bmatrix} \quad \text{and} \quad B(1) = \begin{bmatrix} 0 & 0 & 0 & 0 \\ 0 & 0 & 0 & 0 \\ 0.5 & 0.25 & 0.125 & 0.125 \\ 0 & 0.5 & 0.25 & 0.25 \end{bmatrix}$$

and

$$A(2) = \begin{bmatrix} 0.45 & 0.27 & 0.13 & 0.15 \\ 0 & 0.45 & 0.27 & 0.28 \\ 0 & 0 & 0 & 0 \\ 0 & 0 & 0 & 0 \end{bmatrix} \quad \text{and} \quad B(2) = \begin{bmatrix} 0 & 0 & 0 & 0 \\ 0 & 0 & 0 & 0 \\ 0.45 & 0.27 & 0.13 & 0.15 \\ 0 & 0.45 & 0.27 & 0.28 \end{bmatrix}$$

Using MATLAB we calculate

$$\rho(1) = (0.2, 0.4, 0.2, 0.2)$$

and so we set

$$x^{(1)} = \frac{1}{4}(\rho(1), \rho(1), \rho(1), \rho(1))$$
$$= (.0500, .1000, .0500, .0500, .0500, .1000, .0500, .0500,$$
$$.0500, .1000, .0500, .0500, .0500, .1000, .0500, .0500)$$

and calculate

$$x^{(2)} = (.0500, .1250, .0625, .0625, .0250, .0625, .0312, .0312$$
$$.0750, .1375, .0687, .0687, .0500, .0750, .0375, .0375)$$
$$x^{(3)} = (.0338, .1046, .0604, .0638, .0338, .0821, .0469, .0498$$
$$.0647, .1148, .0643, .0688, .0478, .0765, .0425, .0457)$$
$$x^{(4)} = (.0338, .1103, .0551, .0551, .0323, .0735, .0368, .0368$$
$$.0775, .1338, .0669, .0669, .0534, .0839, .0420, .0420)$$

$$\vdots$$

$$x^{(13)} = (.0291, .0994, .0576, .0607, .0343, .0801, .0456, .0485$$
$$.0660, .1199, .0672, .0719, .0494, .0791, .0439, .0472)$$
$$x^{(14)} = (.0317, .1056, .0528, .0528, .0330, .0764, .0382, .0382$$
$$.0763, .1323, .0661, .0661, .0556, .0874, .0437, .0437)$$

Thus we have

$$x(1) \approx (.0291, .0994, .0576, .0607, .0343, .0801, .0456, .0485$$
$$.0660, .1199, .0672, .0719, .0494, .0791, .0439, .0472)$$
$$x(2) \approx (.0317, .1056, .0528, .0528, .0330, .0764, .0382, .0382$$
$$.0763, .1323, .0661, .0661, .0556, .0874, .0437, .0437).$$

References

[Abd03]   Abdel–Hameed, M.: Optimal control of dams using $P_{\lambda,\tau}^{M}$ policies and penalty cost. Mathematical and Computer Modelling, **38**, 1119-1123 (2003)

[Gan69]   Gani, J.: Recent advances in storage and flooding theory. Advanced Applied Probability, **1**, 90–110 (1969)

[KT65]   Karlin, S., Taylor, H.M.: A First Course in Stochastic Processes. Wiley and Sons, New York (1965)

[LR99]   Latouche, G., Ramaswami, V.: Introduction to Matrix Analytic Methods in Stochastic Modeling. SIAM (1999)

[Mor54]   Moran, P.A.P.: A probability theory of dams and storage systems. Journal of Applied Science, **5**, 116–124 (1954)

[Mor59]   Moran, P.A.P.: The Theory of Storage. Wiley and Sons, New York (1959)

[Neu89]   Neuts, M.F.: Structured Stochastic Matrices of M/G/1 type and Their Applications. Marcel Dekker, Inc. (1989)

[Pia04]   Piantadosi, J.: Optimal Policies for Management of Urban Stormwater. PhD Thesis, University of South Australia (2004)

[Yeo74]   Yeo, G.F.: A finite dam with exponential variable release. Journal of Applied Probability, **11**, 122–133 (1974)

[Yeo75]   Yeo, G.F.: A finite dam with variable release rate. Journal of Applied Probability, **12**, 205–211 (1975)